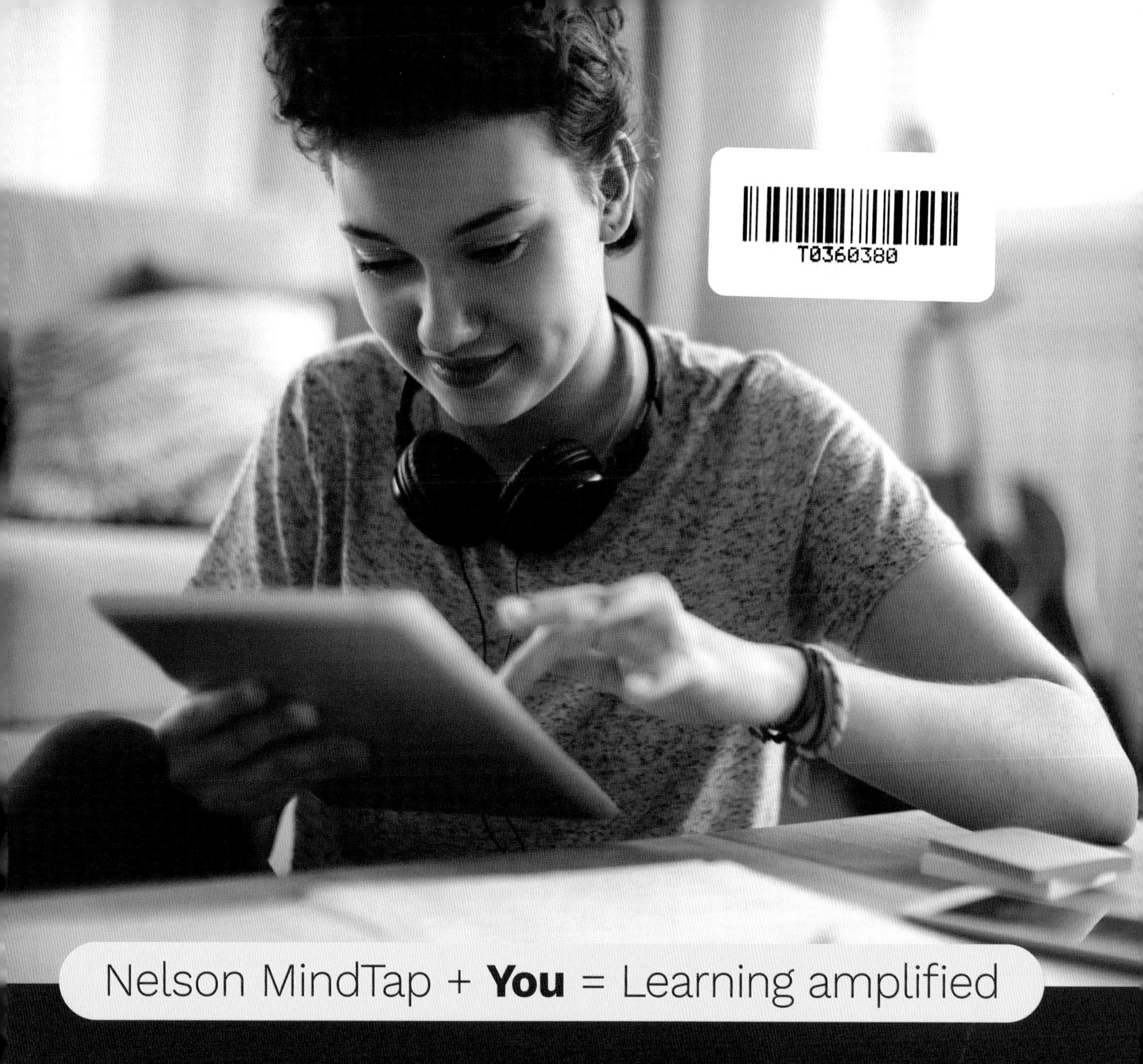

Nelson MindTap + **You** = Learning amplified

"I love that everything is interconnected, relevant and that there is a clear learning sequence. I have the tools to create a learning experience that meets the needs of all my students and can easily see how they're progressing."

— **Sarah,** Secondary School Teacher

Ea.

The cover was designed by Everyday Ambitions, a visual story-telling studio based in Melbourne which has produced work for national Australian brands including Warner Music, Kookai, and Monash University, alongside working with small, independent brands in food, beauty and fashion industries.

In approaching the brief the studio considered: how might we inspire students' imagination across a wide variety of materials, processes, outputs, and end users, and ultimately convey a design that is sympathetic to an interest in the hands-on, practical nature of the subject?

The visual solution was informed by the methodology of design thinking as a problem-solving approach, with inspiration taken from the classic design and graphic works of Bruno Munari and Dieter Rams.

5e

NELSON PRODUCT DESIGN & TECHNOLOGIES

VCE UNITS 1–4

Jacinta O'Leary
and Jill Livett

Nelson Product Design and Technologies VCE Units 1–4
5th Edition
Jactina O'Leary
Jill Livett
ISBN 9780170477499

Product manager: Caroline Williams
Editor: Nick Tapp
Proofreader: Vanessa Lanaway
First Nations' content reviewer: Dr Mark Lock
Text designer: Rina Gargano (Alba Design)
Cover design: Emilie Pfitzner (Everyday Ambitions)
Project designer: Linda Davidson
Permissions researcher: Catherine Kerstjens
Content developer: Morgan Begg
Content manager: Alice Kane
Typeset by: Straive

Acknowledgements
We would like to thank the following for permissions to reproduce copyright material:
Unit 1 opener: stock.adobe.com/Art_Photo
Chapter 1 opener: stock.adobe.com/Pixel-Shot
Chapter 2 opener: stock.adobe.com/Chaosamran_Studio
Unit 2 opener: stock.adobe.com/New Africa
Chapter 3 opener: stock.adobe.com/Peakstock
Chapter 4 opener: stock.adobe.com/FotoArtist
Chapter 5 opener: Alamy Stock Photo/Philip McAllister
Unit 3 opener: stock.adobe.com/Chaosamran_Studio
Chapter 6 opener: stock.adobe.com/Seventyfour
Chapter 7 opener: stock.adobe.com/MIND AND I
Chapter 8 opener: stock.adobe.com/weedezign
Unit 4 opener: stock.adobe.com/anon
Chapter 9 opener: stock.adobe.com/DC Studio
Chapter 10 opener: stock.adobe.com/tgordievskaya
Chapter 11 opener: Izzy Donat
Theory opener: stock.adobe.com/Chaosamran_Studio
Chapter 12 opener: stock.adobe.com/Daniel Laflor/peopleimages.com
Chapter 13 opener: stock.adobe.com/pressmaster
Chapter 14 opener: stock.adobe.com/Sergey Ryzhov

Extracts from the VCE Product Design and Technologies Study Design (2024-2028) are used by permission, © VCAA. VCE® is a registered trademark of the VCAA. The VCAA does not endorse or make any warranties regarding this study resource. Current VCE Study Designs, past VCE exams and related content can be accessed directly at www.vcaa.vic.edu.au.

For product information and technology assistance,
in Australia call **1300 790 853**;
in New Zealand call **0800 449 725**

For permission to use material from this text or product, please email
aust.permissions@cengage.com

National Library of Australia Cataloguing-in-Publication Data
A catalogue record for this book is available from the National Library of Australia.

Cengage Learning Australia
Level 5, 80 Dorcas Street
Southbank VIC 3006 Australia

For learning solutions, visit **cengage.com.au**

Printed in Malaysia by Papercraft.
3 4 5 6 7 27 26 25 24

CONTENTS

About this book ix
About Nelson MindTap........................ x
About the authors xi

UNIT 1: DESIGN PRACTICES

CHAPTER 1: Developing and conceptualising designs 4
The Double Diamond design process 6
Investigating and defining..7
Generating and designing 7
Producing and implementing...7
Teamwork: collaboration in design 9
Advantages of working in a team 9
Working as a team 11
Use of digital technologies in your team 11
Approaches to teamwork 12
Using divergent and convergent thinking in a team................................. 13
Beginning the design process 15
Understanding drawing for product design 23
Developing visualisations (initial design ideas) 25
Team and end user feedback 30
Developing design options 31
Evaluating your design options 32
Optional working drawings for models or prototypes 33

CHAPTER 2: Generating, designing and producing 34
Planning and conducting technical research.. 36
Knowledge of materials 37
Materials testing and/or process trials.. 40
Other possible research areas... 43
Developing physical product concepts... 44
Choosing the preferred product concept.. 48
Working drawing.. 49
Producing and implementing... 50
Production 54
Evaluating.. 55
Assessment.. 56

UNIT
1

UNIT 2

UNIT 2: POSITIVE IMPACTS FOR END USERS

CHAPTER 3: Opportunities for positive impacts for end users 64
Designing to improve lives 66
Products that make a positive impact 70
Case studies of positive impact 80
Case study investigation and report 82
CHAPTER 4: Designing for positive impacts for end users 86
Creating a positive impact by identifying a need 88
First diamond: exploring, researching and defining the end users' needs 90
What does it mean to work technologically 91
Second diamond: generating, trialling, testing and producing solutions 92
Production planning and safety 98
Evaluation 100
Assessment 101
CHAPTER 5: Cultural influences on design 102
Focus on Aboriginal and Torres Strait Islander culture 104
Broad considerations of culture: impact on choices 113

UNIT 3

UNIT 3: ETHICAL PRODUCT DESIGN AND DEVELOPMENT

CHAPTER 6: Influences on design, development and production of products 118
Manufacturing: methods, scales (volume), technologies and contexts 120
Traditional, new and emerging technologies 125
Lean manufacturing 133
Sustainability 136
Greenwashing 157
Planned obsolescence 157
Experimental and alternative materials 161
Summing up 171
Assessment Task for Outcome 1 176
CHAPTER 7: Investigating opportunities for ethical design and production (SAT) 182
School-assessed Task (SAT) 184
The Double Diamond design process 184
Critical, creative and speculative thinking 187
Your SAT 'folio' 192
Factors that influence design 193
An ethical design problem 194

9780170477499

Deciding on an end user 196
Research 198
End user research 201
Qualitative and quantitative research methods 201
Recording, collating and forming information 206
Ethical design of a product 206
The design brief 209
Evaluation criteria 214
Further research 216
Product concepts 217

CHAPTER 8: Developing a final proof of concept for ethical production 228
School-assessed Task (SAT) 230
Design thinking 230
Ethical considerations 231
Testing materials, tools and processes 231
Prototyping 233
Finalising a proof of concept 239
Developing a scheduled production plan 241
Risk assessment and management for safety 244

UNIT 4: ETHICAL PRODUCT DESIGN AND DEVELOPMENT

CHAPTER 9: Managing production for ethical designs 254
Production 256

Chapter 10: Evaluation and speculative design 260
Evaluate a range of products 262
What is R&D? 262
Speculative design thinking 266
The product development process 267
Market research in the product development process 269
Considerations in the use of new and emerging technologies 271
Success or failure of products 273
Key factors that determine the success of a product 276
Evaluating a range of products and data 278
Assessment 283

CHAPTER 11: SAT folio 285
Visual checklist for SAT folio 286
Examples of items for your SAT 288

DESIGN FUNDAMENTALS

CHAPTER 12: Design thinking 313
Design thinking 314
Being creative and innovative 317
Drawing and design 322
Use of models and prototypes 343
Design aesthetics: the Design Elements and Principles 346
Design Elements 346
Design Principles 353

CHAPTER 13: Factors that influence product design 362
What are the Factors? 363
Need or opportunity 365
Function 366
End users 368
Aesthetics 376
Market opportunities 378
Product life cycle 379
Technologies: materials, tools and processes 380
Ethical considerations in design 384
Design specialisations 394

CHAPTER 14: Materials and testing 398
Knowledge of materials 399
Understanding materials 399
Wood 406
Metal 410
Plastics 413
Fabrics, fibres and yarns 416
Materials testing 421
Test suggestions for resistant materials 422
Test suggestions for non-resistant materials 425

Glossary 428
Index 431

9780170477499

ABOUT THIS BOOK

Nelson Product Design and Technologies VCE Units 1–4 5ed is explicitly aligned to the *Victorian Certificate of Education Product Design and Technologies Study Design 2024–2028*, which underwent a major revision from the previous study (2018–2023). This edition provides a comprehensive resource that guides students and teachers through the practical and theoretical demands of this course.

Written by the experienced and trusted author team of Jill Livett and Jacinta O'Leary, *Nelson Product Design and Technologies* includes contemporary case studies of Australian designers and makers, up-to-date folio and production activities, practical advice and expert insights to assist students in completing their SAT and developing the key knowledge and skills necessary to achieve success in both their internal and external assessments for VCE Product Design and Technologies.

Key features

- The book follows the structure of the Study Design, with each Area of Study and corresponding outcome covered in its own chapter.
- Clear explanations of the **Double Diamond** design process and an in-depth exploration of its activities will help students understand this approach and apply it to product design.
- Contemporary and relevant **case studies** of local designers, makers and businesses demonstrate the key knowledge and key skills being explored.
- Samples of **past students' work** provide inspiration and examples for students in completing their own projects.
- Detailed **planners** for each unit give guidance to teachers and students on planning and pacing their approach to the required practical and coursework components of the subject.
- A whole chapter focused on **First Nations peoples' design** perspectives and practices.
- The **language and terminology** of the Product Design and Technologies Study Design is used throughout to help students familiarise themselves with key terms.
- Key knowledge and key skills, key terms and infographics at the beginning of each chapter clearly break down the key concepts that will be covered in the chapter and link back to the Study Design.
- Nelson MindTap icons throughout the student book indicate additional resources that can be found on our online platform, including worksheets, templates, weblinks and video weblinks.
- Signpost and Notes boxes appear throughout the text to provide additional advice and cross-referencing within the text.
- The **valued features** of the previous edition are retained, with the Materials Processes chapter along with updated worksheets and templates now to be found on Nelson MindTap, our online platform.

ABOUT NELSON MINDTAP

Nelson MindTap is an online learning space that provides students with tailored learning experiences.

Access tools and content that make learning simpler yet smarter to help you achieve success in VCE Product Design and Technologies.

Nelson MindTap

A flexible and eay-to-use online learning space that provides students with engaging, tailored learning experiences.

Your Nelson Product Design and Technologies Nelson MindTap includes:

- an eText with integrated activities and online assessments.
- margin links in the student book signpost multimedia student resources found on Nelson MindTap.

For students:

- End-of-chapter revision quizzes
- Worksheets
- Drawing and designing worksheets
- Bonus case studies and worksheets
- Chapter summaries
- Folio and production templates
- Activities to prepare for the external assessment (exam) for Units 3 and 4
- Additional resource: Materials and processes
- Additional resource: Design styles

For teachers:

- Teaching plans (editable)
- Folio and production templates
- Advice for teaching 'unscored' students
- Tips for preparing SACs
- Chapter summaries

9780170477499

ABOUT THE AUTHORS

Jacinta O'Leary has taught Woodwork, Plastics, Junior Electronics and Visual Communication Design in various schools in Victoria. She currently teaches at Virtual School Victoria, where she created courses for Design and Technology in the junior level, VCE Visual Communication Design, VCE Food and Technology and VCE Product Design and Technology.

In 2010 Jacinta was the chief writer for the VCAA review of *VCE Product Design and Technology 2012–2017*. In 2015 and 2016 she contributed to the VCAA Assessment Criteria for the major task for this subject and she has been an exam assessor for many years.

Jacinta has been a board member of the Design and Technology Teachers Association Victoria (DATTA Vic) since 2007, writing and editing DATTA Vic teacher support material for the previous VCE Product Design and Technology course and giving numerous presentations on the subject to both teachers and students.

Jill Livett has been teaching Technology Studies for well over 20 years in state and independent schools. During that time she has taught classes in VCE Wood, Metal and Plastics and Fabrics, Years 7–10 Design and Technology in the areas of Wood, Metal, Plastics and Fabrics, and Systems Technology. She currently teaches Years 5–12 Design and Technology at Overnewton Anglican Community College.

Jill was the Product Design and Technology Examination curriculum vettor and writer for a number of years, and has been on the exam marking team for many years.

Jill has supported the teaching of the Design and Technology learning area by organising and presenting workshops and professional learning days, developing curriculum material and organising conferences for DATTA Vic and the VCAA. She worked for a number of years as DATTA Vic's Executive Officer and Conference Organiser, and held both the President and Secretary positions for DATTA Australia. Jill has taught Technology Education subjects at the University of Melbourne, La Trobe University and Ballarat University, and is currently the Vice President of DATTA Vic.

UNIT 1

DESIGN PRACTICES

Areas of Study 1 and 2

Unit 1, Area of Study 1: Developing and conceptualising designs

You learn about:

- the Double Diamond process and design thinking
  - » importance of divergent and convergent thinking, and creative and critical thinking
  - » strategies and application
- collaborative approaches and roles, use of digital technologies
- design brief structure – end user profile, project scope, constraints and considerations
- Factors that influence design
- evaluation criteria – purpose, writing and use
- research methods – primary and secondary
- drawing systems and their use to develop design ideas.

Unit 1, Area of Study 2: Generating, designing and producing

You learn about:

- experimenting with materials, tools and processes (refine with creative, critical and speculative thinking)
- modelling and prototyping strategies
- choosing product concepts, proof of concept
- developing and implementing a production plan
- working collaboratively to develop concepts and complete a product solution
- evaluating finished product, collaboration and recommending improvements.

9780170477499

Unit 1 Planner

Please note: in this textbook, the size of the diamonds has been adapted to reflect the time you will most likely spend in each.

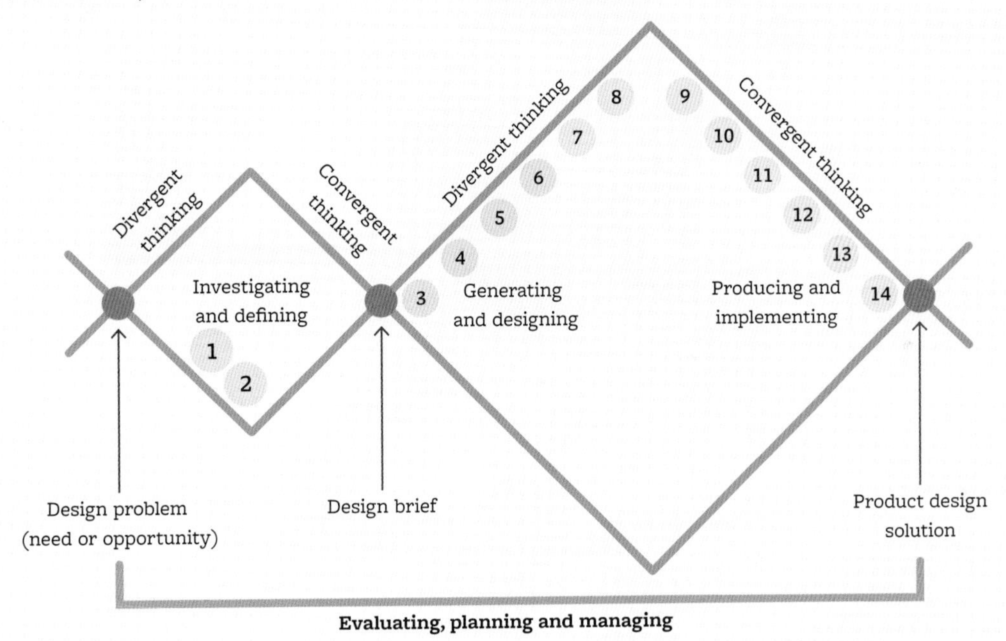

Outcome 1	**Week 1** • Identifying need/ opportunity • Brainstorming ideas • Exploring ideas for what to make, possible end user/s	**Week 2** • Making decisions about end user/s • Researching end user requirements • Investigating existing solutions	**Week 3** Writing the design brief, evaluation criteria	**Weeks 4 and 5** • Developing drawing skills • Exploring ideas in visualisations	**Week 6** Two or more design options developed and evaluated
Outcome 2	**Week 7** Research, testing of materials and trialling processes	**Week 8** Modelling and prototyping	**Week 9** • Final working drawing • Planning	**Weeks 10–13** Production	**Week 14** Finishing and evaluation

Assessment in Unit 1

The following work will be assessed for Unit 1:

- a multimodal record of progress (folio of work)
  - » This is a collection of the work that you have done throughout the design process for Unit 1. Your work can be presented electronically or in hard-copy form, and can include a wide range of tasks in multiple formats. It will also include a record of physical 3D work, such as model-making, prototyping and production work (in the form of images and a journal/ production log)
  - » For ease, this will be called a **folio of work** throughout the textbook. For Unit 1, your teacher will advise you regarding which folio components will be assessed as a team and which will be assessed individually
- practical work
  - » models, prototypes, proof of concept and the finished product.

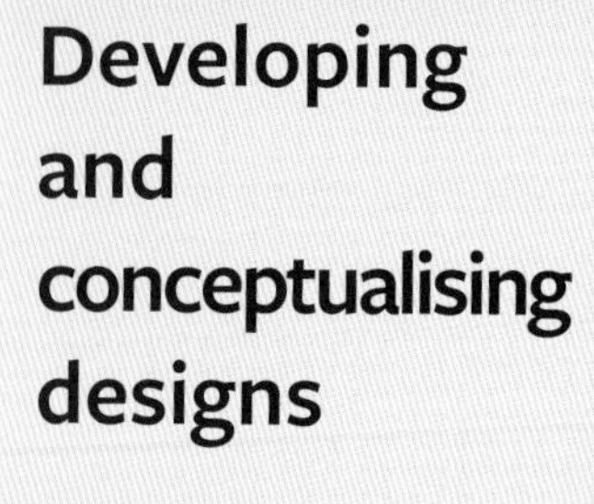

CHAPTER 1

Developing and conceptualising designs

UNIT 1, OUTCOME 1

In this outcome, you will:

- apply design thinking strategies to research, critique and communicate a response to a need or opportunity, and work collaboratively and in teams to develop and propose graphical product concepts that address a design brief.

Key knowledge

- activities and their purposes within the Double Diamond design approach to investigate and define, generate and design graphical product concepts, evaluate and plan and manage
- relationships between the Double Diamond design approach and design thinking strategies
- approaches, roles and responsibilities of working collaboratively and as a team
- methods to support collaboration and teamwork, including use of digital technologies
- elements of a design brief: need or opportunity, profile of end user/s, function, project scope (constraints and considerations)
- importance of design thinking when responding to a design brief
- Factors that influence product design
- purpose of, and methods to develop, evaluation criteria
- qualitative and quantitative research methods using primary and secondary sources
- methods to conduct ethical research and to appropriately acknowledge sources and intellectual property of others
- types of drawings to represent graphical product concepts: visualisations, design options and working drawings.

Key skills

- explain activities and their purposes within the Double Diamond design approach
- describe and apply design thinking strategies to refine graphical product concepts
- construct, justify and use research methods to gather data to investigate and define needs and/or opportunities
- explain and demonstrate approaches, roles and responsibilities to support working collaboratively and as part of teams, including the use of digital technologies
- formulate a design brief that addresses a real need or opportunity, with reference to Factors that influence product design
- develop and use criteria to:
  - » inform and evaluate graphical and physical product concepts
  - » evaluate processes to design and make the product
  - » evaluate the finished product
- use manual and digital technologies to represent graphical product concepts.

Source: *VCE Product Design and Technologies Study Design*, pp. 21–2

9780170477499

DEVELOPING AND CONCEPTUALISING DESIGNS

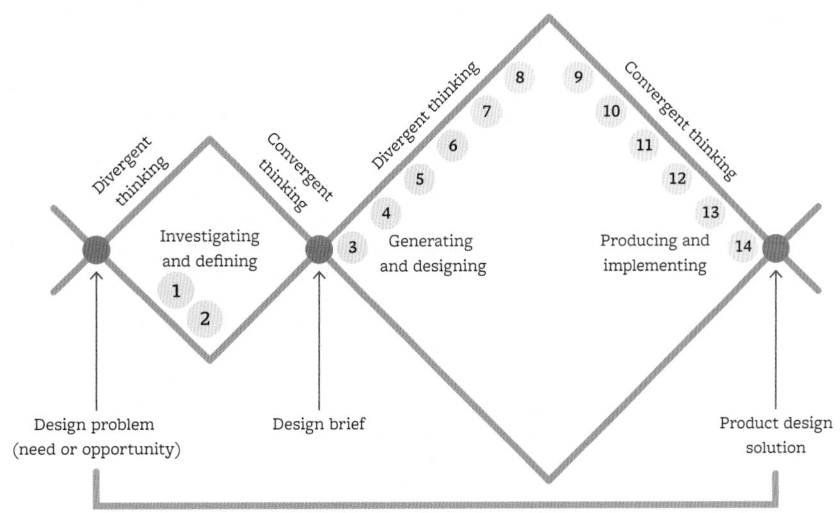

Identify and research a need or opportunity and possible end users

Define the design problem in a design brief

Formulate a design brief

Develop product concepts through drawings and prototyping

Activities:
- Mini design process **(p.6)**
- The Double Diamond interactive **(p.6)**
- The Double Diamond explained **(p.6)**

Case studies:
- Charlwood Design **(p.8)**
- Tim Garrow **(p.33)**
- Jim Hannon-Tan **(p.33)**
- Chapter 1 case study report activity

Worksheets:
- 1.1.1 The Double Diamond design process **(p.6)**
- 1.10.1 Teamwork **(p.14)**

Templates:
- Researching a need or opportunity **(p.17)**
- Product analysis PMI **(p.19)**
- Design brief **(p.22)**
- Evaluation criteria **(p.23)**

Quizzes:
- Chapter 1 revision **(p.33)**

Summaries:
- Chapter 1 **(p.33)**

Nelson MindTap

To access resources above, visit **cengage.com.au/nelsonmindtap**

The Double Diamond design process

In the *Victorian Certificate of Education Product Design and Technologies Study Design 2024–2028*, the Double Diamond design process diagram looks like this:

Activities
Mini design process

The Double Diamond interactive

The Double Diamond explained

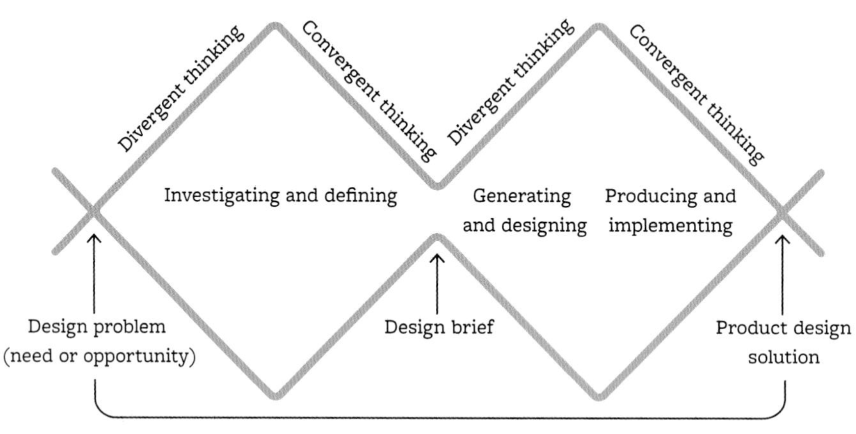

From *Victorian Certificate of Education Product Design and Technologies Study Design 2024–2028*, p. 13

The Double Diamond design process is a very flexible and adaptive process that changes to suit the situation or context and the different ways that designers think and work. So, for the purposes of this textbook and to make it easier to put into action, it has been slightly modified by changing the size of the diamonds. This better reflects the time you will spend in each diamond, and shows the balance of activities in each section.

Worksheet
1.1.1

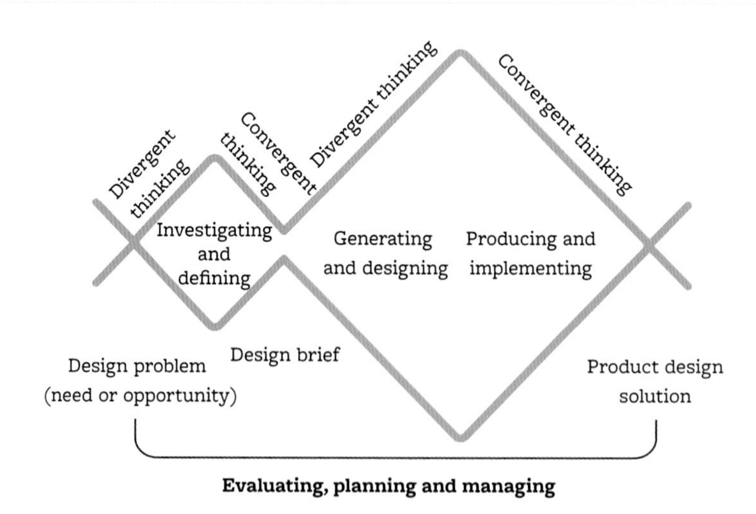

The Double Diamond process describes the process designers use to:

- identify and research a need or opportunity and possible **end users**
- define the design problem through writing a design brief (and evaluation criteria)
- research a wide range of information that can be used when designing
- develop and refine product concepts through drawings and physical modelling/prototyping
- identify the best solution (proof of concept)
- plan and produce the product
- evaluate the end result.

End user
Someone or something that might use the product or solution. Can be a person or a creature (human/non-human)

9780170477499

The Design Council (UK) developed the Double Diamond design process in the early 2000s. It requires two types of thinking, divergent and convergent (see pages 8–9). There are four main phases of this design process.

Investigating and defining

Discover (divergent)

Designers identify and research a need or opportunity. They spend time listening to people and learning about their needs, and consider opportunities that are open for new/improved solutions. Research is also done to identify the problems and benefits of products that are already available.

Define (convergent)

The problem or opportunity is clarified and defined in a written design brief. This outlines why a product is needed, for who, what it is required to do, etc. and its constraints (budget, size, time, etc.). Evaluation criteria are developed from the design brief. They are used throughout the process to guide the design, development and production to achieve a suitable solution.

Generating and designing

Develop (divergent)

During this stage, ideas are suggested, explored and sketched, developed, trialled and modelled (both digitally and physically). Useful research is carried out on materials, functionality, tools and processes. A range of viable options are drawn and the best ideas are prototyped, tested, and refined until the most effective solution is chosen.

Producing and implementing

Deliver (convergent)

Working drawings of the chosen design are developed, and detailed plans are made for production (focusing on efficiency, cost-effectiveness and safety). The product solution is made using the best available materials, tools and skills. Checks are frequently made to ensure the solution meets the quality standards required – this may lead to modifications. When the product is completed, it is tested and evaluated by the designer, manufacturer and end users.

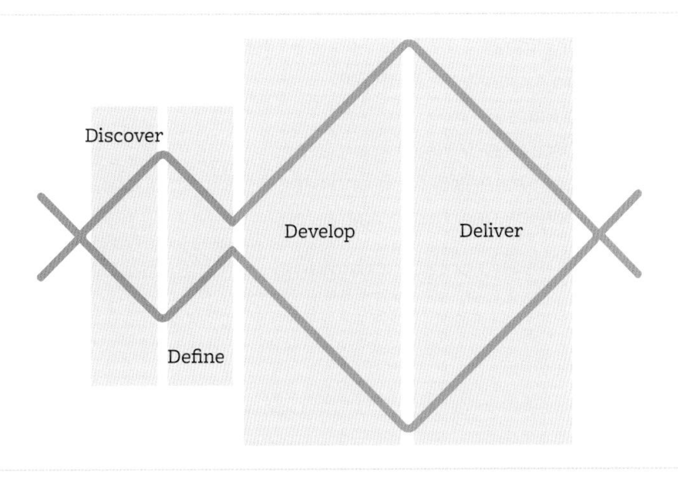

The design process doesn't always follow a linear path, with four clearly separated phases, as described on the previous page. Designers often jump from one phase to another, using different types of thinking depending on the needs of the design situation and the flow of their work.

Core design principles

The Design Council also outlines its four core design principles for problem-solving within the Double Diamond process:

- Put people first. Start with an understanding of the people (end users) using a service, and their needs, strengths and aspirations.
- Communicate visually and inclusively. Help people gain a shared understanding of the problem and ideas.
- Collaborate and co-create. Work together and get inspired by what others are doing.
- **Iterate**, iterate, iterate. Do this to spot errors early, avoid risk and build confidence in your ideas.

Iterate
Repeat, adjust and refine through small improvements

For more information about the Double Diamond design process, look up the Design Council's description of the process on their website.

Weblink
Design Council: Framework for Innovation

 Divergent and convergent thinking

You will have noticed the words 'divergent' and 'convergent' used in the section above. These terms describe different ways of thinking that are used during the design process. Both ways of thinking are essential for successful design.

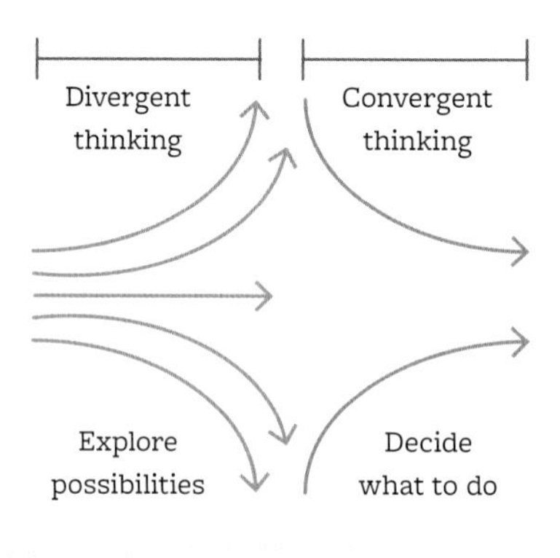

Divergent thinking

This involves being open and exploring a broad range of possibilities. Divergent thinking should increase the number of ideas and possibilities you have to play with, and help develop new ways of thinking about and solving problems. It involves being curious and creative, and taking risks, seeking inspiration, researching broadly, following leads, tips and hunches, and applying ideas and knowledge in new ways.

Activities and approaches that lead to divergent thinking

- Brainstorming and mind-mapping
- Broad exploratory research to gain ideas
- Viewing from different perspectives (points of view)
- Observing, questioning and listening to the end user
- Sketching and visualising ideas
- Using creativity triggers (such as SCAMPER)
- Trialling and modelling ideas
- Trying out new functional and/or visual approaches
- Using functional aspects, materials, tools and processes in an unconventional way
- Learning about and applying new technologies, etc.

Most of these activities involve **creative thinking**.

Case study
Charlwood Design

Chapter 1 case study report activity

Creative thinking
Thinking to come up with new and original ideas and explore new possibilities

9780170477499

Convergent thinking

This involves checking and using the information gained to refine ideas and make decisions. Convergent thinking is critical and analytical. Designers analyse the evidence they've collected to define and clarify the content of a design brief; they weigh up the pros and cons of each design concept, and analyse the results of trials, tests and modelling to choose the most appropriate design function or feature, or the best course of action; and they put together (synthesise) information and ideas to hone their concepts into the best solution for the end user/s.

Activities and approaches that lead to convergent thinking

- Using analysis of initial research to develop a design brief and evaluation criteria
- Critiquing existing ideas/solutions
- Analysing the results of trials and testing
- Modelling and prototyping
- Evaluating ideas and concepts against criteria
- Making decisions about the features, materials, processes, functions that are the most effective/appropriate
- Using evaluation criteria to consider the advantages and disadvantages of prototypes And choosing the best design solution (proof of concept) to produce
- Careful costing and budgeting
- Planning for production
- Deciding on modifications to improve the concept during production
- Evaluating the final product and considering areas for improvement

These activities involve **critical thinking**.

Critical thinking
Thinking that questions, researches, analyses and makes judgements/decisions

Teamwork: collaboration in design

Designers and makers pool their ideas, put forward suggestions and give constructive criticism, agree on certain aspects and share knowledge related to their areas of expertise. Designers often work as part of the team in a business setting. Many companies now employ designers to work in-house, where they have to **collaborate** with the business and its clients to solve problems. The designer brings a wide range of skills and a different perspective to product decision-making, but they need to communicate and justify their ideas clearly. Hopefully, teamwork is an area in which you will improve your skills while working in Unit 1.

Collaborate
To work together cooperatively for a common goal or task

Advantages of working in a team

Teamwork is considered beneficial for many reasons and is highly valued in industry.

- It brings multiple ideas and brains together – people get inspired when working closely with others, sharing ideas, skills and knowledge.
- Team members can bounce ideas off others rather than make all the decisions in isolation. Teamwork in the same room also has the benefit of quick responses.
- When the workload is distributed between a number of people, it can speed things up.

- Through team interaction, more energy and enthusiasm is created, which can improve motivation and relieve monotony.
- Members can support each other.
- When working as a team, everyone in the group gains from the special capabilities and skills of the individuals in the group.
- Working as part of a team gives you practice at presenting – speaking up and explaining your ideas to others.
- Teamwork improves your problem-solving, reasoning, cooperating, speaking and listening skills.

Rip Curl: an example of teamwork

Read through the case study on Rip Curl on page 57. At Rip Curl, all the design and development team of the wetsuit division are passionate surfers who care deeply about surfing and wetsuits. They constantly compare, share, make suggestions and confer about how to create the best solutions. Prototypes are made downstairs from where the designers work, and superb surf is only minutes away, so it's quick and easy to share the information, and trial and test the product to get it right, with almost immediate feedback.

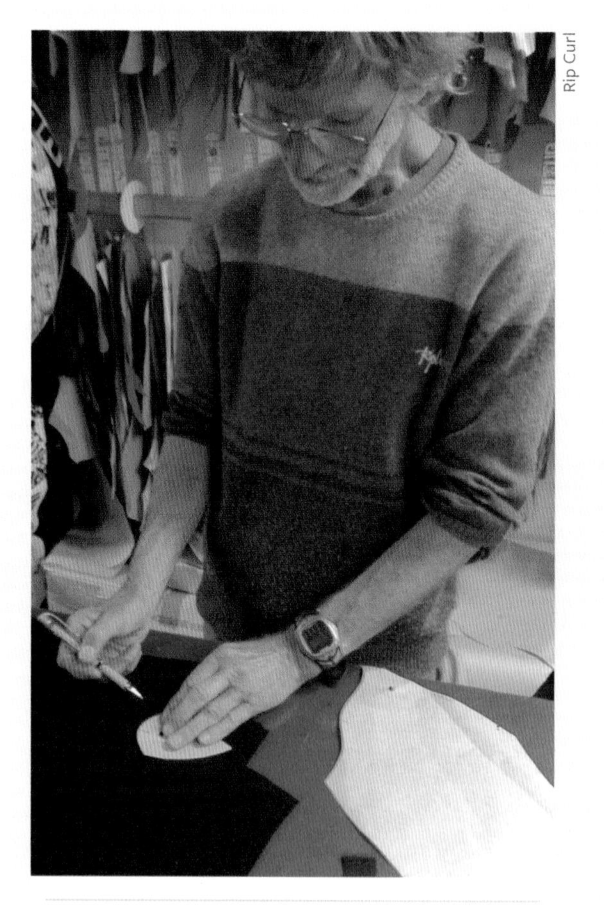

Rip Curl

Rip Curl pattern maker John 'Sparrow' Pyburne at work

Working as a team

There are some strategies used in business that could help your approach to teamwork. Try to:

- share your ideas and advise each other, but also allow for informal discussions as you work through the process
- make sure other class members know what they need to do.

For teamwork to be effective, each member needs to:

- contribute and communicate frequently and effectively
- ask for input from each team member and listen to their ideas
- express their ideas clearly and freely
- make constructive statements (i.e. how something could be improved) rather than simply criticising or being negative.

Use of digital technologies in your team

Many designers are located far from their clients or manufacturers. They use a wide variety of digital technologies to help them effectively communicate design work across an office, between states or around the globe. Read the case study of Rip Curl (page 57) to see how they use digital technologies collaboratively.

Wherever possible, create digital versions of your work in this unit to allow for easy sharing with others. Suggestions include:

- using a cloud-based sharing platform (e.g. Teams or Google Classroom) for sharing and editing documents, or creating a website on your school's system
- creating mind maps and mood boards in Microsoft Word and PowerPoint, Google Slides or with specific software
- sharing research in a Google Slides file with one or two slides per student
- using Illustrator or Photoshop for initial concept ideas, design options and technical 'flats' for textiles products
- using computer-aided design (CAD) to create 2D and 3D technical drawings for wood, composite materials, metal and plastics
- scanning any drawings done by hand and adding a digital table for others to give feedback
- manipulating scanned drawings and enhancing them with software; captions can be added. Learn how to reduce the file size of images
- uploading photos of work in progress into a shared journal or production log
- using a spreadsheet for timelines, costs, calculations, etc. to plan and monitor production done by the team.

Work for submission can be compiled digitally and presented in the order that best demonstrates the design process. To show individuals' contributions, work can be colour-coded.

Approaches to teamwork

In Unit 1, your teamwork can be approached in one of the following ways. You could:

- develop one product as a team
- work individually but collaborate with classmates at various points for ideas and information
- as a team, develop a product range based on a theme, sharing research and giving feedback.

Developing a product as a team

In this approach, a group of students design and make a single product together. The product needs to be substantial enough to allow input from each team member. At the designing stage, there needs to be close collaboration to make sure that each student is thinking in the same direction. This approach means that each team member is required to make a part, or that team members each take responsibility for certain processes or stages within designing and producing. It's important to plan carefully so that no-one is sitting around waiting to do their part, and to ensure that the product can be finished in the time available.

Working individually and collaborating

Each student works on their own product but joins group discussions, shares research and ideas, gives and receives feedback, assists others when creating models and learning construction processes, and collaborates when making decisions. Ideas, skills and information are shared.

Developing a product range as a team

A product range usually includes a number of products that have a similar style or function, or use materials in a similar way. In this approach, each team member gets to be fully responsible for a completed product, while using a team 'theme' or style, or sticking with team requirements in some way.

Students of Overnewton College in Keilor worked in teams to make a range of seating for the school grounds, after agreeing on some design and aesthetic features related to the surroundings.

9780170477499

Using divergent and convergent thinking in a team

Divergent/Creative thinking

Use the following activities to help you think divergently and create lots of ideas while completing the Double Diamond process.

Brainstorming

One of the most important creative techniques for a team is **brainstorming**, where each team member contributes. A brainstorming session is crucial at the beginning of the design process, and it may also be useful at other points throughout the process. Once a few directions have been decided on, team members can start to use other creative techniques to extend the ideas. Team members can use post-it notes to write as many different ideas as possible and 'post' onto a team sheet. This can be photographed and/or transferred to mind mapping software to record the ideas.

Brainstorming
Working as a group to share ideas and explore new ideas, encouraging creativity and problem-solving

Increasing your knowledge and skills

Knowledge of **aesthetics**, and the design elements and principles, will assist you to communicate and refine your visual ideas and concepts. Increasing your drawing skills will also help you to share your creative ideas with your team more effectively. **Technical** knowledge of a range of materials, tools and processes will give you a wider range of choices to be creative.

Aesthetics
The science or philosophy of beauty and taste, which relates to visual and tactile aspects of a product. It is explored through the design elements and principles.

Technical
Relating to the way things work in practice in industry, science or art

Idea sharing

Sharing ideas and researched information helps everyone in the team to grow from each other's efforts. It helps to spark new ideas. Asking for, and following up on, feedback from others could also help you to explore new directions in your designs.

iStock.com/SolStock

Sharing helps you to explore and refine ideas.

Convergent/Critical thinking

The following activities can help a team to think and work convergently to make the best decisions.

Questioning and providing critical feedback

If a team is to be effective, each team member must have a firm grasp of the requirements of the task.

Sharing critical suggestions should focus on the positive and negative aspects of the idea being shared, and areas for improvement or modification. It is important for the person whose ideas are being critiqued to listen carefully and think positively about differing opinions, without reacting defensively.

In a team, it is helpful to ask clarifying questions of each other. In response, members can refer to examples to assist in their justification.

Checking and evaluating

Critical thinking requires you to check and evaluate sources of information for credibility: who is saying what, and for what reason? What are the facts or truth? That is why critical thinking is sometimes known as 'scientific thinking'.

Evaluating requires the team to consider and discard information and ideas that aren't helpful and keep those that best suit the design situation, the time available and the skills within the group. It helps keep the purpose clear, and the work focused on the needs of the design brief.

Making clear and considered decisions

It is important to make good decisions efficiently. Shared research analysis, discussions, checklists and evaluating tools (such as rating concepts according to criteria) help to effectively make shared decisions. Regular meetings with others will give you time and opportunities to learn from each other and make decisions. Adding a time limit to make decisions will avoid wasted time.

Managing the process

In a team, it is important to plan so that every member uses their time efficiently. Allocate jobs that suit the skill set of each person and keep track of work that is completed and tasks yet to be done. Use checklists, spreadsheets and project management software to manage and monitor tasks.

Worksheet
1.10.1

ACTIVITY

As a team:

- identify members of your team and their skills/interests
- discuss and list the collaborative/communication skills needed for effective teamwork
- list the digital tools you can access in your school to assist in your team's design process.

Beginning the design process

Conducting research to investigate and define the need

To begin the design process, you first need to determine the reason for creating something new, or in other words, what need or problem your product is going to address. To do this, you will need to complete a range of steps. Some suggested steps for investigating and defining the need are to:

- brainstorm to identify a need
- conduct initial research, particularly focusing on the user/s and their needs
- set team goals/requirements
- write the design brief
- write the evaluation criteria
- conduct further research.

Identifying the need

As mentioned above, the first step in the design process is finding the reason to create something new. A designer might identify a need or opportunity in a number of ways, including:

- perceiving a need through personal experience. Needs are often identified as a result of some daily activity, or through interaction with family or friends who have identified a need through their own experience, or through direct observation of others
- identifying the need to improve something. The need may arise because a product doesn't function effectively or may come from the desire to improve something ergonomically. New technological developments (such as a new material or construction process) might also provide the opportunity to improve a product or provide an easier method of performing a difficult action
- market-led designs or opportunities. These usually come about when someone identifies a gap within an existing market and creates a product or solution to fill the gap
- creating ethical/sustainable change. Many designers have a desire to create positive change by improving the ethics/sustainability levels of a product. You can read more about this in Chapter 6.

The *Oi* bike bell by Knog

A good example of a market-led product that filled a gap is the bicycle bell called *Oi*, developed by Australian company, Knog. They wanted to create a bell that was more aesthetically interesting and user-friendly than traditional bicycle bells. Go to the Knog website for more information about their innovative bicycle and outdoor products.

Knog

The *Oi* bike bell developed by Australian company, Knog

Researching end user needs

To identify a need, you or your team can:

- observe friends or family to determine any needs they may have
- consider your own peer group and any needs they might have

- reflect on and draw from your own experiences
- listen to people with a special need
- talk to a typical end user about their experiences, needs and requirements
- examine consumer demands through market research (e.g. surveys or questionnaires)
- research and be aware of new technologies that could create new opportunities or needs
- observe or research problematic situations that could be improved with good design
- be aware of social trends
- use speculative thinking and be forward-thinking about what people might need in the future.

FOLIO ACTIVITY

Brainstorm

1 As a team, brainstorm situations that you have experienced, observed or are aware of that might create a need or opportunity for a new or improved product and record this activity (photograph or create a digital version). Your teacher may give you some guidelines about areas to focus on.

2 Decide on your approach (see the heading 'Approaches to teamwork' on page 12), the type of product/s and your end user/s and the areas/situations that you or your group will focus on.

Primary and secondary research

Primary research
Information that is created first-hand

Secondary research
Information collected and published by others

To explore your end users' needs, your team should carry out both **primary research** and **secondary research** (this describes the sources of your information).

- Primary research covers investigations that you have carried out personally. It may include your own photographs, observations, interviews, surveys and questionnaires, etc.
- Secondary research is research that other people have done. It can be carried out by searching the internet, or by reading books and catalogues or any other published materials (in print or digital form) that have been created by another person. This material may include videos and other forms of multimedia.
- Whenever you use someone else's research or content, you need to acknowledge who did the research and where you found it: cite the writer of the article, or designer or manufacturer of the product, the name of the website where you found the content or its URL, etc.

Both types of research are valid and useful.

Quantitative and qualitative research

The research information or data that you collect will be either quantitative or qualitative. These terms describe the type of information or data you have collected, and may guide how you present your information.

- Quantitative data is research information that can be explained with numbers. It is usually collected from a large group of people who answer simple questions to which there are a limited range of responses. Their responses can be grouped and the results shown through graphs, tables and charts. Quantitative data can also be information gained from a secondary source, and may include specifications and measurements that can be directly compared.
- Qualitative research data is detailed, individual and has a lot of depth. It is usually gathered from a small number of people in response to open-ended questions. This data cannot be easily generalised, and is usually presented as written text. Qualitative research may also be in-depth research gained from a secondary source about the background and needs of particular groups of people.

See Chapter 2 and Chapter 7 for more information about different research approaches and data types.

9780170477499

Ethics when carrying out research

It is important that research is conducted in an ethical way, considering the needs and rights of others. There are different ethical considerations when carrying out primary research (research that you directly collect) and secondary research (research that someone else has collected). See more on ethics in research in Chapter 7.

Ethics in primary research

When carrying out primary research, you and/or your team need to:

- ensure that you have **informed consent**. Make sure that the person or people who you are interviewing, in discussions with or observing understand what your research information will be used for and give their consent for it to be used. If you are observing children, you may need to have parental consent
- **do no harm**. Ensure that your research doesn't put any person or animal in a situation that may cause physical or mental harm. For example, don't encourage the use of an existing product in a way that is dangerous, or ask questions that may cause anxiety
- **show respect for privacy**. Don't use the names or photos of people in your research unless you have permission, and respect the confidentiality of people's responses, particularly if they are of a sensitive nature.

Examples of primary research include taking your own photos, taking notes from observations, interviewing or surveying your end user and conducting your own material tests or process trials.

Ethics in secondary research

When carrying out secondary research, you and/or your team need to:

- appropriately **acknowledge the sources** of the research and the creators of any 'works'. Work by others should not be copied and misrepresented as your own work. See page 199 in Unit 3 on acknowledging sources in your folio
- ensure the research is **presented honestly**, **analysed appropriately** and **conclusions are valid** (i.e. research is not taken out of context, modified or misinterpreted).

FOLIO ACTIVITY

Conduct your initial research

Research aspects related particularly to the needs of the end user/s of the product, how it might fit into or improve their lifestyle, details about how they might use the product, and issues such as ergonomics, safety and comfort. Read about the end users Factor in Chapter 13 (pages 368–75) and see what is relevant to your project. Think about how research around the end user can be incorporated into the design of your product or products.

1. Plan actions you can take as a team to research the situation around your end user and their needs. These might include:
   - formal interviews or informal discussions with potential end users
   - surveys/questionnaires
   - targeted observations and note taking
   - collection of images relevant to the end user's interests and needs
   - research into new markets or social trends
   - research into new materials, equipment or processes that might be accessible to your team.

   Make sure you carry out both primary and secondary research.
2. Allocate different research tasks to team members. Make sure you follow the ethical guidelines outlined above when you carry out your research.
3. Collate, annotate and upload your research into a file, folder or drive where it can be accessed by others. As a team, discuss what you have discovered.

Template
Researching a need or opportunity

Research into existing solutions/products

Research into existing products helps you to understand and learn from the problem-solving work of others, and to learn from their mistakes. This research is much more than just providing a photo collage of similar products to the ones you will be designing and making. You need to analyse the advantages and disadvantages of the product you have found. This will provide you with:

- features of products that are useful and that you might want to include in your designs
- features that could be improved, adapted and modified to make a better product.

A Plus, Minus, Interesting (PMI) activity is a useful way to consider the pros and cons of each product. When completing a PMI, make sure you consider most of the following features/factors:

- aesthetics – the visual appearance of the product
- function – how the product works and whether it works well for the type of end user you are designing for
- cost (if available)
- use of materials and processes
- quality of production and finish
- level of sustainability.

Through this research, you will start to build up a range of options and directions you can refer to and be inspired by during the design stage to develop new, innovative concepts for your design situation.

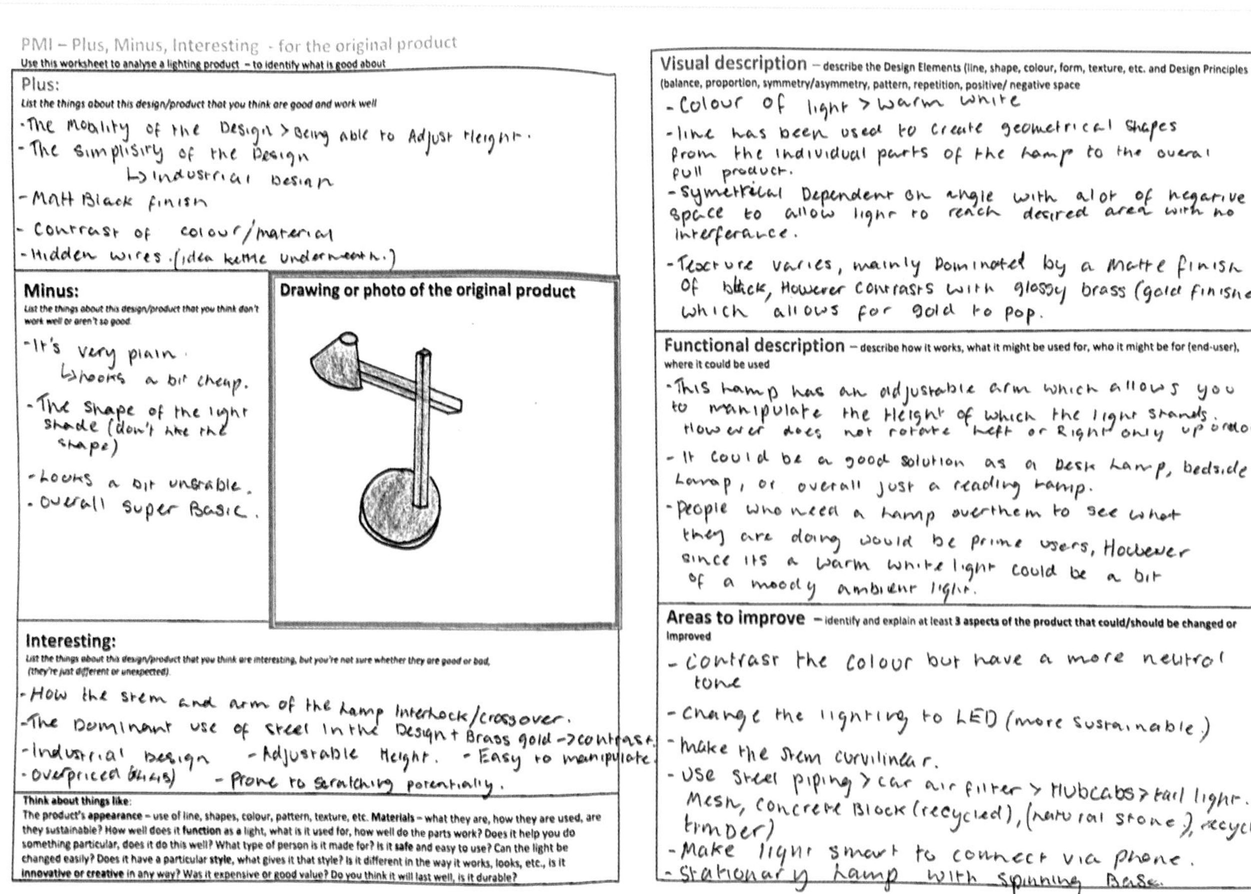

PMI – Plus, Minus, Interesting - for the original product

Use this worksheet to analyse a lighting product – to identify what is good about

Plus:

List the things about this design/product that you think are good and work well

- The Mobility of the Design > Being able to Adjust Height.
- The simplisity of the Design
  ↳ Industrial Design
- Matt Black finish
- Contrast of colour/material
- Hidden wires. (idea kettle underneath.)

Minus:

List the things about this design/product that you think don't work well or aren't so good

- It's very plain.
  ↳ looks a bit cheap.
- The shape of the light shade (don't like the shape)
- Looks a bit unstable.
- Overall super Basic.

Drawing or photo of the original product

Interesting:

List the things about this design/product that you think are interesting, but you're not sure whether they are good or bad, (they're just different or unexpected).

- How the stem and arm of the lamp interlock/crossover.
- The Dominant use of steel in the Design + Brass gold -> contrast.
- Industrial Design
- Adjustable Height.
- Easy to manipulate.
- Overpriced ([illegible])
- Prone to scratching potentially.

Think about things like:

The product's **appearance** – use of line, shapes, colour, pattern, texture, etc. **Materials** – what they are, how they are used, are they sustainable? How well does it **function** as a light, what is it used for, how well do the parts work? Does it help you do something particular, does it do this well? What type of person is it made for? Is it **safe** and easy to use? Can the light be changed easily? Does it have a particular **style**, what gives it that style? Is it different in the way it works, looks, etc., is it **innovative or creative** in any way? Was it expensive or good value? Do you think it will last well, is it durable?

Visual description – describe the Design Elements (line, shape, colour, form, texture, etc. and Design Principles (balance, proportion, symmetry/asymmetry, pattern, repetition, positive/ negative space

- Colour of light > warm white
- line has been used to create geometrical shapes from the individual parts of the lamp to the overal full product.
- Symetrical Dependent on angle with alot of negative space to allow light to reach desired area with no interferance.
- Texture varies, mainly Dominated by a Matte finish of black, However contrasts with glossy brass (gold finished) which allows for gold to pop.

Functional description – describe how it works, what it might be used for, who it might be for (end-user), where it could be used

- This lamp has an adjustable arm which allows you to manipulate the Height of which the light stands. However does not rotate left or Right only up or down
- It could be a good solution as a Desk Lamp, bedside Lamp, or overall just a reading lamp.
- People who need a lamp over them to see what they are doing would be prime users, However since its a warm white light could be a bit of a moody ambient light.

Areas to improve – identify and explain at least **3 aspects of the product that could/should be changed or improved**

- Contrast the colour but have a more neutral tone
- Change the lighting to LED (more sustainable)
- Make the stem curvilinear.
- Use steel piping > car air filter > Hubcabs > tail light. Mesh, Concrete Block (recycled), (natural stone), recycled timber)
- Make light smart to connect via phone.
- Stationary lamp with spinning Base.

A PMI analysis of a lamp carried out by student Noah Veljanovski

9780170477499

FOLIO ACTIVITY

Initial end user and problem/opportunity research

Collect a range of images of existing products related to your design situation. Of those, select and analyse at least three products, using a PMI approach. Make sure you acknowledge the sources, brands or designers of these products in your work.

Template
Product analysis PMI

Writing your design brief

Your design brief will need the following components:

- an end user/s profile
- purpose/function of the product
- project scope (including constraints and considerations).

End user profile

You will need to create a brief end user profile. This will summarise information about them gained from responses to your initial research – from observations, interview questions and market or situation research. It does not need to be detailed with private information, but it should paint a picture of the user/s in relation to their situation and need. If your end user is non-human (animal), either discuss some of these issues with an expert or consider the needs of the animal's owner.

Some suggestions for information to include are:

- gender or age bracket
- living situation (if relevant – e.g. family, housing), whether they live in a city, inner/outer suburb, country/coastal town or on a farm
- income level (in general terms – i.e. high, medium, low) or willingness to spend
- their taste or things they like visually (colours, lines, shapes, textures, objects, architecture, etc.)
- special needs (if relevant)
- values – attitudes to spending, sustainability, quality, function or aesthetics
- passions and lifestyle interests that are relevant.

The end user profile should include some key images and information, which can be presented under suitable subheadings.

Purpose/Function of the product

This section briefly defines why the product is needed, what it is for, and what it needs to do. Make sure that you are fairly general at this stage and focus on the product's purpose. Don't describe how the finished product could look, or specifically define how it will function or work. You need to have room to develop creative solutions to the problem. This section can be short – about one or two sentences.

Project scope

This is a simple statement where you clarify and summarise the details of what your or your team's project is (what you aim to accomplish through the design process). The following questions will help you:

- Who? Briefly summarise who will be using the product (from your profile).
- Why? Why do they need the product? What issues are being solved?
- What? What does the product need to do (what is its purpose)?

- How? How might this benefit the end user/s? How much can they afford?
- Where? Where will the product be used? In what sort of situation or environment?
- When? When might the product be used? How frequently? For a particular event?
- What? What does your team need to deliver at the end of the process?

Constraints and considerations

As part of the **project scope**, you also need to outline the requirements of the situation: the things the end product must have (constraints) and those aspects that require discussion or further research (considerations). These are clear statements in dot-point form. Consider the **Factors that influence design** when writing this section – they will help you to identify the areas you need to include in your constraints and considerations. See more on the Factors that influence design in Chapter 13.

Factors that influence design

Throughout the design process, there are a range of **Factors that influence product design**. These are important areas that designers consider when making design decisions. You also need to identify and consider them through your design process. They cover:

- need or opportunity
- function
- end users
- aesthetics
- market needs and opportunities
- product life cycle
- technologies: materials, tools and processes
- ethical considerations in design.

In your design brief, make sure your constraints and considerations are related to a range of these Factors. For more information about the Factors, refer to Chapter 13.

Your design brief should be *brief*. All parts should fit onto a single A4/A3 page or folio slide.

Design briefs for different team approaches

If developing one product, a single 'team' design brief is all that is required. Your team will need to work together to compile the information gathered from your research and write a team brief that is uploaded into the team folio.

However, if creating an individual product, decide on a short team statement outlining the team direction and requirements, and then complete a more detailed individual brief using the structure above. If developing a product range, the team statement might describe a particular style the team products need to fit within, or a specific range of materials or colours to incorporate, or it might describe a particular product type, environment/setting or functional aspect that links the team products.

Your teacher will give you guidance about the design brief format your team will need to use.

Example of a team design brief for a product range

The following brief is an example of a comprehensive approach to this task. Students should select the aspects that are suitable for their project.

Shutterstock.com/rickyd

End user profile

The end users of our products are young adults who love camping and value sustainability.

Age	16–25 years old, not gender-specific
Living circumstances	Mostly students living at home with family, though some of the older adults in the age group might be living independently.
Income level	Low, the products will need to be low cost.
Interests	Camping and being in nature, some like the challenge of bushwalking (carrying all their equipment with them), but others prefer camping in a place that is accessible by car.
Special needs	Not really relevant, most are fit and active.
Values	Most people in the end user group value the natural environment, and wish to be as environmentally sustainable as possible in their product choices, particularly in relation to material choices. They also appreciate the latest in walking and camping tech.
Tastes, aesthetics	From our research, there were two main interests in terms of colour – some like natural colours that blend into the environment, others like bright colours that are easily visible. Most in the group prefer either natural materials or high-performance outdoor materials.

My individual purpose/function

To create clothing accessories for campers that provide warmth, are lightweight and pack efficiently, and are multipurpose.

My individual project scope	
Keen campers and bushwalkers like to go camping throughout the year, but need products that keep them warm in colder weather. Campers are limited in the amount of clothing and equipment they can pack, so creating multi-purpose products that keep them warm and provide another benefit will save space and weight. The end users want their products to be environmentally sustainable, and the materials used need to function well in cold conditions and provide thermal insulation. The products need to incorporate either natural or bright colours. Our team needs to develop four different clothing accessory products that incorporate a second function, e.g. provide light, storage, GPS tracking, etc. (If the second function is electronic, this can be modelled rather than being fully functional.)	
Constraints and considerations	
Budget	Cost limit of $20 per product
Time frame	Completion date: start of June
Function	Functional requirements: • to keep the wearers warm • include a second function, e.g. can be used to store accessories, have an embedded light, or can be used as a pillow • be lightweight and pack efficiently.
End user considerations	Must be comfortable to wear, and fit the size of a typical young adult.
Aesthetics	Colours should either: • be based on colours in nature or • be bright for greater visibility in the bush.
Ethical considerations	Need to be made of materials that are environmentally sustainable (e.g. are biodegradable when disposed of or extremely long-lasting).
Quality	Need to be made to a high level of quality, and last well.

FOLIO ACTIVITY

Writing your team/individual design brief

Create your team or individual design brief using the sections above:

- Team theme/focus (optional)
- End user profile
- Purpose/function
- Project scope
- Constraints and considerations (indicating specific areas covered).

It can be formatted in a table like the one shown above, or using any other format, provided the sections are clear. Try to keep your brief to one folio page.

Template
Design brief

Evaluation criteria

Evaluation criteria identify the standards your product needs to meet to be successful. They are sometimes called success criteria.

Evaluation criteria can be written collaboratively straight after the design brief is finalised, or after the research is completed. Evaluation criteria help to keep the group work well-directed and to keep the focus on the agreed objectives of the finished solution.

Evaluation criteria have the following features:

- They need to be drawn directly from the design brief, usually from its constraints and considerations, repeating or emphasising what is important.
- They can be written as a question.

9780170477499

- They need to be clear and specific to the product.
- You need to be able to test the product against the criteria to judge whether the product has succeeded in each specific area. The product can be tested by:
  - carrying out physical tests and trials
  - seeking feedback from your team members and end user/s
  - comparing it against other similar products
  - observing the product when it is used, and over time.

Evaluation criteria from the design brief above

Criteria question
Do the products cost less than $20 each?
Can the products be, or have they been, finished by the June deadline?
• Do the products keep the wearers warm (are they thermally insulating)? • Does each product incorporate a second function (is it multifunctional), e.g. can it also store accessories or function as a pillow? • Are the products lightweight and can they be packed efficiently?
Have bright or natural colour schemes been used in the products?
Are the products environmentally sustainable (particularly their chosen materials)?
• Are the products soft and comfortable to wear? • Can they be worn by a range of differently sized end users?
Is the product of high quality and will it last well in harsh outdoor environments?

If developing individual products rather than a group product, each person may need to develop evaluation criteria that are specific to their design situation.

Evaluation criteria

As a team (or as an individual), develop at least **5–8 evaluation criteria**. Make sure that they come directly from your design brief, and that they cover a range of Factors.

Template
Evaluation criteria

Next stage: drawing!

To help inspire you during the drawing stage, you might like to do a bit more research to get additional ideas from other product examples.

Understanding drawing for product design

You will need to produce three different types of drawings to help you develop and visualise your ideas during the design process. The three types are:

- visualisations – sketching your initial design ideas, used to explore a large number of ideas
- design options – two or more polished and realistic drawings, used to show others, to help choose the best solution
- working drawings – detailed technical drawings of your chosen design option, used to guide you through production.

In the Study Design, drawn ideas are called 'graphical product concepts'.

Visualisations

The drawings known as visualisations are the first step in the design and development stage, where you play around with and explore potential ideas for the whole and/or parts of a solution to meet the brief. Graphic organisers, mock-ups and 3D models can be used in this step but mostly you will use visualisation drawings, which are quick sketches, called thumbnails, concept or idea sketches. These drawings can be rough or refined; detailed or simple. Both the sketches and any models are a record of your design thinking in this early stage and show the influence of your research (which you should place nearby on the same page) and annotations. See more on page 25 of this chapter.

Your folio should demonstrate links between your research, visualisations and your design options along with exploration of the design elements and principles (part of the aesthetics Factor).

Design options drawings

Design options drawings are 3D views (realistically drawn for clothing) that show more detail than visualisations and indicate what the whole product will look like. They are rendered and coloured to represent form and textures. Annotations indicate the proposed materials and construction processes; explain how the product will suit the end user and address the brief; and explain aspects that cannot easily be seen in the drawing. See more on page 31 of this chapter.

Working drawings

Working drawings are needed once the concept is finalised. Working drawings (or technical drawings) are used to show exact measurements, shapes, joins and placement of components for construction or manufacture. They need to be drawn to scale and accurately, so CAD can be used – or rulers/templates if drawing by hand. See more in the next chapter.

Graphical product concepts and drawing systems

There are particular systems of drawing that are suitable for wood, metal and plastics products, and others for fabrics products. Within each area, there are further specific systems of drawing that are useful for visualisations, design options or working drawings. The tables below describe the different drawing systems and where they are most appropriately used.

Suitable drawing systems for wood, metal and plastics products

Drawing system	Purpose/ Description	Visualisations	Design options (presentation drawings)	Working drawings
Concept sketches; mock-ups and 3D modelling	To express a lot of ideas quickly for parts or the whole of a product, either as drawings on paper or in 3D form	✓		
Oblique: cabinet and cavalier	A semi-realistic, pictorial 3D view of an object with accurate measurements (front flat and sides, 45° axis), rendered to represent materials		✓	

9780170477499

Isometric	A semi-realistic, pictorial 3D view of an object drawn to scale (30° axis), rendered to represent materials		✓	May be used to support 2D drawings, or for showing details
Perspective	A realistic 3D view of the object, using one, two or three vanishing points, rendered to represent materials		✓	
Orthogonal	Aligned, flat 2D views of at least three sides of the object (to scale). Can be generated from CAD drawings			✓
Hidden details and exploded diagrams	Used to show details not visible in full drawing or to explain construction. Increased size of a section, or parts separated to show joint shaping		✓	✓

Drawing systems for textiles

Drawing system	Purpose/Description	Visualisations	Design options (presentation drawings)	Working drawings
Concept sketches; mock-ups and 3D modelling	To express a lot of ideas quickly for parts or the whole of a product, either as drawings on paper or in 3D form	✓		
Fashion illustration	Fluid drawings, rendered, giving a strong impression of a particular style, showing details such as shoulder seams, buttons etc.	✓	✓	
Descriptive drawing	Clothing drawn on the body – clear, elongated proportions, including details and colour		✓	
Flats	Clothing drawn off the body – 'flat', front and back views, showing all details accurately			✓
Construction diagrams	Detailed diagrams explaining a construction step, showing seams and front/reverse side of fabric			✓

Refer to Chapter 12 for more detailed information on the different drawing systems and student examples.

Developing visualisations (initial design ideas)

This phase of the design process gives you the freedom to really explore and try out a lot of new ideas. Even though you might use existing products as a starting point, the idea is to expand on these, to be creative and experiment with different ways of thinking about the product.

Remember these things:

- Have fun and play with ideas.
- Don't be too critical to start with – your ideas don't have to be practical and 'makeable' at this stage.
- Don't throw anything out – even if you don't like an idea or it is totally impractical, it still shows the flow of your design thinking.

- Experiment with:
  - the visual appearance of the product (using the design elements and principles), using trigger words such as those in the SCAMPER design activity
  - how the product functions and many variations for different parts
  - other things the product might do in combination with its main function
  - different materials and how they might affect your ideas
  - different drawing media, such as pencils, fineliners, watercolour inks, markers, cut paper.

Visualisations are mostly presented as sketches – drawings that are quick and not very polished or carefully done. They can include indications of colour and material texture, but they are not expected to be fully rendered. You can also use other visualising methods to support or extend your sketched ideas, e.g. 3D models, graphic organisers or flow charts.

Annotations

You need to reflect on your visualisations – this is done through the annotations and comments you write around, and on, your drawings (or on photos of models). Your annotations can be descriptive, but they should also show some analysis of what you think will work and what won't (explaining why), what looks good or not so good, and features that can be explored further.

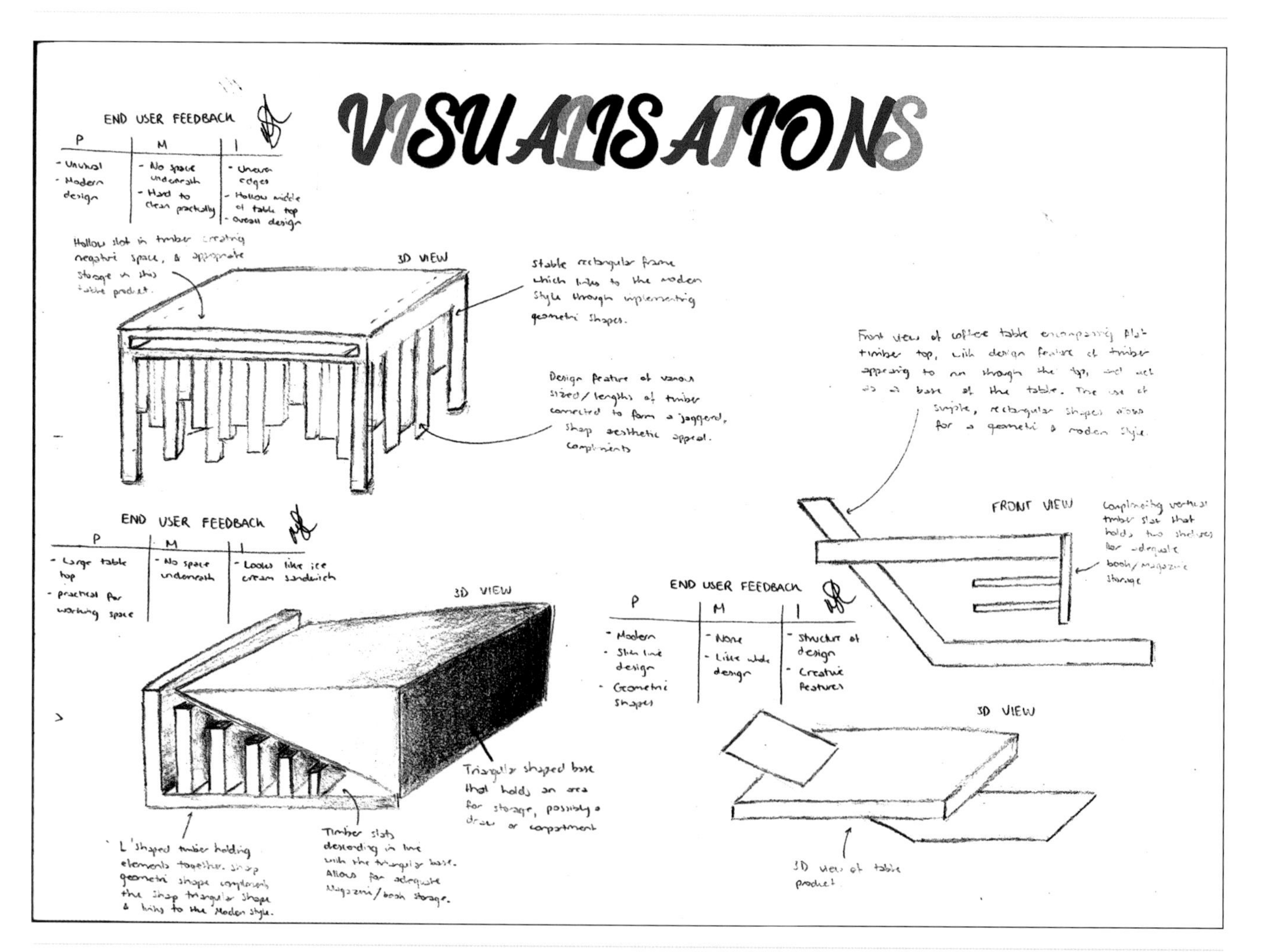

Visualisations created by student Stephanie Gouder

Mood board used as a design influence on visualisations by student Jasmine Lee

Visual mood board for your team's theme

You might find it useful to develop a mood board to clarify your team's visual theme. Find images that reflect the colours, shapes, textures, materials and overall style of your team's theme. Refer to Chapter 12 for more details about what a mood board is and how to create one.

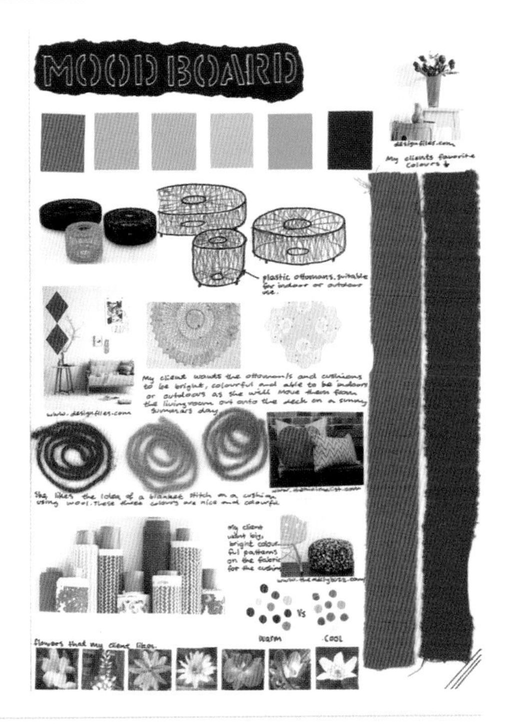

Mood boards by students Demi Spyropoulos and Sophie Allen

Digital visualisations and modelling

Computer-aided design (CAD) enables you to quickly draw and render initial ideas for your product. Use CAD to play around with shapes, materials and the overall form of the design (the visuals) or focus on a functional aspect to try out different ways of making something work. The advantage of this form of visualisation is that you can use your best visualisation ideas as a basis for your more developed concepts in CAD.

Any computer drawing program, such as Adobe Illustrator, can help you play around with visual ideas for textiles products. It is very easy to change shapes, colours and patterning to get a quick idea of those visuals that work effectively and those that don't.

Make sure you record your ideas as you go. Some CAD software allows you to step back in a file's history. Take screenshots or download your work at different stages to show the breadth and progress of your ideas. Include this material in your folio.

iStock.com/EvgeniyShkolenko

Using physical modelling to explore ideas

Many designers find it helpful to make simple physical models during the visualisation stage. They play around with inexpensive materials (paper, cardboard, scrap fabric) to explore shape, form, colour combinations and basic functions. You can use a range of methods and materials to help you brainstorm and expand your visual ideas. You can also create small 3D-printed models to visualise your ideas.

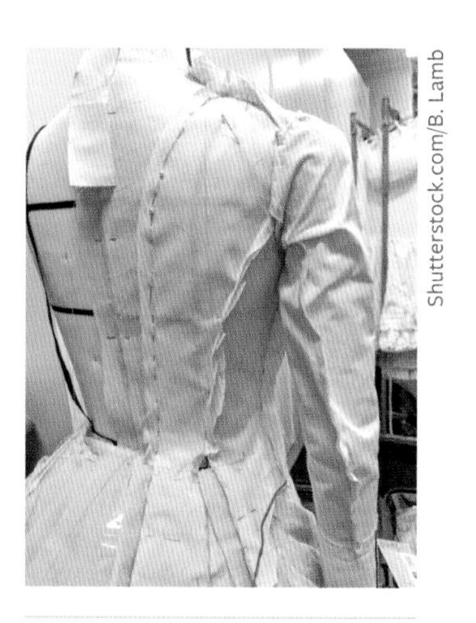

Shutterstock.com/B. Lamb

An example of a physical model

Chapter 2 has detailed suggestions for how to carry out physical modelling in different materials. Read the section on pages 45–7.

9780170477499

Using design elements and principles in your designs

When developing your visualisations, how you use the design elements and principles will determine how your product ideas look. The design elements and principles are the main content of the aesthetics factor and relate to the visual appearance of products.

You need to recognise and become very familiar with design elements (the building blocks of visual design) and design principles (how the elements are put together) and learn to experiment with them.

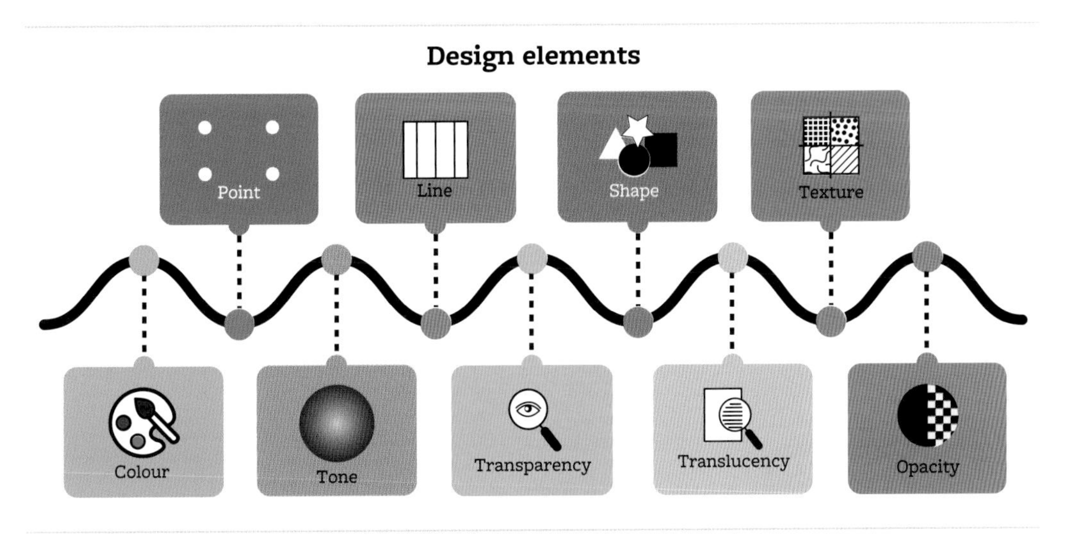

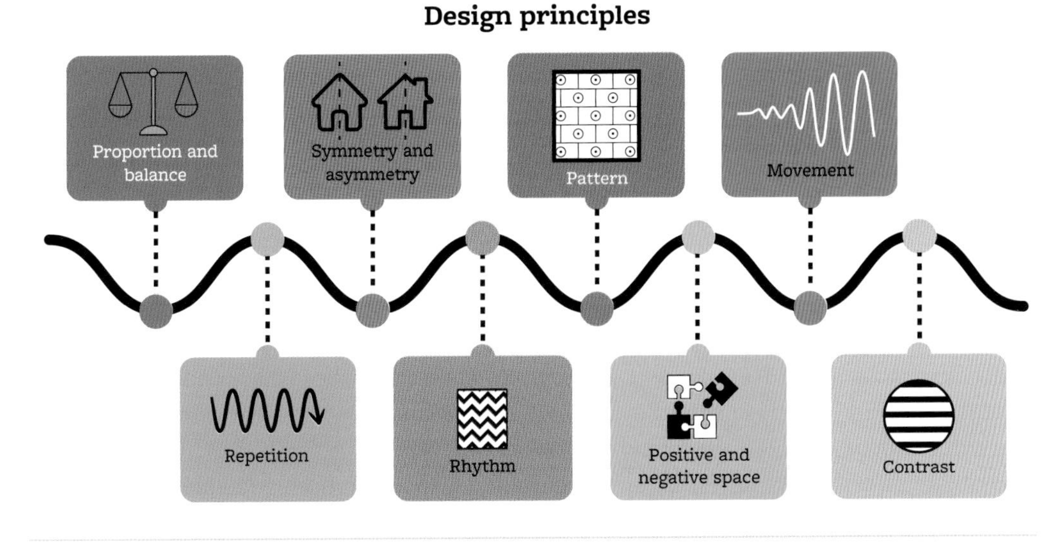

Aesthetic or iconic design styles select and use the design elements and principles in specific ways. You might recognise a product made in the Art Deco, Memphis or any other style by the way they use colour, shape, line, pattern, balance, contrast, etc. because designers of these products have a similar approach to their use. Your team may decide on further research and analysis to find a suitable 'set' of design elements and principles from a known style to follow.

Please refer to Chapter 12 for very detailed information on how the design elements and principles are used with different materials and product types. You may need to experiment and practise using these to develop your skills and your confidence in visual design.

Alamy Stock Photo/Azoor Photo

A Memphis-styled product with its recognisable use of colour, shapes/form and pattern

FOLIO ACTIVITY

Exploring ideas through drawing and modelling

Develop pages in your folio that explore ideas and concepts for the whole product and features or aspects of the product. Use the list of suggestions to experiment with on page 26 to help you come up with a wide range of ideas. When generating ideas, make sure that you refer to the research you've gathered (about your end user and existing solutions) to spark ideas and help you explore different directions.

Team and end user feedback

Remember that working as a team gives you the advantage of having peers to give you feedback on your ideas. This can be a bit daunting at first, but these discussions about your designs can spark new ideas and directions – things you may not have come up with by yourself. It can be useful for someone to look at your ideas with 'fresh eyes'. When you ask a classmate to **critique** your initial ideas, they should focus on both good and not-so-good features and make suggestions for modifications and improvements. Decide on the best places for feedback, e.g. on visualisation pages, next to photos of models and in digital files, and what colours or what method to use. You might keep a special team feedback file.

Critique
To review and appraise; to provide an analytical assessment

9780170477499

Hints for **giving** good feedback include the following:
- Use your evaluation criteria to focus your feedback.
- When giving feedback to your classmates, focus on the ideas in the drawings/model. Don't criticise the person and their visualisation skills.
- Phrase things positively, honestly and constructively.
- Provide suggestions that build on their ideas, rather than going off in a different direction.
- Clarify your suggestions if needed.

Hints for **taking** feedback include the following:
- Take notes, write suggestions down, ask classmates to write or draw their suggestions on post-it notes around your work.
- Don't take suggestions personally. Your classmates are not attacking you when they critique your work; they are considering your ideas.
- If you don't understand a suggestion, ask for clarification and have a discussion.
- Use the feedback to improve your work. Try out new suggestions and directions, and use the opportunity to refocus and refine your design work.

FOLIO ACTIVITY

Team activity: feedback
- Organise and schedule a meeting, at which each member shows and discusses their ideas, and gains feedback from other team members. Make sure that everyone has the opportunity to gather and to give feedback.
- Discuss and record the feedback given.
- Use the feedback as the starting point for creating new ideas or modifying your existing ideas. Add these to your folio.

Developing design options

You or your team now need to develop a range of design options that satisfy the design brief.

For one group product, only one or two very detailed design options per member need to be drawn. For individual products, at least two design options need to be developed.

Your design options need to include:
- a 3D view of each design idea, which needs to be clear, realistic and detailed
- colour and rendering (tone and texture) to indicate materials
- annotations that describe your idea (materials, joining methods, components, etc.)
- annotations that explain how your product design meets the requirement of the brief.

For specific information about the type of presentation drawing required for your design options, go to pages 329–34.

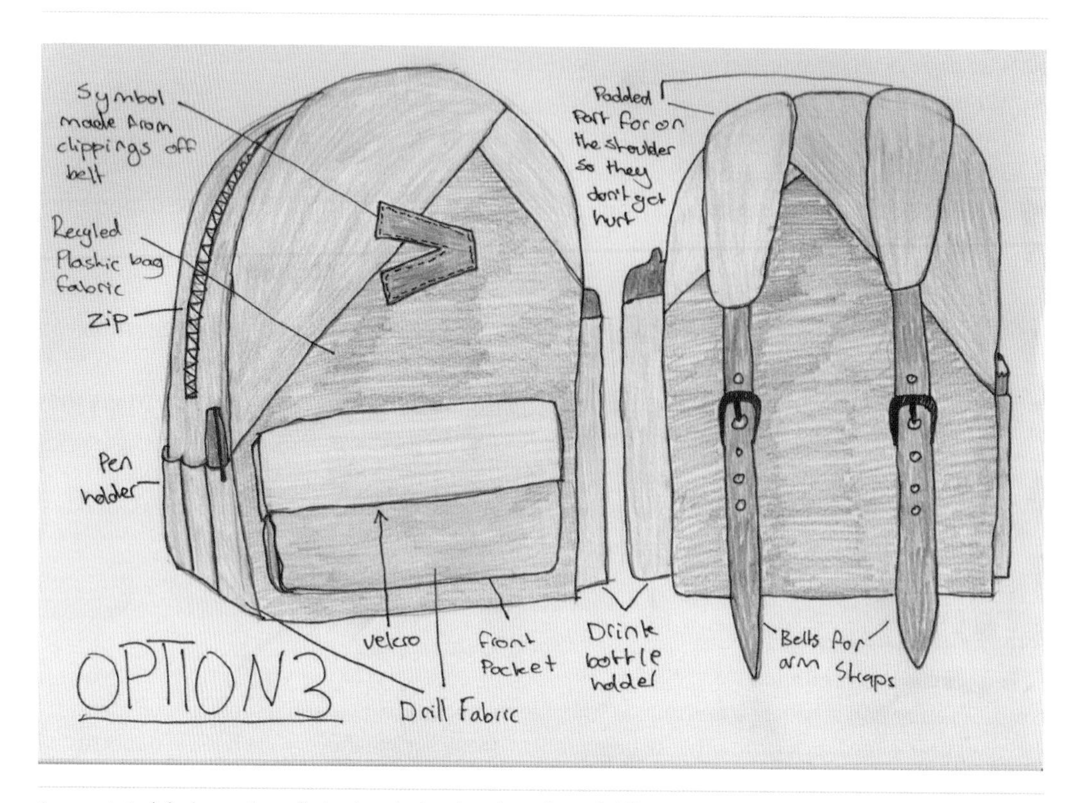

An annotated design option of a backpack showing the points of difference and improved sustainability intention, created by student Elisabeth Roberts

FOLIO ACTIVITY

Drawing your design options

Complete two annotated design options, on either A4 or A3 paper, or using CAD or drawing software. Follow the instructions above and the directions given in Chapter 12 regarding drawing style, content and annotations.

Evaluating your design options

Once your design options are drawn, it is time to go back to your evaluation criteria and choose three or four that can be applied to your design drawings. This is important as some criteria, such as 'Was it made within the $55 budget?' or 'Does it fit the end user perfectly?', cannot be answered from drawings. Apply the criteria and, where possible, include classmates' feedback along with your own responses. You can present this work in a small table next to the design option drawing. The first column will contain the three or four suitable criteria questions and the second column will have a comment (from yourself, an end user, or your classmates), a score, or a rating. Refer back to the 'Team and end user feedback' section in this chapter.

9780170477499

FOLIO ACTIVITY

Analysing your design options

- Use the evaluation criteria you developed earlier to analyse and decide which of your designs best suits the requirements of the design brief. You could rate each of your designs according to each criterion, and total your ratings to give each design a score – this should indicate the best one to choose.
- Show your design options to others and note down their feedback. Use colour to identify different people's feedback.
- Write a brief paragraph comparing the designs, outlining feedback, and identifying and explaining your choice of design.
- Decide which design option or which aspects of both designs you will take into modelling or prototyping.

Case studies
Tim Garrow

Jim Hannon-Tan

Chapter 1 case study report activity

Optional working drawings for models or prototypes

Once the design options are evaluated, you can decide whether to create working drawings to assist in creating models or prototypes, or whether to wait until you have your final proof of concept (your final prototype). You might decide to create rudimentary (simple and basic) working drawings to assist you at this stage. These might come from:

- the most suitable option
- both options (i.e. two different working drawings for prototyping)
- a combination of the best aspects of both design options.

If this seems too difficult at this time in the Double Diamond process, then you can leave it until your final proof of concept has been determined.

Read more about working drawings in Outcome 2 in the next chapter.

The next stage of the process: design development and production

Quiz
Chapter 1 revision

In Outcome 2, you will explore and refine your chosen idea further. Working as a team, you will:

- research materials and construction techniques
- make models and a prototype to improve your design concept
- finalise your concept
- make your product, using tools and equipment safely
- evaluate your finished product using your evaluation criteria and team feedback.

Summary
Chapter 1

CHAPTER 2

Generating, designing and producing

CHAPTER FOCUS

In this chapter, you will be introduced to the following core concepts:

- Testing materials and trialling processes
- Developing models, prototypes and the proof of concept
- Planning, making and evaluating the product and your collaborative process.

UNIT 1, AREA OF STUDY 2: OUTCOME 2

In this outcome, you will:

- work collaboratively and in teams to trial and test, evaluate and use materials, tools and processes to determine your chosen product concept and produce a product through implementing a scheduled production plan, as well as reflect on and make suggestions for future improvements when working collaboratively and as a team.

Key knowledge

- activities and their purposes within the second diamond of the Double Diamond design approach to generate and design physical product concepts, produce and implement, evaluate and plan and manage
- relationships between the second diamond of the Double Diamond design approach and design thinking strategies to refine physical product concepts and product
- materials, tools and processes used in specific design specialisations and the purpose of experimenting and practising with these technologies
- risk management for safe, accurate and efficient use of materials, tools and processes
- strategies to experiment with the physicality of product concepts through prototyping, including use of digital technologies
- methods to test and communicate physical product concepts, such as data from tests and trials, videos and photos
- relationships between product concepts and final proof of concept, and methods to develop a final proof of concept from a product concept
- methods to evaluate the finished product against the criteria described in the design brief
- traditional and/or new and emerging materials, tools and processes to produce a product
- methods to plan to produce a product, including developing a scheduled production plan
- strategies to reflect on collaborative and teamwork activities when designing.

Key skills

- explain activities and their purposes within the second diamond of the Double Diamond design approach
- conduct and evaluate tests and trials using design thinking techniques to propose, critique and justify the chosen product concept
- explain and use a range of materials, tools and processes to experiment with physical product concepts
- experiment with, and document the use of, a range of materials, tools and processes to produce a finished product
- collect and use data to inform and record refinements to develop a final proof of concept and apply a design process
- use criteria to evaluate the production process and determine how well a product addresses the design brief
- reflect on collaboration and teamwork and make suggestions for future improvements when working collaboratively and as a team
- work technologically, collaboratively and as part of a team to manage the activities within the second diamond of the Double Diamond design approach to implement a scheduled production plan to make a finished product safely
- manage risks to use materials, tools and processes safely.

Source: *VCE Product Design and Technologies Study Design*, p. 22

9780170477499

THE DOUBLE DIAMOND DESIGN PROCESS

Experimenting with materials, tools and processes

Working collaboratively to develop concepts

Modelling and prototyping possible solutions

Choosing product concepts, proof of concept

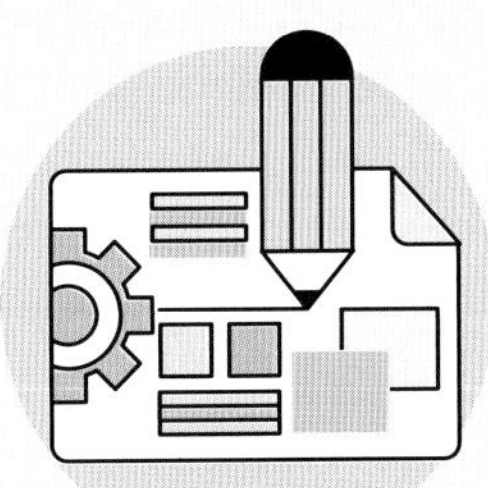

Developing and implementing a production plan to complete a solution

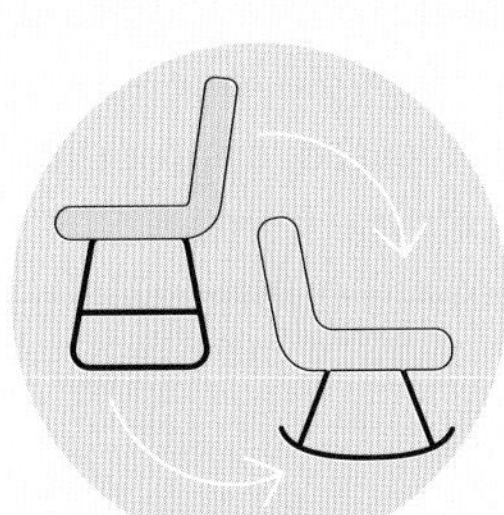

Evaluating finished product, the collaboration and recommending improvements

Templates:

- Materials research **(p.40)**
- Materials testing **(p.42)**
- Research **(p.44)**
- Prototype journal **(p.48)**
- Proof of concept **(p.49)**
- Materials cutting and costing **(p.50)**
- Production steps and timeline **(p.51)**
- Risk assessment **(p.53)**
- Quality measures **(p.53)**
- Journal/production log **(p. 55)**
- Evaluation report **(p.56)**

Quizzes:

- Chapter 2 revision **(p.61)**

Summaries:

- Chapter 2 **(p.61)**

Nelson MindTap

To access resources above, visit **cengage.com.au/nelsonmindtap**

Planning and conducting technical research

In Unit 1, Outcome 1, you and/or your team explored a design problem, defined your design brief and evaluation criteria, and developed a range of possible design solutions as graphical product concepts. In Outcome 2, you will:

- carry out technical research into materials and processes
- refine your design ideas through physical product concepts, i.e. modelling, prototyping and testing or checking them
- develop and define your best solution to the brief (proof of concept)
- plan for production
- safely make your product (recording your progress)
- evaluate your finished product and collaboration process.

While carrying out these tasks, you need to work as a team – gathering knowledge and sharing ideas, giving and taking feedback, and supporting one another through the production process.

As a team, you must carry out further research, using both primary and secondary sources. Research tasks and evidence can be shared, for example, one A3 page of research per individual. Research areas need to be planned, allocated appropriately (group or individual), be relevant, and be done within your required time frame.

See page 198 in Chapter 7 (Unit 3, Outcome 2) for more information on primary and secondary sources.

FOLIO ACTIVITY

Planning research

Use a mind map to explore and plan your research in relation to construction processes, material tests, product features or any aspects being checked in prototypes, etc.

Each student should maintain a personal research file, as well as contributing to a shared research collection.

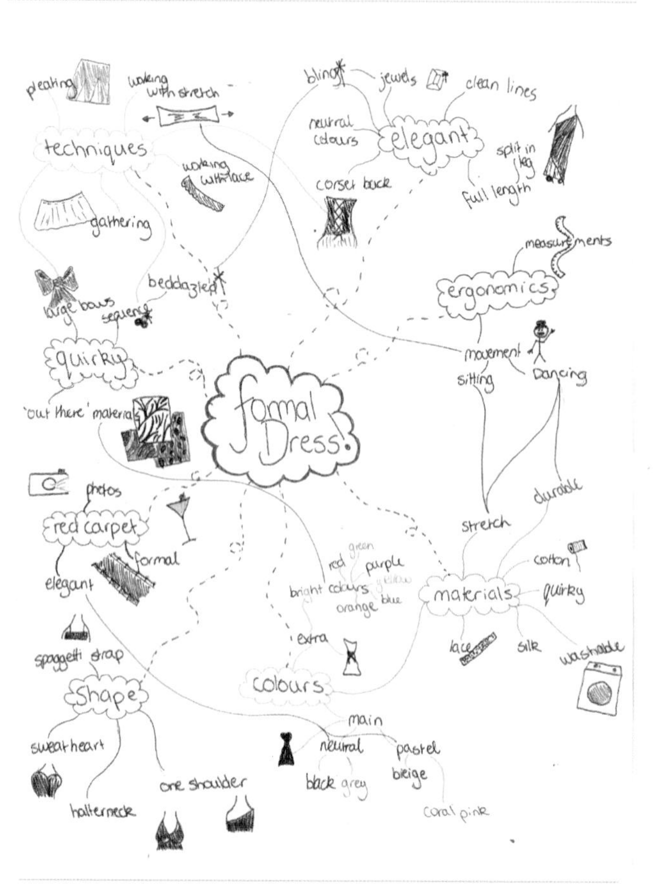

Concept map to organise ideas, by student Shannon Milgate

9780170477499

Data to inform your decisions

Throughout this outcome you will gather data from your tests and trials and evaluations of your physical product concepts (models or prototypes). This can be in the form of photos, videos, charts, graphs and written explanations. This information will be important for communicating with others and for your assessment. Make sure you keep records of all the work you do as follows in this chapter.

Knowledge of materials

In Unit 1, you will gain an understanding of materials and their characteristics and properties through testing/trialling and using materials (primary sources) and through searching the internet, contacting suppliers or through reference books (secondary research).

It may suit your project to experiment with or learn about materials and processes before any modelling or prototyping (primary sources) or you may prefer to start with modelling or prototyping. In some cases, you may need to complete further research. Read about all these topics in this chapter before you decide.

Materials classification

Materials are divided into different categories and classifications determined by their source, raw materials and properties.

Refer to pages 399–421 to find out more about the classification of materials, and to learn about the materials you are likely to use.

Characteristics and properties of materials

Your design situation and the needs of your end user will determine the **characteristics** and **properties** of materials that will be useful or required.

We usually consider the properties and characteristics of resistant materials differently to those made from non-resistant materials – even though some of the terms are similar.

Resistant materials are those that are hard, dense and don't change shape easily unless subjected to mechanical forces.

Non-resistant materials move, stretch or compress with very little force such as that applied by human hands or by wind.

Characteristics
The visual and textural aspects that are inherent in or typical of a material

Properties
Relates to how a material performs and behaves in different situations

Resistant material
Wood, metal, plastic, ceramic, stone, etc.

Non-resistant material
Fabric, fibre, textiles, paper, card, flexible plastics, etc.

Common properties of resistant materials (wood, metal and plastic)

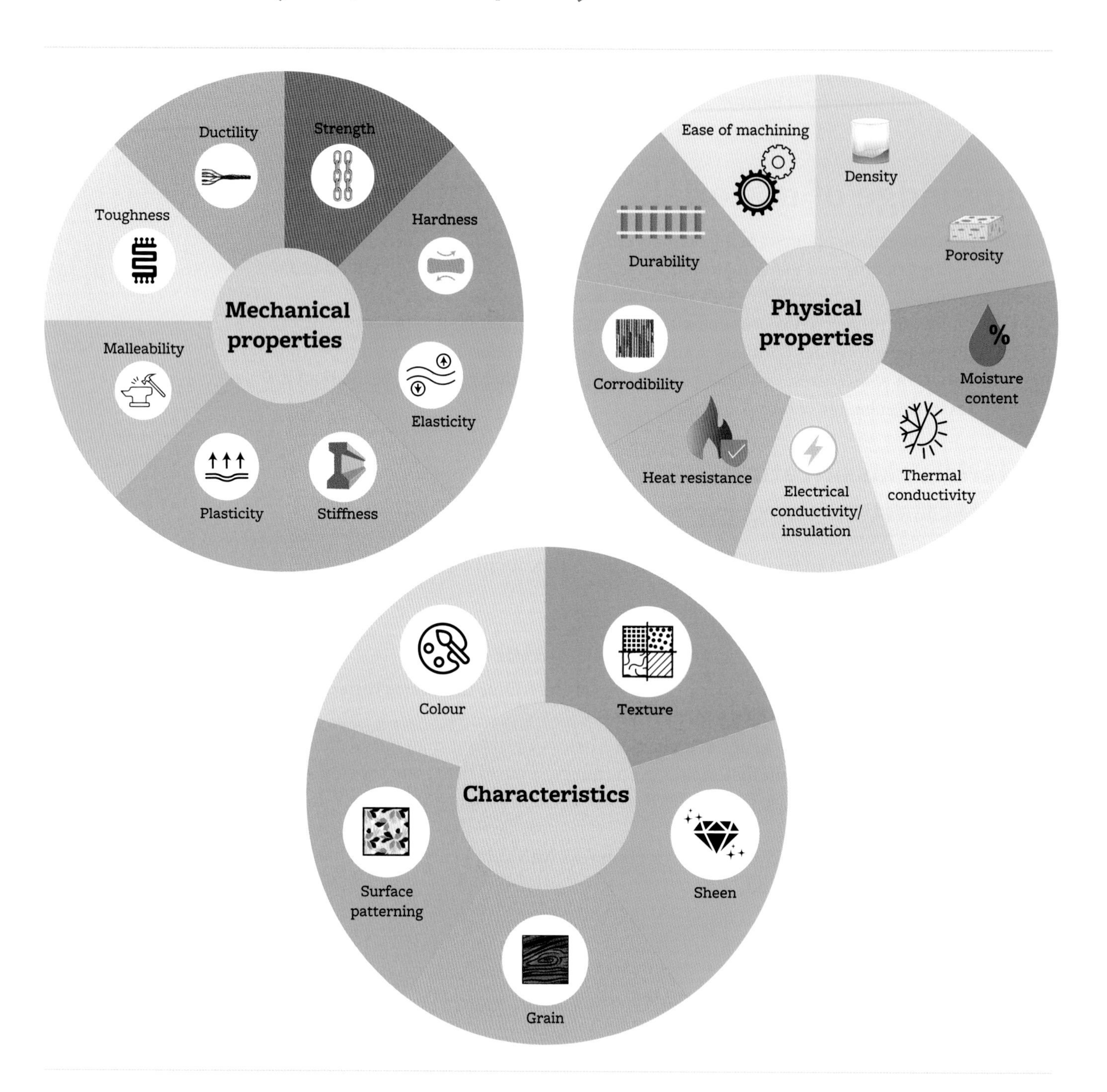

For much more detailed information about the characteristics and properties of wood, metal and plastics, go to Chapter 14.

9780170477499

Common properties of non-resistant materials (fabric, fibre and textiles)

The characteristics and properties of fabrics are determined by three aspects:

- fibre – the hair-like structures that are the source for yarn and fabric
- yarn – the spun structure that fibres are usually turned into
- fabric construction – the way a flat cloth structure is made using yarn or fibres. Fabric can be woven, knitted, felted or extruded.

Variations in each of these aspects determine the final characteristics and properties of a fabric.

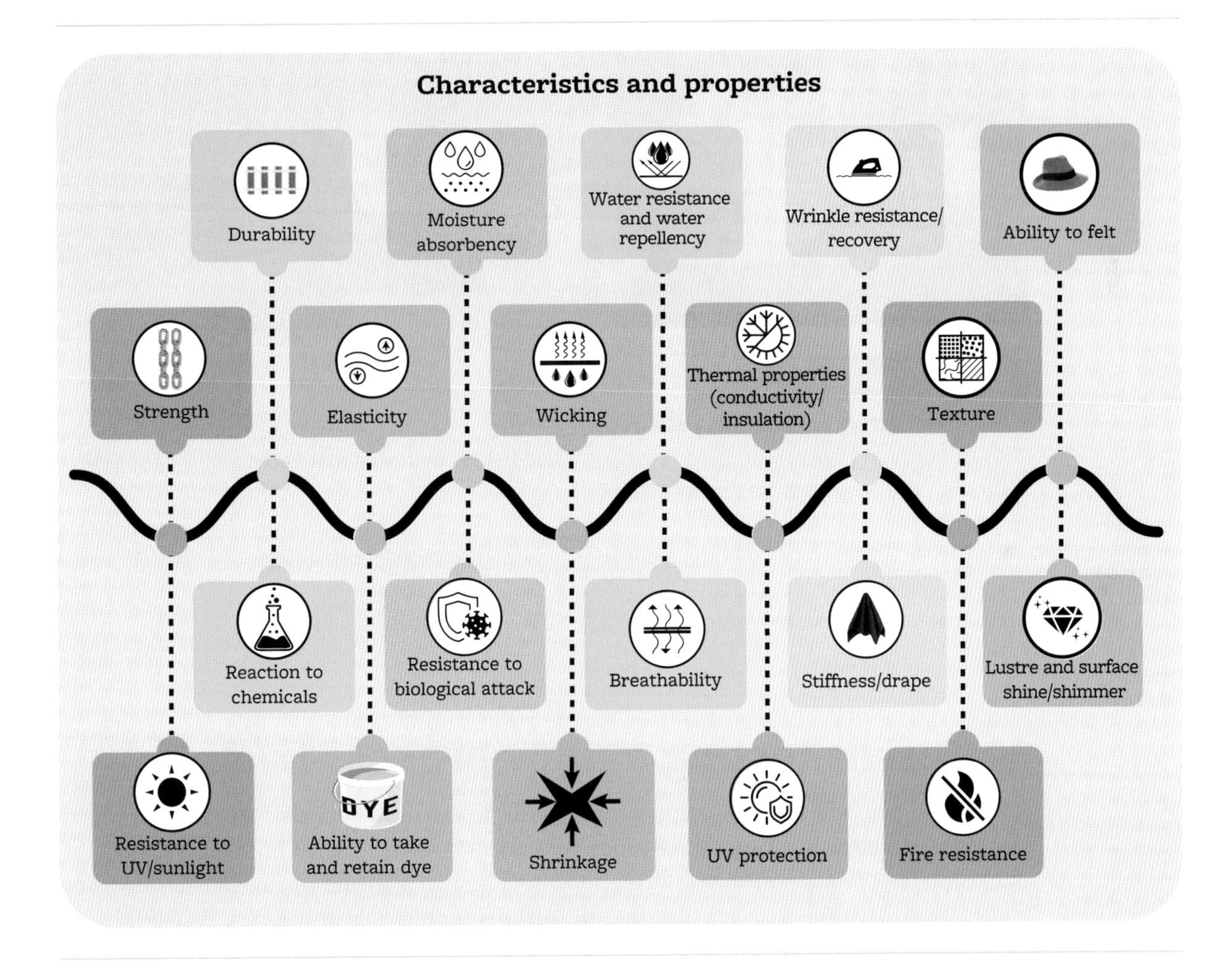

Note that most of the visual impacts of fabrics are determined by processing and by decorative and finishing methods, such as dyeing, printing, brushing, embroidery and laser burning.

For much more detailed information about the characteristics and properties of fibres, fabrics and textiles, go to Chapter 14.

iStock.com/Baloncici

Template
Materials research

FOLIO ACTIVITY

Research activity: materials research

As a team, carry out research into two or three different materials that could be used for your product and are accessible in your school.

Find out:

- specific/scientific name of the material
- the material classification
- characteristics and properties
- advantages/disadvantages
- cost/availability
- sustainability issues.

Materials testing and/or process trials

As a team, you will need to carry out materials testing and process trials to help make decisions about the most appropriate materials and processes for your product. Pages 421–7 in Chapter 14 provide you with information about how to plan and run a materials test, and give many suggestions for tests that investigate different material properties. Trials and tests can be carried out as a group or individually (with results shared) and the results and their analysis need to be added to shared files, folders or drives.

The purpose of **testing materials** is to investigate the properties of a limited range of possible materials and to determine which would be the best for your design situation. You need to plan so that you isolate and test only the property being investigated.

9780170477499

When testing materials, make sure your tests have the following:

- **consistency** – only the aspect that is being tested (the variable) should change. All other parts of the test should be controlled/the same
- **good planning** – make sure you have everything you need and have planned the test set-up and all of your steps before the test starts
- **repeatability** – if you did the test again, would you get the same results?
- **accurate measuring** – work out exactly what you are measuring and comparing. Part of your planning involves gathering accurate and appropriate measurement tools
- **record of results** – make sure you not only record any numeric results in a table, but also record your observations during the test. Make sure you take photos during your test
- **safety** – ensure that all the steps in your test are safe, and don't create the possibility of causing any injury.

The purpose of carrying out **construction process trials** is twofold:

- to choose the most appropriate process for the design situation
- to increase your skill and accuracy when performing the process.

Processes could include cutting, shaping and joining methods, decorative and finishing techniques, or might involve the use of new technologies.

For more detailed information about planning, carrying out and reporting on a materials test, go to pages 421–2.

New technologies

You can focus your process trial on a new technology (one that is available in your school, or that you could **outsource**). Processes that could be explored include:

- CAD designing and virtual modelling (if you haven't done this before)
- CNC routing – cutting and shaping timber or plastic
- 3D printing – for shaping plastic, good for small components (some expensive printers shape other materials)
- plasma cutting – for metal
- laser cutting – for timber, plastic, fabric, paper/card.

Outsource
To hand over to a more experienced person/group or to pay for a technology that is not available to you

Find out:

- the design possibilities created by these technologies
- how they might benefit your design and production
- the sorts of skills and files required.

Most new technologies require some form of digital/CAD skills. Trial the skills and processes required to develop CAD/computer files, and then learn how to set up and 'drive' the machine.

iStock.com/Laurence Dutton

Research reports on tests and trials

When reporting on a trial or test, the following sections need to be covered:

1. Heading – state the property/aspect being tested or the focus of your trial
2. Identification of the materials included in the test or the production process being trialled
3. The aim of the test or trial – explain how you will use the information gained from the test or trial ('How does this relate to our product?')
4. Testing/trialling procedure – step by step
5. A diagram or photo of the set-up
6. Numerical results (for a test) – in table and graph form
7. Observations – in writing
8. Analysis of results and observations ('What do the figures tell us about these materials?')
9. Consequences for material or process use ('How will this affect the decisions we make?')
10. An evaluation of the testing procedure ('Was this an accurate test?' 'How could it be improved?' 'What sort of equipment would improve the accuracy of our results?').

All sections can be reported on briefly, and make sure you support your report with images and diagrams where helpful.

Template
Materials testing

FOLIO ACTIVITY

Research activity: materials testing and process trialling

- In pairs or as a team, choose a test or trial that is relevant to your design situation/product and assign roles.
- Plan and gather the materials and equipment required for the test/trial.
- Follow all safety precautions as instructed by your teacher.
- Carry out the test/trial. Record your results and observations as you go (make sure you take photos).
- Complete a report detailing your planning, results and conclusions.

Other possible research areas

At any stage, you may need to complete further research. This may need to occur before, after or during your material testing, trialling of processes or modelling/prototyping. Read this section for possible areas where further research might be needed.

Ergonomics

Ergonomics is an area concerned with how well a product fits and is suited to the human body. It focuses on the comfort of the end user, the product's ease of use, and ensuring that no strain or injury is caused by using the product. Anthropometric data (human body measurements of populations) helps designers to make good ergonomic decisions. Designers don't usually use average sizes, but instead make sure that their products suit at least 90 per cent of all users. In situations where products are designed in small, medium and large sizes, anthropometric data can be used to determine the percentage of users that will fall into each size category. For example, 30 per cent of users might need a small size, 30 per cent a medium size and 40 per cent a large size. Specific ergonomic and anthropometric data might be needed for people in particular groups – for example, young children, the elderly or people with special needs and disabilities. This data can be used to ensure that products are safe and comfortable for these users.

 See more on ergonomics in Chapter 13 under the Factor 'end users'.

Different information is used depending on the nature and use of the product, as shown in the table below.

Useful data for different products

Can opener	Pot plant	A shirt
• Size of the hand • Position of the hand when gripping • Amount of pressure the hand can apply • The extent to which a hand can comfortably turn and twist	• How much weight someone can lift without strain (related to a fully loaded pot plant) • Position of hands to comfortably grip a pot plant (related to the size and comfort of the rim edging)	• The ease of reaching, opening and closing fasteners such as buttons or snaps • The size of different parts of the body, such as neck, arm/wrist, chest • The ease of fit to allow the shirt to be easily put on and taken off

Shutterstock.com/Imagentle

Identify the ergonomic information you will need to successfully design your product/s so they fit the body and cause no strain. This might be specific anthropometric data about the size of relevant parts of the body, or information about how the body performs (e.g. stretch or reach information). You may also ask potential end users to trial different existing products or components to give feedback on their comfort, ease of use and possible areas of body strain.

Components (parts for your product)

For many products, you may need to use components or parts that have already been manufactured. To make good decisions about components:

- research the different options available and how they work/operate
- consider which component/s will be the best option for your team's product
- find suppliers and investigate costs.

Possible components

For fabric products	For wood, metal and plastics products
• Fasteners (zips, buttons, buckles and tabs, eyelets and snaps, snap tape) • Velcro • Iron-on hemming and interfacing • Reflective tape • Wearable digital components and conductive thread	• Handles • Hinges • Brackets and supports • Castors • Electronic components • Fasteners (nails, screws, rivets, nuts/bolts, etc.)

FOLIO ACTIVITY

Other research

Refer back to your initial research mind map and decide what further research needs to be completed.

Remember to **identify the source** of any secondary research you use:

- Copy the website name and source URL for any images and content found online. If you found the information on Google Images, Pinterest or a retail site, you will need to dig deeper to find the author, brand, group or creator's name.
- Provide the title, author, publisher and year of publication of any reference books you use.

Template
Research

After all your research is done, meet and discuss how your research might impact the design decisions and directions you will take next. Annotate each piece of research with a suggestion for how it could be used.

Developing physical product concepts

Models and prototypes are considered to be physical product concepts even if computer-generated, i.e. with CAD. They are used broadly by designers, when designing, to visualise and check their ideas in a physical form. When you see an object in a 3D form, you perceive your design differently and see things that weren't obvious in drawings. Models and prototypes can also be used to physically test how the product idea works. You will probably find aspects that need to be changed and improved.

9780170477499

Models and/or prototypes (actual or virtual) can be developed:

- for each of your design options, to help decide which option is the most effective and solves the design situation in the best way
- after the preferred option is chosen – to refine and improve this design.

It may suit your project to develop your physical product concepts (modelling or prototyping) first and then do your trials and tests.

Modelling

Designers often use a model, in combination with drawings, to trial ideas and to get early feedback when presenting initial ideas to clients/end users. You can start your physical 3D concepts by experimenting with shapes and proportions in miniature (or full-size) models. These can be made of paper, cardboard, old plastic containers, timber scraps or fabrics. Make sure you photograph your models and include them in the design development section of your folio, with analytical comments/annotations.

Shutterstock.com/Chaosamran_Studio

A designer checks a plywood model of a chair for proportions and structure.

Modelling for resistant materials (wood, metal and plastics)

Simple, rough models help you explore shape, colour and proportion. A well-made model will represent the size, proportion and form of the object, along with other qualities, such as texture, surface finish, colour and pattern. They can be used to explore and critique ideas and to try out functional aspects.

Some models are constructed from solid material blocks, and show the outward appearance and external details only. They are made of easily worked materials – such as cardboard, polystyrene foam or clay – or other materials such as aluminium, acrylic and plywood, and allow for quick and easy evaluation. Changes are easily made if required, and only a small amount of time and resources have been committed to their production.

If you are competent in the use of CAD software, you may prefer to create virtual concepts (models or prototypes). Be sure to take a screenshot or image download for each version to add to your folio.

Modelling for non-resistant fabric products

Ways of modelling product ideas in fabric include:

- draping a small artist's mannequin with different fabrics, of different proportions or lengths, using pins or paper clips, to gain a 3D sense of the finished design
- creating models from cut and folded paper
- cutting out the negative space of a garment silhouette from thick card and placing it over different fabric samples, to get an idea of how the fabric might suit the style, or to check different fabric combinations.

Take photos to add to your folio.

For other fabric products, you could create miniature bedroom or lounge furniture to trial soft furnishings for shape, colour combinations and fabric.

See the examples of paper models developed with the use of DALL-E in Chapter 8 (Unit 3, Outcome 3).

Starting a toile (left) made of check gingham by student Tamara Bottomley; includes markings in ballpoint pen for adjustments to finalise pattern pieces for one of her design options (right).

Prototyping

The purpose of a prototype is to trial or test a design concept to see if it functions, looks good, and achieves the design brief requirements. Prototypes can also be used to trial and develop skills in the production methods required to make the product. Prototypes help designers to refine their ideas before committing to production.

9780170477499

Prototypes are versions of a design concept that are either virtual (viewed in CAD) or physical (can be touched by hand).

If physical, they:

- can be made from lesser quality materials than the end product
- can be a scaled version (a smaller model) of the end product
- usually have some level of functionality
- may be made up for some details or components only, not necessarily the whole product.

Prototype materials may include those shown in the table.

For products made from resistant materials	For products made from fabric
Paper or cardboard	Paper
Radiata pine	Calico
Mild steel	Checked cotton (works like a grid)
3D printing filament	An inexpensive knit fabric
Recycled materials	3D printing filament (for non-fabric components)
Old plastic containers	Pre-used or recycled materials

You can work as a team, in pairs or individually to develop a prototype. Refer to pages 233–40 for more detailed information about prototypes.

iStock.com/andresr

A design team discussing their prototypes

Template
Prototype journal

Developing your prototype

Either here or after the next section, develop a prototype of your best design/s. Work as a team, in pairs or individually to develop your prototype. Select the most appropriate materials, and decide as a team how much detail your prototype will contain (which details are the most important to trial, and what is of lesser importance and can be left out or mocked up).

Journal section of your folio

When prototyping, take photos or download CAD files of your concept development in a journal or log. Add these to your team or individual folio with a discussion of what you have learnt, the successful features of the prototype and ways it could be improved.

Choosing the preferred product concept

Other class members should be involved in choosing the most suitable product concept (to be finalised and produced). For detailed information about how to select a concept, see pages 239–40. Rate your evolving concepts against your evaluation criteria – give a score out of five (or ten) for each model or prototype, then add up and compare the scores. Constructive comments (involving more critical thinking) are valuable here – add them in the columns, or with post-it notes stuck onto each of the models or prototypes. These comments can be used in the written justification explaining why one product concept was chosen and the other/s were not.

Write three or four of your evaluation criteria in a small grid with columns for feedback. This grid can be placed right beside an image of each of your product concepts, and feedback with a score can be written directly on the sheet.

Evaluation criteria	Class member 1	Class member 2	Class member 3	Add extra columns as needed
1 Do the colours match the team theme?	Too green! 2/5	Love it! 5/5	Neutral 3/5	
2 Does the product function as required?	Clever! 5/5	Bit too complex! 2/5	Yes, clever! 5/5	
Add extra rows as needed				
Total scores	**/10**	**/10**	**/10**	
Final score for Product Concept 1			**/30**	

Justification

For individual products, each member writes a justification of their chosen product concept using group feedback. If the team is working on one product, then the team should write a justification collaboratively, with a comment from each member.

Make it clear which drawing, image or prototype is the chosen product concept and how it satisfies your criteria. Also describe why the other options weren't as effective.

Your proof of concept

Once the chosen product concept has been selected, you and/or your team need to decide if it needs further refining. If it does, team discussion is important – feedback and new ideas will help improve your product's design. Decide whether a new prototype is required. Any new changes need to be documented. Once it is all finalised, it is called your proof of concept.

FOLIO ACTIVITY

Selecting your chosen product concept and proof of concept

In the journal section of your folio, record responses and comments about design options, models and/or prototypes, and decisions that are made. Record the final proof of concept using drawings, photos or CAD images, with explanations of how it fulfils the brief and why it is the most suitable solution.

Template
Proof of concept

Working drawing

Whether you are working as a team or an individual, the next stage in the design process is to create the final working drawing of the product you or your team will make. This may have been completed before prototyping but, if changes were made during the prototyping process, the working drawing will need to be updated.

A working drawing is a clear and detailed technical drawing, with a focus on accuracy and construction information. The working drawing is used to guide your final production, and should include all the details you need to make your product. If the product is complex, you may need to include enlarged hidden details, or an exploded diagram to show how the parts go together. Accurate measurements are crucial, particularly if different people are constructing separate parts of a product that need to fit together.

The drawing systems required for working drawings are:

- for wood, metal and plastics products – create orthogonal drawings (with hidden details and exploded views to explain further)
- for textiles – create 'flats' to provide visual details of your final design (supported by construction drawings of stages or production details, if needed).

If you have developed your design option drawings using CAD, working drawings will be easy to generate. Just ensure you have updated your CAD files to your latest product design.

Go to pages 335–42 to learn more about working drawings, and particularly to page 335 for wood, metal and plastic products, and page 338 for fabric and textile products.

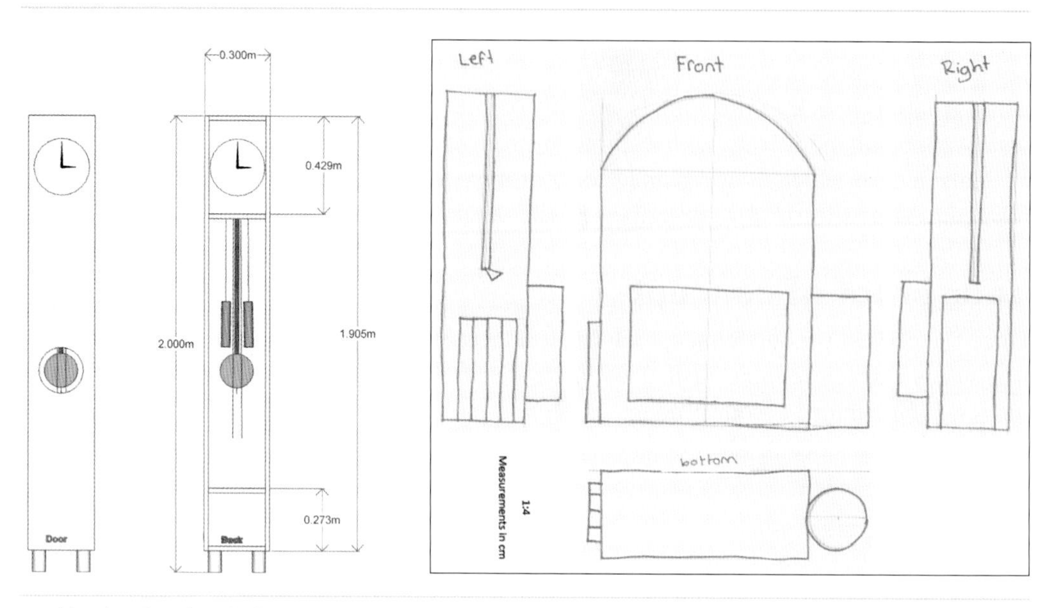

Working drawings for a hall clock (left) by student Clark Fitton, and for a backpack (right) by Elisabeth Roberts

FOLIO ACTIVITY

Working drawing

Use the directions on the pages from Chapter 12 suggested on the previous page and your teacher's instructions to help you complete clear and informative working drawings of your final design. This will help guide your production. Place these drawings in your digital or hard-copy folio. For a team product, it is worth printing out a copy of the drawings for all team members.

Producing and implementing

Your scheduled production plan

Production planning involves creating a number of planning documents/tools. These will help organise and manage time and resources, so that your end product will be completed safely, on time.

For Unit 1, you will need to complete:

- materials cutting and costing list
- combined timeline/production steps including safety precautions
- risk assessment
- quality measures (may be combined with the production steps).

Materials cutting and costing

Template
Materials cutting and costing

A materials cutting and costing list details all of the materials and components you need to source to make your product. Most of the detailed information for this will come from your working drawing. How you complete this task depends on the type of product you are making and the materials involved. When manufacturing a product from wood or metal, your materials list is sometimes called a 'bill of materials'. For fabric/textiles, you need to include fabric and 'notions' (components for joining and fastening fabrics).

Use a table or spreadsheet to help organise the materials, components or notions you need, the sizes or amounts of each, and their cost.

For more information about developing your materials cutting and costing list, see page 249.

Timeline with production steps and tools/equipment

Template
Production steps and timeline

In Unit 1, you can complete a combined scheduled work plan (with production steps) *and* timeline, or do these two tasks separately if you find it easier.

Use the template below (in landscape format) to identify the main production steps, the more detailed tasks for each step (in dot points), any tools and equipment required, and safety guidelines for each step. On the far right, use the columns for the production weeks or sessions to show when work needs to be completed (add extra columns/ sessions as needed). If your team is going to make the same product, allocate jobs to each team member and shade their sections in different colours, as shown in the Column for Session 1. Make sure each person is occupied during every session. Use your school calendar to assist with adding dates for each session.

Example of a scheduled production plan for a timber product

Step	Tasks	Equipment	Safety	Session							
				1	2	3	4	5	6	7	8
1 Select and mark timber	• Select timber from racks • Mark main lengths	• Tape measure • Pencil • Try square	When moving timber, ask for assistance for heavy lengths, carry safely, support timber wholly on the bench								
2 Making 3D printed parts	• Finalise .stl files • Set up and check 3D printer • Send files to printer • Monitor and troubleshoot	• 3D printer • Computer • Filament	• Keep hands away from heated nozzle • Ensure doors are closed when printing • Check filter regularly								

Team members and production tasks colour key

To learn about timeline and production steps/work plan in more detail, and to see wood, metal and plastics, and textiles examples, refer to pages 241–3.

Risk management and your risk assessment

Risk assessments are carried out in industry and in many businesses to identify possible hazards, and to minimise the risk of incidents related to those hazards. A risk assessment covers the first three stages of the risk management process.

The last stage requires monitoring and checking that all safety controls are in place and are working as required during production, and fixing them if they are not.

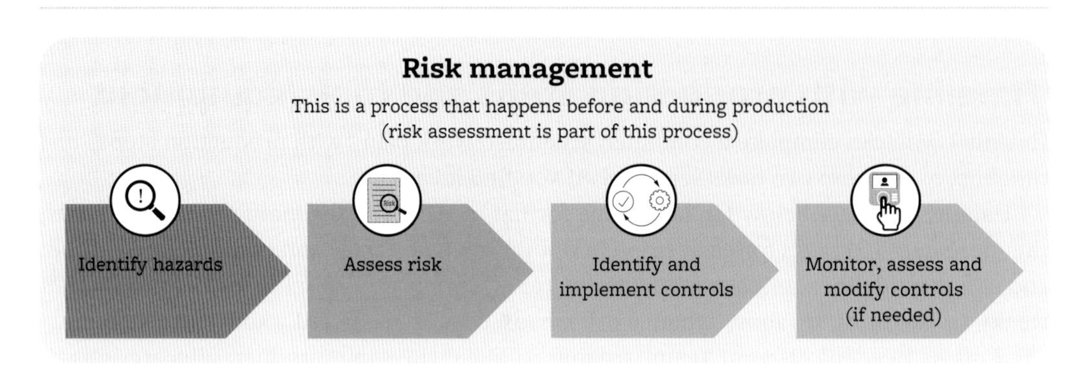

Risk assessment

Hazard
A thing, action or behaviour that could cause harm

Weblink
Safe Work Australia

A risk assessment covers the first three steps in the risk management process above. First you need to identify any **hazards** involved in your production. The Safe Work Australia website outlines the steps to take for managing risks.

A risk assessment is usually written in a table with five columns, as follows.

1. **Purpose of the tool or process**, naming the machine and/or process being used in your production
2. **Identifying hazards** – these are the parts of the machine or process that might cause an injury. Types of hazards include: a sharp blade or cutting edge or point, a rotating/ spinning part, an area where parts of your body or clothing might be caught or trapped (entrapment), electricity cord or supply, heat, dust, noise, bad positioning of the work area, flying objects (being ejected from the workspace), liquid that might splash or burn, a cord or part of the machine that might be a tripping hazard
3. **Identifying the possible injuries** that might occur if something went wrong with different aspects of the machine. Types of injuries include: lacerations (cuts and abrasions), amputation, broken bones, eye injury, bruising, burns, stab wounds, scalping, degloving (stripping skin from hands), respiratory problems, hearing loss, eye strain, electrocution, spinal injuries
4. **Calculating the levels of risk** – a combination of the likelihood of an incident happening, and the consequence or seriousness of a resulting injury, from minor to serious, including illness or death. This is usually given as a rating out of 5 for each area (from 1 = very low to 5 = extreme). Use the chart below to help you calculate risk
5. **Identify and implement controls** – outlining precautions, safety controls and/or personal protective equipment needed to eliminate or minimise the risk.

9780170477499

Risk matrix

Likelihood	Consequences: Insignificant (1) No injuries	Minor (2) First aid treatment	Moderate (3) Medical treatment	Major (4) Hospital	Catastrophic (5) Death
Almost certain (5)	Moderate (5)	High (10)	High (15)	Catastrophic (20)	Catastrophic (25)
Likely (4)	Moderate (4)	Moderate (8)	High (12)	Catastrophic (16)	Catastrophic (20)
Possible (3)	Low (3)	Moderate (6)	Moderate (9)	High (12)	High (15)
Unlikely (2)	Low (2)	Moderate (4)	Moderate (6)	Moderate (8)	High (10)
Rare (1)	Low (1)	Low (2)	Low (3)	Moderate (4)	Moderate (5)

Source: SafeWorkPro

If your calculated level of risk for a machine or process lies in the orange or red sectors, modify or substitute it for something safer.

As a team, identify three or four significant pieces of equipment or production processes that might have some hazards associated with them and require a risk assessment. Allocate one machine or process to each team member to complete. Collect and share your risk assessment planning across the team.

Template
Risk assessment

You can read more about risk assessment in Chapter 8 on pages 247–9.

Quality measures

Quality measures are a brief outline or description of the standard (and/or accuracy) you expect to achieve in major construction processes.

Examples of quality measures for textile products might include:

- Stitching seams:
  - All stitching will be of correct and consistent stitch length.
  - Thread will be a suitable colour (could be a contrasting colour or a similar colour to the fabric, depending on your design).
  - No skipped stitches.
  - No gaps or missed bits of seams.
  - No bunched or tangled thread on the underside.
- Inserting a zipper:
  - Zipper sits perfectly flat.
  - Top edges of zipper align with top edges of fabric on either side (or wherever situated).

Examples for resistant materials (wood, metal or plastics) might include:

- Joining materials:
  - No gaps bigger than 0.2 mm (or your own measure).
  - No visible lumps of glue or welds outside the join.
- Finishing edges:
  - No sharp edges.
  - No scratched grain or surface.

In Unit 1, choose two or three major construction processes for which to create quality measures. These can be drawn up in a table with two columns. You could also add a third column with hints or tips on how to achieve this quality, which might include set-up

Template
Quality measures

procedures, techniques or specialised tools or equipment to help with accuracy. You can add any new tips or techniques you learnt during your construction process trials. This could involve:

- practising/trialling the process if it is unfamiliar
- identifying and using the most accurate/suitable tools
- checking measurements before cutting
- using aids for holding work in the correct place, such as basting, pins, jigs and guides
- checking machinery set-up before use
- checking for consistency and accuracy and redoing if it isn't done well.

You can read more about quality measures in Chapter 8 on pages 250–1.

FOLIO ACTIVITY

Scheduled production plan

Complete:

- a materials cutting and costing list
- a timeline with production steps (either individually or as a team, depending on your product/s)
- a risk assessment (if shared, make sure you have a copy for any hazards)
- quality measures.

Place these in your digital or hard-copy folio. Again, for a team product, print a copy of the planning documents for all members of the team.

Production

During production, it's important to check that your work follows team goals in terms of aesthetics, quality and safety. Plan as a group; share tasks, skills and knowledge and take photos of each member's contribution. Follow the risk management process and apply the safety controls planned in your risk assessment for every step.

During production, make sure you:

- work safely by checking and following all safety controls/guidelines
- think about quality – use the techniques you learnt in your process trials to be as accurate as possible
- communicate with others – talk about what you are doing, ask for help if needed, check in on your classmates regularly
- regularly refer to and follow your planned steps and timeline – make a note in your journal when this changes
- use your time effectively, and make sure you 'pull your weight' (do your fair share of production work).

Recording production: journal/production log

A journal or production log is a visual and written record of the work you have done to make the product. It is something you need to do individually – it is a record of *your* work, and provides insights into what *you* have learnt. A journal is most effective if it is completed weekly, while you are working on your product, so that your memories are fresher and more accurate (don't leave it until after production is finished). The journal consolidates the names of tools and processes used, gives opportunities to reflect on what you've done and on what you've learnt through your experience, and to plan what to do next.

In your journal, you need to:

- outline what you have done using photos and a written description
- explain any safety measures/controls you have used
- reflect on what you have learned about materials, equipment, machinery and processes and name them all correctly
- describe any modifications to the product or process that have been made
- plan the next stage of production.

For Unit 1, it is also helpful to comment on your contributions as a team member and whether your collaboration and the collaboration of others was effective or not. A template is provided in the Nelson MindTap resources, but you could use other digital methods to record your progress – for example, a blog, a series of short videos or a set of production instructions with hints and guidance (such as relevant instructions you found on the Instructables.com website).

FOLIO AND PRODUCTION ACTIVITY

Journal/Production log

Complete a journal or production log, using any format you choose, provided it is done regularly during production and covers most of the content dot point list above. This task needs to be done individually by each team member.

Template
Journal/production log

Evaluating

Product and team evaluation report

When your product is finished, you need to judge whether your efforts have been successful and the product fulfils its purpose. Consider how you are going to test or check that the product satisfies the evaluation criteria and the requirements of the design brief. Choose evaluation methods that suit each criterion. They could include:

- observing end users using the product and asking for their feedback
- getting feedback from classmates or experts (e.g. your teacher)
- trying the product out yourself
- placing or using the product in context (in the situation where it would normally be used)
- checking timelines and budget details.

Your questions should be focused on the needs of the end user and the requirements of the design brief (as expressed through the evaluation criteria). You may wish to add a few more areas if you don't think all aspects of the product are being assessed.

Record responses from these product trials related to evaluation criteria.

Your product won't be perfect and most likely won't match your original design or final prototype. Identify areas/suggestions for improvements that have come out of the evaluation process, or that you have noticed. These might relate to:

- how your product could better satisfy the end user's needs (as identified in the team design brief)
- the sustainability of the product and the materials used
- the quality of the product (and how carefully/accurately materials and processes were used during production).

Also comment on how well your collaborations went:

- Did class members share ideas and communicate effectively?
- What feedback did you receive and was it helpful and timely? How did you manage to give helpful timely feedback to others?
- Was planning effective and production completed efficiently? Were goals clear and did members know what they needed to do?
- Did you contribute well to collaborative tasks and efforts?

Consider improvements to the ways in which your class members collaborated and shared the design and production process.

FOLIO ACTIVITY

Evaluation report

Put all the information from the section above together into a clear report, with:

- appropriate headings
- a table listing the evaluation criteria, how they were checked, and the feedback and responses given
- an assessment of collaborations and how classmates' feedback helped
- areas for improvement identified:
  - » for the product
  - » in how the class members worked as teams when required.

Template
Evaluation report

Assessment

In Unit 1, you will be assessed on the following work:

- a multimodal record or folio that contains:
  - research on specific end user needs
  - design brief, including an end user profile, intended function, project scope (including constraints and considerations)
  - evaluation criteria
  - visualisations (with optional modelling and/or prototypes)
  - design options (with optional modelling and/or prototypes)
  - technical and practical research
  - proof of concept, with explanation and working drawing
  - scheduled production plan, including:
    - » materials cutting and costing list
    - » timeline
    - » work plan/production steps with quality measures and safety controls
    - » risk assessment
  - record of production/journal
- mock-ups/prototypes
- finished product
- evaluation of the product and process.

Rip Curl

Rip Curl: wetsuit development

The global headquarters of Rip Curl is in Torquay, where it all began. The Rip Curl business was founded on wetsuits and they are still one of the core products, along with surfboards and watches.

Peter Coles is the Global Research and Development/Production Manager in the wetsuit division. All the wetsuit design and development happens in Torquay, and Peter's role is to assist with the development of new ranges of wetsuits. As well as working closely with the design and development team in Torquay, he looks after logistics, research and development (R&D), sourcing and production.

Alamy Stock Photo/G. Scammell

Teamwork

Getting a new wetsuit right and ready for market involves a whole chain of people – it is a great example of collaboration. Apart from Peter, the team in the wetsuit division consists of the chairman, a product manager (responsible for sales in the Australian region) and product managers in the northern hemisphere. John Pyburne, or 'Sparrow', is the pattern maker, the head wetsuit designer is Jay Abbot, and together they look after wetsuit design and graphics, but there is significant input from the whole team. The team needs to approve concepts and ideas at various stages and sign off before development continues.

Field trips for new ideas

Field trips for Rip Curl designers might be as far afield as Indonesia, Manhattan, Sydney or Tokyo, where they do a lot of in-store checking. They check competitors' products and other types of sportswear, looking for new ideas that they might be able to bring into the wetsuit division, such as colour combinations, interesting seam lines, garment shape, construction techniques, finishings and trims. The wetsuit designs completed in Torquay are aimed at a world market. Each design has to satisfy users in different parts of the world.

Designing and pattern drafting

The designers often do hand sketching for initial ideas, and draw by hand straight onto a mannequin to see where the seam lines of a design lie on the body. Adobe Illustrator is used to refine and digitise those ideas.

The design is then marked out (drawn) onto an existing wetsuit for the pattern maker. Sparrow, the pattern maker, who has been working at Rip Curl for years, will create the pattern pieces on large pieces of brown paper (his workroom is filled with shaped pieces). He also cuts out the pattern pieces from the designated fabric. A prototype is made in the 'factory' workroom next door and given to a staff member to wear and test in the surf. This can be as immediate as that afternoon or the next day.

After the prototype has been used in the surf, feedback is given about:

- the fit – wetsuits need to be tight to be effective, but not too tight
- the neck – is the collar too high and choking, or too low and letting in too much water?
- flexibility in the elbows and knees
- the ease of getting it on and off.

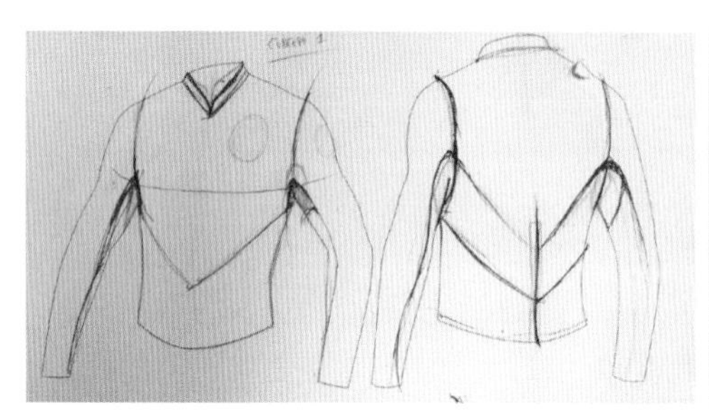
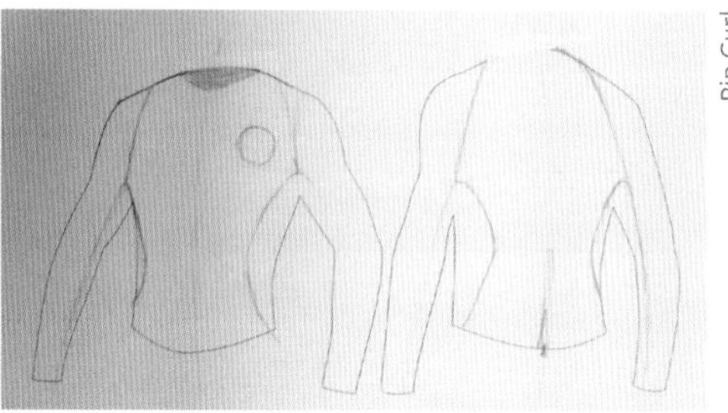

Rip Curl

Hand sketches for initial ideas

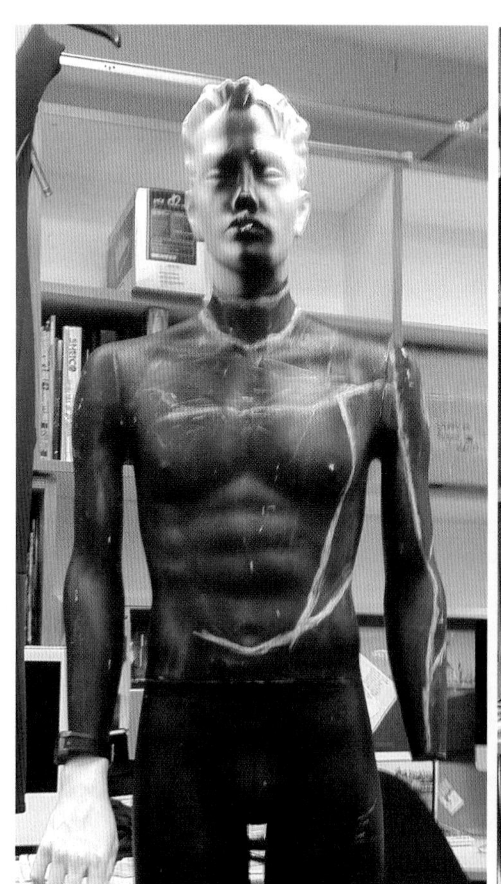
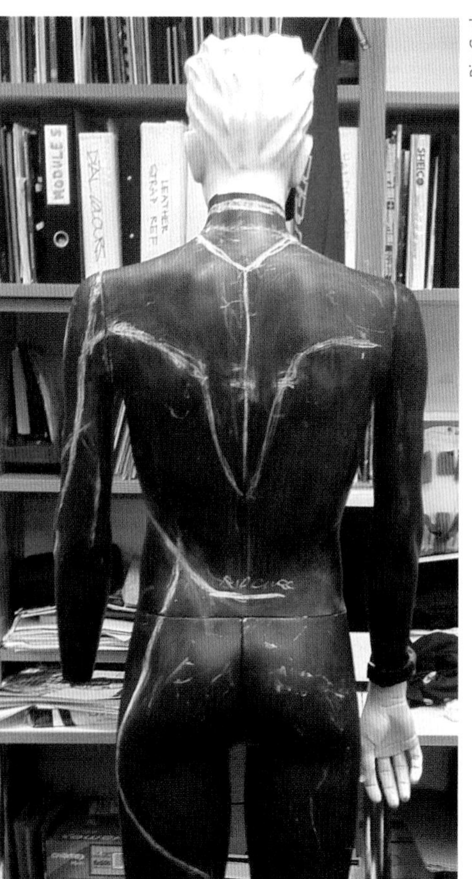

Rip Curl

Lines are often drawn straight onto the mannequin in the Rip Curl design room.

Digital pattern grading

This initial prototype, usually made to a medium size, may be adjusted in minimal or numerous ways after it has been tested. Once the final prototype has had all its 'imperfections' ironed out, it needs to be graded (sized). Patterns are graded digitally using the Lectra CAD System, which is also used by the factories in Japan or Taiwan where production takes place.

To turn the paper patterns into digital form, the pattern pieces are taped with masking tape, one at a time, on a giant digitising board. The board has electronic grid lines (which are not visible). The computer operator goes over the outlines with a cordless digitising mouse and the shape is plotted accurately into the computer. For precision, only half of the symmetrical pieces are 'scanned' and the computer program creates a mirror image of the exact dimensions for the other half.

Rip Curl has worked out its own grading system for wetsuits over many years to get the right fit for quite a range of different body proportions, limb lengths and sizes. Because a good, tight fit is important for wetsuits, the grading system is more complicated than for a product such as a T-shirt, where it might only be necessary to get the correct length and width measurements.

Grading patterns for wetsuits is not just a simple matter of enlarging or reducing (scaling up and down) all the pieces from the base medium size; it needs careful consideration of how all the pattern pieces interlock. Wetsuits don't have a straightforward back, they might have one piece that fits over the shoulder without seams. In this case, an increase in upper arm width for one grading may require a change in the shape and dimension of the shoulder shape in other sizes.

Once the pattern is graded, 'fit samples' (equivalent to final prototypes) have to be constructed to check that the design 'works' and all the bits fit together in every size. When all the graded pattern pieces have been finalised, the designs are emailed to the factory in Japan or Taiwan, where the majority of the wetsuits are made. This is a much faster system than was used at Rip Curl up until the early 1990s, when paper patterns with all their graded sizes for one design were rolled up and sent by normal post.

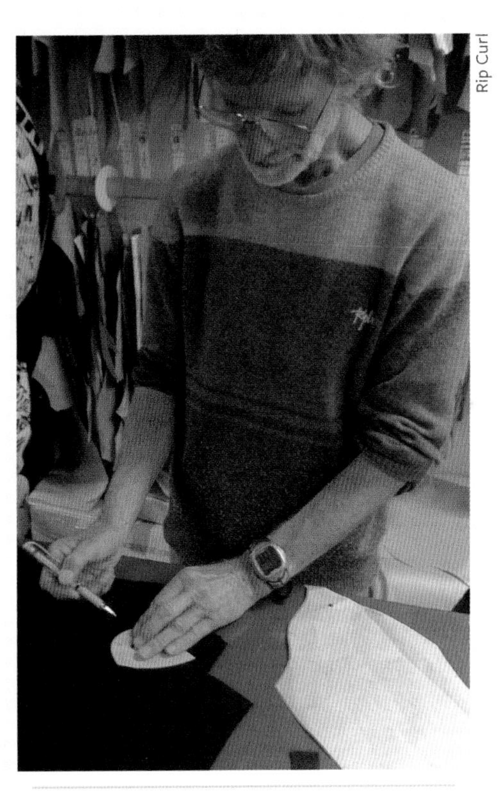

Rip Curl

Pattern maker John 'Sparrow' Pyburne cuts out pieces from neoprene.

Weblink
Rip Curl Wetsuit Factory Tour

The machinists

Rip Curl rash vests are machined in Torquay. It is important to have continual production work, so that the machinists are skilled up to make the prototypes and 'fit samples' when they are needed. Rash vests involve minimal labour, so it makes sense that they are the product to be made in Torquay, where labour costs are much higher than in Asia. To see inside the Rip Curl 'factory' in Torquay, watch the video 'Rip Curl Wetsuit Factory Tour' on AustraliaSurfingLife's YouTube channel.

All repairs for Rip Curl wetsuits Australia-wide are done in Torquay. This is beneficial as it also provides work for the machinists and it allows the company to check where wear and tear might occur (other than from accidents such as dog bites).

Rip Curl

Various cutters are used on the neoprene.

The material: neoprene

The most expensive component of a wetsuit is the raw material, which is neoprene. Neoprene (chloroprene) has been used extensively in the footwear industry and in the automobile industry (as trims of doors and windows, car tyres and gaskets in the engine block) for decades. The price of neoprene is highly dependent on how the global automobile industry is tracking in terms of sales. When automobile sales are strong, suppliers' prices tend to spike. Unfortunately, Rip Curl has little control over this, as the demand for automobiles is far greater than for wetsuits!

The Lectra CAD system that Rip Curl uses can determine the yield for different fabrics, or in other words, how much 2-millimetre or 3-millimetre neoprene is required for a wetsuit design. If the team can see on a prototype or via the Lectra system that the patterns are not interlocking as well as they could, they will go back to the designer. Even though the design looks great, they might suggest adding a seam somewhere to get an extra 5 per cent efficiency in the use of fabric. They might suggest that instead of having a bump on the elbow, it could be tucked in a little bit to improve the yield.

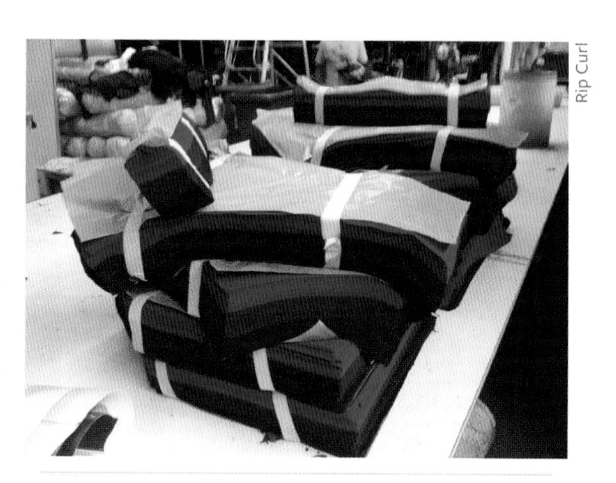

Rip Curl

Layers of pattern pieces that have been cut from neoprene

Research and development (R&D)

Rip Curl works very closely with neoprene suppliers in Asia (where the bulk of neoprene comes from) for continual improvement and development.

E1 neoprene was introduced in about 2000; it was roughly 2 millimetres thick, weighed 200 grams per square metre (gsm), and stretched at 100 per cent at 1 kilogram of force. The neoprene fabric is constantly evolving, and Rip Curl is always working with the suppliers to push the material to the next level: to dry fast, weigh even less, to stretch more and to provide more warmth. Suppliers work with the parameters and present several versions before it is 'right'. Multi-thickness wetsuits are now the norm, with thicker panels on the chest and legs, and thinner ones on the arms, to get warmth at the core and freedom for the limbs. Customers choose from all sorts of thicknesses to suit the conditions – the temperature of the water, the air and their own body. In 2023 the E7 Flashbomb Heat Seeker (which generates heat when stretched) was the latest incarnation of the world's most popular wetsuit materials used for the top-end range.

In-house R&D

Sheets of the 'new material' are sent to Torquay and wetsuit samples are made up. Rip Curl will test the materials internally with lab equipment and, more importantly (being near Bells Beach and being keen surfers), staff will wear the wetsuits out surfing. They will closely document the number of hours in the surf and do an appraisal on the material. In-surf test results are stacked up against the lab tests before a decision is made to commercialise the material or not.

The R&D performed at Torquay headquarters with the company's own lab equipment includes the following:

Rip Curl

A modulus machine for stretching neoprene

- UV is applied from a tungsten filament lamp to test the effect of sunlight on fabrics.
- A Rip Curl–engineered machine tests the endurance of fabrics in salt water. First, a sample is fixed at either end between two moving parts. They swish back and forth in a salt-water mix for a set amount of time or number of movements to continually elongate (and return) the fabric. This machine can be used to see how screen prints, laminated fabrics, glues and trims withstand this treatment.
- A modulus testing machine checks the elasticity of different laminates and fabric.
- Real-life testing in the ocean is also undertaken.

Summary
Chapter 2

Quiz
Chapter 2 revision

Quality control

It is important to have quality control at the factories in Asia. Once the quality checklists and standard operating procedures (SOP) are in place, there is usually less need for stringent checking. However, quality control is always an issue as there are so many variables, such as the correct temperatures for bonding, etc. Variations can occur from batch to batch, so the product always needs to be checked.

Team feedback

Throughout the entire process of design, prototyping, pattern grading and production, staff feedback is highly valued. Most of the staff have years of experience in surfing and in manufacturing, and are enthusiasts who have a tactile understanding of the fabrics from working with them and an intimate and experienced knowledge of what is required in the water. World champion surfer Mick Fanning has been working with Rip Curl for years, testing out the latest wetsuits and giving feedback. This expert knowledge and feedback from the team is what sets Rip Curl above new companies in terms of 'fit for purpose' and quality.

To see more about Rip Curl wetsuits, go to the Rip Curl website. To see real-life testing of Rip Curl's wetsuits by some world champions, watch the video 'Behind the Scenes: The Making of Rip Curl's E-Bomb E7 Limited Edition Wetsuit' on the company's YouTube channel.

Jacinta O'Leary spoke to Peter Coles in January 2017 and again in 2023.

Weblinks
Rip Curl

Behind the Scenes: The Making of Rip Curl's E-Bomb E7 Limited Edition Wetsuit

UNIT 2

POSITIVE IMPACTS FOR END USERS

Areas of Study 1, 2 and 3

Unit 2, Areas of Study 1, 2 and 3

In this unit, you will be introduced to the following core concepts:

- designing to create positive impact
- developing products for specific/special needs
- cultural influence on design (particularly that of Aboriginal and Torres Strait Islander peoples).

You will also review some concepts from Unit 1:

- the use of the Double Diamond process
- writing a design brief and evaluation criteria
- drawing systems
- testing and using materials and processes
- developing models, prototypes and the proof of concept.

Unit 2, Area of Study 1: Opportunities for positive impacts for end users

You learn about:

- role and work of designers addressing needs and/or opportunities that support positive impacts for end users
- Factors that influence designing for positive impacts for end users
- critiquing products that have positive impacts for end users
- communicating research using a variety of multimodal forms.

Unit 2, Area of Study 2: Designing for positive impacts for end users

You learn about:

- research methods to gather primary and secondary research to create a profile of an end user and describe their needs
- ethical research methods when gathering data
- materials, tools and processes used in a variety of design specialisations
- generating and recording graphical and physical product concepts (including prototyping) and to develop a final proof of concept for an inclusive product
- using criteria to evaluate product concepts, final proof of concept and finished product.

Unit 2, Area of Study 3: Cultural influences on design

You learn about:

- cultural influences on product design including those of Aboriginal and Torres Strait Islander peoples
- Factors that influence cultural needs and opportunities of end users
- quantitative and qualitative research methods.

Source: *VCE Product Design and Technologies Study Design*, p26

9780170477499

Unit 2 Planner

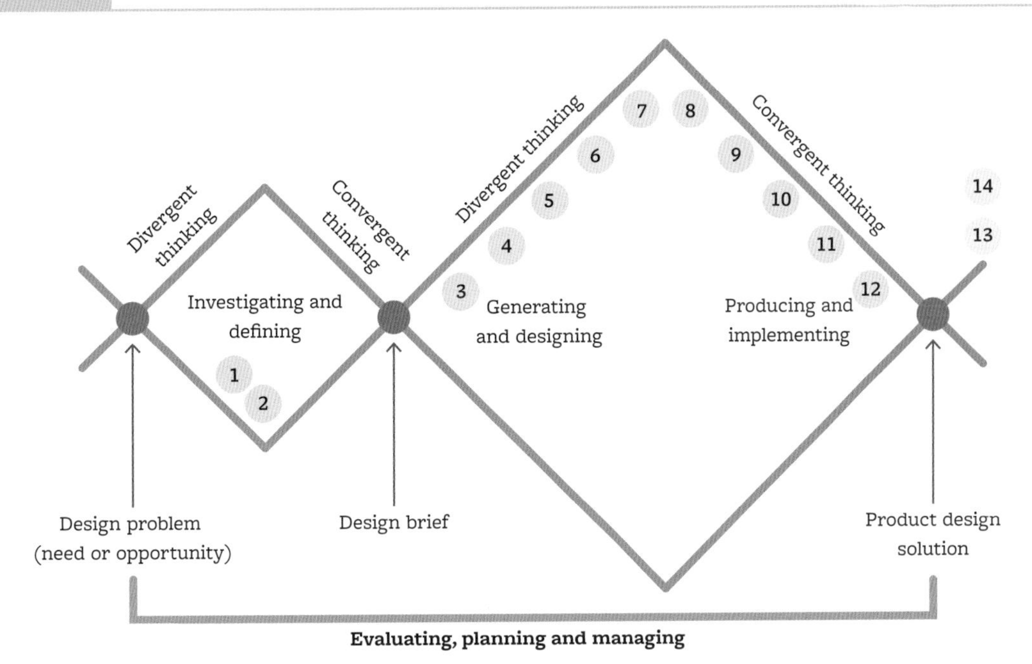

Outcome 1	**Week 1** • Investigate designers' approach to addressing need, brainstorm need/opportunity • Explore areas for 'positive impact' opportunities, possible end users	**Week 2** • Research, analyse and write report on designers making a positive impact • Research end user requirements, investigate existing solutions for Outcome 2	**Week 3** • Write the design brief, evaluation criteria • Visualisations • Modelling
Outcome 2	**Weeks 4 and 5** • Materials and process research • Test materials • Trial skills/joining methods/material combinations, etc.	**Week 6 or earlier** • Two design options • Select the preferred design concept and justify	**Weeks 7 and 8** • Prototype and design refinement • Working drawing
	Week 9 Planning and production	**Weeks 10 and 11 or as required** Production	**Week 12** Finish and evaluate
Outcome 3		**Week 13** Research Indigenous case study and other cultural influences	**Week 14** Case study report

Assessment in Unit 2

The following work will be assessed for Unit 2:

- **a multimodal record of progress (folio of work)** of research, as well as the development and conceptualisation of products addressing a need or opportunity related to positive impacts for the end user/s
- **practical work**
  - » models, prototypes, proof of concept and the finished product
- **case study analysis** or **research inquiry** of a designer and end user/s that explores the influence of culture in product design.

CHAPTER 3

Opportunities for positive impacts for end users

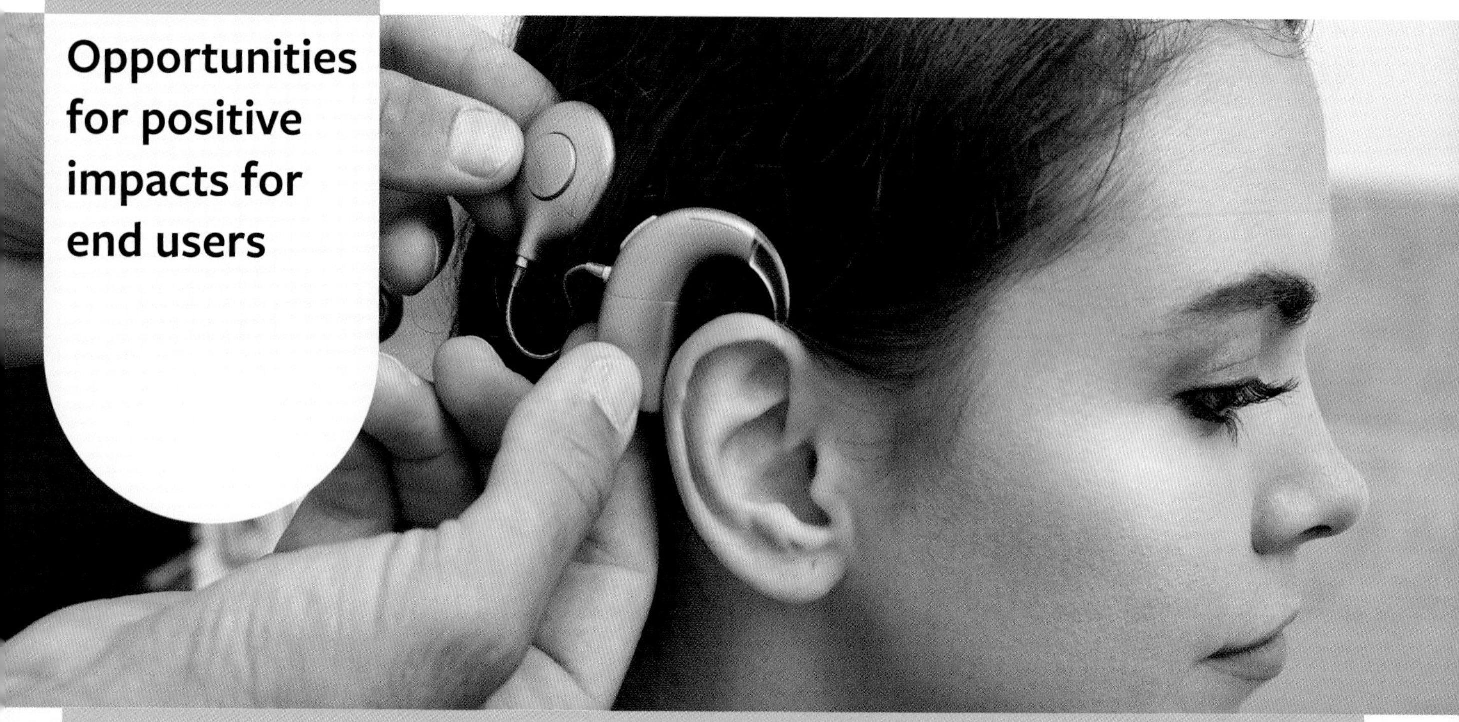

UNIT 2, AREA OF STUDY 1: OUTCOME 1

In this outcome, you will:

- investigate and critique products, using the Factors that influence design, to make judgements about the success or failure of products to support positive impacts for end users.

Key knowledge

- role and work of designers from a range of industries and specialisations whose products address needs and/or opportunities that support positive impacts for end users
- market needs and opportunities for products that support positive impacts for end users
- methods used by designers whose products address needs and/or opportunities that support positive impacts for end users
- use of the Double Diamond design approach to apply design thinking to critique products that have positive impacts for end users
- methods to communicate research using a variety of multimodal forms, such as written reports, audio files, video files, mood boards, diagrams, charts and/or drawings
- Factors that influence designing for positive impacts for end users
- criteria used to evaluate processes to develop inclusive products that support positive impacts for end users as well as to evaluate the inclusive products.

Key skills

- investigate practices of designers and critique their products
- research and discuss market needs and/or opportunities for products that support positive impacts for end users
- use manual and digital techniques to collect, collate, interpret and communicate findings of research
- use Factors that influence product design to develop evaluation criteria to make judgements about a product's ability to support positive impacts for end users
- work technologically to manage the Double Diamond design approach to consider positive impacts of products for end users.

Source: *VCE Product Design and Technologies Study Design*, pp. 24–5

9780170477499

DESIGN THAT IMPROVES LIVES

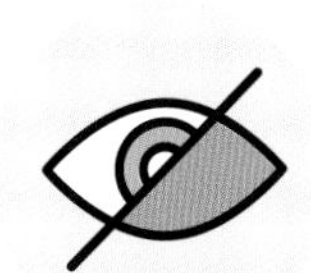

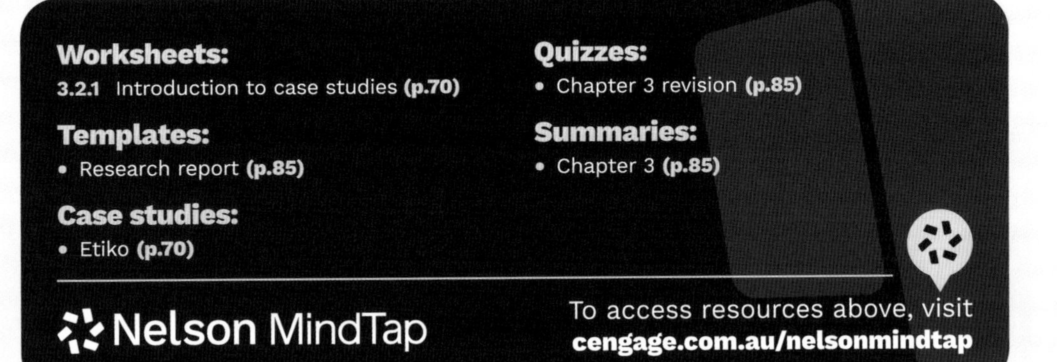

Designing to improve lives

Many designers want to improve the lives of others. They use their skills, knowledge and the design process to create products for people (or creatures) in need. They identify opportunities or situations where a new or modified product will have a positive impact on the lives of people and other living creatures. They can achieve this in several ways.

Universal and accessible design

Universal/accessible design refers to designing products and environments that all people can access and use, no matter what their level of physical and/or mental capacity. It is a term frequently used in building design, but is also strongly relevant to product design. It is about ensuring that public buildings, spaces and products can be used by everyone, not just the 'average' person. It is also about creating specific products for certain end users. It considers the diverse needs of people of varying:

- ages – including the very young and very old
- sizes – including the very small and very tall/large
- abilities – including those with disabilities, reduced mobility, etc.
- incomes – including the very poor
- races, religions and cultures – including minority groups
- genders – including those identifying as non-binary.

Designers aim for their products to be easy and instinctive to use, safe and comfortable, appropriately priced, and not to offend or exclude. In public spaces, there is a greater necessity for products and spaces to be universally accessible so that people with limitations can still move and function safely in those spaces and not be restricted or excluded – for example, by providing an 'easy-to-access' lift for wheelchairs next to an entrance that was originally built with steps only.

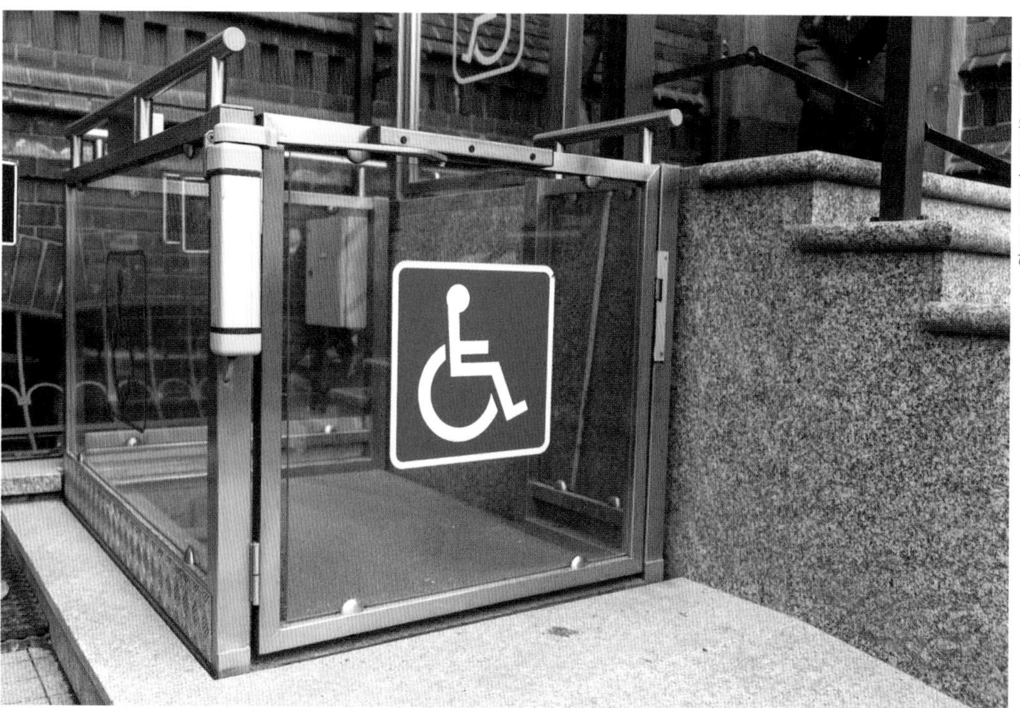

Shutterstock.com/Sergey Granev

9780170477499

7 Principles of Universal Design

The 7 Principles of Universal Design were developed by architects, product designers, engineers and environmental designers brought together by the Center for Universal Design at the University of North Carolina, USA.

Principle 1: Equitable use

The design is useful for people with diverse abilities.

Principle 2: Flexibility in use

The design accommodates a wide range of individual preferences and abilities.

Principle 3: Simple and intuitive use

Use of the design is easy to understand, regardless of the user's experience, knowledge, language skills or current concentration level.

Principle 4: Perceptible information

The design communicates necessary information effectively to the user regardless of the location's conditions or the user's sensory abilities.

Principle 5: Tolerance for error

The design minimises hazards and the adverse consequences of accidental or unintended use.

Principle 6: Low physical effort

The design can be efficiently and comfortably used with a minimum of fatigue.

Principle 7: Size and space for approach and use

Appropriate size and space are provided for approach, reach, manipulation and use – regardless of user's body size, posture or mobility.

Adapted from the Center for Excellence in Universal Design website. Universal design case studies can be found at the website and by doing an internet search on 'universal design case studies'.

Weblink
The 7 Principles

Design to aid specific needs

Designers often identify a specific problem that people have and design a solution specifically for those people. This could be a problem related to:

- age:
  - diseases and conditions related to old age that limit movement, grip, reach, sight and hearing, problems related to dementia/loss of mental capacity
  - related to the young, such as limited physical ability, and limited ability to read and to make considered decisions
- specific illnesses and/or disabilities, permanent injuries, genetic conditions or chronic health conditions
- mental health issues
- brain injuries and dementia
- poverty and/or limited access to resources such as food, water, electricity and housing.

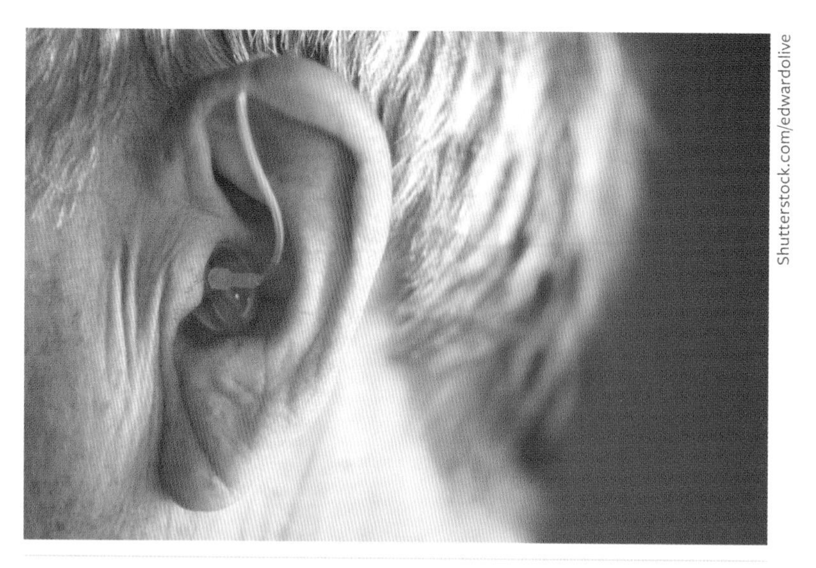
Shutterstock.com/edwardolive

Designing for those with a hearing impairment

Design for animals (non-humans)

Products and specially designed spaces are needed to improve the lives of the creatures that live with us on this planet. These include:

- domestic animals – pets and livestock
- captive animals
- wild animals – on land, in the sea and in the air.

For all animals, designers help to create products that focus on their comfort, health and safety.

- For pets, products focus on the basic needs above and the pets' engagement or enrichment.
- For livestock, the focus is on keeping them safe, contained and content, and on maximising productivity.
- For captive animals, designers consider how to mimic the creature's natural environment, provide enrichment and promote breeding (especially for endangered species).
- In the wild, many products and systems are developed to monitor and repair the damage human impact has caused to animals and their habitat, including after bushfires and floods.

Jill Livett

Nesting boxes to replace tree hollows lost through bushfires or land clearing

9780170477499

Getty Images/Parks Australia

This bridge is a design response to the annual migration of millions of red crabs on Christmas Island.

Sustainable design

Many designers seek to protect our world and ways of living by developing products that lessen our negative impacts on the environment. It is one way they can have a positive impact. Sustainability is defined as 'meeting the needs of the present without compromising the ability of future generations to meet their own needs'. Sustainable designers aim to develop products that effectively satisfy the needs of the end user without further damaging the world around us. They consider the impacts of a product at every stage of its life, and create positive change by reducing environmental, social and economic harm. They apply practical strategies, such as:

- developing long-lasting products
- designing multifunctional products
- designing products that solve an environmental problem
- selecting materials that are recycled, locally sourced and/or processed in ways that cause minimal harm
- reducing the amount of materials used
- minimising the use of water and energy during manufacturing and when the product is used (or by using renewable energy sources)
- minimising and repurposing waste from the manufacturing process
- eliminating the use of harmful toxins in production and ensuring the safety of workers and end users
- ensuring the materials in the product can be reused/recycled at the end of its life.

Sustainable Development Goals (United Nations)

The United Nations has developed 17 Sustainable Development Goals. The basic aim of these goals is to improve the lives of those who live on this planet, and to provide practical approaches to all forms of development. In the words of the UN Sustainable Development web page 'The 17 Goals', they are 'an urgent call for action by all countries – developed and developing – in a global partnership. They recognize that ending poverty and other deprivations must go hand-in-hand with strategies that improve health and education, reduce inequality, and spur economic growth – all while tackling climate change and working to preserve our oceans and forests.'

Many of the areas discussed above and the product examples that follow illustrate how designers create products that aim to address a number of these goals.

Case study
Etiko

SUSTAINABLE DEVELOPMENT GOALS

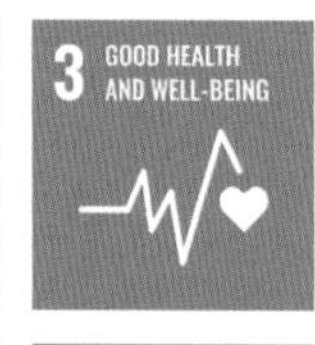

7 AFFORDABLE AND CLEAN ENERGY

8 DECENT WORK AND ECONOMIC GROWTH

9 INDUSTRY, INNOVATION AND INFRASTRUCTURE

10 REDUCED INEQUALITIES

11 SUSTAINABLE CITIES AND COMMUNITIES

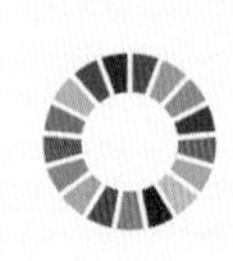

The 17 United Nations Sustainable Development Goals

Weblink
UN Sustainable Development Goals

For more information about each Sustainable Development Goal, refer to the UN website.

Products that make a positive impact

For Unit 2, Outcome 1, you need to examine a designer's process and product to see if they achieved their aim – to create products that have a positive impact on people and the environment. The following case studies are examples of products you could investigate.

Worksheet
3.2.1

Introduction to case studies

Read **at least three** of the case studies in the following pages and use them as a starting point for your Outcome 1 research task. You could also source other case studies of products developed to have a positive impact on those with special or specific needs.

9780170477499

FOLIO ACTIVITY

Discussion activity

In small groups, discuss:

- Who are the end users in each case study, and what is their need?
- Consider which need the product is specifically designed to address (either refer to the sections on pages 66–9 explaining where designers can have a positive impact, or decide which of the 17 Sustainable Development Goals is relevant).
- Are there aspects of the end users' situation that could make it more difficult to create or use a solution?

The Shoe That Grows

Kenton Lee travelled to Nairobi, Kenya, after graduating college. He lived and worked at a small orphanage with some incredible kids. One day, he noticed a little girl walking next to him. As he looked down at her feet, he was shocked to see how small her shoes were. They were so small that she had to cut open the front of her shoes to let her toes stick out. That was the day that Kenton thought:

What if there was a shoe that could adjust and expand its size?

What if there was a shoe that could grow?

And the idea for The Shoe That Grows™ was born.

It took Kenton and a group of friends back in their hometown in Idaho more than six years to take this idea and turn it into a reality. But they finally did it; they made The Shoe That Grows – a shoe that grows five sizes and lasts for years.

They started the nonprofit organisation Because International™ to serve as the structure to get The Shoe That Grows to as many kids as possible who are in desperate need of footwear to protect their feet.

How do these shoes make a positive impact?

- Over 1.5 billion people suffer from soil-transmitted diseases worldwide.
- Without shoes, children are especially vulnerable to soil-transmitted diseases and parasites that can cause illness and even death.
- Children who get sick miss school, can't help their families, and suffer needlessly. And since children's feet grow so quickly, they often outgrow donated shoes within a year, leaving them once again exposed to illness and disease.

The Shoe That Grows can be donated for $20, and fundraising has allowed many of these shoes to be given to children in need who can't afford to buy them.

Because International now mentors entrepreneurs developing products that make a positive difference through the Because Accelerator. This program gives them the skills and opportunity to develop and commercialise innovative ethical/sustainable products.

Courtesy of Because International

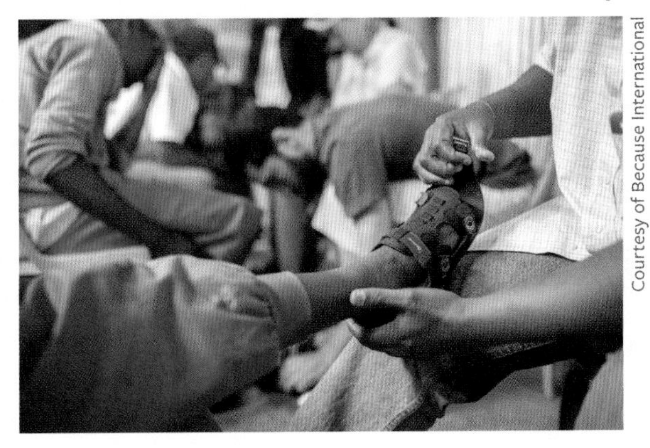

Courtesy of Because International

CASE STUDY

A Liter (Litre) of Light

Liter of Light is a global, grassroots movement committed to providing affordable, sustainable solar light to people with limited or no access to electricity. In many places in the world, millions of people live in makeshift housing without windows and no access to daytime light. Through a network of partnerships around the world, Liter of Light volunteers teach marginalized communities how to use recycled plastic bottles and locally sourced materials to illuminate their homes, businesses and streets.

Liter of Light has installed more than 350000 bottle lights in more than 15 countries and taught green skills to empower grassroots entrepreneurs at every stop.

https://literoflight.org

The original inspiration

In 2002, the Brazilian mechanic Alfredo Moser had a light-bulb moment and came up with a way of illuminating his house during the day without electricity – using nothing more than plastic bottles filled with water and a tiny bit of bleach.

So how does it work? Simple refraction of sunlight, explains Moser, as he fills an empty two-litre plastic bottle. 'Add two capfuls of bleach to protect the water so it doesn't turn green [with algae]. The cleaner the bottle, the better,' he adds.

Wrapping his face in a cloth he makes a hole in a roof tile with a drill. Then, from the bottom upwards, he pushes the bottle into the newly-made hole. 'You fix the bottle in with polyester resin. Even when it rains, the roof never leaks – not one drop. An engineer came and measured the light,' he says. 'It depends on how strong the sun is but it's more or less 40 to 60 watts,' he says.

Alamy Stock Photo/RZAF_Images

The simple concept behind Liter of Light: a plastic water bottle protrudes from a roof.

Source: Gibby Zobel, 'Alfredo Moser: Bottle light inventor proud to be poor', BBC News, 13 August 2013

https://literoflight.org/

Elements needed in a 'Day light kit' to transform a plastic bottle into a household light

Using this as inspiration, the Liter of Light organisation aims to develop a range of solar lighting solutions for homes with minimal access to electricity. As well as the plastic bottle solution, other innovative products use cheap solar panels to provide inexpensive lighting solutions in homes and community spaces at night. There are now chapters of the Liter of Light organisation in 15 countries, and they have improved the lives of over 353 600 homes worldwide.

Liter of Light's open source technology has been recognized by the UN and adopted for use in some UNHCR camps.

https://literoflight.org

9780170477499

Bamboo Labs

Bamboo Labs is an award-winning social enterprise in Ethiopia working to address the need for accessible mobility in Addis Ababa. Using the large supply of bamboo in Ethiopia, the organisation produces wheelchairs and bicycles made out of bamboo for the local market. They also have plans to produce bamboo bike frames for export.

Founded by a young Ethiopian entrepreneur, Abel Hailegiorgis, the company began its work in 2019 by addressing a need for improved mobility for wheelchair users in Addis Ababa. With some seed funding from an international non-governmental organisation (NGO), Abel was able to set about designing wheelchairs made from bamboo. Since then Abel has gained experience working at the Bamboo Bike Club in Munich and more recently participated in a three-month training and mentorship programme organised by the UN's Global Green Growth Institute where Bamboo Labs won one of the Green Mobility awards.

Samuel Bekele/Bamboo Labs

Bamboo Labs founder Abel Hailegiorgis

Why bamboo?

Ethiopia possesses the largest area of bamboo in Africa, estimated at around 1.75 million hectares and amounting to approximately 70 per cent of all bamboo grown in Africa. As a resource bamboo is highly renewable, sustainable, and easy to grow. With a tensile strength stronger than steel and a compressive strength comparable to concrete, bamboo also has the properties to absorb shocks and vibrations.

Bamboo Labs want to contribute towards bringing about a more equitable, safer and greener mobility plan for the country. Using locally sourced bamboo, it creates jobs and strengthens the local economy. The company is also advocating for more sustainable modes of transport and a reduced carbon footprint.

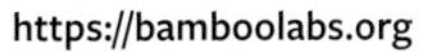

https://bamboolabs.org

Bamboo Labs

A Bamboo Labs bamboo wheelchair

Bamboo Labs

A Bamboo Labs bike

CASE STUDY

LifeStraw

LifeStraw is a personal water filter designed to remove bacteria and parasites from water. It allows individuals to access clean and safe drinking water in areas where water sources may be contaminated. The LifeStraw was initially developed to prevent Guinea worm disease, but, after being improved over many years of trialling and on-the-ground testing, it now functions as a more effective generalised filter in many situations. The filter is lightweight, portable and easy to use, making it a valuable tool in countries where clean water isn't easily accessible, and in emergency situations and for disaster relief. It is also useful for hikers, campers and travellers. Additionally, for every LifeStraw product purchased, a school child in need receives safe drinking water for an entire school year. This gives an additional benefit of providing access to clean water to children in developing countries.

From the LifeStraw website:

'At LifeStraw, we believe everyone deserves equitable access to safe drinking water. We design beautiful, simple, and functional products that provide the highest protection from unsafe water based on where and how they will be used. We also take our responsibility to people and the planet seriously. For every product sold, a child in need receives access to safe water for an entire year. We have been partners in the fight for the eradication of Guinea worm for 25 years, and we actively respond to emergencies across the globe. We are Climate Neutral Certified B Corp and from product to packaging, we measure and minimise our environmental impact. We strive to support underrepresented communities through our actions, our products, our marketing and our communications. We live, work and create with impact in mind. We fight for good and always err on the side of action.'

https://lifestraw.com

Alamy Stock Photo/Mike Greenslade

Alamy Stock Photo/Keith Homan

Alamy Stock Photo/Mike Greenslade

Alamy Stock Photo/dpa picture alliance

LifeStraw and other water-filtering systems in use

9780170477499

Lego Braille machine

Braille is a written language using a system of raised dots that can be felt with the fingertips and used to read and write by people who are blind or have low vision. Shubham Banerjee, 13, created a Braille printer out of Lego pieces for a school science project and, with the help of his family, turned his concept into a start-up company that gained financial backing from the tech company Intel Corp.

Shubham came up with the idea after researching Braille online and realising that printers for the blind cost $2000 or more. Concerned it was too expensive for most people to afford, Shubham wanted to make an inexpensive version.

He asked his dad for a Lego robotics kit that cost about $350 and built a model in about a month for a science fair in early 2014. He shared the plan online in an open source format so that anyone could build it.

'He wanted to make it very cheap and DIY, do-it-yourself,' Neil Banerjee, Shubham's father, said. From there, newspapers picked up on the project and he went on to win at the county fair level.

Children who were blind (and their parents) started contacting Shubham soon after, asking if he could make a printer they could buy off the shelves.

Shubham made a prototype using parts from a desktop printer and an Intel chip with wi-fi and bluetooth. He showed what he created to Intel.

'They were really impressed,' Banerjee said. Banerjee helped his son set up a company with Shubham as the founder. They called it Braigo Labs, a combination of Braille and Lego. A few months later, Intel announced they would back Shubham's company.

'He's solving a real problem, and he wants to go off and disrupt an existing industry. And that's really what it's all about,' Edward Ross, director of Inventor Platforms at Intel, told the Associated Press.

Lori Grisham, USA Today Network

Alamy Stock Photo/Tribune Content Agency LLC

Shubham Banerjee and his Lego Braille printer

CASE STUDY

Adaptive clothing: Christina Stephens

If you have some form of physical limitation (e.g. if you are wheelchair bound, have prosthetic limbs or are limited in your hand movements), finding clothing that is comfortable and easy to manage is very difficult. Most adaptive clothing (clothing made for those with physical limitations) focuses on function with little thought given to making the wearer look and feel stylish. Founder of the Christina Stephens clothing company, Jessie Sadler, works with Carol Taylor, a quadriplegic fashion designer, to give the people they design for a choice in clothing – helping them to feel confident and empowered in clothing that looks as good as it feels. The drive for this work comes from personal or family experience, which gives powerful insights into the needs and wants of their customer base. The company's aim is to bring inclusive and adaptive clothing to mainstream fashion retailers. They create pieces that are not only discreetly functional and fit for purpose, but are also widely appealing and bought by customers with and without disabilities. Practical approaches used in their products include:

- a focus on clean lines, flattering cuts and colours, minimal trims, and basic pieces that can be worn with ease
- fasteners – easily accessible positions, reducing the number needed, some use of magnets for fastening (removing the need for buttons, zips, etc.)
- reducing fabric and seams in the seating area for wheelchair wearers, choosing fabric in these areas that are high in comfort
- clothing shapes that work in a seated position (e.g. shorter jacket length, flowing, non-waisted styles), and that allow for greater arm movement
- other clothing shapes designed for different abilities.

Getty Images/Mark Nolan

Label founder Jessie Sadler with designer Carol Taylor and other members of the team at Christina Stephens

The company has a strong environmental and ethical focus, and uses the following sustainability measures:

- using remnant ('dead-stock') fabrics throughout its collections, diverting unused production fabrics from landfill
- manufacturing to Global Organic Textile Standard and using STANDARD 100–certified bamboo viscose fabrics to its own and individual specifications
- providing beautiful, practical and low-impact packaging designed for post-packaging use
- using all natural materials in labelling (including clothing labels, tags and strings)
- partnering with textile mills and wholesalers who are committed to sustainability and fair trade
- designing and manufacturing in Australia and with reputable and accredited offshore partners.

Weblink
Global Organic Textile Standard

OEKO-TEX® STANDARD 100

9780170477499

Products for those with autism

Designers are recognising the special needs of people with autism. One aspect that benefits many is the feeling of pressure on their body. Research has found that children and adults with autism find weighted products provide a sense of deep pressure touch (DPT) stimulation. DPT is thought to have a calming effect, and may help reduce anxiety, improve sleep and increase focus. Weighted products provide a sense of security and comfort for individuals with autism, which can help them feel safer and more grounded in their environment. The use of weighted products also helps to improve body awareness and can be used as a tool for self-regulation during times of high anxiety or stress. There are a range of products on the market that provide this kind of relief – the following are two examples.

OTO chair

The James Dyson Award–winning OTO chair uses a set of inflatable cushions to hug the person sitting in the chair. The cushions expand from the sides, emulating the feeling of being body-hugged and helping people with special needs overcome sensory overload.

The OTO chair was designed by Alexia Audrain, who learned more about the special needs of people on the autism spectrum while she studied cabinetmaking and designing. 'Noise, light or physical contact can be a real challenge in everyday life [for people with autism]', says Audrain. 'To compensate for this sensory disorder, autistic people regularly feel the need to be held very tightly or to be hugged.' This form of deep pressure therapy can have a calming effect and reduce anxiety while improving the person's sense of body awareness.

The award website says: 'The OTO chair has inner walls that inflate and create deep pressure on the user's legs and chest. The aim of the chair is to be used autonomously … users respond to their own sensory needs. OTO is easy to use. The remote has + and – pictograms buttons that inflate or deflate cells … Thanks to its cocoon shape, OTO offers privacy and gives a feeling of safety for the user. The upholstery of the chair provides a muffled acoustic that helps the user to focus on their own body and isolate themselves from the outside world.'

www.jamesdysonaward.org

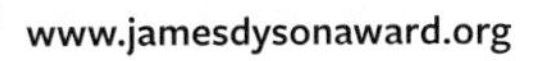

Coralie Monnet

The OTO chair

Harkla sensory products

Children and adults with autism and/or a sensory processing disorder (SPD) have a heightened experience of the sense of touch. Clothing can be difficult: the texture of some fabrics, seams, labels, etc. may cause frustration and constant irritation. In other situations, some forms of body compression or enclosure can be very comforting, particularly when sensory inputs from the world feel overwhelming. Specialist clothing has been developed, in which fabrics are carefully chosen, the number of seams is reduced or they are placed on the outside, buttons and zips are replaced with other closures, and a firm fit can provide compression.

The Harkla company creates products to give children with special needs a more positive experience of their environment. Their aim is to 'help parents create the ideal environment for their children to thrive in'. The company, founded by Casey Ames, has developed creative, fun and useful products such as:

- sensory swings (the compression swing and the pod swing) – these are spaces that provide rhythmic movement and a place to be enfolded (reducing external inputs)
- the Harkla hug – an inflatable 'peapod' that can be used in play or to become a safe sanctuary. The hug provides deep touch pressure that is calming and comforting.

Both of these products are meant to reduce anxiety – they imitate a hug and induce a sense of calm and safety. The company focuses on 'creating products that help improve sleep and comfort at home, focus and behaviour in the classroom, and growth and development at therapy'.

Harkla products are part of a larger service to families. The company also supports parents by providing online information, podcasts and webinars, courses and training. Find out more about Harkla at the company's website.

Weblink
Harkla

Harkla

The pod sensory swing

The Harkla hug

For more information about clothing for people with special needs, search the internet for 'adaptive clothing'.

CASE STUDY

Life Saving Dot

The bindi delivering much-needed iodine to impoverished women in India

Women in India traditionally wear a bindi, a small dot between the eyebrows, for religious purposes or to show they're married, but it's grown in popularity among all women. Talwar Bindi's Life Saving Dot has an even higher purpose. The Life Saving Dot is coated with iodine and delivers the recommended amount of 150–220 micrograms of the nutrient daily to poor women in India, where approximately 350 million people are at risk of iodine deficiency. A lack of iodine can cause a

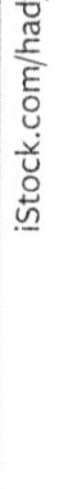

iStock.com/hadynyah

A woman wearing Talwar Bindi's Life Saving Dot

9780170477499

number of health problems, especially during a woman's pregnancy. Iodine can be absorbed through the skin, and the Life Saving Dot is a particularly low-cost nutritional supplement – it costs only 10 rupees, or 16 cents, for a packet of 30 bindis. It has been distributed to women across rural India through health camps and clinics in several villages.

https://theindexproject.org/award/nominees/1349

Artificial reefs

Reefs are important ecological infrastructure that supports a wide variety of marine creatures. A reef is a shallow strip or ridge in the sea, ocean or other water body that rises to or near the water surface. They are hard structures on which plants and coral larvae can find surfaces to take root in and that create shelter for fish life and crustaceans. Reef structures also affect water movement: they can reduce the energy level, which forms calm water and thus good shelter. It also creates extra turbulence at some spots and may increase concentrations of small animals such as plankton. Subsequently, fish and other marine animals find shelter and food on the reefs. So a healthy reef forms a complex habitat for marine flora and fauna and usually shows great biodiversity. Because of the diversity of life found in the habitats created by corals, reefs are often called the 'rainforests of the sea'. About 25 per cent of the ocean's fish depend on healthy coral reefs.

Reefs are being destroyed in greater number due to human intervention and extreme weather events. In the past, artificial reefs were made from local plant material or recycled waste (sunken boats, tyres, etc.) but these were either short-lived or caused further environmental issues due to the release of toxic materials. Scientists and engineers are working together to design long-lasting and ecologically effective artificial reefs. They have explored a wide range of materials, technologies and forms to make structures that are quickly inhabited by sea life. Concrete balls with hollows were found to create good habitat for plants and coral, and the hollow structure provides shelter for fish and mammals. According to the EcoShape website, 'Experience with artificial reef balls in New South Wales has shown that after two years the richness of the artificial reef exceeded the natural one close to the artificial reef.' Other new reefs are being made from ceramic, geo-textiles, bio-rock and 3D printed equipment – thorough testing and trialling continues.

To find out more, research the case studies on the EcoShape website.

Alamy Stock Photo/David Kilpatrick

A breakwater built from artificial reef balls

Shutterstock.com/Angela N. Perryman

Close-up of reef balls

Weblink
Artificial Reefs

Case studies of positive impact

For this outcome, you need to investigate a range of products (at least two or more) to see how successful the designers and producers in different specialisations are in addressing and improving the lives of others.

Designers working in different specialisations

Designers can work in one or more of the following specialisations to create effective solutions:

- engineering
- furniture and furnishings – including homewares
- health and medical
- agriculture
- industrial design – including casings for electronic products, and wearable technology
- jewellery
- music products
- sports, travel and recreation
- textiles – clothing and non-clothing products.

Of course, designers are not limited by these categories – they work within and beyond them to create solutions focused on the needs of end users. For example, depending on their expertise, designers might develop:

- sensory gardens for those who are sight and hearing impaired
- adaptive clothing for those with physical disabilities, clothing for specific psychological or physical needs
- toys or furniture for children with autism, physical disabilities or learning difficulties
- enclosures or housing for animals (domestic pets or farm animals, native animals, aquatic creatures, etc.)
- enrichment toys for captive animals
- universal/accessible housing fittings and fixtures
- kitchenware and utensils/tools for those with physical impairments
- inexpensive lighting for those with limited access to electricity
- technological tools that improve access to education for remote or impoverished young people
- musical instruments that are accessible to those with a range of disabilities
- tools for reading and writing for those with physical disabilities and learning difficulties
- interactive jewellery for those with communication difficulties
- walking and domestic aids for the elderly and those with reduced mobility
- sporting equipment for those with physical impairments
- products for indigenous/cultural groups
- any product that reduces its environmental impact when compared with existing alternatives.

9780170477499

Working together

Designers usually work with a range of skilled specialists to create an effective solution. They often call on people with expert knowledge in areas relevant to their project, such as:

- health professionals, biomedical engineers and researchers (physical and mental health)
- scientists and/or engineers, researching or working in the following areas:
  - materials, manufacturing processes and electronics
  - environment and sustainability
  - animal behaviour
  - manufacturing systems
  - ergonomics
- manufacturers of component parts.

Designers may also draw on information from articles and research publications carried out by specialists in different fields (secondary research) to help them understand their end user, or for specific knowledge relevant to design and production.

Sometimes, a non-design person such as a doctor or sportsperson comes up with an innovative concept, but may not know how to turn it into a workable solution. A designer might be asked to work with that person – to use their specialised design knowledge and skills to convert the concept into a well-planned product that can be manufactured. For an example of this, refer to the Cobalt Design case study on pages 179–80, where designer Steve Martinuzzo discusses working with the originator of KeepCup.

KeepCup. Design by Cobalt Design

The completed KeepCup *Brew*

Research methods to investigate need

To develop an effective solution for special end users, designers need to put a lot of effort into their research. This is because the needs and circumstances of those who they are designing for are outside their own experience – they have less knowledge to draw from. Designers need to use the information collection 'tools' that are used in the first phase of the Double Diamond to explore the needs and opportunities of end users, including:

- observation
- questioning potential end users and those with specialist knowledge of the situation (through interviews, surveys, focus groups etc.)
- secondary research – searching out the research of others.

In some situations, an end user may have limited or no ability to communicate. In these cases, a designer would need to discuss the details of the situation with others – usually the carer of the end user, or someone with specific knowledge of their needs. Designers should not make assumptions about what is needed based on their own experience (unless their experience is relevant). **Innovative and creative solutions come from looking at a situation with new information and different perspectives.**

Data collection and communication tools

Designers can use a wide range of manual and digital tools to carry out their research.

Collecting information

- When observing, they may use digital cameras to record observations, so that they can be reviewed and checked. They might also take observation notes, either handwritten or digital. Sometimes the presence of cameras might change the person or creature's behaviour if they are conscious of being observed – in these cases, cameras are hidden (with care to ensure privacy laws aren't infringed).
- GPS tracking devices might also track and collect data on the movement of animals or creatures. Consent would need to be given for tracking humans.
- Audio recording apps can be used for recording interviews and converting them to text.

Communication

- Designers and other researchers can use a range of computer software to record and process data, such as word processing, database and spreadsheet software. They can use charts, tables, graphs and infographics to process and display their research data to make it easy to understand.

Case study investigation and report

What is critiquing?

As part of this outcome, you will be critiquing the practices and products of designers as they aim to improve the lives of others. Critiquing is the process of evaluating or analysing a design or product, in order to identify its strengths and weaknesses and provide constructive criticism. You need to carefully examine the design or product, consider how it works, and see if it meets your expectations and the expectations and needs of those who will use it. To do this, you will create a set of criteria. These criteria should cover a range of the Factors that influence product design – these are listed and described in Chapter 13.

Critiquing using the Factors that influence design

When creating criteria to critique the practices and products of a designer, you need to consider many of the Factors that influence design, which are listed in the table on page 83. You can use the questions in the right-hand column, which come directly from the scope of each Factor. Choose from the questions listed for the first five **Factors** (need or opportunity, function, end users, aesthetics, ethical considerations); you may also consider the last three Factors (optional). Choose the questions that best suit your case study.

 You can find more detail on the Factors in Chapter 13.

9780170477499

Useful questions for critiquing

Factors that influence product design	Questions related to the scope of each Factor as it relates to product design
Need or opportunity	• What is the product for? Where will it be used and how will it be used? • Has the designer identified the needs of the end user well? Have they used direct contact, research and/or experience to help create their solution? • Does the product effectively satisfy the end user's specific needs?
Function	• How is the product meant to function? Does it function effectively and fully?
End users	• Who are the intended human and/or non-human users of the product? • How does the product improve their welfare and quality of life? - Particularly related to aspects of safety, accessibility, universal design, comfort, social and physical needs, and the use of ergonomics and anthropometric data. - Also considering culture and religion, emotional and sensory appeal, demographics, and trends. • Can the designer's solutions to these aspects be improved?
Aesthetics	• What design elements and design principles are used to appeal to the end user?
Ethical considerations in design	• Has the product been designed and made ethically? - Has the designer considered the individual and public 'good'? - Has the designer worked towards social goals such as belonging, access, usability, and equity for the disadvantaged? - Is the product sustainable? Does it encourage environmental protection, social inclusion and economic access and growth? Does it reduce negative impacts for both present and future generations? Is the material long-lasting or can it be reused or biodegraded after the product is disposed of? If it is recycled material, can it be upcycled again and again? • How has the designer considered their legal responsibilities? - Intellectual property (IP) - Australian and International (ISO) standards, regulations and legislation (including OH&S) • Are the products produced safely and safe to use?
Market needs and opportunities	• Has the designer used a creative approach to develop their solution? • Have they designed innovatively and worked entrepreneurially (created business/commercial opportunities)? • Does the product fill a niche that was previously unfilled?
Product life cycle	• Has the designer considered and minimised the negative impacts of the product at each stage of its life (including sourcing of materials, manufacture, distribution, use and disposal/reuse)?
Technologies: materials, tools and processes	• What properties and characteristics of the chosen material make it suitable for the product? • From what you can see, has the product been manufactured using suitable technologies?

It may be very hard to find information or evidence about some of these areas. When investigating, you might need to make an informed guess from what you have read or what you can see, and explain why you think the designer's or manufacturer's choices are suitable or unsuitable.

Critiquing the designer's use of the Double Diamond process

For products to be designed well, the designer needs to work through a process that strategically goes from initial recognition of a need or opportunity through to a completed product that is ready to use. When analysing and critiquing how effective a designer's process has been, consider how well the designer has used components of the Double Diamond process. Use the following questions related to the stages of the Double Diamond process to see if you can find evidence of strategic design thinking. Be cautious, however, of making hasty judgements about a designer's process, as evidence of their methods and process development is often hard to find.

If you can pick up information that answers just a couple of the questions in the table, it will give you some indication of the designer's use of the Double Diamond process.

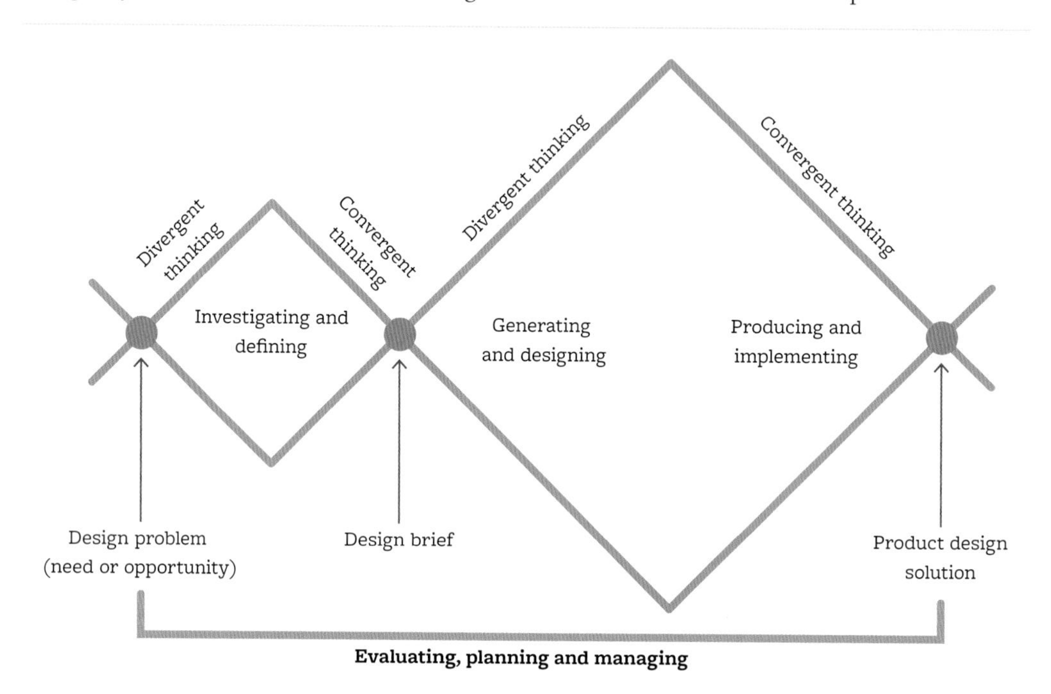

Double Diamond stage	Questions
Identifying the need or opportunity	Has the designer identified a real need? Have they investigated the needs of the end users using a range of research 'tools'? Have they had direct contact with the end users? Have they consulted experts in the field?
Defining/design brief	Did the designer clearly define the end users' needs and their situation? (Be aware that it is unlikely you will find information about a design brief.) Did that clear focus guide the rest of the process?
Design development	Did the designer take time developing a range of concepts? Did the designer research and seek expert advice? Did they create prototypes and test them with input from end users? Was the design improved through many iterations?
Design production and solution	Was the product made using appropriate materials and production technology? Was the final solution evaluated with end user feedback? Has the product undergone further modifications?

9780170477499

FOLIO ACTIVITY

Research report

- Choose at least three case studies to research. These can include any of the products described in pages 71–9, other appropriate case studies in this textbook, or case studies of products you have found elsewhere.
- Use the questions on pages 83 and 84 to guide your research – choose questions related to:
  - » Factors that influence design
  - » the designer's use of the Double Diamond process.

  These will become your criteria for assessing the success of the products and the designers' process.
- The case studies in this textbook can be a starting point for your research, but you will need to carry out more in-depth research into each product using the links provided. You won't be able to find information to answer all of the questions, but research as many as you can.
- When writing your research report, use the Factors and the Double Diamond process to structure your report. You may choose to use tables and/or develop a rating system or symbols to communicate how well each designer/product has addressed each area.
- If possible, ask a typical end user that the products are designed for to give you feedback. This will mean your report is more accurate and valid.

Template
Research report

Summary
Chapter 3

Quiz
Chapter 3 revision

CHAPTER 4

Designing for positive impacts for end users

UNIT 2, AREA OF STUDY 2: OUTCOME 2

In this outcome, you will:

- design and make an inclusive product that:
  - » responds to a need or opportunity of an end user/s
  - » addresses positive impacts in relation to belonging, access, usability and/or equity.

Key knowledge

- quantitative and qualitative methods to gather primary and secondary research, such as focus groups, interviews and product reviews to formulate a profile of an end user/s and describe their needs or opportunities related to positive impacts
- ethical research methods when gathering, representing and using data
- design thinking strategies to examine positive impacts for end users
- methods to plan, record and implement the Double Diamond design approach using a scheduled production plan
- materials, tools and processes used in a variety of design specialisations to make inclusive product designs
- methods to generate and record graphical and physical product concepts, including prototyping, and to develop a final proof of concept for an inclusive product to support positive impacts
- methods to develop criteria to evaluate product concepts, final proof of concept and finished product.

Key skills

- conduct and use research, including data from tests and trials, as well as design thinking strategies to:
  - » formulate a profile of an end user/s and describe their need or opportunity
  - » generate, refine, evaluate and critique product concepts
  - » justify a chosen product concept and develop a final proof of concept
- use criteria to evaluate processes used to develop product concepts and final proof of concept and make a finished product
- plan and record implementation of design process: development of product concepts, final proof of concept and making of product
- use materials, tools and processes to safely make a product that supports positive impacts for end users in relation to belonging, access, usability and /or equity
- work technologically to implement, manage and document the Double Diamond design approach: development of product concepts and final proof of concept and making of product.

Source: *VCE Product Design and Technologies Study Design*, pp. 25–6

DESIGNING FOR POSITIVE IMPACTS FOR END USERS

What is the opportunity or need?

Identify the problem

Defining

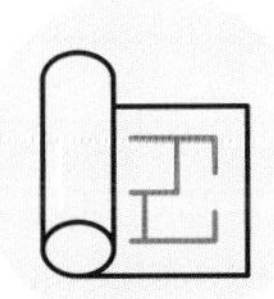

Generating and designing

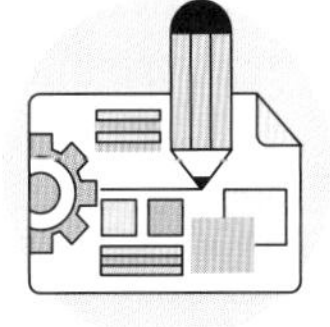

Modelling, prototyping and testing

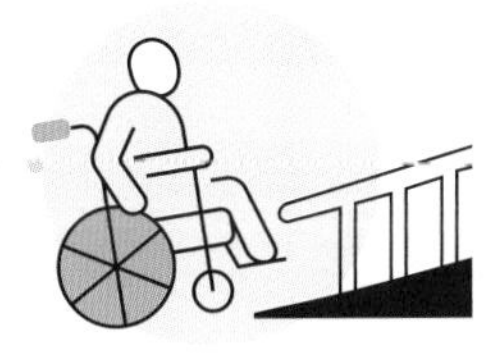

Producing and implementing

Templates:

- Researching a need or opportunity **(p.90)**
- Positive impact design brief **(p.90)**
- Evaluation criteria **(p.90)**
- Materials research **(p.94)**
- Materials testing **(p.94)**
- Research **(p.95)**
- Choosing the best design **(p.96)**
- Mock-up or prototype analysis **(p.98)**
- Proof of concept **(p.98)**
- Production steps **(p.100)**
- Timeline **(p.100)**
- Materials cutting and costing **(p.100)**
- Risk assessment **(p.100)**
- Journal/production log **(p.100)**
- Evaluation report for positive impact **(p.101)**

Summaries:

- Chapter 4 **(p.101)**

Nelson MindTap

To access resources above, visit **cengage.com.au/nelsonmindtap**

Creating a positive impact by identifying a need

Your research for Outcome 1 may have inspired you to create your own product that will make someone's life better. For your design task this semester (Outcome 2), you need to identify, design and make a product for someone in need (either a human or a non-human).

To identify an area of need:

- choose one of the case study situations, and consider these questions:
  - Can you design another solution for that need?
  - Does it inspire you in some way?
- choose one of the product examples listed on page 80
- consider the situation of someone you know who has specific needs
- draw from your own experience
- identify a group of people in the community who might have particular needs and discuss their situation with someone 'in the know' (a person who has experience caring for or working with these people)
- focus on improving the lives of non-humans
- improve a product so that it has a positive ethical/sustainable impact
- choose a product that isn't designed well to be used by all, and modify it to make it accessible
- choose an area of significant need overseas, in a developing country (perhaps your school has a social service connection to a place of need).

Your product can be:

- an entirely new idea/creation or approach to a need
- an improvement on an existing product.

Try to choose a product type that will allow you to work over two or more material categories or specialisations (listed on page 80). In Unit 2, you have less time to focus on the making of the product, as more time is expected to be spent on researching the need and refining the design. Choose a small product to design that won't be too complex to make.

iStock.com/Weekend Images Inc.

There are many design opportunities related to safety, access and providing products appropriate to the stage of understanding of small children.

9780170477499

Shutterstock.com/Janelle Lugge

Bushfires can destroy the habitats of many tree-dwelling animals. Products can be developed to improve access to safe nesting spaces for animals affected in this way.

OXO kitchen tools: design that grew out of experience

CASE STUDY

Over 30 years ago, OXO founder Sam Farber designed the first OXO peeler for a pair of hands he knew very well. Sam and his wife Betsey were cooking together when Betsey's mild arthritis made using an old-fashioned metal peeler a struggle. They knew there had to be a better way.

They created the now-iconic OXO handle – with its distinctive ergonomic form and signature non-slip grip – and paired it with a sharp stainless-steel blade. OXO kitchen tools were originally designed for people with arthritis, but they found widespread popularity because of their soft, easy-to-grip handles and overall ease of use.

iStock.com/shutswis

An OXO peeler, with its well-known grooved, soft-grip handle

First diamond: exploring, researching and defining the end users' needs

FOLIO ACTIVITY

Templates
Researching a need or opportunity

Positive impact design brief

Evaluation criteria

Design folio activities

- Complete some initial research to help you decide who or what you are going to design for (the end users), what their significant need is, and what type of product you might explore.
- Write these initial ideas up as a product proposal of no more than **two** sentences. This will help direct your initial research. The proposal should identify:
  - who the product will be designed for (what type of end users, and their need)
  - the aim/purpose of the product.
- Create a research plan for gaining insight into the specific needs of the end users. This should incorporate:
  - secondary research – finding background information about the end users and their area of need (this might require discussions with experts)
  - primary research – observing, interviewing, talking to the end users, and experts/carers, focus groups, etc.
  - research into existing solutions/products for the area, identifying good features and areas for improvement.
- Carry out that research to help you define the requirements and special needs of the end users you are designing for.
- Write a design brief that includes:
  - an end user profile that summarises the information you have collected from your research, and outlines their requirements in the area of the product you are going to create
  - the function or purpose of the product – a sentence that clearly defines what your product needs to do
  - the scope of the task
  - its constraints and considerations.

  For more information about writing a design brief, refer to page 210.
- Develop a range of evaluation criteria for your design situation to:
  - evaluate the product/solution
  - evaluate the processes used to create, refine and produce the product.

Ethics when gathering end user research

In a situation where your end users have special needs, it is even more important that you behave in an ethical way when collecting information for your design situation. This is especially important if your end users have limited capacity or are not able to communicate effectively.

- As the designer, you have a responsibility to treat your end users with respect and maintain a positive relationship with all people that you relate to in the situation.
- You need to ask for consent when carrying out research – whether to conduct an interview or take observations. If the end users are not able to give consent (they may be very young, or incapable) it must be sought from parents, guardians or carers.
- You need to be honest about what you need the research for.
- You need to be fair and transparent in the way you present your information. Make sure you don't present or twist information in a way that suits your needs rather than the needs of the end users.
- You need to put the needs of the end users before your own. The purpose of the process is to create a product that benefits the end users and makes their life better/easier. If the process causes discomfort to the end users, find other ways of completing research tasks. Also refer to the section on ethics in design on page 385.

9780170477499

- You may need to gather information from those who care for the end user or who have expert knowledge about their needs. Make sure you explain the purpose of your research clearly, and consult with them throughout the process.

iStock.com/SDI Productions

It is important to be respectful and sensitive when interviewing a person with a disability.

What does it mean to work technologically?

Working technologically means using all the technical and technological tools and strategies available to create the best possible outcome. The 'tools' may be planning or thinking tools, or they may be physical and practical tools.

Thinking and planning technologically

Thinking and planning technologically involves using 'tools' to help you collect information, make decisions and plan a course of action that will get you to the end result as efficiently as possible. It could involve:

- using computer software (such as spreadsheets) to record critical research
- being thorough in considering and defining the design situation, and recording documentation clearly
- extending your drawing skills to visualise your ideas effectively (both hand and computer drawing)
- using flow charts and other planning tools to keep you on track
- testing and trialling your ideas to check which is the best – using accurate measuring tools, photographic evidence, etc. to make decisions.

Producing technologically

Producing technologically involves using the most appropriate skills and tools available to create trials and prototypes and to construct your final solution. It will involve:

- searching out and testing materials to decide which are the most suitable for the task
- trialling materials, equipment and solutions that are not usual or conventional

- selecting the best equipment, which will work efficiently and effectively; checking accessibility
- extending your technical skills to enable you to use a wide range of materials and equipment
- searching for alternative solutions if something fails.

Second diamond: generating, trialling, testing and producing solutions

After writing your design brief and evaluation criteria, you need to explore:

- design ideas
- materials
- processes and equipment
- other areas of research to help you develop a successful solution.

Methods of generating and recording design ideas

There are three main forms of **design drawings**:

- visualisations – sketches of initial ideas (of the whole products or focusing on parts) annotated with discussions of positive and negative features, and descriptions of how the ideas work. The purpose of these drawings is to explore and expand many different ideas, to consider what is possible
- design options drawings – detailed and fully considered design solutions, drawn with colour and rendered, showing materials and dimensions (if appropriate) and annotated to explain how they satisfy the needs of end users
- working drawings – detailed technical drawings of the final solution, to guide production.

These drawings can be hand-drawn or generated digitally. You need to use the drawing methods and techniques that suit the materials and form of the product you are developing.

Refer to Chapters 1, 7 and 12 to find out more about drawing styles and methods.

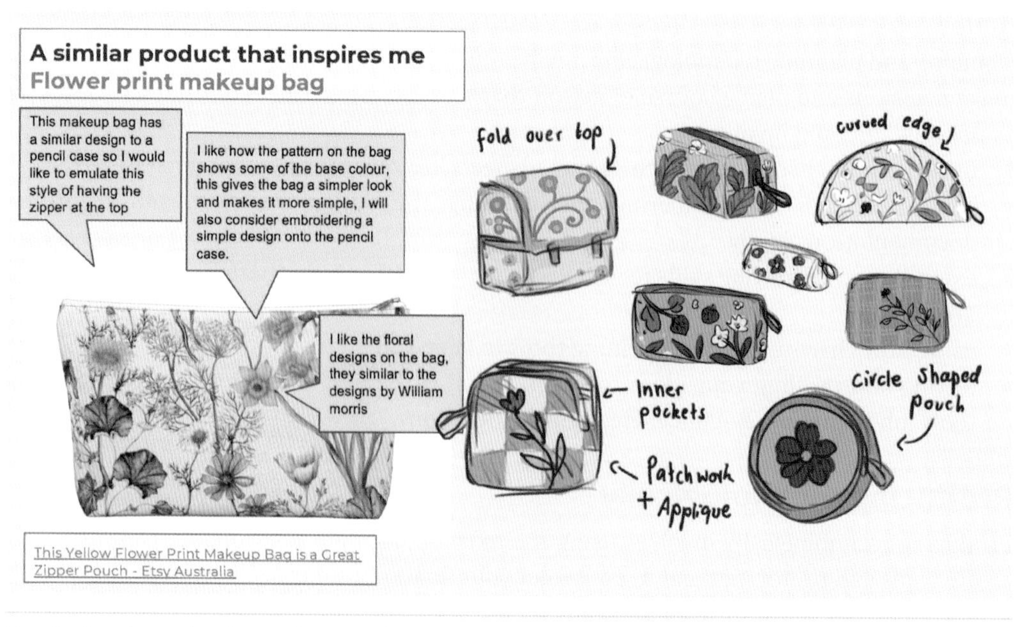

Visualisations by student Rayan Al-Mosa, using Procreate and inspired by William Morris wallpapers

9780170477499

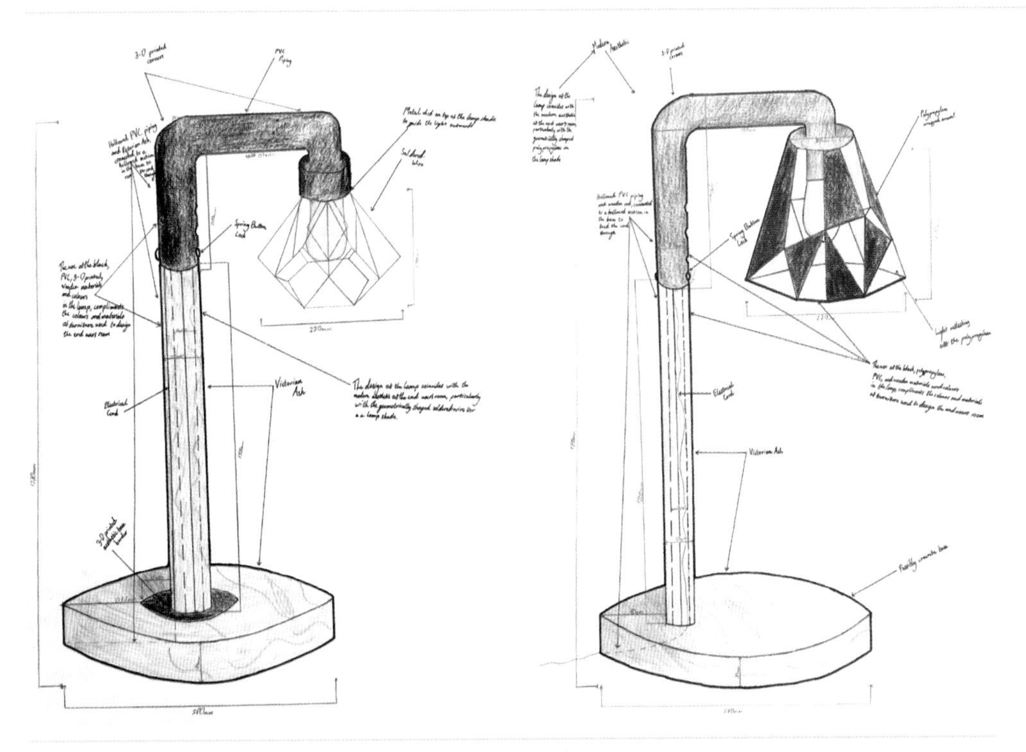

Design options for a lamp created by student Samuel Nastasi

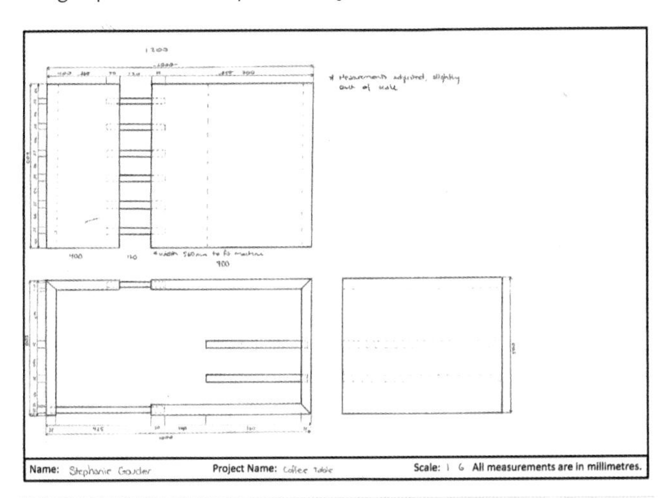

Working drawing for a small table designed by student Stephanie Gouder

FRONT
BACK
Erin - dress
Drawn with Inkscape

A 'flat' technical drawing for a sailor-style dress, made by student Erin Tinker using Inkscape

Modelling is also used to generate and test ideas. Models can be quick ideas put together fairly roughly with paper, cardboard, fabric, foam, etc. to trial basic shapes or functional features. Record these ideas with photos that are annotated to record your analysis of positives and negatives.

For more detail about modelling, refer to pages 45–6.

You will need to complete a range of visualisations, design options, modelling and a working drawing as you explore and finalise your design solution.

Solution-focused technical/practical research

To develop the best solution, you need to research areas that are relevant for your design situation. This research will help you to choose the most suitable materials, processes and equipment for the construction of your product. Depending on the type of product, you may also need to research commercially available components, and functional aspects.

Materials

As part of your design process, you need to explore and research materials that could be used for your product, so that you can choose the ones that are most suitable. You need to become familiar with a range of materials used in different design specialisations (for information about materials categories and design specialisations, refer to pages 37–9 and page 80). Consider exploring and using materials from at least two categories, e.g. timber, fabrics, plastics, biomaterials and/or metals.

Your evaluation criteria might give you some direction regarding the material qualities your product will need (e.g. strength, flexibility, low weight, ability to be washed/cleaned, ability to withstand weather, etc.).

When researching a range of materials, for each one you need to investigate:

- the characteristics and properties of the material
- its appearance
- where the material comes from, and how easy it is to obtain
- its uses
- the cost of the material
- how sustainable it is. Are there environmental issues related to its sourcing, processing and use? Can it be reused or recycled?
- what makes it suitable or unsuitable for the product.

Template
Materials research

Use the Nelson MindTap Materials research template provided for Chapter 2.

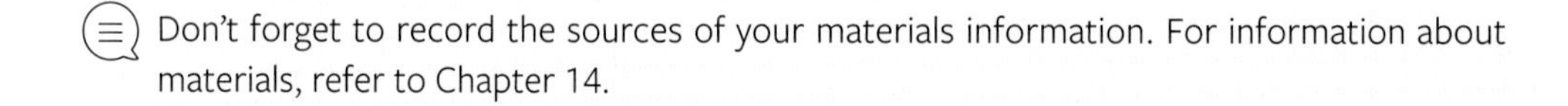

Don't forget to record the sources of your materials information. For information about materials, refer to Chapter 14.

You could carry out a **materials test** to investigate the performance of a small range of materials that could be used in a specific situation. A materials test is like a scientific experiment, where a controlled testing procedure is carried out on the materials. The test results are recorded and analysed to identify the most suitable material for your design situation. There are examples of appropriate materials tests on pages 422–7, with an explanation of how to carry out the test and report your findings. A materials test report usually includes the following sections:

- the aim of your test – where you identify what you are testing for and why
- some background information about each material
- a description of the test procedure and set-up
- a record of results – in tables, graphs or charts
- a statement on the validity of the results
- an analysis of the results and your conclusions regarding material suitability.

Template
Materials testing

Use the Nelson MindTap Materials testing template provided for Chapter 2.

9780170477499

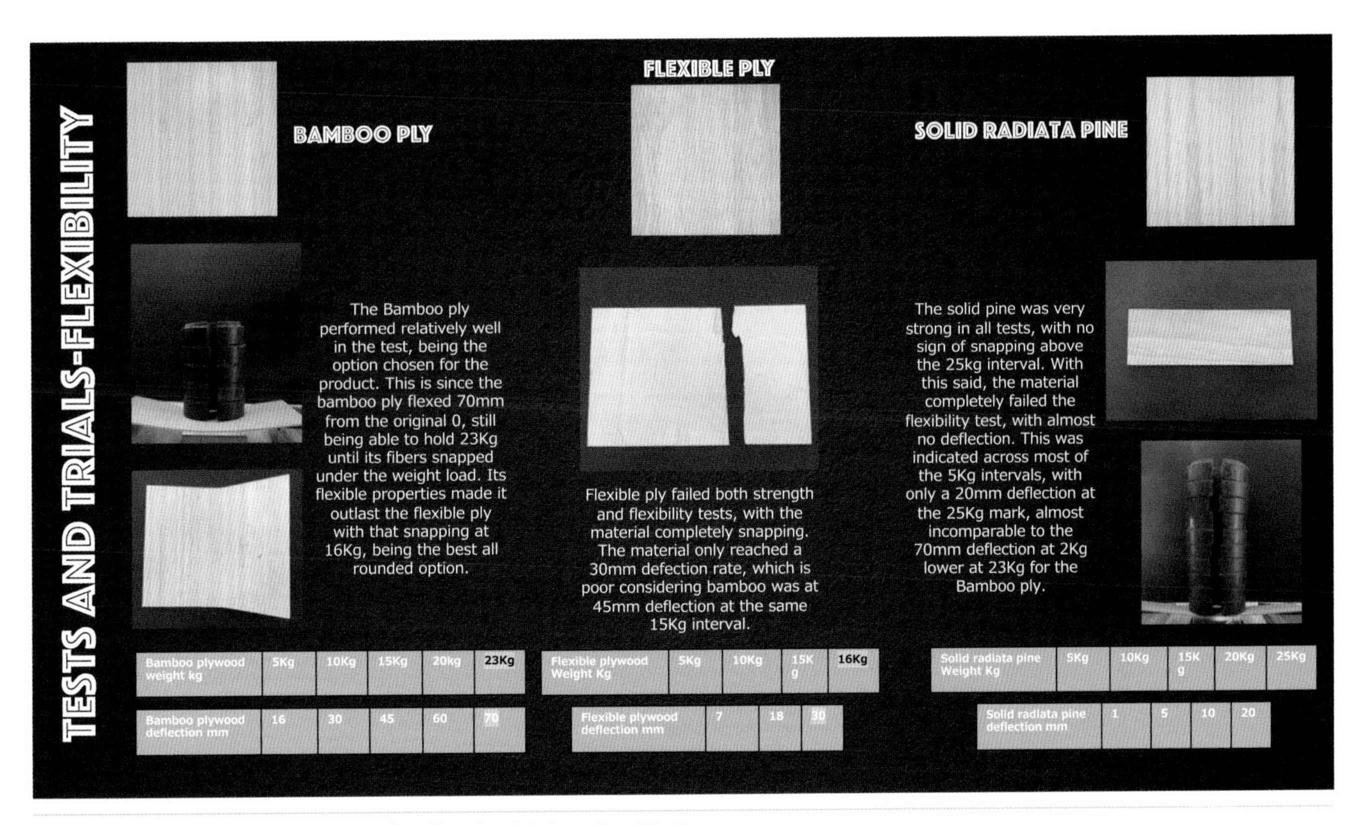

Part of a materials flexibility test, carried out by VCE Unit 3 student Ella Costanzo

Other research

You need to select and carry out other research (such as trials) to make informed decisions, and to develop effective solutions for your product.

Other areas you could investigate in this way include:

- trialling shaping and joining processes or methods
- investigating and modelling functional aspects related to your product
- researching components that could be useful
- trialling finishing processes
- exploring and trialling tools, equipment and machinery – what is accessible and what skills are required?
- researching existing products – testing and/or analysing their positive and negative aspects (if this hasn't already been done).

Use the Nelson MindTap Research template provided for Chapter 2.

Template
Research

When carrying out research, be clear what you need to find out to help you make design decisions – have a definite aim. For more detail about research areas, review the section in Chapter 2 on pages 36–44.

Nelson MindTap has a wide range of information about construction processes relevant to different materials. You can select a range of these processes to trial.

Lighting Research

Type of LED Light:
Wire LED Fairy Lights

Source of Power:
The wire LED fairy lights are battery powered.

Advantages:
- It is sustainable as the product is battery powered.
- It creates a mood light to add to the ambiance or aesthetic of the room because the light source is dim.
- The wire lights are safer to work with than other mains powered wire lights because it is low voltage.
- The LED lights use a minimal amount of power, therefore, it lasts longer.

Disadvantages:
- The wire lights are a weak light source and therefore will not be a main source of light.
- The batteries need to be changed when the lights fails to work.

How will you hold or contain the LED lights?:
The wire LED fairy lights will be contained in a fake plastic light bulb.

How will the parts be connected?:
The wire LED fairy lights in the fake plastic light bulb will be placed in the socket of the recycled lamp. The battery pack will in the small drawer in the base of the lamp. The wire light will go from the battery pack in the base of the lamp, through the tube-like body of the lamp, and into the fake plastic light bulb that will be in the socket of the lamp.

Type of LED Light:
Low Powered LED Light Bulb

Source of Power:
The low powered LED light bulb can be powered by battery or mains power.

Advantages:
- The LED light bulb use minimal power, therefore, the batteries can power them for a longer period of time.
- The light bulb can be battery powered, unlike other light bulbs.
- The LED light bulb is safer to work with than mains powered light bulbs because it is low voltage.

Disadvantages:
- The LED light bulb is a weaker light source and therefore will not the the main light source, compared to a desk lamp.
- The batteries have to be changed when the light bulb does not function properly.

How will you hold or contain the LED lights?:
The LED light bulb will be held in the socket of the recycled lamp.

How will the parts be connected?:
The low powered LED light bulb will be placed in the socket of the lamp.

Which LED lighting system will best suit your design and why?:
The wire LED fairy lights will best suit my design as it fits the sustainability requirements of the assessment, it fits the end-user's mood light requirement as the wire lights create a dim light source that will add to the aesthetic of her Hampton styled front room, it is a safer option to work with as they are low voltage lights, and they use minimal power so the batteries last longer.

Research into low-voltage lighting components carried out by VCE student Julia Nastasi

FOLIO ACTIVITY

Design development

You need to complete the following tasks and include them in your design folio:

- visualisations – a variety of sketches to explore possible ideas:
  - » for the whole product, and possibly for parts of the product, to sort out details
  - » considering aspects of appearance and function
  - » ensuring they are annotated (with lots of comments describing your ideas, and noting their good features and how they will satisfy the needs of the end users)
- models – simple 3D experiments with shape, form and function (made from paper, cardboard, fabric, etc.). Photograph and comment on your ideas. These can be created as part of the visualisation process or after your design options are drawn
- at least three design option drawings:
  - » of complete concepts
  - » coloured and rendered
  - » showing major dimensions (for products other than clothing)
  - » annotated to describe features and to comment on how the design addresses the needs of end users
  - » rated (on a scale of 0–5) using evaluation criteria
- research into materials and at least two other areas:
  - » materials – background information, including characteristics and properties
  - » materials test
  - » construction, decoration, components and/or finishing methods
  - » machinery and equipment (including safety concerns and control measures)
  - » aspects of function
- identify the most suitable solution (from design option drawings and models). Explain and justify your choice, using feedback and rating comparison.

Template
Choosing the best design

9780170477499

Mock-up or prototype

Creating a 3D physical mock-up or prototype of the chosen design helps to refine your concept. It can be actual or computer-generated. A mock-up is usually made to quickly analyse the visual appearance of an idea, whereas a prototype is often more detailed and is used to test both the visual and functional aspects of a concept. You can make a mock-up or prototype of a concept/ product idea either full size or to a smaller scale. It can be made from a cheaper material than intended for the final solution, or from something like paper, cardboard or corflute. It needs to be functional enough to test whether the design will work or not.

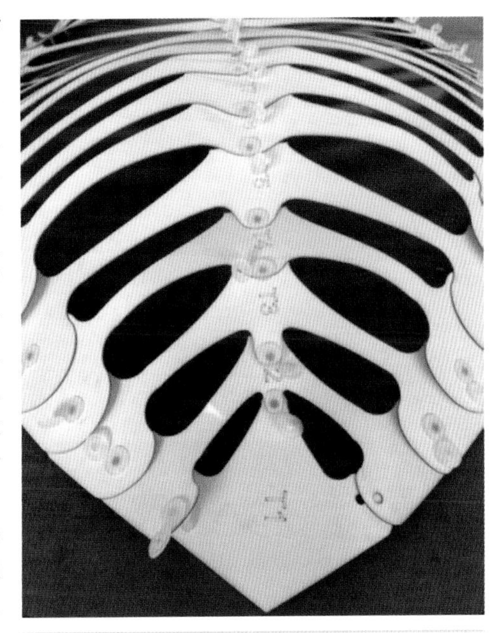
Cardboard mock-up, full-sized, by VCE student Ella Costanzo

Photograph and analyse your mock-up or prototype, considering areas that worked well and identifying areas for improvement. Trialling the mock-up/prototype will give you useful practical information about how well your design will perform. A physical version of your concept is also more useful when seeking feedback from other people in the class, your teacher, and possible end users. If your trialling and feedback points to areas for significant improvement, change your mock-up/prototype – develop a new, 'improved' version. Again, photograph and comment on your changes and how this version satisfies the end users' needs better, and record this analysis in your folio (multimodal record of progress).

 You can repeat this process until you are satisfied that you have developed the best possible solution.

Proof of concept: justification

The final solution that you have trialled, modified and improved is called your 'proof of concept'. It is the end result of your trialling and testing, and the evidence that you have a design solution that is as resolved as possible and will work. The proof of concept can be a detailed, annotated drawing, a CAD design with comments or a physical prototype that is photographed and explained. Refer to your design brief and any feedback you've gained through the development process to explain why your design is the most suitable, and to explain any improvements to your original design option drawing.

Working drawing

This solution needs to be drawn ready for production. Working drawings illustrate the details of your concept and provide direction to construct your final solution. Complete a detailed technical drawing (a working drawing) of this solution.

For textiles products, this is called a 'flat'.

For wood, metal and plastics products, the most useful drawing method to use is orthogonal.

 Refer to pages 335–42 for information about working drawing techniques.

 FOLIO ACTIVITY

Concept trialling, modifying and recording

Complete the following tasks and document your progress in your design folio.

- Develop a 3D mock-up or prototype of your chosen design option.
- Trial, test and get feedback on your concept, identifying areas for improvement. Photograph your mock-up/prototype and record your analysis and feedback.
- Modify your mock-up/prototype. Record improvements, analysis and feedback on this version.
- Repeat until you are satisfied you have the best possible solution.
- Proof of concept – photograph or provide detailed drawings of the final version of your design idea, your 'proof of concept', and write a justification to explain why this is the best solution for your end user/s.
- Complete a working drawing of this proof of concept.

Templates
Mock-up or prototype analysis

Proof of concept

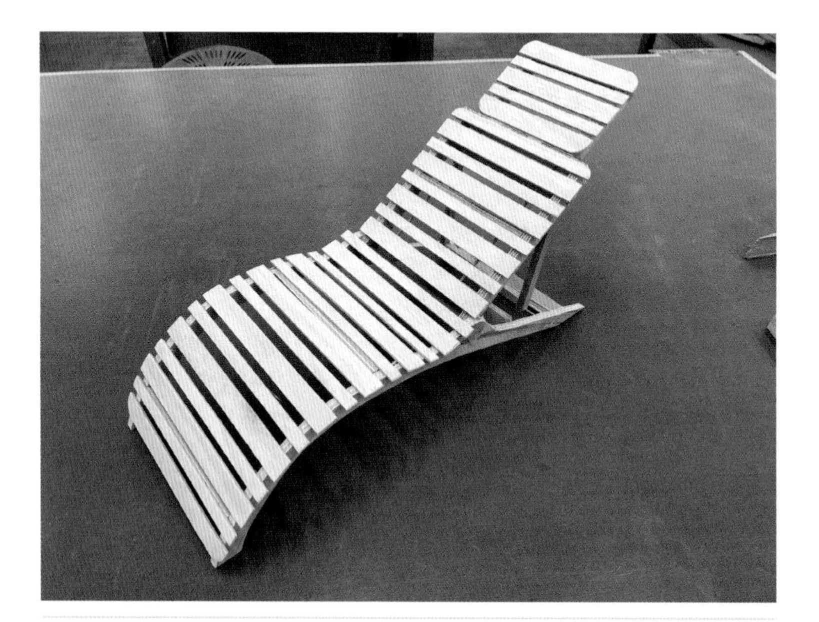

A scale model by VCE Unit 3 student Charlee Turner trialling the adjustable back mechanism and the patterning of slats

Production planning and safety

A series of production planning documents need to be developed – these go together to create your scheduled production plan. The following are important 'tools' to help you work efficiently, effectively and safely:

- **materials cutting and costing list**. Review the instructions on pages 50 and 249, and use that information to complete a cutting and costing list to identify the materials you need to produce your product
- **production steps with quality measures and safety controls** – a list of all the steps required to make your product, and guidelines for safe work, as in this example:

Step no.	Step heading	Step description	Tools and equipment	Quality measures	Safety controls	Time required
1	Pre-production	Move material from storage ready for marking or laying out	Ladder	Select materials without damage or faults	Awareness when moving materials, using trolleys, lifting equipment	

9780170477499

- **timeline** or **flow chart**. Complete either a timeline table or a flow chart to indicate a logical flow of work through production:
  - **timeline table** – a table with the step heading on the left, the weeks or sessions marked across the top, and the times you plan to finish each step shaded (easy to do using either word processing or spreadsheet software, or specialised timeline or Gantt apps)
  - **flow chart** – shapes and arrows are used to describe or plan the flow of production and the points where decisions are made, inputs are added and quality checks are carried out. Check online to find out how they are created. The table shows some of the simple symbols you could use in a flow chart:

Flow chart symbol	Name	Function
	Start/end	An oval represents a start or end point
	Arrow	A line or arrow is a connector that shows relationships between the representative shapes
	Input/Output	A parallelogram represents input or output
	Process	A rectangle represents a process or production step
	Decision	A diamond indicates a decision, and the workflow may loop or fork in a different direction depending on the answer

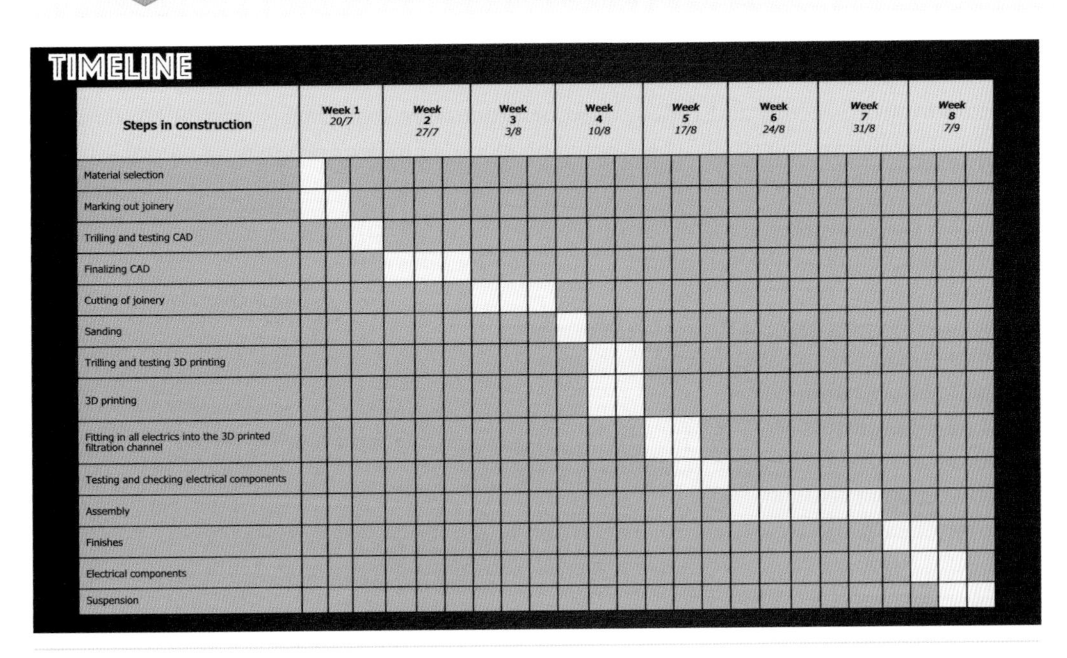

Timeline created by VCE Unit 3 student Ella Costanzo

- **risk assessment** – a table or chart where you:
  - list significant steps, processes or machinery
  - for each, identify its hazards and the possible injuries that might result from the hazards
  - assess the level of risk (1–5 for both the likelihood of an incident occurring and the seriousness of a possible injury)
  - list a range of safety controls/measures that should be used to keep safe.

Refer to pages 50–4 in Chapter 2 and in Chapter 7 for detailed information about the components of your scheduled production plan.

FOLIO ACTIVITY

Planning

Using the instructions above, and following directions to other references in this textbook, create the following planning documents for your final production (this needs to be done before you start making):

- production steps with safety controls and quality measures
- materials cutting and costing list
- timeline or flow chart
- risk assessment.

Use the templates provided on Nelson MindTap for all these components of your production plan.

Templates
Production steps

Timeline

Materials cutting and costing

Risk assessment

Journal/production log

Production

Now that you have refined your final design, prepared detailed drawings and completed your production planning, it is time to make your product. During production, regularly refer to your planning documents to guide your construction stages and to help you use your time efficiently. It is fine to change the content and order of production steps as you go in response to any changes to your situation during production, or if you can see a more efficient way of doing things. Just make sure you explain any changes or modifications in your production record or journal.

Production record or journal

It is important to record and reflect on your production experiences as you go. Refer to page 257 for information about what should be recorded in your production record/journal.

Use the production record/journal template provided for you on Nelson MindTap or develop your own journal format.

PRODUCTION ACTIVITY AND JOURNAL

Making your product

Use your time effectively to make your product. Make sure you work safely, using all the safety controls you described in your work plan.

Each week, complete a journal to:

- describe what you have done to complete each step – in photos and words
- explain what you have learnt about materials, processes and equipment
- outline the safety measures you have used
- describe any modifications you have made, explaining your reasons for these changes
- plan your next steps.

Evaluation

When you have completed your product, you need to evaluate it using your evaluation criteria and feedback from your end users. Remember that your product is meant to make a **positive impact** on the lives and wellbeing of your end users, and this should be the focus of your evaluation. If possible, get feedback from a range of possible users who may have

9780170477499

different perspectives. If your end users have difficulty communicating (if they are too young, old, incapacitated or withdrawn):

- observe how they use the product, and their reaction to it (e.g. do they show signs of being pleased with the product, do they want to keep using it, etc.)
- talk to their carers or to other people close to the end users
- ask for feedback from experts in the area.

If possible, take photos of the products being used (make sure you ask permission first).

FOLIO ACTIVITY

Evaluation report

Create a clear evaluation report, with:

- appropriate headings
- a table listing the evaluation criteria, how they were checked, and the feedback and responses you received
- an overall assessment of how well your product performed and whether it provided the **positive impact** you intended (explain why/why not)
- areas for improvement identified:
  - » for the product
  - » for the process of researching, designing, planning and producing.

Template
Evaluation report for positive impact

Assessment

In Unit 2, Outcomes 1 and 2, you will be assessed on the following work:

- a multimodal record or folio that contains:
  - research of products that have made a positive impact (Outcome 1)
  - research of specific end user needs
  - design brief, including an end user profile, intended function, project scope (including constraints and considerations)
  - evaluation criteria
  - visualisations (with optional modelling and/or prototypes)
  - design options (with optional modelling and/or prototypes)
  - technical and practical research
  - proof of concept, with explanation and working drawing
  - scheduled production plan, including:
    - » a timeline, a work plan/production steps with estimated time, tools and equipment
    - » quality measures
    - » materials cutting and costing list
    - » risk assessment
  - record of production/journal
- mock-ups/prototypes
- finished product
- evaluation of the product and process.

Summary
Chapter 4

CHAPTER 5

Cultural influences on design

UNIT 2, AREA OF STUDY 3: OUTCOME 3

In this outcome, you will:

- research and discuss how a designer and end user/s are influenced by culture.

Key knowledge

- cultural influences on product design including those of Aboriginal and Torres Strait Islander peoples
- factors that influence cultural needs and opportunities of end users
- quantitative and qualitative research methods to collect and record information about culture of designers and end users, including ethical considerations.

Key skills

- examine a variety of products through a cultural lens
- use ethical research methods to extrapolate information from a variety of sources to connect and make meaning between a designer's culture and the products that they design
- investigate cultural factors in relation to end user needs and/or opportunities
- discuss examples of cultural influences in the design of products from the perspectives of both designers and end users.

Source: *VCE Product Design and Technologies Study Design*, pp. 26–7

9780170477499

Templates:
- Indigenous case studies **(p.113)**
- Cultural influences on product choices **(p. 115)**

Case studies:
- Valentino's resort range **(p.113)**
- Bryan Cush **(p.113)**

Summaries:
- Chapter 5 **(p.115)**

Quizzes:
- Chapter 5 revision **(p.115)**

To access resources above, visit **cengage.com.au/nelsonmindtap**

Weblinks
Help Students to Learn About Aboriginal and Torres Strait Islander History and Culture: Australian Curriculum

Teaching Aboriginal and Torres Strait Islander Culture: DofE Victoria

In this chapter, the authors have made every attempt to cover the topic respectfully. Teachers or students embarking on any projects or lessons related to Indigenous culture should do their best to check material beforehand by contacting elders or traditional owners from the local area on which the school stands. More information including recommended reading, contact details for Koorie Education Offices and protocols can be found on Victorian Government websites. Links are provided in the Nelson MindTap resources for this chapter.

Focus on Aboriginal and Torres Strait Islander culture

Aboriginal and Torres Strait Islander peoples' culture is widely recognised as the longest continuously existing culture on the planet. Their cultural knowledge and practices have existed for and continued to develop over at least 60000 years. Indigenous knowledge of the land, plants and materials, food, medicine, creatures and astronomy has grown and adapted to local conditions and has been moulded by the culture of the many individual Indigenous nations. It has been passed down over many generations through stories, songs, artwork and object making.

Alamy Stock Photo/Adisha Pramod

The coolamon is a multipurpose vessel that could be used to carry water or bush food or to cradle a baby.

Aboriginal and Torres Strait Islander cultures' deep connection to the land is often reflected in the way products are created and used by Indigenous Australians. Connection to Country is central to any design. Available materials in the local area inform the different uses of tools, weapons, clothing and textiles and consequently are manifested in songlines, daily activity and ceremonies. First and foremost, the aim is to protect natural resources for the future. The following are just a few ways in which Indigenous people connect their culture and traditions with their contemporary life.

- In Indigenous communities, there is usually a strong emphasis on community and sharing, which can influence purchasing decisions and the use of products within families and groups.
- In health care, traditional medicines and healing practices may be used in conjunction with Western medicine, and certain plants and animals may hold cultural significance and be used in ceremonies or for food.
- In some parts of Australia, modern buildings use both Western and traditional construction methods. For example, building materials such as steel and concrete may be used to build the structure of a house, and traditional elements, such as the use of natural materials like bark or woven fibres in roofing, or cultural symbols or artwork, may also be incorporated into the design. This approach of blending Western and Indigenous construction methods can create homes that are functional, durable and also reflective of Indigenous culture.

9780170477499

Indigenous Australian cultures are diverse – there are many different Indigenous nations within Australia and the Torres Strait Islands. This diversity needs to be recognised and respected when considering the influence of culture on product design.

Aboriginal and Torres Strait Islander peoples' cultures can influence product design in a variety of ways. For example:

- Connection to Country (land and environment) and respect for it is shown in everyday life and all actions. See more on the next page.
- The use of natural materials and techniques, such as basket weaving, can also be a source of inspiration, and the activity facilitates storytelling and the passing of wisdom.
- Product design may also be informed by the cultural significance or meaning of certain objects or symbols in Indigenous cultures.
- Social practices also influence the design of objects – how the objects are used, what they are used to do and where they are used has an impact on their forms and the materials used. Objects are adapted to local needs and resources.
- Traditional visual patterns and designs from these cultures can be incorporated into the appearance of a product.

When developing products that may be used by Indigenous communities, designers, engineers, architects and planners need to listen to and respect Indigenous cultural perspectives to develop solutions that are appropriate and useful.

Cultural appropriation

When recognising the deep importance of cultural influences on design, the concept of cultural appropriation is important to understand and avoid. Cultural appropriation is where elements of one culture are taken and used by members of another culture without proper understanding, respect or permission. This often happens when people within a dominant cultural group take elements from a minority culture – and in the process, the elements lose their significance and cultural power, and can be stereotyped. It can also be exploitative: profits do not benefit those to whom the cultural ideas belong. Use the link on Nelson MindTap to investigate the Australian and International Indigenous Design Charters, which outline 10 protocols that lead to appropriate Indigenous design collaboration.

Weblink
Indigenous Design Charters

Traditional Indigenous Australian design

Traditional Aboriginal and Torres Strait Islander nations created remarkable solutions for everyday living that solved practical problems. They also developed visual languages and symbols, and oral stories that communicated their culture, values and knowledge of Country.

Ways of doing and making

There are widely varying Indigenous traditions of crafting objects and products that assisted in collecting, growing, managing, hunting and gathering food. These include tools such as spears or harpoons, spear throwers (woomeras), knives, boomerangs, bowls and baskets, and fish and eel traps. Many of these objects were separated into 'men's and women's business' depending on the roles men and women performed in each Indigenous nation.

There is a strong tradition of boat making, using a wide variety of materials, such as hollowed logs, peeled cambium layers from a tree, and bound bark. Many groups also used the materials that surrounded them to make shelters.

Some Indigenous nations changed the landscape to assist in the trapping of fish and eel, forming channels, pools and 'gates' to guide and trap their aquatic food. Specially shaped baskets were woven to successfully trap the larger fish and eels, while allowing the smaller, younger creatures to escape – a sustainable measure to ensure the safety of the next generation of wildlife. Evidence of these practices can be found in the World Heritage-listed Budj Bim area, near Warrnambool.

All of these technologies were developed over many generations, through many iterations, by trialling and testing, adapting and modifying. These objects developed differently in different areas of Australia and the Torres Strait Islands, responding to the local environmental conditions and moulded by the cultural practices of each nation.

Newspix/David White

Eel traps designed to trap adult eels and allow young eels to escape (sustainable practices ensuring the survival of the next generation of eels)

For more information, the ABC has a series of excellent videos on Aboriginal and Torres Strait Islander technologies: *Deep Time History of Indigenous Australians*. Of particular interest are the videos on the:

- cultural significance of possum-skin cloaks
- ingenuity of Indigenous fish traps
- history of Indigenous watercraft
- diversity of Indigenous watercraft.

Other TV programs have been developed that explore the creativity and inventiveness of Indigenous Australians, including *First Weapons* on the ABC and *The First Inventors* on NITV/10.

Weblinks
First Weapons

The First Inventors

Possum-skin cloaks

In Victoria and New South Wales, Aboriginal peoples have a rich history of making and using possum-skin cloaks – both as a practical garment and as a way of telling their stories. The cloaks are made from joined rectangles of tanned possum skins that have images or patterns cut into the inside surface (which also help to make the skin flexible). Before the mid-19th century, Aboriginal people used these cloaks to keep warm, to carry their young, as a drum stretched over their knees, and as a wrapping when someone died.

In 1999, Gunditjmara women Vicki and Debra Couzens and Yorta Yorta women Lee Darroch and Treahna Hamm were moved by two possum-skin cloaks in Melbourne Museum's collection – the only remaining cloaks in Australia. To make new cloaks, they set about learning the old skills and developing new ones. Sourcing the possum skins was a problem as these animals are protected in Australia, so they shipped the skins from New Zealand, where possums have become a pest. These were tanned differently, and were very soft and pliable, so the old cutting method for creating designs was not appropriate. Instead, they used pyrographic equipment to scorch lines onto the inside surface of the cloaks.

9780170477499

The artists reproduced the original designs on the cloaks – the lines and shapes that symbolised their Country and Koorie life, the rivers, lakes, eel traps, meeting places, etc. It was through the experience of making the cloaks, and discussions with their elders about the cloaks, that the women made connections with the 'Old Ones', their ancestors, and grew to recognise and understand the symbols used on the cloaks.

AAP Image/JULIAN SMITH

Aboriginal artists Treahna Hamm (left), Lee Darroch (centre) and Vicki Couzens (right) wear traditional decorated possum-skin cloaks, made for an exhibition at Melbourne Museum's Aboriginal Cultural Centre, 2006.

Case studies: contemporary Indigenous designers

Nicole Monks: Indigenous artist and designer

CASE STUDY

Nicole Monks is a multidisciplinary creative of Yamaji Wajarri, Dutch and English heritage, living and working on Worimi and Awabakal Country (Newcastle). Nicole's work is strongly influenced by her cross-cultural identity. She uses storytelling as a way to connect the past with the present and future. Her works take a conceptual approach that is embedded within story and aims to promote conversation and connection.

As an award-winning designer and artist, Nicole crosses disciplines to work with furniture and objects, textiles, video, installation and performance. Across these varied forms of contemporary art and design, her work reflects Aboriginal philosophies of sustainability, innovation and collaboration. Nicole is well known for her success as a solo and collaborative artist and founder of design practice blackandwhite creative as well as the public art company mili mili.

The *Marlu* furniture collection was inspired by a trip back on Country, visiting Nicole's 93-year-old 'Auntie' Dora Dann, and reminiscing about Dora's childhood, including Nicole's great grandmother's renowned kangaroo tail stew. 'From the beginning of making the kangaroo tail stew, we pay respect to the kangaroo, and then the hunter and we are grateful to eat together.'

The focus of the *Marlu* collection was about 'how the furniture made you feel and about honouring my great great grandmother's story, not just about the aesthetics or function of the furniture.'

Marlu, meaning kangaroo, tells the story of this lived experience through a collection of furniture pieces:

- *Wabarn-Wabarn* – representing the kangaroo, in particular the mother caring for the young in the pouch. The chair has a pelt blanket attached to it and is like a big hug
- *Walarnu* – or boomerang, representing innovation and hunter, sitting here, feeling empowered
- *Nyinajimanha* – representing coming together, eating and creating story. It has a low group setting to ground us and create community.

For Nicole, *Marlu* represents the importance of knowledge transfer and the role of memory and lived experience in this process.

Prue Aja, www.prueaja.com

A model wearing Nicole Monks' textiles in a collaboration with Cara Mancini at Pandanah Australian Indigenous fashion week

In Nicole's furniture collection, you'll see the use of seamless joins, simplified lines, forms and a minimalist style. All works promote sustainable practice and mindful consumption that reflects a caring for Country. The pieces are made to order, in limited production. The materials are found on, or represent, Country (gold plate, kangaroo pelts and native timbers).

Source: http://nicolemonks.com/

Courtesy of Australian Design Centre. Photograph: Boaz Nothma

Nicole Monks sitting in her *Wabarn-Wabarn* (bounce) chair made from hemp canvas and kangaroo leather

9780170477499

Courtesy of Australian Design Centre. Photograph: Boaz Nothma

Nicole Monks' *Nyinajimanha* table and chairs, which form part of her *Marlu* collection

Courtesy of Australian Design Centre. Photograph: Boaz Nothma

Nicole Monks' *Walarnu* chair, made from black powder-coated steel and kangaroo fur

Stories from Layers of Blak

CASE STUDY

The Layers of Blak design program and exhibition was co-ordinated by the Koorie Heritage Trust. Eleven Victorian First Peoples artists/designers developed their individual skills and professional practice, while fostering broader design collaboration. The work of these First Peoples designers was showcased in an exhibition from 1 October 2022 to 19 February 2023 in Melbourne and continues in regional galleries as a touring exhibition during 2023–2026.

Nikki Browne

In the lead-up to being involved in the Blak Design program, Nikki Browne (Bidjara) was working outdoors with a school group at Healesville Sanctuary when a small branch of blue-gum leaves tumbled from a tree and landed on her. It amused her, and helped determine the direction she wanted to go in with her designs. As an artist, she has always loved the extraordinarily long leaves of the blue gum, which can extend up to 45 cm. How though, she asked herself, could she incorporate them into a piece of wearable art?

The creations she ended up pursuing – one featuring leaves, the other the spines of sea urchins – are reminiscent of shawls, but also of traditional possum-skin cloaks in the way they extend protectively across the heart-centre. Embedded in them is Nikki's long-standing passion for connecting with Country, with issues of sustainability and protecting wildlife being of particular interest. Healing, medicine, culture and connection to the environment are threaded through the work.

'And those big gum leaves: I have always loved them and collected them since I was a little girl growing up in Cockatoo. Mum's best friend had this tree with these enormous gum leaves, I was fascinated by them. I used to paint them and give them as bookmarks for Christmas and birthday presents. They are ridiculously oversized, and a lot of people wouldn't realise they grow this big.'

Suggestions that it would be impossible to use them gracefully in a piece of jewellery goaded her on and an exciting journey began, with many technical skills and techniques being passed on to her during Blak Design. With another strong interest in the history and effects of industrialisation, Nikki has used old cogs from a sewing machine she had dismantled – she loves recycling when she pulls things apart to see how they work. The cogs, made of an ultra-tough metal alloy, have been used as a brace from which the copper-plated leaves dangle.

'The cogs reflect the churning of industry since colonisation,' Nikki says. 'Industry is needed – but it is wrecking Country. I'm interested in how can we continue industry but in a better way, for example by getting people out of logging and into new, sustainable industries that look after Country. Industry at the moment is killing Country and not protecting us.'

Her use of sea urchin spines follows a similar train of thought and the metal cogs connecting them have been heated so that the material is reminiscent of the iridescent swirls of an oil slick – suggesting the effects of waterway pollution injuring or killing sea fauna. Unlike the large blue gum leaves, the spines jangle together musically when they dangle, their dried surfaces hardening to an almost ceramic type of material. 'I imagine the music under the sea and fresh waterways, yabbies and eels moving and making sounds under the water, just like when the wind blows and the trees move and you get those whistling sounds,' Nikki says. 'And just because we can't see something [in the ocean depths] doesn't mean we shouldn't look after it and protect it.'

Both her neck-arrangements are reminiscent of shawl-like forms, a reference to the traditional usage of possum skins and other adornments used around the upper torso. 'I like the idea of wearing Country as an artform,' Nikki says. 'I am trying to get away from the tokenistic blakfella situation, where you turn up and they expect you to "look" a certain way. For me, it is about a contemporary way of telling my story.'

When she interacts with school groups, Nikki often begins by talking to them about deep listening – to the birds and trees, to connect with the space. 'For some Elders and older Aboriginal people there are two veins of thought: one is that we don't share and we keep our culture to ourselves, because it was stolen and we need to keep it sacred. The other is to share it so that we do have this opportunity of people loving our culture and understanding what it is. To me, sharing culture is creating this wave of understanding and respecting of our Country's history: you see it now, there is more culture being woven into our everyday lives through business, advertising and education, I feel like there is a curiosity and awareness, there are more arts, there are more Aboriginal people and culture being received really well. People want to know and I love that sharing of culture, at the same time understanding and respecting cultural lores, that there are some elements that really do need to be kept private and sacred for our future generations.'

To see examples of Nikki Browne's work and an interview with the artist, visit the Layers of Blak page on the Koorie Heritage Trust website.

Weblink
Layers of Blak catalogue, Koorie Heritage Trust

Source: *Layers of Blak* catalogue, Koorie Heritage Trust, 2022, pp. 42–50, https://kooriheritagetrust.com.au/wp-content/uploads/2022/09/Blak_Jewellery-Layers_Of_Blak_Catalogue.pdf

9780170477499

Tracy Wise

Walking along Laverton Creek one day, Tracy Wise, a proud Barkindji Ngiyampaa Maligundidj woman, saw an eel, it was not in the creek, but high above her in the sky – what looked to be an image of an eel as the clouds. It was slowly floating there for a while, then it eventually melted away, and it put her in mind of using her creative talents to make something eel-related, such as an eel-trap. 'It got me wondering about the traditional people who may have lived along the creek and caught eels for their meals,' Tracy says. 'So, I went on a whole new journey of learning about the eel and the traditional people of this Country.'

Tracy's explorations with creativity in recent years have been rich and rewarding, with a strong spiritual foundation underpinning her interests and connecting her to nature and Country, walking along the lands of the Boon Wurrung people (as Tracy later discovered), where she usually goes for the bird life. 'I am a real cloud person,' says Tracy, 'they mimic the ocean ... the swirls are like rip curls, they are connected in that sense to the sea with its circular currents; and clouds, of course, produce water.'

Embarking on Blak Design, she pursued the eel theme with great curiosity, with her other desire being to make jewellery – something she'd always wanted to do. She went into it with great passion, continuing to see eels in clouds, as well as in road signage and on roads, and in stencilled artworks on the road along Flinders Lane to warn motorists of a hump and to slow down. 'As a spiritual person I observe everything through my travels, so I was very open to what was appearing before me,' she says. 'There was a lot of trial and error when I got started but the concept sat in the space of knowing and creating and using the skills being learnt. Then, my creative flow was unlocked thanks to the talented co Blak Designer Lorraine Brigdale, who shared her knowledge with me; there and then I started weaving an eel trap with natural grass material found on the campus and wire. I'm a natural wire weaver.' The act of weaving, she says, is a spiritual practice that brings deep connection.

Jahgany by Tracy Wise (Barkindji Ngiyampaa Maligundidj)

Tracy's work reflects the cycle of an eel's life. This entails the creature moving between the sea, estuaries and freshwater. It starts at the egg stage, with spawning happening in the sea, then progressing to larvae, to 'glass' eel (baby) and elver stages, to juvenile and adult eels making their habitat in freshwater. Freshwater eels can live from 25 to 35 years, with the female living longer than the male.

Tracy's work includes small swivels, doubling as eggs, larvae forms and more mature eel shapes. She has interwoven brass wire with natural grass, and the necklace incorporates an eel-trap shape that can be adjusted to trap the neighbouring eel inside. Many components were made with the lost-wax process, casting wax-modelled shapes into sterling silver.

Tracy's research for the pieces was underpinned by reading stories about eel-related practices among various groups. The Wurundjeri and Boon Wurrung people, for example, have traditionally gathered along the Birrarung (Yarra River) to mark the luk (eel) season, while the Gunditjmara World Heritage-listed Budj Bim Cultural Landscape includes the famous eel trap system at Tae Rak (Lake Condah) for trapping, storing and harvesting kooyang (eel), a practice that has been dated back thousands of years. And the Bundjalung people have a creation story of the giant eel Jahgany, who created three islands along the Clarence River.

A highlight for Tracy was when N'arweet Carolyn Briggs AM visited the Blak Design workshop and, looking at Tracy's work, explained how the eel was part of her own Boon Wurrung culture; Tracy encouraged her to try on the eel jewellery, which she did. 'I had been worried and concerned about being culturally inappropriate by using the eel and when I found there was a Dreamtime story, it made me feel better that I could do the eel trap; then Aunty Carolyn came along and it all made sense. It was meant to be.'

Source: *Layers of Blak* catalogue, Koorie Heritage Trust, 2022, pp. 132–3, https://kooorieheritagetrust.com.au/wp-content/uploads/2022/09/Blak_Jewellery-Layers_Of_Blak_Catalogue.pdf

Weblink
Layers of Blak catalogue, Koorie Heritage Trust

The stories of all designers can be found in the *Layers of Blak* catalogue, published by the Koorie Heritage Trust, in which all designers discuss their deep connection to their culture.

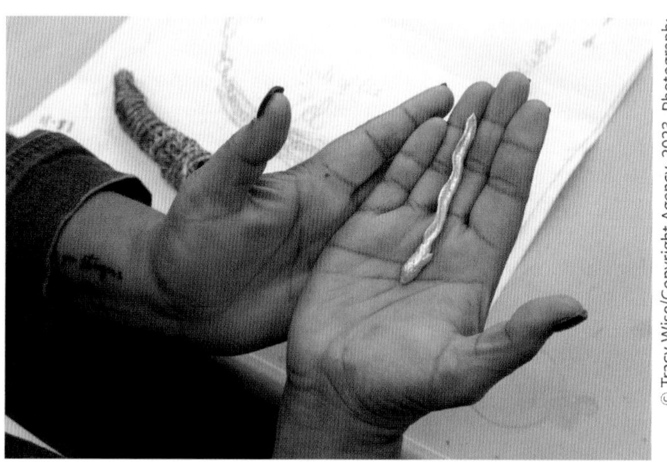

Work in progress by Tracy Wise, Blak Design workshop 2022

FOLIO ACTIVITY

Research activity: Indigenous case studies

The weblinks on Nelson MindTap include a number of alternative case study sources and leads for your research.

1 Read at least two of the case studies above or others that you have sourced.
2 Discuss the following questions.
   a How do the designers talk about their Indigenous culture? What is their connection to culture? What aspects are important to them?
   b How do the designers discuss their connection to the people close to them who link them to their cultural heritage?
   c Many designers discuss the importance of Country. What specific aspects do they focus on?
   d How does this affect the designers' decisions in relation to sustainability, and how they use materials and resources?
   e Do they discuss areas that are restricted or that are private to those within their culture?
3 For each case study, identify two specific examples where the designer's Indigenous culture has influenced some aspect of design (e.g. visual elements, materials, methods of construction) and explain the cultural connection using images and words.
4 **Acknowledge all your sources.**
   a Explain why it is important for you to recognise the sources of all the images and information you use.
   b Discuss why you need to source information about cultural influences directly from the people concerned.

Template
Indigenous case studies

Case studies
Valentino's resort range

Bryan Cush

Broad considerations of culture: impacts on choices

All over the world and across Australia, different cultures have their own practices and traditions, and these have an impact on how products are both designed and used. Consider some of the following points when researching and discussing the influence of culture on the products that people value, choose and use.

- Culture is related to the things that are valued by a specific group. It is formed by how the group thinks about the world around them (their worldview), how they value and treat people (cultural behaviours), and the stories, objects and artforms they produce that reflect their worldview.
- Culture may relate to traditions that are passed down through many generations, through stories, artworks and objects, and social customs.
- Cultural groups differ in the way they perceive and value things.
- Different cultures may have certain areas that are 'taboo', or discouraged.
- Culture can be formed within a variety of settings or groups:
  - national, regional and ethnic groups, language-based groups, communities/localities, religious groups, social and family connections, age groups, specific interest groups (e.g. sport, arts, music, the environment, etc.).
- Culture can also be fluid and may change over time, and within different settings.
- Some cultures are very group-focused; others are very individualistic.
- Culture influences the objects that people choose and how they use those objects or products.

When researching the influence of culture on product choice and use, it is important to talk directly to people within that culture. You need to find out what it is about the culture that influences how different products are valued and used. You will need to carry out some background research, ask 'good' questions, and listen carefully to their responses. If you are interviewing someone from a different culture to your own, you might need to check that you are not affected by your own cultural biases or ways of thinking.

With research into a very small group of people (a small sample), it is unwise to make any broad general conclusions. Share your information with others in the class to see if you notice any similarities and trends in your findings.

Shutterstock.com/Rawpixel.com

Template
Cultural influences on product choices

FOLIO ACTIVITY

Research activity: end user choices influenced by culture

Carry out this primary research with at least two people (one might be yourself). Select people who come from different cultural groups, and possibly different age groups. You might like to create a table that compares the results of your investigations and discussions.

1 Identify a cultural group that the people you are researching are part of, and provide some background information about their connection to that culture. Remember that nationality/ethnic culture is just one of many cultural groupings.
2 What are the key aspects that unify people within their cultural group? You could discuss:
   » regional location, worldview, values that unite, spiritual beliefs, behaviours, core interests, visible signs (e.g. clothing), taboos.
3 Discuss how these aspects influence the specific products your research subjects choose to use and buy. Ask for examples of situations in which aspects of their culture have affected the way they think about specific products.
4 Are there any products that would be 'off limits' or offensive? Explain.
5 On a scale of 1–10, ask your research subjects to estimate how strongly their culture affects their product choices in different areas, e.g. clothing and jewellery, furniture, tools, sports and entertainment, food, housing and interior decoration, etc.

9780170477499

6 Ask the research subject to list a range of features and characteristics that products should have if they are to be valued by people within their culture. Compile another list of features and/or characteristics that would make a product unacceptable or cause it to be rejected within their culture.

7 **Write a report** that summarises the information you have gathered. Cover the following sections:
   » Aim of your research
   » Brief description of your research subjects and their cultural background
   » Your research findings (the information gathered from the questions above)
   » Your analysis and conclusions.

Quiz
Chapter 5 revision

Summary
Chapter 5

Assessment

In Unit 2, Outcome 3, you will be assessed on the following work:

- a research report that uses case studies and/or primary research to investigate:
  - the influence of culture (particularly Aboriginal and Torres Strait Islander culture) on product design
  - the influence of culture on users' product choices.

UNIT 3

ETHICAL PRODUCT DESIGN AND DEVELOPMENT

Areas of Study 1, 2 and 3

Unit 3, Area of Study 1 (SAC 1): Influences on design, development and production of products

You learn about:

- technologies and sustainability approaches in industry/manufacturing, experimental and alternative materials

Unit 3, Area of Study 2 (SAT): Investigating opportunities for ethical design and production

You learn about:

- the Double Diamond design process
- creative, critical and speculative thinking
- Factors that influence design
- ethical research on needs and opportunities
- use of quantitative and qualitative research
- how to write a design brief and evaluation criteria
- drawing systems to create graphical product concepts.

Unit 3, Area of Study 3 (SAT): Developing a final proof of concept for ethical production

You learn about:

- refining and evaluating product concepts into prototypes (virtual or physical)
- use of drawing systems as required
- materials – characteristics and properties, experimentation
- construction processes – degrees of difficulty and trials, justifying tools, equipment and machinery
- gathering user feedback – with both quantitative and qualitative research
- selecting a final proof of concept
- roles and components of a scheduled production plan – how to record modifications.

9780170477499

Units 3 and 4 breakdown of scores

This pie chart shows a breakdown of how students are scored for Units 3 and 4. To gain a pass or S for Satisfactory for each unit, you need to achieve an S or Satisfactory for each outcome.

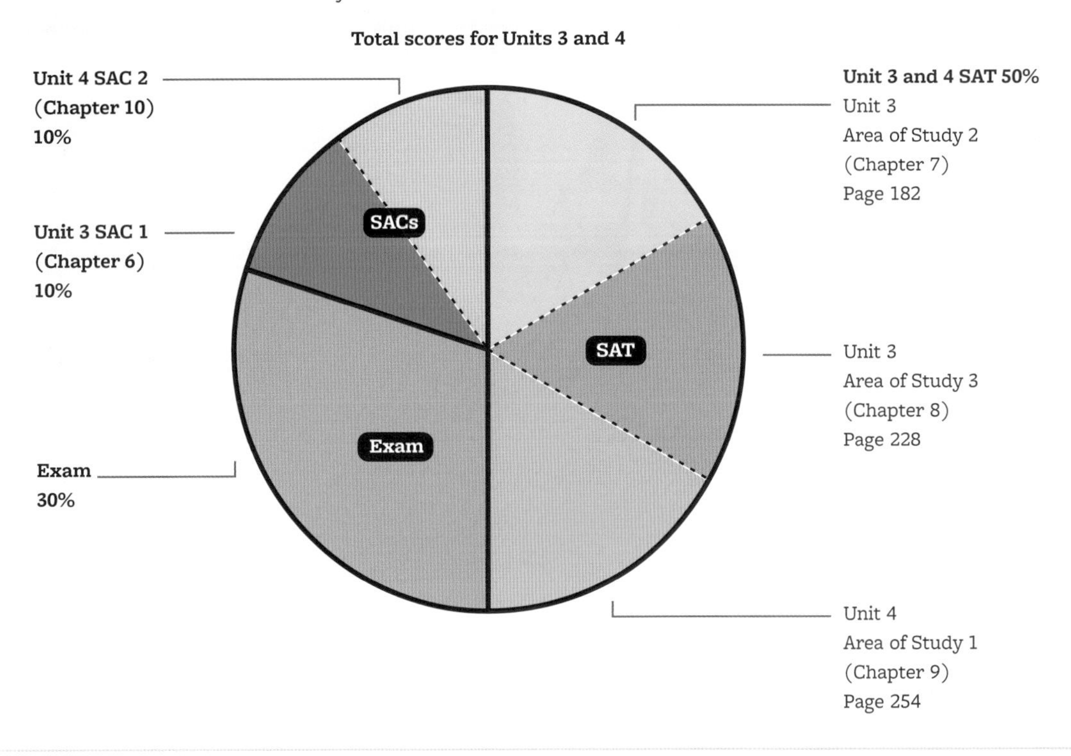

Unit 3 SAC

Area of Study 1: Influences on design, development and production of products

The SAT

Unit 3, Areas of Study 2 and 3: In response to a design brief, you create graphical product concepts and refine them through prototypes and experimentation. A final proof of concept allows you to plan your product's production.

Unit 4, Area of Study 1: You will follow the scheduled production plan to make the product designed in Unit 3. These three areas of study make up the School-assessed Task (SAT) – your major project.

Unit 4 SAC

Area of Study 2: Evaluation and speculative design

SAT Assessment Criteria

It is important to check the SAT Assessment Criteria published every January for the current year, to see how the SAT work is assessed. All work for assessment must be completed within the current school year only.

CHAPTER 6

Influences on design, development and production of products

UNIT 3, AREA OF STUDY 1: OUTCOME 1

In this outcome, you will:

- be able to critique examples of ethical product design and innovation within industrial settings.

Key knowledge

- methods of manufacturing in low-volume and high-volume production settings
- technologies used in different scales of manufacturing: one-off, low-volume, high-volume, continuous production
- relationship between lean manufacturing and flexible and responsive manufacturing
- sustainability frameworks and strategies:
  - » 6Rs: rethink, refuse, reduce, reuse, recycle, repair
  - » circular economy
  - » Cradle to Cradle approach
  - » Design for Disassembly (DfD)
  - » Extended Producer Responsibilty (EPR)
  - » life cycle analysis/assessment (LCA)
  - » triple bottom line
- benefits and issues for the producer and consumer, and associated environmental, economic and worldview issues with planned obsolescence (style, technical and functional)
- technologies and their impacts on processes used in production: artificial intelligence (AI), automation, computer-aided design (CAD), computer-aided manufacture (CAM), computer numerical control (CNC), laser technology, rapid 3D prototyping, and robotics
- experimental materials and processes and their sustainability and worldview impacts: bio-products including mycelium, innovative polymers used for 3D printing, composite metals, repurposed plastics
- alternative materials and their sustainability and worldview impacts: vegan leather instead of animal hide, and bamboo instead of hardwoods.

Key skills

- describe methods of manufacturing processes in different production settings
- compare technologies used in different scales of manufacturing regarding their viability in different contexts and influence on productivity
- discuss sustainability frameworks that influence design, manufacturing and marketing in industry and critique examples of product design in relation to these frameworks
- analyse the impact of planned obsolescence on sustainability and other ethical considerations
- compare and critique impacts of the use of technologies in production processes and experimental and alternative materials for both consumers and producers.

Source: *VCE Product Design and Technologies Study Design*, p. 29

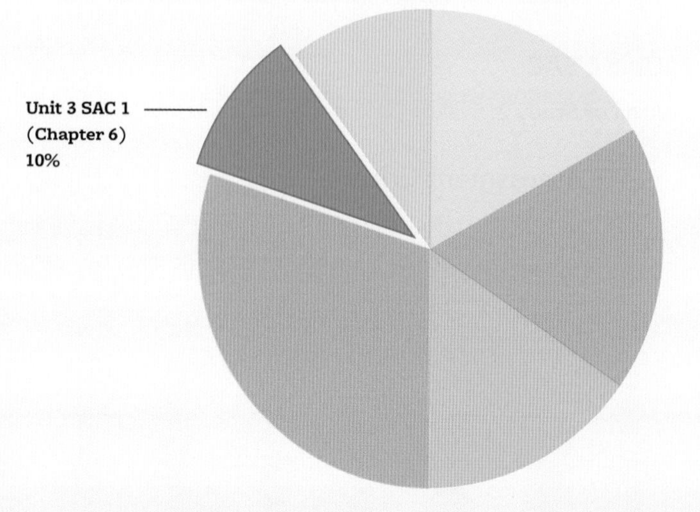

INFLUENCES ON DESIGN, DEVELOPMENT AND PRODUCTION OF PRODUCTS

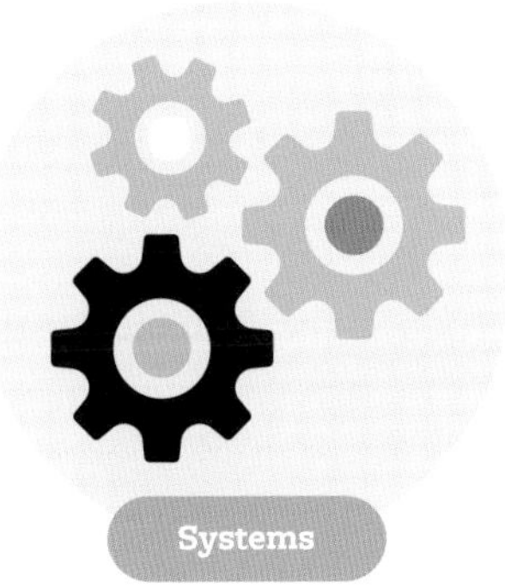

Systems

Technologies

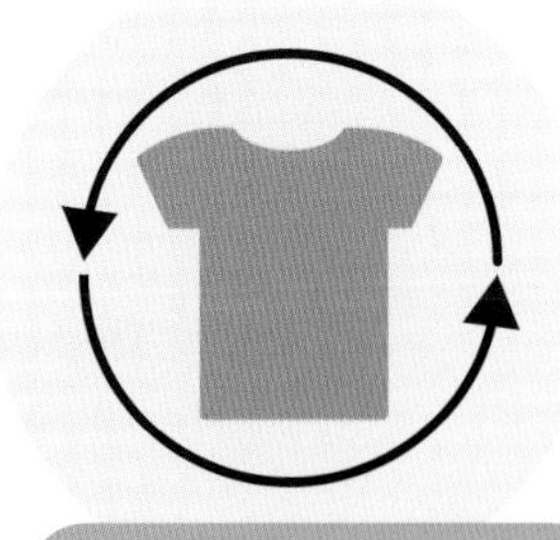

Sustainability approaches in industry/manufacturing

Worksheets:

- 6.1.1 Exploring one-off production **(p.121)**
- 6.1.2 Low-volume production and continuous production **(p.123)**
- 6.1.3 Aspects of all manufacturing scales **(p.124)**
- 6.1.4 Supply chain issues **(p.125)**
- 6.2.1 Technologies – traditional and new/emerging **(p.126)**
- 6.2.2 Researching the use of AI in manufacturing **(p.128)**
- 6.2.3 Researching the use of lasers in manufacturing **(p.129)**
- 6.2.4 Robotics and scale **(p.130)**
- 6.2.5 Researching CNC processes **(p.132)**
- 6.2.6 Prototyping **(p.133)**
- 6.3.1 Lean manufacturing **(p.136)**
- 6.3.2 Revision activity on technologies, scales and contexts **(p.136)**
- 6.4.1 Circular economy **(p.139)**
- 6.4.2 Sustainability and terminology **(p.143)**
- 6.4.3 Triple bottom line **(p.144)**
- 6.4.4 The 6RS **(p.146)**
- 6.4.5 LCA (Life Cycle Assessment) **(p.148)**
- 6.4.6 C2C (Cradle to cradle) **(p.149)**
- 6.4.7 DfD (Design for Disassembly) **(p.152)**
- 6.4.8 EPR (Extended Producer Responsibility) **(p.153)**
- 6.4.9 Sustainability strategies **(p.156)**
- 6.6.1 Obsolescence **(p.158)**
- 6.6.2 Types of obsolescence and impacts **(p.161)**
- 6.7.1 Research polystyrene and mycelium for packaging **(p.163)**
- 6.7.2 SAT activity - 3D printing possibilities **(p.164)**
- 6.7.3 Researching plastics **(p.166)**
- 6.7.4 Researching composite materials **(p.167)**
- 6.7.5 Vegan vs animal leather **(p.168)**
- 6.7.6 Bamboo vs hardwood **(p.170)**
- 6.7.7 Impacts for producers and consumers of alternative materials **(p.170)**
- 6.8.1 Impacts of technologies **(p.174)**

Bonus case studies:

- Bryan Cush **(p.135)**
- Schamburg + Alvisse **(p.180)**
- Marc Pascal **(p.180)**
- Alessi **(p.180)**
- Jim Hannon-Tan **(p.180)**
- Chapter 6 case study report activity

Chapter summary:

- Chapter 6 **(p.177)**

Quizzes:

- Chapter 6 revision **(p.177)**

To access resources above, visit **cengage.com.au/nelsonmindtap**

Manufacturing: methods, scales (volume), technologies and contexts

Scales of manufacturing

Production occurs at a number of different scales:

- one-off creations
- low-volume production
- mass (high-volume) production
- continuous production.

The suitability of the scale depends on the type of product, the demand for the product and the production facilities.

One-off manufacturing

One-off products are suitable in contexts where a client requires a unique product – either a small article or a large, complex product. With smaller products, items are essentially managed by a single skilled craftsperson, using some hand skills, some machinery and outsourcing certain processes (usually these require more expensive machinery, or they are dangerous, repetitive or time-consuming construction processes that can be finished safely and quickly in a factory or bigger workshop). The craftsperson adds the special individual touches. In a large item such as a luxury yacht, a team of skilled tradespeople are involved, working under a chief designer or architect. The boatyard becomes a mini factory.

In one-off situations, clients may select special rare or expensive materials, depending on their budget.

One-off manufacturing can be costly in terms of labour and materials, but usually results in a high-quality, **bespoke** or **custom-made** product. Many people like a product that has been touched by human hands.

Bespoke or **custom-made**
Designed and made to suit a customer for a specific situation

Examples of one-off items include:

- a piece of jewellery
- a **couture** dress or a special outfit
- an entire boat, plane or vehicle
- custom furniture
- prosthetics made for an individual person's disability.

Couture
The French word for tailoring and dressmaking; refers to 'high-end', exclusive fashion garments (sometimes called 'haute couture')

Sometimes a prototype is considered a one-off product if it has not yet been mass produced (manufactured in identical multiples), although it may have been made with similar technologies. A final prototype can function as well as a finished product, but it may be called a 'one-off' as a decision was made for various reasons not to take it to full production. In some cases, it may also be made from different, cheaper materials than those used for the finished production run. Most often it will not be offered for sale, but may be used by employees of the company.

9780170477499

Technologies used in one-off products

Depending on the context, whether a product is wholly handmade, has some construction outsourced or involves a team of workers, technologies used could include:

- hand-operated machines
- CAD
- 3D printing
- laser cutting
- hand tools
- some 'factory' technologies.

ACTIVITY

Exploring one-off production

1. Consider the possible products you could design and make for your SAT, which will be a one-off product.
2. For each possible product, make a list of the 'technologies' that could be used for its one-off production.
3. Explain briefly what each technology will do.

Worksheet 6.1.1

Low-volume production

Low-volume production is often called 'job-lot' or 'batch' production; it is when a small number of products are made identically. It is used in contexts where designers or companies are working with new or specialised products that have a niche (e.g. boutique clothing) or an uncertain market. After testing the market and ironing out any manufacturing issues, these products may be put into high-volume production. It is also a suitable scale for 'limited edition' sets that are often of very high quality, or 'upmarket' products for affluent consumers.

In Australia, companies may choose to manufacture onshore (locally or within the same area) in low volume, or offshore (e.g. in China or South-East Asia), where manufacturers will only take orders of thousands of units to make it worthwhile. In low-volume production, designers and manufacturers can respond quickly to changes in market demand, especially if the production is onshore (local), and adapt the designs accordingly. They can also easily do repeat runs or incorporate new technologies. Low volume allows stringent quality checking.

The costs of low-volume production are higher than for mass production due to higher costs for materials (without the benefit of economies of scale) and manufacturing 'downtime' (time spent adapting machinery for changed designs). However, it usually brings financial benefits as initial set-up costs can be low if production is not highly automated, and waste is minimised by only producing what is likely to sell. Customers are also often willing to pay a little bit extra for items that are special or 'different'.

See more under the heading 'Flexible and responsive manufacturing' (page 135).

Fairfax Syndication/Edwina Pickles

Mud Australia founder Shelley Simpson (right) working with a team member in the Mud factory in Sydney

Technologies used in low-volume production

Technologies used depend on the context, whether a product is new to market or has a small target market, what degree of quality is required and also the complexity of the product.

Low-volume production may occur in a small area within a large factory or in individual workstations, or it may take place on an assembly line for a limited amount of time. It is not for handmade or single, unique products but takes place in a factory setting, making use of all the technologies available. Some products, such as clothing, may include some skilled 'hand-work'.

Technologies used could include:

- CAD
- CAM
- CNC
- automation
- robotics
- laser or 3D printing.

Alessi is an Italian 'industrial arts complex', located in the northern Lombardy region. The area is renowned for the number and quality of the links between schools, studios and manufacturers in the design field. Alessi uses a low-volume scale of manufacturing. You can read more about them in the case study on Nelson MindTap and by researching online.

Mass (high-volume) production

Mass production is the fast, automated production of thousands or millions of identical items. It is suitable in the context of worldwide need, where products are made to suit a large number of people without any customisation (special or individual adjustments). Production at this scale is less flexible than low-volume production as it cannot be changed quickly, but it gives the greatest **economies of scale**: there is minimal downtime and materials can be purchased in bulk at lower cost. However, some mass production, especially of clothing, occurs in unautomated factories. These are usually found in countries where labour is cheaper and there are fewer (expensive) regulations to comply with regarding environmental issues and workers' wellbeing.

Economies of scale
Cost savings made when the per unit cost is low as materials can be purchased in bulk, and manufacturing set-up costs are shared over many products

In automated factories, production lines or assembly lines 'build' the products step by step in an order that produces the end product economically. Automation and robotics are highly accurate and reliable and have reduced human error, which can be costly.

However, mass production requires careful planning as it is time-consuming and expensive to set up and is risky financially – manufacturers need to be sure of their market. It uses skilled and unskilled workers, but workers need to be vigilant for any problems or errors that need fixing to avoid masses of faulty products and/or bigger problems further down the line. It also requires real estate for the storage (or a stockpile) of parts, materials and finished products, which adds to the cost.

Shutterstock.com/panpote

9780170477499

Continuous production

Continuous (volume) production occurs 24 hours a day, seven days a week. It is suitable in contexts where there is a great need for products with few variations, such as in the food industry, oil and gas refineries, mining and metals, and pharmaceuticals. It is important for materials/products that need to be consistent and are subject to change due to heat, time or chemical variabilities and would be harmed by any pauses. Operating continuously not only prevents any unwanted changes in heat- or time-sensitive materials; it also capitalises on the setting-up costs, which are very expensive.

Continuous production is similar to mass production, is costly to set up, is highly automated and is usually run by computer-aided manufacturing (CAM) and automated/robotic systems. Conveyor belts move parts and products between storage facilities and different technologies. It 'flows' with no interruptions and needs to be monitored so that any problems are quickly detected. It requires a managerial workforce to operate at all times, in rotating shifts.

Technologies used in high-volume or mass production and continuous production

Technologies used include:

- CAD to feed CAM
- CNC
- automation
- AI
- robotics
- laser
- 3D printing.

Weblink
Satisfying Assembly Line GIF

ACTIVITY

Low-volume production and continuous production

1 Consider the same possible products you looked at in the previous Activity, 'Exploring one-off production', and for each possible product, make a list of the technologies that could be used for its low-volume production. Explain briefly what each technology will do and how it will benefit the production run.
2 Outline the differences between low-volume and continuous manufacturing.

Worksheet
6.1.2

A comparison of scales of production

One-off production	Low-volume or batch production	Mass, high-volume or continuous production
The people who make the products are multiskilled, and the entire product is overseen by one person.	Workers require specific skills, or may work on only one part of the product (piece workers).	Workers need to be trained in using the technology to detect any errors and to do maintenance. There is less need for human labour.
This type of work may be hands-on, varied, and interesting and rewarding for the creator when complete.	This work may be repetitive, boring and dull. Factory workers may only see one part of the product that they produce. CAD or CAM operators may only see virtual versions of the product/s.	
One item is produced at a time, and each product is individual (no two products are exactly the same).	Tens, hundreds or thousands of identical products are made.	

One-off production	Low-volume or batch production	Mass, high-volume or continuous production
Costs are usually high, as materials are purchased at full price, and some parts or aspects require individual input or special details. Changes can also be expensive due to time added or purchasing of extra materials.	The cost of each unit is lowered, due to 'economies of scale' – a cheaper price for buying large quantities of materials. However, the bigger the volume, the higher the financial risk as a big capital investment is required. It also carries the risk of having a high inventory (parts, finished products and/or equipment) if sales slow down, there are transportation problems or some parts become unavailable (owing to worldwide shortages, etc.).	
Takes a long time to produce but adjustments to the design, materials or functional aspects can be made easily if required.	Is a fast and reliable way to produce a batch of products that are identical, but can also be adjusted easily after each batch if changes are needed. However, this might add time.	Once set up, the average time taken to produce each unit is reduced massively. The output quantity is consistent.
The quality of the product depends on the skill of the creator/maker. Mistakes/errors can reduce quality.	The quality of the product depends on the accuracy of the technologies and systems and vigilance of the staff that run them. Quality is often high in low volume as it is easily checked and fixed (or the technologies quickly adapted) for the next batch if there are any issues.	The quality is decided at the start of manufacturing and remains consistent and steady, whether it be high or low quality, and depending how often the quality and the system are checked. Depending on the target market, this kind of product may be of high quality due to safety requirements, or it may be of a lower, 'suitable' quality, fit for its purpose. Automation reduces the chances of human error.
There may be wait times for materials, parts and components, especially if these are rare or exotic. A craftsperson will keep their own store of bits and pieces to reduce waiting.	Suppliers of parts or components, or providers of special treatments (supply chains) are often situated locally, which reduces transportation waiting times. Longer waits will occur for imported parts.	Can be affected by delayed transportation of parts or whole products (supply chains and outsourcing) if manufactured overseas and if waiting on parts from various places all over the world.

Worksheet
6.1.3

ACTIVITY

Aspects of all manufacturing scales

Draw up your own table to separate the information in the table above into categories according to what it relates to for each scale of manufacturing.

- Workers – satisfaction and skills required
- Time
- Costs
- Quality
- Technologies used.

Make a brief comment for each dot point for each scale, or score out of 5 (1 = lowest, 5 = highest).

9780170477499

Examples of different scales of manufacturing for a chair

One-off for ...	Low-volume for ...	Higher volume for ...
• the Prime Minister • the chairperson of a board • a play, TV or film • a special commission or gift	• an upmarket restaurant • the foyer of a five-star hotel • the office of a large company • a large house used for entertaining	• school chairs • seats in a theatre • food courts • regular household chairs

Impact of world events on supply chains

Worksheet 6.1.4

The information here is not essential in this unit but could be of interest to you.

Regardless of the scale of manufacturing, a well-managed supply chain helps businesses in many ways, such as by reducing waste (overproduction) and costs such as storage and freight. There are many issues to take into account, particularly now that so many countries are interconnected in trade and economics (as a global economy).

War, geopolitical issues, natural disasters, pandemics and economic challenges are good reasons why businesses should avoid a single source of materials and components. The following are examples of events that have affected availability of materials and disrupted supply chains in manufacturing on a global scale since 2018:

- COVID-19 – from early 2020 until 2022, lockdowns and illnesses due to the pandemic reduced the amount of available workers, manufacturing output and deliveries worldwide.
- Low interest rates, intended to protect economies during the pandemic, enabled some segments of society to borrow large sums to purchase or build more housing, resulting in a shortage of housing, and also contributed to high demand for building materials.
- The Russia–Ukraine war, which began in 2022, destroyed much infrastructure and created a shortage of energy and commodities (e.g. oil, gas, wheat, corn, barley and fertilisers, aluminium, nickel) and blocked off many trade routes.
- Natural and man-made disasters included:
  - » fires in Victoria and NSW in 2019 that burnt huge areas of land as well as small towns. Other large fires occurred in Portugal and California in 2021 and 2022 and Hawaii in 2023
  - » a giant container ship that became stuck in the Suez Canal for six days in 2021, blocking all other ships transporting goods and materials between the Americas, Europe and Australasia
  - » floods in 2022 that destroyed rural towns, agriculture and other businesses in Australia and almost one third of Pakistan, where a lot of textile manufacturing was based
  - » a giant landfill dump (17 storeys high) in Delhi, India, which caught fire in 2022 due to extremely high temperatures, halting the work of thousands of 'waste-pickers'
  - » record summer temperatures in Europe in 2022 and 2023 that destroyed or reduced many crops and slowed down manufacturing
  - » droughts in 2022 that dried up rivers used to transport cargo, e.g. the Rhone in Germany, the Mississippi in the USA and the Yangtze in China, creating greater transport costs
  - » a massive earthquake along the border of Syria and Türkiye (formerly Turkey) in 2023 that killed almost 60 000 people and damaged infrastructure, blocking movement of exports.

Traditional, new and emerging technologies

Technologies are the tools, machinery, equipment, systems and software involved in production.

- **Traditional technologies** are those that are still used that have been around for hundreds of years, those that are cultural and passed on through generations and those that might be specific to an area because of the materials available there. These technologies include hand tools, hand machinery or simple electrical machinery used in a straightforward system. Many are still in use and are valued for their simplicity, low impact and skilled, individual control.

- **New and emerging technologies** are completely new technologies as well as those that have been around for decades but continue to develop and evolve, becoming ever more sophisticated and complex. They also tend to be safer and more efficient than traditional technologies. Occasionally an innovative technology comes along and upends all previous methods of production.

Worksheet 6.2.1

Automation

Automation is the automated control of technologies and machines, usually by computers. The aim of automation is to increase production efficiency and accuracy by reducing human input and therefore human error. Software is used to control mechanical, electrical or computerised actions that were previously performed by humans. It has automated feedback loops and checks to reduce human involvement even further. It is most useful for repetitive, complex or unsafe processes. It works with many other new and emerging technologies described on the following pages.

Shutterstock.com/Es sarawuth

Automated storage and retrieval in a modern warehouse

Artificial intelligence (AI)

Artificial intelligence (AI) can be used to improve overall productivity and operational efficiency in manufacturing. AI involves complex computer programming that uses machine learning (ML) to make decisions on manufacturing processes that are efficient and economic by analysing data quickly without human interaction. These decisions are implemented via other **4IR** technologies.

In manufacturing, AI can:

- create and optimise a design in CAD from scratch by using algorithms (using data to create sets of software instructions and calculations), which engineers can test and check, replacing months of design work
- move parts around the factory and respond to changes in the environment so no humans are needed
- collect a large amount of operational data and make decisions in response to the data
- use automated image recognition to perform quality checks and inspections to reduce errors/defects
- find and identify patterns that are unseen by humans, particularly of small errors resulting in defective products that need to be discarded, known as the scrap rate
- predict maintenance that will be needed and perform it
- predict delays or issues with **supply chains** and identify bottlenecks (crowded slow spots in the production line)
- forecast and monitor completion/delivery dates of products and parts
- monitor facilities in real time.

4IR
The Fourth Industrial Revolution, where technologies and data are interconnected and responsive to each other

Supply chain
Every step involved in getting a finished product to consumers, from sourcing raw materials, outsourcing of components, labour, transportation at all stages and finally to retail outlets

9780170477499

Fanuc is a Japanese company that manufactures different technologies such as CNC machines, robots, lasers and many others, using AI technology. Take a look at their factory floor and AI in action. For more on AI, read the article '17 Remarkable Use Cases of AI in the Manufacturing Industry' on the Birlasoft website.

Weblinks
17 Remarkable Use Cases of AI in the Manufacturing Industry

Enjoy Effortless Outings with the Award Winning GlüxKind AI Stroller

AI in products

Artificial intelligence can also be incorporated into products. A well-known example is self-driving cars. Cameras and sensors feed information into 'neural networks' (like neural pathways in the human nervous system) to build up a vast amount of data for machine learning (ML). The algorithms build information on obstacles such as trees, traffic lights, pedestrians and other cars as well as data from Google Maps. All the technology involved makes up the artificial intelligence that replaces human decisions and actions in a fraction of a second to control the car.

See the self-driving child's pusher *Ella* by GlüxKind on display at the Consumer Technology Association (CES) tech show in Las Vegas, USA, in 2023 via the GlüxKind website. The pusher is packed with sensors, cameras, motors and AI. It can stop automatically when it detects obstacles or goes out of the owner's reach, drive itself when empty and keep up with parents as they stroll.

AI can also be used in film editing, drones, video games, security and surveillance, and home devices, to name a few.

Alamy Stock Photo/Phillip Bond

A Google Street View car collecting data in the form of photographs to feed into AI systems

Suitable scales, context and influence on productivity of automation and AI

Automation and AI require a big investment and are mostly suited to high-volume production. They reduce the need for human labour. This reduces the incentive for companies to use offshore manufacturing in other (usually developing) countries where labour costs are lower. Use of AI and automation in manufacturing helps to decrease errors and therefore increase sales, decrease lost sales and reduce the workload of managers and workers. Use of AI in products can increase their safety and comfort. For more on this topic, read the article, 'Fashion's new fakes: How AI will change what you wear' by Janice Breen Burns on *The Age* website.

Worksheet
6.2.2

ACTIVITY

Researching the use of AI in manufacturing

1 Research a product for which AI is used in its manufacturing. Answer **one** of the following:
   a In approximately 150 words, identify the product, explain how AI is used in its manufacture, the scale of manufacturing and how the use of AI suits the manufacture of that product.
   b Insert an image of the product or a rough sketch into PowerPoint and annotate it to show how AI is used in its manufacture, and how this benefits that production line.
2 Research a product that incorporates AI into the product. Make a presentation using **one** of the methods suggested in a or b above.

Laser technology

Laser is an acronym of '**l**ight **a**mplification by **s**timulated **e**mission of **r**adiation'. In simple terms, a laser is an intense beam of monochromatic (single wavelength) light that is concentrated through a special lens. The first common use of lasers was in the supermarket barcode scanner, introduced in 1974. This was followed by laser disc players, the CD player (1980s), laser printers, fibre-optic communications and the use of lasers in surgical medicine. Lasers can be harmful to human eyesight and are classed according to the degree of hazard they pose.

Lasers are used in manufacturing for:

- cutting complex shapes and curves, and/or many layers at the same time, accurately and in a clean way, giving ultra-high quality edges that reduce extra 'finishing' techniques required – used in plastics, fabrics, timber and less commonly in metals (due to costs and the high power required)
- body scanning for measurements
- taking measurements of distances, levels, positions and speed
- surface treatments such as embossing, imprinting, engraving (to identify materials or brands or QR codes) or hardening metals
- drilling in tiny layers until the drill 'hole' breaks through – this can be done in metals, polymer and rubber
- detecting faults such as puckering of seams in clothing or irregular weave/prints of fabrics
- welding using low heat, used for tiny and/or thin materials, and welding remotely, which increases safety
- rapid prototyping – in additive manufacturing, laser accumulates the material layer by layer in a similar way to 3D printing, by sintering (heating) metal powder to produce metal components, directly from CAD. This method of manufacturing also allows easy repair of parts.

Suitable scales, context and influence on productivity of laser

Laser technology is now commonly applied in production. It is used in one-off, low-volume and high-volume productions and is easy to automate, reducing human error. It is extremely speedy, accurate and

Shutterstock.com/alterfalter

Laser technology is widely used in production.

9780170477499

precise to one thousandth of a millimetre. It improves safety in the factory. It uses less energy and takes up less space than conventional machinery, so can reduce costs. A laser beam does not become blunt, and thus need sharpening, or wear out like traditional cutting machines with blades.

Laser technology allows a level of detail and complexity in designs that might have been avoided in the past because of safety, accuracy and time issues. It also improves product quality and productivity, and can reduce pollution and material consumption. For example, hardening metal through laser surface treatment can make it more resistant to wear and corrosion, increasing its life.

ACTIVITY

Researching the use of lasers in manufacturing

1 Research how lasers are used in manufacturing products from your SAT materials category and explain their benefits and any problems with their use.
2 How does a laser cut compare with results from previous cutting methods or the use of hand machines?

Worksheet
6.2.3

Robotics

Robots, like lasers, are speedy, accurate and safe. They are automated machines, and can be programmed to work long hours, completing repetitive tasks. The use of robots has eliminated a lot of dangerous and repetitive jobs once performed by workers in manufacturing. Robots don't tire out and can be programmed to work non-stop. They perform processing activities, such as welding and painting, material handling, assembly and inspection. The name 'robot' is also applied to complex computers that mimic human movement, but with much more available movement and reach. This makes them suitable for sorting (if equipped with sensors) and pick-and-place (PnP) processes, which involve moving objects from one place to another.

Alamy Stock Photo/Roger Bamber

Robots are used in car manufacturing and speed up the assembly line.

Suitable scales, context and influence on productivity of robotics

Robotics requires a big investment, but is suitable for both high- and low-volume production lines. It allows procedures that were previously considered unsafe to be included in the construction of products. It reduces costs and improves the quality of products. Robots are extremely suitable for high-volume production of complex products such as cars. However, robots can also easily be programmed for smaller production batches, making them flexible and responsive to consumer demand.

Worksheet 6.2.4

Robotics and scale

Explain why robotics is more suited to high-volume production of complex products.

Computer-aided design (CAD)

In computer-aided design (CAD), designs can be developed from scanned sketches or drawings, or they can be created from scratch. When adjustments are made to the design, the CAD program calculates all the required dimensions. There are many CAD drawing programs, such as TurboCAD, AutoCAD®, Autodesk, Intercad, Creo, SolidWorks and CATIA. Textiles and fashion designers may adapt the use of computer drawing software (such as Adobe Illustrator) for developing their designs, but there are also specialist CAD software programs that are used in the textiles and fashion industry. All of them allow 3D visualisation or 'virtual prototypes', which reduces the costs and development times that physical prototypes entail. Some programs allow simulation in a detailed real-world context, to the extent that a physical prototype is not needed. CAD files are linked quickly and easily to computer-aided manufacturing (CAM).

iStock.com/Laurence Dutton

CAD allows designers to create 3D visualisations of their products.

Suitable scales, context and influence on productivity of CAD

CAD is used in all scales of production, from one-off to high-volume. CAD allows changes to designs to be implemented quickly and accurately, avoiding errors in calculations. Designs can easily be emailed for immediate feedback, adjusted and stored electronically. This brings many advantages compared with designs that were historically drawn and stored on paper and sent by mail.

Computer-aided manufacturing (CAM)

Computer-aided manufacturing (CAM) uses digital information (from CAD designs) to directly drive machines and manufacturing systems. Companies design and create instructions in CAD that work in conjunction with CAM, which controls and automates technology such as lasers, routers, knitting and embroidery machines, machine tools, etc.

Almost any modern factory would use a form of CAM, where CAD files can be updated quickly to customise the product.

9780170477499

Suitable scales, context and influence on productivity of CAM

CAM is suitable for both low-volume and high-volume manufacturing as the digital designs can be easily updated or customised. It may also be useful in creating parts or the whole of a large, complex one-off product. CAM allows designs to become physical in a short time. It increases accuracy by removing opportunities for human error, reduces material wastage and overlaps with the benefits of CAD.

Computer numerical control (CNC)

Computer numerical control (CNC) is the automation of machine tools and the use of CAD/CAM programs in a series of highly complex steps to create parts and components. CNC coordinates numeric information to control position, movement and speed of machining parts that are calculated from designs and sent directly to CNC machinery. It is mostly used to control drilling and milling tools that cut shapes, 'holes' and 'trenches' in various positions and at different angles. It controls the size and shape of pieces, the specific cutting tools used and how they move. CNC has dramatically reduced the number of machining steps that require human action, as well as making considerable improvements in consistency and quality due to its high degree of accuracy.

Shutterstock.com/Red ivory

CNC-like systems are now used for any process that can be described with a series of movements and operations. These include:

- laser cutting, flame and plasma cutting, sawing
- welding, friction stir welding, ultrasonic welding
- bending
- edging, lathe spinning and router cutting
- spinning, pinning, gluing, fabric cutting, sewing, tape and fibre placement.

Suitable scales, context and influence on productivity of CNC

CNC speeds up manufacturing of products and components. It is suitable for high-volume manufacturing as it can quickly produce large quantities of parts or products that are highly accurate and therefore of high quality. Its computer programming can be easily modified, and can cater for complex designs that would be difficult to achieve otherwise.

It is also suitable for low-volume manufacturing of small quantities of parts or products thanks to the flexibility of its programming (it can be changed easily) and its ability to adapt to changing customer needs.

Customised for customers

It is important to note that most of the technologies outlined in this book exist in factories. When the term 'customers' or 'clients' is used, it refers to small companies or designers who enter into a business agreement with the factory to manufacture a product (i.e. a contract with the manufacturer to create a product). It does not refer to retail customers or clients making individual purchases.

Worksheet
6.2.5

ACTIVITY

Researching CNC processes

1 Type any one of the CNC processes listed, with the prefix CNC, into the search bar in YouTube to find out more. Choose a process or processes that would be applicable to your material category, watch the video and briefly explain it.
2 Outline or draw the process that occurs from CAD to CAM, or CAD to CNC.

Rapid 3D prototyping

Commonly known as 3D printing, but also known as additive manufacturing, rapid prototyping is a speedy process used for building physical 3D prototypes and/or parts. It allows much more checking than a virtual prototype, which cannot be tested physically. Prototypes are used to test the feasibility, functionality, strength, ergonomics, sizing, feel and suitability of materials, plus more.

Prototypes in general can be made by either additive or subtractive manufacturing:

- **Additive manufacturing** consists of adding layers or building up a material and reduces the need for joining materials.
- **Subtractive manufacturing** consists of removing material from a solid piece by cutting, drilling or carving.

To create a rapid prototype or 3D-printed object, a digital file created on CAD is sent to a special printer, where it is 'built'. There are 3D printers that:

- lay down successive layers of heated liquid plastic filament
- use UV or laser light to solidify liquid resin (good for objects with fine detail)
- use laser sintering to melt powdered metal or plastic.

The main benefit of this process is the greatly reduced time needed to produce a physical prototype – from months to overnight. Previously, if a prototype needed changes, modifications and retooling (setting up the machining) would add extra months.

Most 3D printing uses some kind of plastic filament (meltable thread of different sizes) depending on the properties required – for example, a 'wood' filament that mimics wood, which is 10 per cent wood dust mixed with PLA; or metal impregnated with plastic for metal parts. See more about 'innovative polymers' on page 163.

Solid metal parts can also be printed, but this is an expensive process.

It is a developing field, and research continues to find other materials suitable for 3D printing and applications, e.g. to mimic human tissue and for building structures.

Weblink
Stop Printing Crap

- Search YouTube to see the largest 3D-printed building in the world, built in Dubai in 2022.
- Watch the 'Stop Printing Crap' video on the One Army website before you decide to print 'something'.

9780170477499

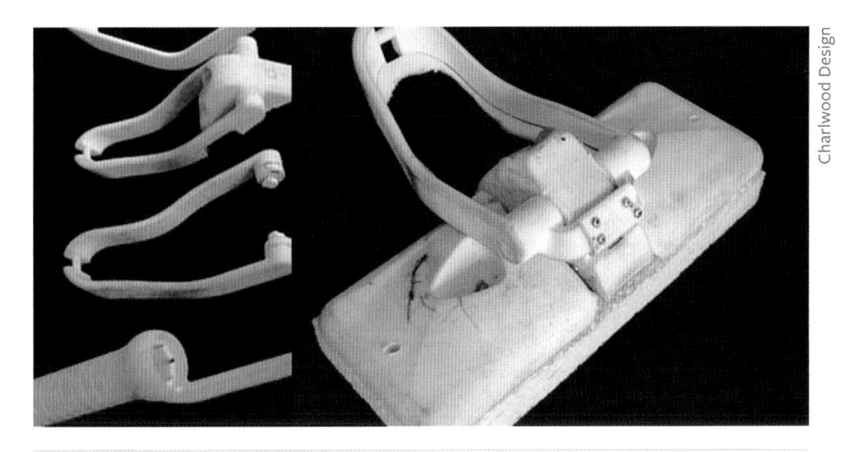
Charlwood Design

Rapid 3D prototyping at Charlwood Design, creating the Oates mop

Suitable scales, context and influence on productivity of rapid 3D prototyping

Rapid prototyping speeds up the entire product development process in high-volume production – the time from initial concept to finished saleable items. A prototype of a design can be 'printed' literally overnight, examined, trialled and tested, and adjustments made almost immediately.

As the cost of 3D printing comes down, it has more applications in both low-volume and mass production, for high-quality identical parts in relatively small numbers. It allows more design freedom as only the digital files need to be changed. Constraints are the high costs of both the equipment and suitable materials, and its slower speed compared with other methods, such as injection moulding.

3D printing also has domestic uses, such as to replace broken parts of household items, or to create one-off objects such as jewellery, prosthetics (artificial body parts) and/or sports equipment.

Prototyping

1 Outline three purposes of prototyping.
2 Outline two benefits of rapid 3D prototyping compared with previous methods.

Worksheet
6.2.6

Lean manufacturing

Lean manufacturing is a manufacturing practice that aims for reduced cost per unit, while maintaining and improving quality. It was derived from the Toyota Production System (TPS) developed by Toyota car manufacturers in Japan after World War II. Based on total quality management (TQM), it raised the quality of Toyota products and identified eight **waste** areas. They are:

- overproduction (too many items produced)
- unnecessary transportation
- storage of inventory (materials and products that won't be used for long periods)
- motion (unnecessary movement of people and machines)
- defects
- over-processing (extra features)
- unutilised talent/skills
- waiting times.

The eight wastes of lean manufacturing

Unnecessary transport

Inventory storage

Unnecessary motion

Waiting

Over-production

Over-processing

Defects

Unutilised talent

In lean manufacturing, the cost of inputs (resources, labour, etc.) for any reason other than adding value for the end customer are considered to be wasteful, and thus a target for elimination. Value is defined as any action or process that results in an outcome that a customer would be willing to pay for. The goal is to reduce waste as much as possible – not just waste of materials, but waste of time and/or labour and skills and any unnecessary costs.

It aims for a smooth production flow with processes that run very efficiently, using a Plan–Do–Check–Act method to solve problems. Measurements such as sales and/or orders, rather than opinions, are used to make decisions and improvements, i.e. on quantities to produce, the product's design or the flow. It is about getting the right materials, components and labour to the right place at the right time in the right quantity to achieve perfect workflow. 'A place for everything and everything in its place, safe, clean and ready for use.' Software to manage this is important and even the layout of the factory floor is organised to promote a good flow. Software also keeps track of supply chains to ensure smooth flow of parts and materials as needed. Workers are carefully trained and their knowledge is respected, valued and put to use effectively. Looking for improvement in all areas is ongoing and all staff are encouraged to contribute ideas. In some situations, 'lean' contributes to more sustainable manufacturing due to the elimination of waste, but this isn't always the case.

Weblink
What is Lean Manufacturing?

For more details, visit the 'What is Lean Manufacturing?' page on the TXM website.

9780170477499

Flexible and responsive manufacturing

Flexible and responsive manufacturing (sometimes termed 'agile manufacturing') is easily achieved through 'lean' manufacturing strategies; it allows manufacturers to change production with little notice and without complicated or expensive changes in the technological systems used. It means that manufacturers can cater for more customised products from small niche businesses and can respond quickly to changes in trends (or sales) by updating styles. Lean manufacturers only make products for projected sales for the very short-term future, not for the next few years. This means they don't need large warehouses to stock parts ready for manufacturing or to stock finished products waiting to be sold (an inventory). It saves the cost of warehousing and frees up space for new parts and/or components that are needed when changes are made to a product. However, if there are delays in the supply chain for materials or components, manufacturing may be disrupted. The use of CAD, CAM and robotics improves a manufacturer's ability to be flexible and responsive. Production can be quickly changed through programming rather than through the costly changeover of tooling and production machinery and systems.

In a competitive environment, a company can stay ahead and uphold their reputation by offering advantages, no matter how small, in the speedy delivery or performance of their product to meet a current need.

Case study
Bryan Cush

Chapter 6 case study report activity

Examples of lean manufacturing approaches or methods

Robotics in manufacturing and rapid prototyping has assisted lean manufacturing in having a very fast product development process, i.e. getting the product from initial idea to finished product ready for sale quickly.

To eliminate anything that is an unnecessary cost and does not give value to the consumer, companies consider these strategies:

- low-volume production
- purchasing parts and materials only needed for current orders (avoids overproduction; reduces unsold stock; saves on storage)
- reducing waste in matters/resources related to production flow such as precision cutting of materials (computerised cutting layouts)
- having 'looser' tolerances to allow for and eliminate errors
- using electricity, gas, water and any other resources in an efficient manner
- eliminating extra unnecessary functions or parts on the product, or paints or finishes that aren't essential
- reducing 'wait' times for staff and managing labour efficiently
- training staff to use machinery correctly to increase accuracy (less mistakes)
- focusing on continuous improvement, for which staff are encouraged to make suggestions
- frequent quality checks so that imperfect components and products are removed from the system, and problems are identified and fixed quickly
- using IT systems, electronic tags and/or mobile phones on the factory floor for fast tracking and communication.

Weblinks
Case Studies: Lean manufacturing success stories that adopted Katana software

Stella Soomlais home page for sustainable practices

Worksheet
6.3.1

ACTIVITY

Lean manufacturing

Create icons and cartoons or redraw them from the infographic for the areas of waste. Read about the eight areas of waste on page 133 and add annotations explaining each of your icons.

CASE STUDY

Lean manufacturing at Zara

Worksheet
6.3.2

Although automotive and product manufacturers have used an agile or lean manufacturing approach for many years, the Spanish fast-fashion company Zara became known as the first clothing company to implement it. Lean enabled a fast turnaround of only 15 days from concept to production, which otherwise took months. By updating clothing styles continuously, the brand gained popularity quickly. Women knew they could find something different and fashionable and that it wouldn't be seen on a thousand other women. Zara invests a lot into finding out what customers want and are buying on a weekly basis, and feeding this information to the designers. New styles are delivered to its stores twice a week on order from each store. However, Zara only ever holds a small amount of stock. The use of radio frequency identification (RFID) enables Zara to 'see' the supply chain; meaning the company can see which products are selling, how many and in what sizes, and what needs replacing in stores. Stores do not get more products than they need to sell.

Weblinks
How Zara Used Lean to Become the Largest Fashion Retailer

Zara's Business Operations and Strategy: How and why they worked

Zara owns its own factories in Europe, with highly skilled staff. This cuts down the time and the cost of transportation and avoids quality issues that other companies have when manufacturing in Asia.

This tight control of stock allows Zara to respond to the market quickly and reduce dead stock, and gives them invaluable competitiveness.

To learn more about Zara's approach, read the following articles:

- 'How Zara Used Lean to Become the Largest Fashion Retailer' on the LinkedIn website
- 'Zara's Business Operations and Strategy' on the Tough Nickel website.

Weblink
The Lean LEGO Story: Building blocks of continuous improvement

As well as Zara, companies known to have implemented lean manufacturing include Lego. You can read more about them in the article 'The Lean LEGO Story: Building blocks of continuous Improvement' on the KaiNexus website. You may also be interested to complete research on fast-fashion company Shein, which has adopted convenient aspects of lean manufacturing to produce millions of super cheap clothes, but has has been accused of neglecting quality construction and materials and ignoring negative environmental impacts.

Sustainability

Sustainability can be defined as 'meeting the needs of the present without compromising the ability of future generations to meet their own needs' (United Nations, 1987, *Our Common Future* – also known as the Brundtland Report). This statement is commonly understood to be about:

- making the best choices for the planet's health by protecting the natural environment and using resources wisely
- safeguarding human rights and human/social/community wellbeing
- meeting people's needs using ethical economic systems.

9780170477499

Many design strategies address these three dimensions, which may be called:

- the three pillars of environmental, economic and social sustainability
- the three Ps of planet, profit and people

  or
- the 'triple bottom line'.

All strive to reduce negative impacts and generate positive impacts across these interconnected 'pillars' or dimensions. Read more about these and other strategies that designers and manufacturers can use to guide their work on the following pages. First of all, make sure you are familiar with the following relevant terminology.

Terminology linked with environmental sustainability

Important terms that are linked with sustainability include:

- **climate change** – changes to our weather patterns and natural systems due to the increased production of **greenhouse gases** that trap heat in Earth's atmosphere. Increased heat in our environment also reduces the effectiveness of natural systems to sequester (trap) carbon
- **carbon dioxide** – a greenhouse gas, produced by the burning of **fossil fuels** for energy (used in homes and in manufacturing and transportation) and by the destruction of forests (for timber materials and land clearing for agriculture) and other habitats (for plant materials). More carbon dioxide in the atmosphere contributes to more rapid climate change

Greenhouse gas
Carbon dioxide, methane or nitrous oxide, which are all naturally present in Earth's atmosphere and trap the sun's heat to sustain life, but which have increased out of proportion over the last two centuries, increasing global temperatures

Fossil fuel
A carbon-containing compound, formed over millions of years, that releases greenhouse gases such as carbon dioxide when burnt

Alamy Stock Photo/Marc Anderson

Growing forests take carbon dioxide from the atmosphere and store it in living things. Cutting forests down contributes significantly to carbon dioxide emissions.

- **finite resource** – a resource that doesn't replace itself (i.e. it is non-renewable) or replaces itself much more slowly than it is consumed. Examples of finite resources are minerals, metals and fossil fuels such as oil, coal and natural gas. The opposite of this is a **renewable resource** – one that is unlimited or can be replaced easily
- **waste** – read statistics on waste in Australia in the 'National Waste Report 2022' on the Department of Climate Change, Energy, the Environment and Water website
- **landfill** – rubbish that is dumped in deep pits or huge piles (tips or dumps). Much of what goes into landfill could be used in another way. Organic waste in landfill emits gases as it decomposes (rots) and consequently contributes to a build-up of toxic air or spontaneous fires. Toxic chemicals can leach out of other waste products into surrounding

Weblink
National Waste Report 2022

soil, the water table and air. Landfill sites, once they are filled, are difficult to use for other purposes because they are usually contaminated and structurally unstable

- **incineration** – the burning of waste, which is sometimes used to create energy but also emits a lot of carbon dioxide and toxic gases
- **ecological footprint** – measures and describes the impact of our activities on the environment.

Sustainability strategies and frameworks

Frameworks or strategies for improving sustainability of products follow different guiding principles, which often overlap. A brief explanation of some of them follows.

Circular economy

A circular economy has three principles: to 'circulate' products and materials, to reduce waste and pollution and to preserve nature. Ideas about a circular economy first appeared in the 1960s. As the global population and its wealth increases, so too do the use and consumption of materials and resources, and the creation of waste and pollution. A circular economy is seen as a system to combat these issues, as opposed to the linear system, which is also known as 'take, make, waste'.

The concept was defined in more detail by the Ellen MacArthur Foundation in 2010. It has a lot in common with the Cradle to Cradle (C2C; see page 148) approach: both want to eliminate waste, reuse resources and regenerate nature. The circular system aims to work with materials from two cycles: 'technical' and 'biological'.

The **technical** cycle is about keeping materials (such as metals and plastics) that come from finite sources in circulation by reusing or recycling them.

The **biological** cycle is about materials (such as cotton and wood) that are renewable and can be composted or biodegraded when they are disposed of, meaning they are returned to the Earth.

The circular economy also encourages prolonged use of products, rather than the constant manufacturing of new products. This design approach allows products to be shared, durable and/or repairable so they can be passed on to others for reuse.

Source: Downes, J. (2022), 'Framework for understanding, measuring & communicating waste prevention'. Prepared for the Australian Government Department of Climate Change, Energy, the Environment and Water. BehaviourWorks Australia, Monash University.

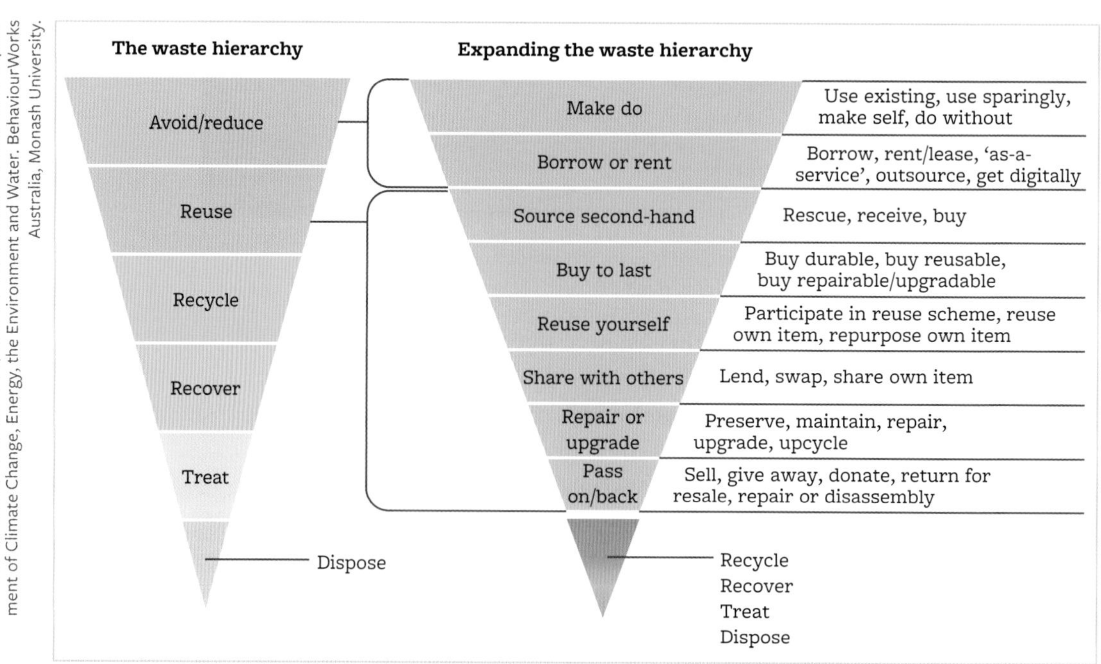

Waste hierarchy for sustainability. The left-hand side shows the simple waste hierarchy, with the most preferred strategies at the top. The right-hand side expands these preferred strategies and describes ways to achieve them.

9780170477499

The circular economy has a hierarchical approach to sustainability, with the reuse of an intact product being the most preferred, and recycling of individual materials being the least preferred. This correlates to the 2022 'Framework for Understanding, Measuring and Communicating Waste Prevention', prepared under Australia's National Waste Policy and Action Plan.

The circular economy and clothing

The circular economy has become more important to implement in the clothing industry. The amount of clothing manufactured in the world more than doubled between 2000 and 2022, and much of this clothing was sent to landfill after a few wears. According to consumer magazine *Choice* in 2021, only 7 per cent of clothing is recycled and Australians on average buy more than 25 kilograms of clothing, and throw away a similar amount, each year. At the same time, however, the industry generated $27.2 billion for the economy and employed 490 000 people – so it is an important industry. The Australian Government aims to reduce the environmental impact of clothing and has launched an initiative to develop a National Clothing Stewardship Scheme.

Shutterstock.com/Ernest Rose

To reduce the impact of our fashion choices on the planet, it is essential to get more use out of each piece of clothing. Designers can take this into account and, with knowledge, consumers can purchase fashion that is repairable, reusable or able to be upcycled. The best approach is to buy less and buy clothes that are durable and timeless.

Weblinks
Is Australia Waking Up to its Textile Waste Problem?

Clothing Textiles

The Impact of Textile Production and Waste on the Environment (infographics)

What is a Circular Economy?

Ellen MacArthur on the Basics of the Circular Economy

A New Textiles Economy: Redesigning fashion's future

The Jeans Redesign

Redesigning the Fashion Industry: The story of the jeans redesign

To read the article, 'Textile waste and what to do with it' visit the Choice website.

To read more about the Australian government's initiative to develop a National Clothing Stewardship Scheme, visit the Department of Climate Change, Energy, the Environment and Water's website.

The European Parliament has put forward many important reasons why circularity is important for the clothing industry in its publication News European Parliament.

Read articles and watch videos regarding a circular approach to production of denim jeans, fashion in general and more at the Ellen MacArthur Foundation website and on YouTube.

ACTIVITY

Circular economy

Draw one or more cartoons or diagrams to:

a show the waste hierarchy in a circular economy, i.e. the best way to be sustainable

b help you remember the difference between technical and biological cycles or materials.

Worksheet
6.4.1

CASE STUDY

Radical Yes

Who are they?

This vibrant Melbourne company designs and makes shoes, sunglasses and handbags. They are dedicated to creating flat shoes for women in ranges of sneakers, sandals and boots. Their name captures the ethos of the company but also provides them with a licence to push against the norms in the fashion industry, particularly in relation to speed and waste.

Radical Yes

Radical Yes *Grace* shoes and *Big Time Mini* tote bag

Their end users and target market

Radical Yes founder Kerryn Moscicki has developed a keen understanding of the target market of her 'flat' shoe ranges. They are confident and intelligent women, who know what they believe in, women who are moving through the world from a grounded base. They like to be grounded, not constricted or restricted by their shoes and/or high heels. They are busy women with a lot to do, and that could span activities as diverse as going to a rock show or a dance rave, visiting friends and shopping, or going to and from work. Movement emanates from them, whether it's walking, dancing or bike riding. They may also be inner-city public transport users who need flat shoes for their commute. They are practical and want to move in comfort and style.

Radical Yes has two customer archetypes, called Grace and Zoe, each representing women in different stages of the life cycle. Zoe is older, probably a mother, and busy juggling a demanding career and child rearing. She generally has less time in her day and tends to drive her car more. Grace is younger and is busy with her career and social life; she probably doesn't have children and is more likely to walk than drive a car. Kerryn designs her shoes with both women in mind, and this clarity about who the company's customers are and how they live also informs all parts of their business, from branding to merchandising.

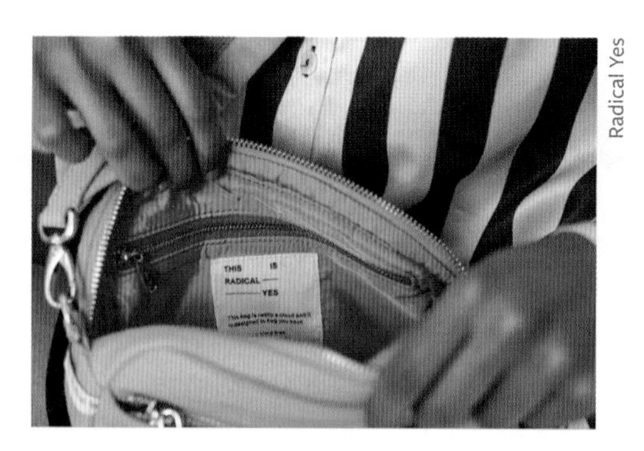

Radical Yes

The company slogans are a variety of phrases around the idea of 'movement', such as 'Hasten slowly'. They even grade their shoes as 1 km, 3 km or 5 km, based on their suitability for walking these distances.

The team

The team at Radical Yes consists of Kerryn and her partner Leo as well as a small team of in-house staff and contractors. Kerryn completed a Bachelor of Business while working in retail. Her love of fashion and retail, particularly the connection she had with customers, led her to pursue a career in the local fashion industry. She began with a role in apparel production before moving into product development in eyewear. This was at a time when Australian companies were beginning to work closely with factories in China, and Kerryn's role was focused on managing this relationship with Chinese vendors. This proved invaluable when she came to launch her own shoe company. Her partner in life and business, Leo, studied media and iterative design and communications before becoming a graphic designer, working mainly in advertising.

The skills, knowledge and experience that Kerryn and Leo had gained during the earlier phases of their careers, in design and production and in branding and graphic design, enabled them to start Radical Yes. It has also informed the roles that they undertake in the day-to-day running of their business. Kerryn manages product design and development, and Leo oversees art direction and marketing. They work closely together on the direction of the company and both have an equal say in the final shoe designs that are produced,

9780170477499

largely basing their decisions on what has or hasn't worked in the past and, critically, on feedback from their customers.

Kerryn and Leo are supported by an in-house team of seven, including retail staff at their shop in North Melbourne and back-office staff who manage e-commerce, inventory, merchandising and customer service, and operate their warehouse in nearby Kensington. They also rely on a team of contractors who specialise in CAD design, trend forecasting, digital advertising and search engine optimisation (SEO).

Low-volume production

After 10 years working for large-scale wholesalers (producing in the thousands of units), Kerryn wanted to produce things more slowly, with less urgency and in a more considered way. She wanted a 'radical' approach.

Sometimes only 100 to 200 pairs of shoes are made in one style. The shoes are made in small quantities for several reasons. One reason is being able to control things almost as they are happening, and make changes if and when needed. A small batch can be manufactured and brought to market, then customers wear the shoe, and all feedback is taken on board to feed into, and make changes in, the next batch of production. Quality, including fit and comfort, can also be improved on. They number their shoes like software by including a version number to indicate the batch and show the progression as they improve the shoe with each iteration – for example, their *Journey 3.0* boots.

As Kerryn explains, 'We are radical – because we let things percolate and take the time to educate our customers. They have been invited into the language of our design world and into our products, so they understand what we are doing'.

Customers also love that there aren't millions of the same shoe out there on the street.

Radical Yes uses factories in China that are small, family-owned businesses aligned with their values. The proximity of China to Australia allows transport by sea and lets them visit the factories as required. Radical Yes checks the factory conditions for things such as good lighting, cleanliness, safety set-ups, break times for workers – similar to Australian conditions.

Radical Yes

Production of Radical Yes *Saturn Returns* high-top sneaker

Design and production process

All shoes are designed by Kerryn and her team of collaborators in Melbourne. The first phase of the design and production is sketching and ideation, which can take between 2 and 4 weeks, depending on whether it is a new design or riffing off existing designs. It often involves reviewing previous collections and determining what they want to put in the new design, informed by trend forecasting and customer feedback. The designs are initially drawn by hand and then transferred into a CAD file.

In the second phase, Kerryn and her team submit the technical drawings and specifications to the factory in China in what is called a 'tech pack'. From time to time, the team may also supply the factory with an 'input sample' to help guide the overall direction of the shoe they are trying to create. This might consist of examples from previous Radical Yes shoes, and could also include a sample from another brand to demonstrate a particular aspect of the design – for example, a heel or the shape of the toe – that she wants to adapt. From this the factory can build a **last**.

Last
A solid form in the shape of a human foot, around which shoes are made

Radical Yes

Work-in-progress sketches and colour swatches from the Radical Yes design process

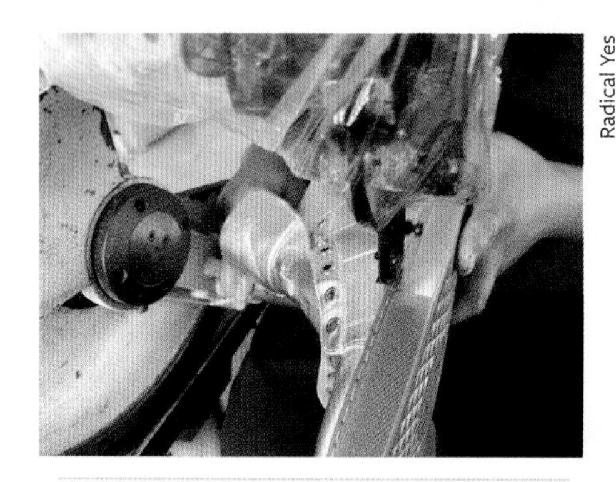

Radical Yes

Outsole stitching on a Radical Yes *Saturn Returns* high-top sneaker

The next stage is marked by the arrival of the first prototype, called 'Prototype 1'. The Radical Yes team use this prototype to do fittings and check that the pattern is the right shape and that the shoe looks right aesthetically. The 'P1 feedback' is then sent back to the factory.

While the next prototype is being created in China, Kerryn begins working on the colours. She creates drawings of the shoes in colours, which are sent to the factory to incorporate in the next prototype. This is called the 'colour up' stage.

When the final sample, or 'Prototype 2', arrives from the factory, Kerryn completes a final check of the shoe's design, including the colours, before production of the final shoes can begin.

Kerryn explains that, unlike larger companies that require at least one more stage to approve a final sample before production can begin, Radical Yes is able to go from prototype to production because of its low-volume approach.

Ideas come from ...

Kerryn and her team explore all sorts of areas to get ideas and colour schemes, such as a 90s nostalgia theme inspired by films, colours, style icons and popular products (e.g. Swatch watches) of the 90s era. The brand is also hugely inspired by its customers – how people use the products 'IRL' (in real life) forms an important part of the inspiration for new designs.

You can see the company's website for available shoes, sunglasses and bags.

Weblink
Radical Yes

Text written by Jacinta O'Leary. Interview with Kerryn Moscicki by Caroline Williams in September 2023.

Triple bottom line or 3BL

'Triple bottom line' is an expression, coined by John Elkington in 1994, that covers the three areas, or 'pillars', of sustainability and is now taught in many business schools. The three dimensions or areas of sustainability – environmental, economic and social – are also referred to as the three Ps: planet, profit and people.

In business, the term 'bottom line' refers to profits. Triple bottom line is a framework that encourages companies not only to focus on profits but to also consider social and environmental impacts – how to improve people's lives and the wellbeing of the planet. It promotes taking the full cost of doing business into account and being transparent (honest and clear) about it – all of which can enhance a company's reputation among its workers and consumers and also, often, its investors.

In manufacturing, water, air and land are crucial resources that are readily accessible, but belong to all humanity and should not be polluted or destroyed by companies prioritising profits. It can be challenging to gauge a company's impact on society and the environment, but its financial performance is easily measurable. Regrettably, this often results in financial gain being prioritised over social and environmental wellbeing.

9780170477499

Adopting the 3BL strategy may require a company to invest heavily, and may result in lower profits compared to other strategies that are more profitable but less socially or environmentally responsible. However, this approach can lead to a positive reputation and contribute to the long-term success of the business. Note that many companies shift production to less strictly regulated countries (offshore) to avoid the costs of complying with environmental regulations.

Here is a breakdown of what is involved for organisations in the three areas:

Worksheet
6.4.2

Planet – looking after our natural resources:

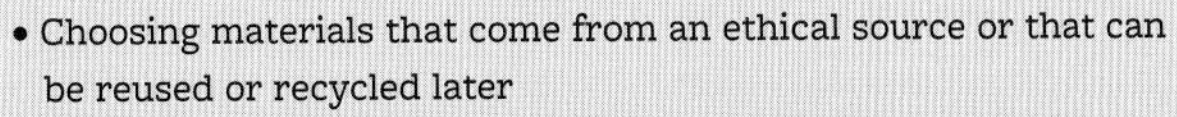

- Choosing materials that come from an ethical source or that can be reused or recycled later
- Reducing energy and water use where possible by using new and efficient technologies and efficient methods to distribute products (e.g. by train or boat rather than by air)
- Managing waste and pollution – following environmental regulations
- Reducing the consumption of resources (particularly those that are non-renewable)
- Reducing the distruption to, and destruction of, natural ecosystems
- Considering the needs of non-human creatures, plants and organisms

Profit – staying in business ethically:

- Managing and taking responsibility for long-term financial viability
- Avoiding unethical or unfair methods of making profits
- Honestly and accurately calculating and predicting overheads (e.g. operating costs, transport) and limiting financial waste
- Paying bills, taxes, costs and wages fairly and in reasonable time frames
- Ensuring quality and price are consistent across all regions
- Connecting with other 'ethical' businesses in the supply chain
- Investing in research and development that contributes to sustainability for the benefit of all

People – supporting the health of all people, workers and communities:

- Considering, and improving where possible, the living conditions, health (mental and physical) and social impacts on consumers, societies and workers throughout the whole life cycle of a product, from extracting raw materials, processing and manufacturing to use and disposal of the product
- Giving back to the community where possible with grants or public infrastructure, by investing in environmental programs or by being philanthropic (donating to causes)
- Ensuring workers receive a living wage, don't experience undue pressure to work long hours or in unsafe conditions, and have opportunities to be trained – which all contribute to creating loyalty and therefore retaining workers, and reduce the cost of hiring and training new employees
- Ensuring products are safe and not damaging to consumers' mental or physical health.

Weblinks
Triple Bottom Line: Sustainability in business

My Perspective on Profit

Corporate Social Responsibility (CSR) Explained with Examples

Investopedia Dictionary

- Watch a video animation explaining the triple bottom line.
- Watch the entertaining video 'My Perspective on Profit' by One Army in their 'Story Hopper' project, which makes educational videos.
- Another framework you might come across is corporate social responsibility (CSR). Many companies follow this to enjoy benefits beyond selling products. The Investopedia explainer on CSR includes information about **ISO** 26000: 2010, a set of voluntary standards (guidance for practical actions) meant to help companies implement corporate social responsibility.
- You can read a lot more about sustainable and ethical business practices, with examples, at Investopedia, by searching terms in their A–Z list.

ISO
International Organization for Standardization, which develops and promotes global standards

ACTIVITY

Triple bottom line

1 Draw your own graphic (Venn diagram or other type) to explain the triple bottom line.
2 Which one of the three elements of the triple bottom line was previously considered the 'bottom line'?
3 Explain how the bottom line has evolved into the triple bottom line and what the 3BL encourages businesses to do.

Worksheet
6.4.3

The 6Rs

The 6Rs are six words – rethink, refuse, reduce, reuse, repair, recycle – that can help designers, manufacturers, retailers and consumers make more sustainable choices around the design, use and purchasing of physical products. They can help us think about:

- how to reduce use of resources, particularly materials and energy (electricity, gas or from renewable sources), by firstly reducing what we buy or create
- ways in which products are used, maintained and disposed of.

All over the world, most products end up in landfill, many are incinerated (burnt), many litter the natural environment and only a small number are recycled and reused. When products go to landfill or are incinerated, all those materials used to produce them are gone and, in effect, wasted. More energy and materials are required to replace them. The disposal methods that create the least impact are those in which materials can be biodegraded (composted) or reused without any lessening of quality.

When applying the 6Rs, an important question that can be asked to consider a product's impact is, 'How soon will this product end up in landfill?' The longer it takes for a product to end up in landfill, the better for the environment; conversely, the shorter the lifespan of a product before it goes to landfill or is incinerated, the bigger the negative impact on the environment. The 6Rs have a similar waste hierarchy to a circular economy. See page 138.

9780170477499

6Rs: Rethink, Refuse, Reduce, Reuse, Repair, Recycle

Rethink

Rethink first whether a product is necessary, and secondly, how the product can be made, used and/or disposed of.
Question: Can the product be made differently in a more sustainable way?

Refuse anything that is not necessary, such as a whole product that can never be reused or that will never break down once disposed of, and that contributes to a negative environmental impact.
Also refuse a product made with materials (or that has parts made with materials) that cannot be reused, nor will break down.
Questions: Is this product really needed? Does it need to be made in this way or with these materials? Will it shortly end up in landfill?

Refuse

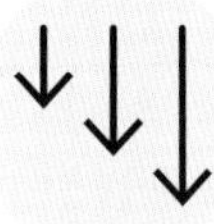

Reduce

Reduce the size of the product or the amount of materials used. Reduce the amount of energy, such as electricity, that is required when using the product. Reduce the amount of water required for its use - for example, when washing it.
Question: Can the size be reduced and the product be just as effective for its purpose? Can the product be designed to require less electricity or less water during its use?

Reuse a product and its materials (including parts or mechanisms) for a different purpose.
Question: Can I turn this product or some of its materials into a different product? Can this product be designed and made to last for many uses?

Reuse

Repair

Fix a product to extend its life. Design and make products that can be repaired, so that components can be replaced to keep the product in good working order or the exterior treated to keep it looking good. Allow consumers the 'right to repair' instead of mandating brand-certified repairers.
Question: If the product has a weak or vulnerable point, can it be repaired or fixed by the user?

Turn a material into a new material - for example, melt plastic bottles to form a different plastic. Make use of materials that have been recycled. Design/make a product from recycled materials.
Question: Can this product's materials be recycled or can I use recycled material?

Recycle

Note: most environmental groups say that a big issue with the emphasis on recycling is that it does not encourage producers or consumers to reduce their consumption. Instead it creates a false sense of security that waste can be easily dealt with by recycling, which is currently not the case: less than 10 per cent of all the plastics ever produced have been recycled.

Read more under the heading 'Negatives of repurposed plastics' on page 165.

Worksheet
6.4.4

ACTIVITY

The 6Rs

1 Define each of the 6R strategies, then divide them into what is more applicable for consumers and what is more applicable for you as a student designer/maker.
2 Draw an image of the type of product you intend to make for your SAT and annotate ways you could achieve each of the 6Rs (as a consumer or a designer).
3 Discuss the statement in the orange box on page 145 with classmates, and share your ideas and knowledge on why the 'sense of security that waste can be easily dealt with by recycling' is problematic. Bring the waste hierarchy idea (see page 138) into your discussion.

Life cycle analysis/assessment (LCA)

Life cycle analysis/assessment (LCA) is an internationally recognised, standardised technique (ISO 14040 series) for scientifically measuring a product's impact on the environment and/or human health over its total life cycle. The purpose of an LCA is to compare the impacts of two or more similar products (or products that have the same function) over the five life cycle stages to find out which product is more sustainable. An LCA can also measure whether sustainability improvements to a product have made significant reductions to its impact. However, a true certified LCA is costly, resource-intensive and time-consuming as it collects complex data for each stage of the life cycle, which needs interpretation. Refer to the table for the stages in a product's life that are assessed in an LCA.

LCA: Stages of a product's life that are assessed

Stage of a product's life	Description	Possible impacts
Sourcing, extracting and/or processing of raw materials	Taking/harvesting materials from the ground (mining), forests, crops, animals, petroleum, etc. Converting raw materials into forms that can be used for manufacturing by: • mining, melting and purifying metal • farming, harvesting and processing plant and animal fibres for fabric • logging, sawing, drying and dressing timber • purifying, heating and combining chemicals • processing petrochemicals into plastics	Degradation of land, air, rivers and oceans, which may affect the future economy of the area and impact on indigenous people, animals and plants Carbon dioxide emissions from the generation and use of energy from finite resources Pollution from toxic waste, herbicides and pesticides Social and health impacts on workers, nearby communities and their culture
Product manufacture	Making the products (including components that might be used in production)	Consumption of materials, energy and water (finite resources) Creation of waste and pollution, and possible degradation of surrounding land, air and water, which may have social/health impacts on the workers and their communities
Transport (through all its life stages)	Moving materials/products: • from their source to where they are processed; to the manufacturers; to distribution centres and retail locations • at end of life, when they are collected and taken to disposal sites Transporting materials/products by sea, by air, by rail or by road	Carbon dioxide emissions due to the use of fuel and the creation and maintenance of vehicles and infrastructure for transportation Creation of significant pollution and greenhouse gases

Stage of a product's life	Description	Possible impacts
Product use including repair and maintenance	Products may be static/inert but many function in different ways that: • require energy and fuel • need to be maintained • require regular replacement of parts such as filters • need to be washed regularly, which may require energy to heat water and/or use of detergents • may require drying machines • discard microfibres or molecules when used and when washed	Consumption of energy, fuel, cleaning/maintenance materials and/or water to make the product function or to keep it in usable condition Waste created when discarding parts Microplastic fibres from synthetic clothing that are shed during washing, and which end up in waterways and, inevitably, in the food chain
Product disposal	Disposing of the product at the end of its useful life, by methods such as: • reuse/repurposing • recycling • landfill • incineration	Reuse or repurposing of materials/products – this has the least impact Recycling of materials, which requires energy for processing and doesn't solve the issue of their consequent disposal Waste and pollution; increased area needed for landfill, which can emit toxic fumes or be a fire hazard; impacts on the health of people, animals and plants in nearby areas Incineration creates toxic air particles that spread

ISO 14040 and 14044 for LCA

You can read more about ISO 14040 (principles) and 14044 (requirements and guidelines) regarding LCA that belong to the 14000s 'family' or series of standards related to environmental management at the ISO website.

The ISO standard specifies that to achieve Type 1 classification of a product's LCA, a third party such as an environmental scientist is required to certify the process. The company can then market their product as having a certified LCA.

A true LCA (structured and scientific) assesses the impact of a product and compares it with other products made for the same purpose. Detailed research is carried out on all stages of a product's life – the sourcing and processing of materials, production processes and systems, transportation, product use and disposal. It accurately measures the **inputs** (materials, energy, fuel, water, etc.) and **outputs** (waste, pollution, emissions, etc.) at each stage.

Some LCAs also include an analysis of related human behaviour in their 'goal and scope' – e.g. how we interact with and use the product, or how we perform the activity – to form comparisons. For example, drying hands in public bathrooms might include a study of how much paper towel is used or how long people use the electric hand-dryer for.

Weblinks
ISO 14040:2006

KeepCup LCA

Australia Post Packaging LCA

Environmental impacts differ depending on the product

The level of environmental impact differs at each stage of an LCA, depending on the type of product being made. For example, a pair of jeans has some negative impact when the cotton they are made of is grown, cleaned and processed (large amounts of pesticides are used in the production of cotton). A pair of jeans can also have a significant environmental impact during their 'use' stage due to repeated washing and the amount of energy, water and detergents used. The level of impact depends on whether the jeans are washed in hot or cold water, as hot water greatly increases the amount of energy required, and whether they are line dried or machine dried. Many jeans are discarded due to changing fashions and end up in landfill.

On the other hand, the environmental impact of a piece of furniture is more evenly shared between the materials sourcing, production, transportation and disposal stages (though this varies depending on which materials are used, how they are processed, the distances travelled, how long the item lasts for and whether any of the materials can be recycled). Very few of the environmental impacts of a chair occur during its 'use' stage.

The length of a product's life also has an effect on its environmental impacts. For products that have a short life (i.e. those that need to be replaced frequently), the negative environmental impacts of the sourcing and disposal of materials are usually more significant than the impacts of its use. On the other hand, if a product lasts a long time (such as a vehicle or an electrical appliance) and uses fuel or energy when used, it may impact on the environment more during its 'product use' stage than during its materials processing, production and disposal stages.

Worksheet
6.4.5

Weblinks
Recycled Polyester Doesn't Fix Fast Fashion's Over-production Problems

The Principles of Life Cycle Assessment (LCA)

ACTIVITY

LCA

1. Create a flow diagram using cartoons or icons to show a non-scientific LCA of a product of similar type to the product you intend to make for your SAT. Include all the stages as shown in the table on pages 146–7.
2. Textiles students: research the use of polyester and recycled polyester in clothing, considering its LCA, its positives for clothing and its negatives for the environment. One article by 'Good On You' (a group, founded in Australia, that contributes to UN Sustainable Development Goal 12) is titled 'Recycled Polyester Doesn't Fix Fast Fashion's Over-production Problems'.
3. Watch the video 'The Principles of Life Cycle Assessment (LCA)' until you see the diagram of the stages involved in an LCA. Draw your own diagram based on materials you will use in your SAT and draw cartoons to help you remember the stages for your material type.

'Cradle to Cradle' concept (C2C)

Cradle to Cradle® (C2C) is a design framework created by an American architect, William McDonough, and a German chemist, Michael Braungart, and outlined in their 2002 book *Cradle to Cradle: Remaking the way we make things*. It is about eliminating the concept of waste – the idea that all waste can be reused or composted as food for biological and technical systems.

While LCA looks at a product's impact from **'cradle to grave'** (raw materials, processing, manufacturing and use through to its final disposal in landfill or incineration), **Cradle to Cradle** encourages a circular approach. To achieve this it also encompasses a Design for Disassembly (DfD) approach.

In the cradle to grave model, resources are consumed, and when it is no longer used the product is disposed of by methods that render all its materials useless – a process known as the 'take, make, waste cycle'. The Cradle to Cradle concept emphasises a plan for the end of life of a product, so that the materials continue to be reused in some way. It emphasises 'upcycling' rather than 'downcycling':

- **upcyling** – producing materials or products of equal or greater quality after each round of recycling
- **downcycling** – creating materials or products of lower quality, which eventually end up in landfill.

C2C aims to ensure that all waste created during all stages of manufacturing is useful in some way. Like the circular economy, it takes inspiration from nature's processes, where

9780170477499

materials are viewed as nutrients in a healthy, safe system. The systems referred to are either biological or technical:

- **Biological materials** (or biomaterials) are biodegradable and can be composted or consumed. They include cotton, linen, silk, wool, timber, paper and a tiny number of bioplastics (see more on bioplastics on page 161).
- **Technical materials** are reusable or recyclable, and include most metals, some chemicals and, to a lesser degree, oil-based plastics.

It is important, however, if they are to be used effectively once recycled, or to biograde completely, that any of these materials are separated and 'clean', i.e. not mixed or contaminated with different materials.

C2C also aims to:

- protect clean air and reduce harmful emissions by using renewable energy sources where possible, i.e. solar, wind or tidal energy
- safeguard clean water and healthy soils
- respect human rights and treat people fairly.

Weblinks
Cradle to Cradle

Get Certified

Embracing Circularity: Eagle Lighting receives Cradle to Cradle certification

Introduction to Cradle to Cradle video

- Read more about Cradle to Cradle on the EPEA website.
- Read more on how companies can get C2C certification on the Cradle to Cradle Products Innovations Institute website.
- For a case study on how Eagle Lighting was helped to be C2C certified, find 'How we helped Eagle Lighting' on the Thinkstep website.

C2C

1. Explain how C2C differs from the previous thinking of 'cradle to grave'.
2. With your SAT product, suggest a possible choice of materials that could allow your product to be C2C certified.
3. Watch the video 'Introduction to Cradle to Cradle' on YouTube.
   a. Explain to your classmates three things you learnt about C2C. You could use a short PowerPoint.
   b. Explain why continued recycling does not always work.

Worksheet
6.4.6

Design for Disassembly (DfD)

Design for Disassembly (DfD) uses the design stage to consider how parts and components of an entire product are joined and how they can be easily separated at the end of its life. It can also be applied to buildings. The intention is to minimise the loss of value at disposal (or demolition) and reduce the resources required to create new products, which can also reduce production costs. It has a lot in common with C2C and the circular economy and is also known as design for recycling or recovery (DfR).

The DfD design approach incorporates fast and easy methods to take apart a product for two purposes:

- to enable easy repair and replacement of parts or components to extend the product's life
- to enable the reuse or recycling of materials or components upon disposal.

To make reuse and recycling streamlined and efficient, the number of different types of material used has to be minimised. This is because materials can only be recycled when they are not contaminated or mixed with other materials. DfD must allow for the easy separation of materials that are incompatible (not matched) for recycling. This also requires labelling for easy identification of materials.

Extracting materials from DfD products

It is important to note that disassembling a product needs to be cost-effective. If disassembly takes too long or can only be done manually, this reduces the likelihood that the disassembly will happen unless it is going to yield a highly valued, rare or precious material.

Materials with different properties, such as magnetic and non-magnetic materials, can easily be separated by automation. Plastics with different densities can also be separated, and it is easier if the differences between those densities are bigger. Metals are generally easier to recycle, but it depends on whether they are 'plated' or alloyed. The different metals may not be compatible with each other, and the resulting 'mix' may not suit the purposes of the original metal. For example, mixing copper, tin, zinc, lead or aluminium with iron and/or steel can reduce their reliability.

Fabric products can also be disassembled and some fibres, such as cotton, wool and polyester, can be shredded and respun into useful yarns to be used in sports clothing, jackets, bags, furnishings, insulation or wadding. Currently this does not happen on a large scale but it is a growing industry.

Methods that allow for easy disassembly and are 'designed into' the product are shown here and on the following page.

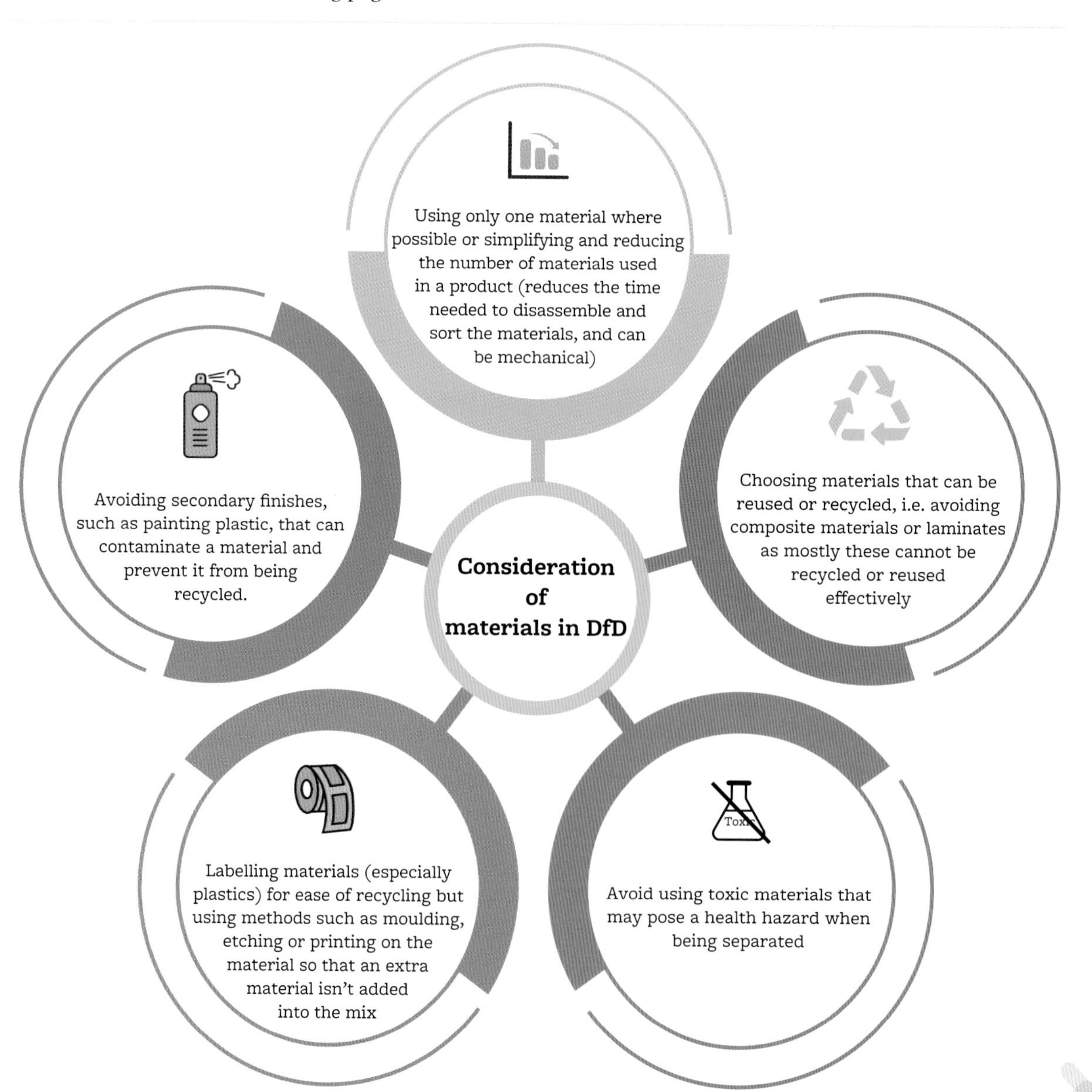

9780170477499

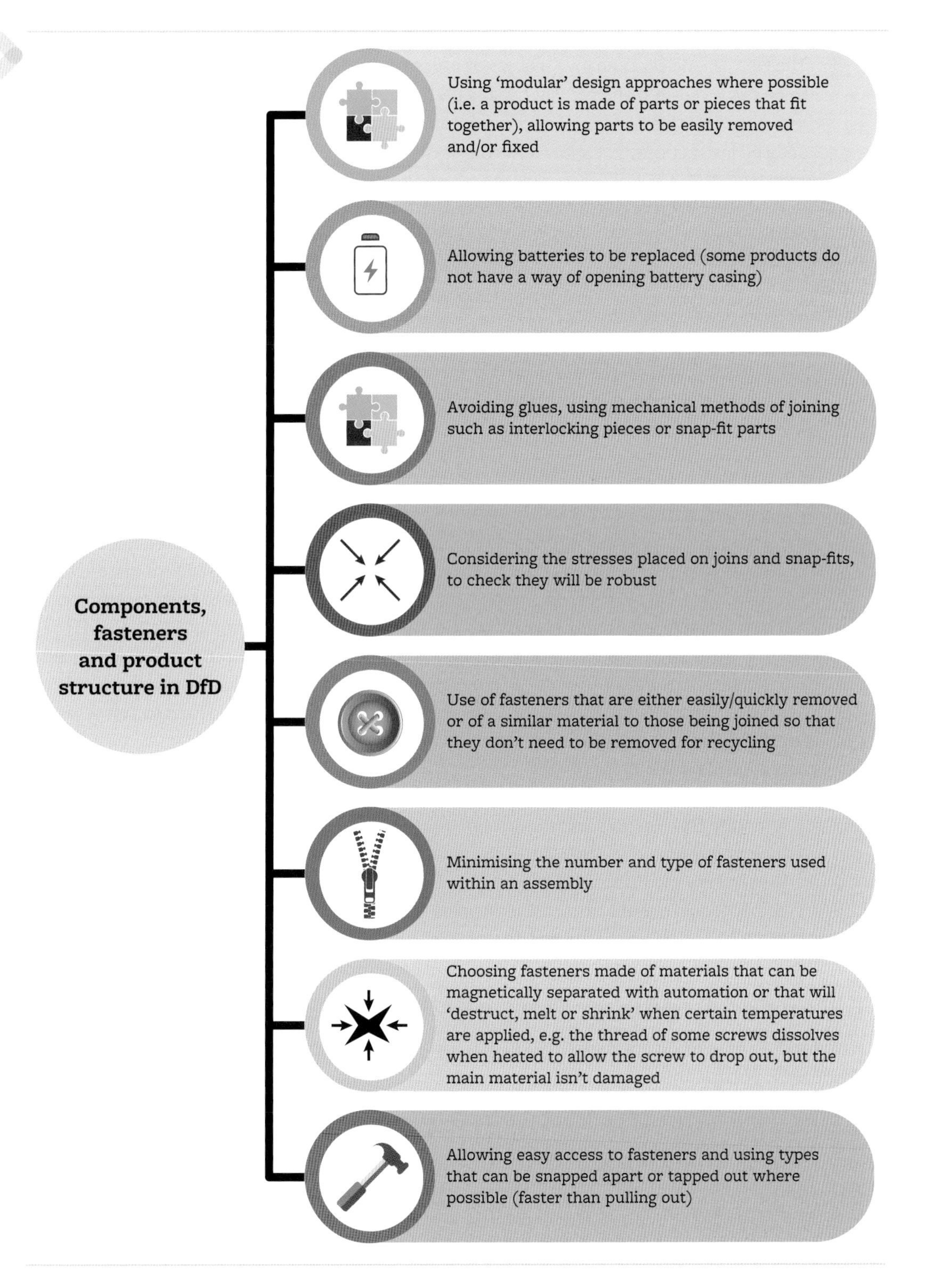

When using DfD, designers need to consider the environmental conditions that a product will be exposed to, such as moisture, humidity and high or low temperatures. The integrity and safety of the product should not be compromised by parts or fasteners that degrade in certain conditions.

Weblinks
Design for Disassembly Guidelines

BlockTexx

- Read more about DfD guidelines on the Engen website.
- Read about BlockTexx, a Queensland company of world renown turning textile waste into a commodity, at the company's website. BlockTexx uses AI to identify different fibres.

The Herman Miller company (makers of office furniture) designs all of its office chairs to be easily repaired and recycled. Many parts can be disassembled and replaced, using minimal tools, to extend the chairs' useful life. The company provides customers with detailed instructions that show how to disassemble for recycling – all parts are screw or snap joined. There is also a market for refurbished Herman Miller chairs.

Worksheet
6.4.7

DfD

Complete some of these activities.

1 Select three or four fasteners in your material category that you could possibly use for your SAT product, and research the materials they are made of and their properties. If put into mass production, how easy would they be to remove when your product is being disposed of? How could all the fasteners be collected? Which is the most suitable for the product? Which is the most sustainable? Justify which one you would choose and why.
2 Choose between two and four materials that you could use and research their ability to be reused or recycled once retrieved from a used product.
3 Most sports shoes are made of 50 different materials. Research your favourite sports shoe to see if you can find out all the materials used. Draw a rough diagram of this shoe and annotate it with ideas on how it could be made with fewer materials and with a DfD approach to its construction.
4 Watch the video 'Design for Disassembly' on YouTube and choose one product (from the video or from any product). Draw a diagram explaining how the product could/can be disassembled. Include some of the DfD design approaches explained in the video.
5 Watch the video 'How to Design for Disassembly and Recycling' on YouTube to see several approaches for DfD described and sketched in the first 3 minutes, and disposal considered in the remaining minutes.
   a Choose 2–4 of them that you have not covered already and do your own sketches. Annotate the sketches and show them to your classmates to see if they can understand the method.
   b Explain how the disposal of a product with Extended Producer Responsibility (EPR; see below) could benefit from C2C principles.
   c What is the difference between down-cycled and up-cycled materials, particularly plastics?

Weblinks
Design for Disassembly

How to Design for Disassembly and Recycling

Extended Producer Responsibility (EPR)

Extended Producer Responsibility (EPR), or Product Stewardship, is a strategy where producers (companies) are made responsible for managing the environmental impact of their products, particularly at the end of a product's useful life. They are expected to 'take back' the product from consumers when it is no longer useful and put it back into the manufacturing of new products. This strategy was developed to reduce the amount of waste and pollution going to landfill, limit the use of water and energy and lower greenhouse gas emissions during the manufacture of goods. It is a strategy that most often incorporates DfD, considers the LCA of products and contributes to the aims of a circular economy. Ideally, producers design more environmentally friendly products.

EPR or Product Stewardship accepts we all have a role to play in protecting the environment. The responsibility is shared between those involved in designing, manufacturing, selling (importing) and disposing of products, along with governments and consumers. However, only some countries around the world have EPR policies enforced by governments. In Australia, EPR is called a 'scheme' and it is usually voluntary or industry-led, but it is mandatory for some products, such as petroleum-based or synthetic oil.

Australia has a Product Stewardship Centre of Excellence. Read more and find case studies at the centre's website.

Weblink
Product Stewardship Centre of Excellence

Shutterstock.com/Ned Snowman

ACTIVITY

EPR

1. Choose a case study from the Product Stewardship Centre of Excellence and explain it on one PowerPoint or Google Slide.
2. Watch the video 'What is EPR (Extended Producer Responsibility)?' by Globalrec or 'Extended Producer Responsibility: How can it help recycling?' by Waster on YouTube. Outline where finance has an impact on EPR, and two other things you learned.
3. Explain how EPR differs from DfD.
4. Copy this table, explain and give an example of each of these words (durability, repairability, reusability and/or recyclability) in terms of EPR. Insert images of two products (and acknowledge their IP) to refer to.

	Durability	Repairability	Reusability	Recyclability
Explanation				
Example				
Images of products				

Worksheet
6.4.8

Weblinks
What is EPR (Extended Producer Responsibility)? by Globalrec

Extended Producer Responsibility: How can it help recycling? by Waster

EPR strategies for producers

Producers can use strategies such as:

- designing products for longevity
- avoiding the incorporation of parts that cannot be replaced or that need constant replacement, such as throwaway filters
- using DfD approaches such as labelling of different materials and modularity to enable reuse and/or recycling
- aiming for 'clean' manufacturing by:
  - designing products that produce less waste, use fewer resources, have less toxic components and contain more recycled materials that are in turn recyclable
  - using alternative energy sources such as wind and solar to reduce emissions
- using efficient transportation for all stages of a product's life
- limiting packaging
- educating consumers on:
  - how to maintain or repair a product with minimal negative environmental impact
  - where and how to dispose of the product safely when they are finished with it
- offering free 'take back' systems for disposed products.

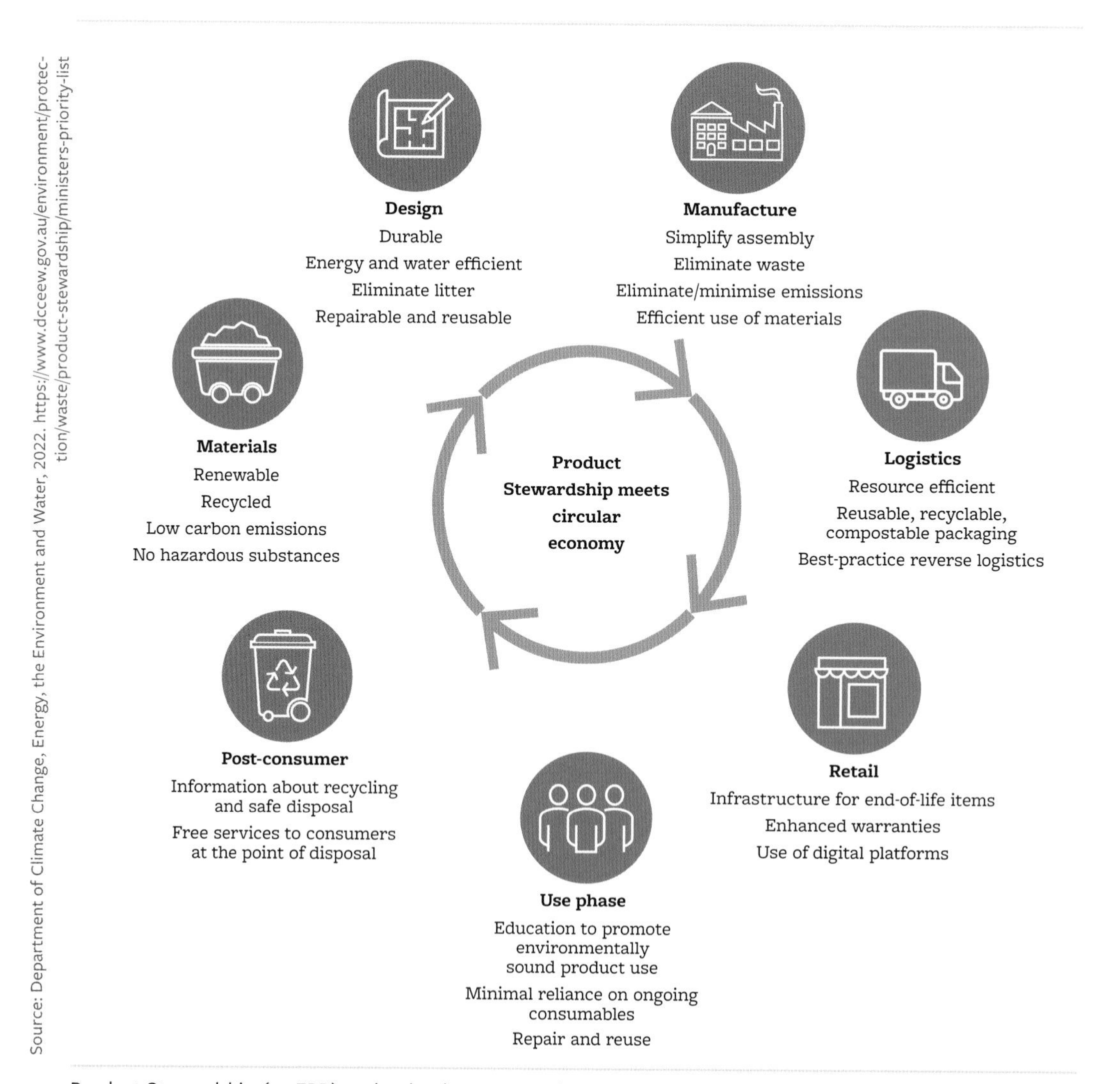

Source: Department of Climate Change, Energy, the Environment and Water, 2022. https://www.dcceew.gov.au/environment/protection/waste/product-stewardship/ministers-priority-list

Product Stewardship (or EPR) and a circular economy have a lot in common.

9780170477499

Role of government

There are many voluntary Product Stewardship policies and arrangements being undertaken across Australia, such as the collection and recycling of mobile phones, mercury-containing lamps, newspapers and PVC products. Many people feel that this approach should be enforced through legislation for many more products. One big issue is tracking the producers of products that are sold online from other countries, as selling in this way makes it possible to avoid having to comply with policies. Another issue is creating policies that don't cause a backlash, such as illegal dumping due to high costs.

Every year the Australian Government updates and publishes a list of products to be prioritised for EPR schemes. There are EPR programs currently in place for things such as tyres, TVs and computers, and batteries. To find out more and see the most recent lists, go to the Department of Climate Change, Energy, Environment and Water (DCCEEW) website and search for the Minister's Priority List for Product Stewardship.

Weblinks
Minister's Priority List

Extended Producer Responsibility

To find out more about the OECD's efforts, go to the EPR page on their website.

Influence of the sustainability models on design, manufacturing and marketing

Influence on design

It is said that 80 per cent of a product's impact on the environment and human health is decided at the design stage. All the strategies listed encourage careful choices at this stage:

- 6Rs – considers what is essential, carefully chooses materials and how to extend a product's life
- circular economy – considers how to avoid waste and encompasses C2C and DfD
- triple bottom line – takes social, environmental and economic sustainability into account
- LCA – chooses materials based on scientific research; considers all possible impacts (energy, waste, etc.) at the design stage
- C2C – chooses reusable or biodegradable materials so that products do not go to landfill; focuses on long-term useful products rather than fads
- DfD – restricts the number of materials, blends, laminates and paints in a design to allow easy recycling; gives access to battery compartments for replacement; incorporates snap-lock fit systems into the design for easy disassembly
- EPR – follows DfD principles; designs quality products with the ability to be repaired or modified in the future; considers how materials can be reused in future products.

Influence on manufacturing/production

Any strategy that is adopted for improved environmental impacts when designing a product will also give careful consideration to manufacturing methods, amount and sources of energy/water used and the waste/emissions created:

- 6Rs – aims for quality production for longevity
- circular economy – chooses efficient technologies
- triple bottom line – ensures production is of high quality; invests in safe conditions for workers and minimal impact on the environment
- LCA – considers carbon footprint (when choosing energy source) and chooses methods to reduce water use and general wastage

- C2C – favours the use of less toxic chemicals (in dyes, etc.); puts waste to other uses where possible
- DfD – eliminates glues; labels different materials for easy identification
- EPR – aims for quality production methods and reuse of waste; reuses or recycles returned parts and products.

Influence on marketing

When marketing products, a company can appeal to a consumer's desire to purchase sustainable products by:

- using scientific support (e.g. certified LCA, C2C and EPR) to verify sustainability status
- promoting products as having the potential for their materials to be reused or having been made from recoverable/recyclable materials that won't go to waste or end up in landfill (e.g. circularity, DfD, C2C or EPR)
- labelling to clearly identify the manufacturer, along with instructions on the repair, reuse or return of used products (e.g. DfD, C2C or EPR)
- making transparent, trackable supply chains available to consumers.

Transportation

Getting a product to consumers (transport) is another stage that can be carefully considered to reduce environmental impact. All companies that are interested will:

- develop efficient distribution systems with the aid of software (i.e. avoid backtracking and doubling up of transport)
- use transport that carries the most goods (or weight) with the least environmental impact, e.g. use sea instead of air, or rail instead of smaller vehicles.

Shutterstock.com/StockStudio Aerials

Worksheet
6.4.9

ACTIVITY

Sustainability strategies

1 Create a chart that shows the overlapping areas of the different sustainability strategies.
2 Work in pairs. Each student creates marketing slogans for each of the sustainability strategies on separate pieces of paper. Pair up with a classmate and see who can match the slogans with the strategy correctly in the shortest time. Discuss your results. Which slogan is hardest to understand? Explain which strategy is your favourite.

9780170477499

Greenwashing

Greenwashing is a negative term, indicating that an organisation is trying to gain 'green' or environmental credibility without deserving it. An organisation might create the impression that a product, a service or their brand is environmentally responsible when it is not (e.g. by deceptive packaging, manipulative or selective use of data, or meaningless 'green' certification), or it might focus on a small positive action of the organisation in an effort to divert attention from more significant negative impacts. Others make claims about sustainability that are not backed up by any evidence.

Greenwashing can mislead consumers about a product's attributes or an organisation's environmental practices in terms of sustainability. Unfortunately, this tends to make consumers suspicious of all sustainability claims.

An example of greenwashing is stating that packaging or products are compostable at home, when in fact many products labelled this way require mechanical and chemical assistance on an industrial scale to be composted. Likewise, there are no systems of collecting compostable and biodegradable plastics, so such products are put in the normal rubbish system and end up in landfill regardless.

Weblink
The Big Compost Experiment

Read more about this topic in the article 'The Big Compost Experiment' on the Frontiers website.

stock.adobe.com/HollyHarry

Planned obsolescence

What is obsolescence?

Obsolescence is when something (a product) becomes obsolete (no longer needed). Reasons for obsolescence may include that the product:

- is no longer useful or usable
- is out of date
- has broken parts that cannot be replaced
- cannot be used with current technology
- is replaced by another product that is more efficient.

Planned obsolescence is sometimes called 'inbuilt obsolescence' as it is built into the design, so that the product becomes obsolete within a few months or a few years. It is often driven by cost cutting in construction or material choices, resulting in products of lesser quality, which have shorter lives. Manufacturers have a vested interest in making a product that has a limited life: a consumer will then buy another one to replace it. However, this type of approach can be blamed for some very negative impacts on the environment and society. Consequently, more and more designers want, and are being encouraged, to think sustainably and to follow a strategy such as those outlined under 'Sustainability strategies and frameworks' in this chapter. Consumers also make demands for more sustainable products.

Worksheet
6.6.1

ACTIVITY

Obsolescence

1 Explain what 'obsolete' means.
2 Choose one sustainability strategy and explain how it would lead to a more positive outcome than a product with planned obsolescence.

Benefits and issues with planned obsolescence

	Benefits	Issues
For consumers	• Products are cheap/affordable • Products are easily replaced • Consumers can feel 'up to date' • Products can be more efficient and labour-saving	• Products don't last • Products cannot be repaired, or are difficult to repair • Replacement costs can be high • Can become 'out of date' quickly
For producers	• More products sold means economic benefits for manufacturers and retailers • Business can keep evolving • Staff can be 'up-skilled'	• Need to invest in re-skilling and new technology (both costly) • Constant attention to the product development process
Associated environmental, economic and **worldview issues**	• Increased employment and wealth for the community • Increased level of hygiene through use of disposable products (e.g. medical, food) • Many products make life easier and more enjoyable for people	• Increase in waste – more rubbish in landfill • Contributes to pollution of air, water and land • Uses up resources – materials, water, etc. • Requires a lot of energy

Worldview issues
Things that a person may believe to be fundamental aspects of life, such as what is good or right and what concerns different groups of people

There are three ways in which a product becomes obsolete:

- functional
- technical
- style.

Functional obsolescence

Functional obsolescence is what results when manufacturers choose lower quality materials and construction processes, knowing that the product will be less durable. One reason for it is the belief that there is no point creating a high-quality, long-lasting 'casing' or structure for a product because the underlying technology will be constantly evolving. The predominant reason is cost – materials that are more durable are often more expensive to source, to work with during production and to transport. In the past, washing machines and refrigerators had an effective service life of 20 years or more. Now they are expected to last for 5 to 10 years.

9780170477499

Some products are designed in a way that makes them difficult to maintain and repair. Clothing may be stitched carelessly and seams placed too close to the edge of the fabric so that the fabric rips and is deemed not worth fixing. A household item such as a toaster might have a plastic knob that breaks off and costs more to repair than a new toaster; or the workings of a watch (or the battery casing of a product) might be totally sealed during manufacture so that it can't be repaired once it breaks down or when the batteries go flat. Some products require specialist tools to pull apart. This shortens the life of a product. Such products become disposable – they cannot be repaired and must be thrown away if something goes wrong.

In terms of future sales and the company's reputation, a certain level of quality and durability must be maintained, otherwise customers will not feel they are getting value for money. It is a fine balancing act.

Disposable products or single-use products are in their own category but could be considered to be subject to 'functional obsolescence' as they are not durable, and can't be repaired or reused.

Shutterstock.com/Denis Galushka

Technical obsolescence

Technical obsolescence occurs when products become obsolete once a new technology or design feature/function is developed. Examples of new technologies that change products and make others obsolete are the microchip, small batteries, artificial intelligence (AI), sensors, LEDs, cloud computing and data, 3D printing, optical fibre, wi-fi and electrical powering of vehicles and machinery. Innovation, when adopted in new products, can create obsolescence in older products.

A product may still function well, but may need to be replaced because of a technological advancement, i.e. the systems that operate them are no longer supported, or they are replaced by newer systems or infrastructure (e.g. 3G, 4G and 5G networks), and therefore a new product is needed for the added features or capabilities. The casings of some electronic or digital devices are made cheaply, as noted above, because the technology within them is likely to be surpassed within a very short time.

Style obsolescence

Style obsolescence relates to the changing nature of trends and fashions. Products go through regular changes in appearance and style, which encourages people to replace the older-styled product with something more fashionable.

Different types of products have fashion cycles of differing lengths. Clothing fashions change every six to 12 months, but for products with a longer life – such as furniture – fashions and trends change more gradually and over a longer period. Trend, fashion and colour forecasting are tools that some designers use to make their designs current and desirable but that also make them go out of date quickly.

The impact of planned obsolescence on sustainability

Despite the benefits from planned obsolescence, there are environmental, social and economic costs:

- **environmental** – more products are created, which means more resources are used, more energy is required for production and a lot of waste is created. Millions of discarded products need to go somewhere, and that is usually landfill
- **social** – people often buy more than they need due to the pressure to keep up or due to 'never to be missed' prices. The results can mean people's houses are crowded with 'stuff' while their credit card debts build up and add stress to their lives. Communities living near sites of environmental degradation can have their health, daily life and income impacted
- **economic** – there is a cost in the need to replace things regularly with something new. The low cost of a lot of products headed for obsolescence may deter people from purchasing higher quality, more expensive products that would last longer. The cost to the environment is often ignored or left to be dealt with at a later time. This can have a negative impact on communities and may make it difficult to continue to profit from certain practices that environmental damage makes unsustainable.

Weblink
No More Cheap

A video by One Army called 'No More Cheap' explains the consequences of 'cheap' products and suggests another way.

9780170477499

ACTIVITY

Researching types of obsolescence

1 Find two examples of products for each of the three types of obsolescence and explain why they are obsolete or will become obsolete. Be sure to acknowledge the IP.
2 Outline one benefit of each product for the consumer and one for the producer.
3 What are some of the worldview issues relevant to these products? Refer to the table on page 158.
4 a Watch the video 'Top 10 Products that are Designed to Fail' (10 minutes) on YouTube.
  b Choose one product that is mentioned in the video for each type of obsolescence. Choose products that would be suitable for Product Design and Technologies (i.e. not textbooks, printer ink or light globes). Notice that some of the language is different, e.g. 'shoddy construction' belongs to functional obsolescence.
  c Explain the product, which type of obsolescence you think would bring about its diminished usefulness and why you think so.

Worksheet
6.6.2

Weblink
Top 10 Products that are Designed to Fail

Experimental and alternative materials

The world of materials changes almost daily. New materials are constantly being developed by scientists, industry and individuals all over the globe. Some of these developments are radical and innovative, while others are incremental.

For information about a wide range of new and innovative materials, visit the Material District website and database. Many of these materials use waste resources, or reinvent the use of traditional materials.

Weblink
Material District

Experimental materials

Bioproducts

A bioproduct is a type of material derived from living organisms or their by-products. Examples of bioproducts relevant to product design are biomaterials and bioplastics made from renewable resources such as soybeans, corn or potato starch.

Positives of bioproducts

- They are usually made from renewable sources and require less energy to make.
- If made entirely from plants or natural materials, many of them can be composted at the end of life.

Negatives of bioproducts

- Biodegradable polymers (bioplastics) use plants, which may have been sprayed with pesticides. These toxic chemicals can remain in the finished product. Biodegradable polymers actually have an extremely long biodegradation rate and are expensive to produce.
- Some bioproducts are more durable or are mixed with other materials to make them last longer and are therefore not compostable or recyclable.

Mycelium

Mycelium is the mass of white thread-like root matter of fungi. It lives mostly underground. In edible fungi, it produces the fruit buds that we eat as mushrooms. It creates a network that has:

- digestive qualities to break down organic matter in soil into smaller molecules to feed itself
- bonding qualities that 'glue' the organic matter together into a felt-like structure.

Mycology or 'myco' is a branch of biology that studies the structure, function and behaviour of fungi, and how they interact with ecosystems. There are millions of species that live on air, water or land. In nature, fungi play a vital role in decomposing plant matter such as fallen branches.

Mycelium composites

Mycelium can be used in laboratories to 'grow' biocomposites. A substrate, which is a replica of soil and matter, is inoculated with fungal spores and left to grow. Various materials can be used in the substrate, including agricultural waste and shredded textiles, but they need to be sterilised. Woody materials that are high in cellulose are favourable as they can contribute to high tensile strength in the composite.

The fungi digest the material and bond it into a new material. Depending on the type of fungus and the growing conditions, long strands like bootlaces or flat sheets like mats can be 'grown'. These are then cooked or dried to stop growth. The new material is light, strong, fire-resistant, static-resistant, sound-insulating and water-repellent.

Mycelium can be used to create materials similar to leather (Mycoworks, Mylo) and lightweight packaging (Mycofoam).

Positives of mycelium composites

- They are compostable by adding water, but this is dependent on any polymers added and the conditions (although not a great deal of research has been done on some mycelium composites).
- They can reduce waste by regeneration, i.e. turning something like fabric scraps into a new material.
- Production uses less energy than synthetic composites, as only space, water, containers and darkness are required.
- They are strong but lightweight.
- They can be shaped into almost any form.

Negatives of mycelium composites

- They may not have suitable properties, such as durability or elasticity, to replace plastics or synthetic composites until further research and investment occurs.
- They are not yet widely available as they are still in the early stages of development.
- They can be slow-growing – depending on the species of fungus, it may take between two weeks and two months to grow a new mycelium material.

Shutterstock.com/Y.P.photo

9780170477499

See the case study 'Fungi Solutions' on page 174.

Weblinks
Biohm home page

Fungus: The Plastic of the Future

The Making of BMW Materia: The Garden of Possibilities

- Read about Biohm in London at the Biohm website. From the home page, scroll to 'BioMaterials', click on the title and scroll to watch a short video explaining what Biohm is doing with mycelium in building construction. Under 'Materials', go to 'Mycelium' and scroll down to click through the great qualities of the product.
- Watch an 11-minute video on turning fungus into plastic in the Netherlands.
- Take a look at the Materia lab inside BMW Designworks to see the work being undertaken to create new materials for cars such as the iX model.

Research polystyrene and mycelium for packaging

1. Research and compare the use of polystyrene and mycelium-based foam material for packaging. Outline the main advantages of mycelium-based packaging.
2. As a class, consider purchasing a mushroom kit, research how to create a substrate, and watch it 'grow' and record your observations.

Worksheet
6.7.1

Weblinks
Ray of Hope

Videos of Biomimicry in Action

Biomimicry – the ways in which nature solves problems – has long been a source of inspiration for innovation in material developments. Visit the Ray of Hope Prize Gallery for information about the latest innovations in biomimicry.

Innovative polymers used for 3D printing

3D printing creates an object by building up layer by layer of 'thread' or melted plastic with directions from a digital file. This technology continues to improve and meet different needs. See previous heading on rapid 3D prototyping on page 132.

Innovative polymers that are being used or developed for 3D printing include:

- biodegradable polymers – designed to break down when they are exposed to heat, moisture or enzymes. They can be disposed of in an environmentally friendly way
- conductive polymers – designed to conduct electricity. They can be used to print sensors and antennas and other objects that are able to perform electrical functions
- shape memory polymers – designed to create structures that can remember their original shape and return to it when exposed to heat, moisture or other conditions
- self-healing polymers – designed to repair themselves when damaged or broken, so that the product is more durable and able to withstand wear and tear
- polymer-bonded rare-earth magnets – using a composite of polymer and magnetic filler to create strong magnets that are protected from corrosion and can be customised in shape
- bio-compatible polymers that are safe to use in medical and surgical situations
- fibre-reinforced polymers (e.g. using carbon fibre) that increase the strength of the finished printed piece.

Positives of innovative polymers in 3D printing

Like all 3D printing, additive manufacturing using polymers can require less material than subtractive manufacturing, reduce waste, and create customised products on demand with desired properties.

Negatives of polymers in 3D printing

There are significant sustainability issues with the polymers used for 3D printing:

- Polylactic acid (PLA) plastic filaments are labelled as being compostable or biodegradable, as they are made from renewable sources such as corn or potato starch and sugarcane. However,

Weblink
What is PLA? (Everything You Need to Know)

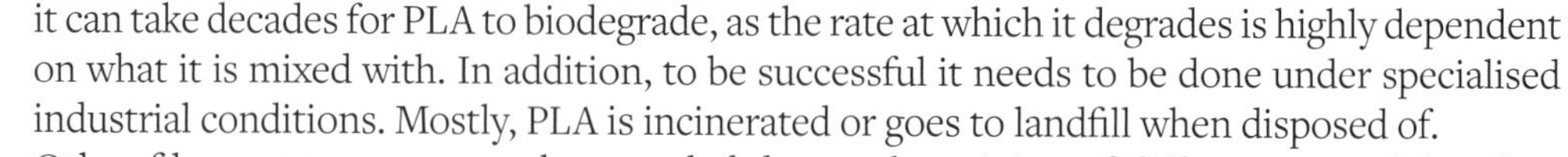

it can take decades for PLA to biodegrade, as the rate at which it degrades is highly dependent on what it is mixed with. In addition, to be successful it needs to be done under specialised industrial conditions. Mostly, PLA is incinerated or goes to landfill when disposed of.
- Other filament types cannot be recycled due to the mixing of different materials. They too usually end up in landfill or are incinerated.

Read more on PLA at the TWI Global website.

Worksheet
6.7.2

ACTIVITY

3D printing for your SAT

1. Identify and list two possible parts of your intended SAT product that could be 3D printed.
2. Research different polymers available to you (or your school) and their suitability for the two parts.
3. Research possible online services for 3D printing if this technology is not available at your school (or home).

Plastics

Plastic is a versatile material that is valued for its many properties, which include flexibility, waterproofness, washability, light weight, ability to be easily coloured, food safety and mouldability. One very important property that plastic has is durability: it takes hundreds of years to break down. However, there is growing global concern about the increasing amount of discarded plastic and microplastics in the environment and their impact on our waterways, oceans, and both marine and land animals.

Microplastics are the result of plastics degrading and breaking down into very small pieces – smaller than 5 millimetres. Microplastics have been detected in remote areas of the world, even in areas where there are few humans. A significant amount of plastic ends up in landfills or is burnt, releasing molecules into the air that have been found in the milk, bloodstream and faeces of mammals, including humans. Research in Italy on human breast milk in 2022, published in the journal *Polymers*, found the presence of microplastics composed of polyethylene, PVC and polypropylene, all used in packaging. The researchers could not analyse particles smaller than 2 microns, or millionths of a metre. Scientists fear that smaller, nano-sized plastics are likely more prevalent and toxic but currently have no reliable way to measure them.

Shutterstock.com/luckakcul

9780170477499

- Read more on microplastics in the article, 'Microplastic Found in Human Breast Milk' on the Star Health website.
- A half-hour video by Patagonia called 'The Monster in Our Closet' talks about the huge problem of synthetic clothing.
- Watch the video 'Backstage Paradise' about plastics in the sea, focusing on the Maldives, on the Story Hopper website.
- Watch a TED-Ed animation on the history of plastics called 'A Brief History of Plastic'.

Weblinks
Microplastic Found in Human Breast Milk

The Monster in Our Closet

Backstage Paradise

A Brief History of Plastic

Repurposed plastics

Repurposed plastics are plastics that have been recycled or recovered from waste and used for a new purpose. They can be mixed with bitumen for road surfaces or used for packaging materials, textiles, construction materials and industrial products. To be processed, they need to be collected, transported to a special facility, sorted and processed into a usable form. Sometimes they are combined with other materials to create composite materials.

Positives of repurposed plastics

- Using repurposed plastics can help to reduce the amount of plastic waste that ends up in landfill and the environment.
- They can reduce the need for processing of virgin plastic.

Negatives of repurposed plastics

- There is a lack of infrastructure and funding for plastic recycling programs.
- At present, only about 9 per cent of plastics are recycled globally (World Wildlife Fund).
- There is a lack of awareness and education about plastic recycling and the need for separating different plastic types because they cannot be recycled when mixed.
- No significant efforts are being made to reduce the use/purchase of plastics.
- The belief that discarded plastics can easily be recycled or reused leads to complacency.
- There is a lack of regulation for controlling/limiting the use of plastics in commercial enterprises.
- The value (and quality) of recycled plastic is low compared with virgin plastic, and demand for repurposed plastic products is also low.
- Recycling plastic consumes energy and has a high cost.

Shutterstock.com/Dorothy Chiron

Weblinks
Is Bioplastic the 'Better' Plastic?

Biodegradable and Bioplastics Explained

- Watch the short video, 'Is Bioplastic the "Better" Plastic?' on the DW Planet A YouTube channel.
- Watch the video 'Biodegradable and Bioplastics Explained', an entertaining look at the big issues of so-called biodegradable plastics, on the One Army YouTube channel.

Worksheet
6.7.3

ACTIVITY

Researching plastics

1 Research products that could be made from repurposed plastics. Keep in mind that the repurposed plastic may not be of similar quality to 'new' plastics and may not be considered 'food safe'. Make a one-page presentation on two products and explain why repurposed plastics would be a suitable material for them.
2 Search the internet for infographics on microplastics. Choose one that appeals to you and roughly redraw it with labels to explain where microplastics come from and where they end up.
3 Research the following:
   a the first plastics made of milk
   b the introduction of petroleum-based plastics
   c the use of plant-based bioplastics
   d possible fungus plastics for the future.
4 Create a diagram to illustrate this timeline, annotating the positives and negatives of each type of plastic. You may find the video 'Is Mycelium Fungus the Plastic of the Future?' on the Undecided with Matt Ferrell channel on YouTube helpful.

Weblink
Is Mycelium Fungus the Plastic of the Future?

Composite metals (metal matrix composites)

Composite metals, or metal matrix composites (MMCs), combine two or more metal elements or have another material added such as a ceramic or a fibre. This is done to give improved properties, such as increased strength, hardness or stiffness, lower weight, wear resistance and corrosion resistance.

There are two main types of composite metals:

- **reinforced composite metals** – made by adding a reinforcement material, such as fibres or particles, to increase its strength and stiffness
- **dispersion-strengthened composite metals** – made by adding a second metal or a non-metal element in the form of fine particles or a thin layer to improve strength, hardness and wear resistance.

Positives of composite metals

Composite metals are used when a combination of strength, stiffness and wear resistance is required. They are typically lighter than solid metals, so using them can reduce the weight of structures.

Negatives of composite metals

Recycling composite metals is problematic due to the need to separate out the pure metal. This can be complex and costly, and globally it occurs at a very low rate. Mostly, composite metals go to landfill when disposed of. Read about the issues with mixing metals under the heading 'Extracting materials from DfD products' on page 150.

9780170477499

Shutterstock.com/teh_z1b

Many contemporary building exteriors feature composite metals.

Composite timber (engineered wood)

Composite timber is made from a combination of sawdust, wood chips and wood shavings joined with a plastic binder (e.g. polyethylene or polypropylene).

Positives of composite timber

Composite timber is suitable for outdoor applications as it is more durable than natural timber, more resistant to moisture, rot and insects, and requires less maintenance.

Negatives of composite timber

Composite timber is expensive and is not biodegradable. When disposed of, it goes to landfill or is incinerated. Some of the binding agents (i.e. glues) used may cause health problems.

ACTIVITY

Researching composite materials

1. Research a composite material that you can purchase a sample of.
2. Once you receive the sample, compare it with a raw material and briefly describe the differences. If it's not possible to purchase a sample, insert images of a composite material and a raw material into a document (acknowledging IP) and annotate the differences.
3. Justify which material you believe would be more sustainable, including whether you believe it is the most suitable for a product similar to your SAT product.

Worksheet
6.7.4

Alternative materials

Vegan leather instead of animal hide?

Traditional leather from animal hide has long been a luxury favourite used to make jackets, bags and shoes.

Positives of animal hide leather

Animal hide leather is extremely durable and long-lasting, is mildly flexible and moulds to the wearer, can be coloured and is water-resistant, windproof and warm. When it is disposed of, it usually biodegrades, depending on how it has been treated or coloured. Many leathers are a by-product or co-product of the meat industry.

Negatives of animal hide leather

On the negative side, it comes from animals, which require breeding and pasture (more land is required to breed animals than to grow crops), and its processing, called tanning, requires a lot of harsh chemicals. These chemicals impact the environment and the health of tanning workers.

Vegan leather

Vegan leather can be made from two types of source material:

- petroleum-based plastics – most commonly, polyvinyl chloride (PVC) and polyurethane (often called synthetic leather, faux leather or pleather)
- plant-based materials.

Petroleum-based vegan leather does not use animal hide, but it produces toxic pollution and waste that are harmful to both animals and humans. Since 2003, Greenpeace has been warning that PVC is the most hazardous plastic in existence. It is rarely recycled and it does not break down in the natural environment.

Plant-based materials are more environmentally friendly and, as might be expected, are biodegradable, though this depends on what is added in creating the 'leather' to stop it breaking down while in use. Often the plant mixture has a polyurethane or petroleum-based coating, which renders it unable to biodegrade or to be recycled at the end of its life.

Examples of vegan leather from plant-based materials are:

- Piñatex – a mix of pineapple leaves and plastic polymers, which is not biodegradable
- Desserto – made from the leaves of the nopal cactus, and 'partially biodegradable'
- Demetra, by Gucci – includes 'bio-based polyurethane', viscose and wood compounds
- Pellemellah (Italy) and Veerah (India) apple peel leather – 50 per cent waste apple peel and 50 per cent polyurethane
- Myclo – made from mycelium, the long, white thread-like roots of fungi
- Mylo – from the mycelium of shiitake mushrooms
- Muskin by Grado Zero (Italy) – made from a large parasitic mushroom. It is soft but very delicate and absorbs moisture, so needs a wax coating
- Reishi by Mycoworks (USA) – made from mycelium and wood fibre.

Other plant sources for vegan leather are cork, coconut husks, kombucha starters, mango, algae/seaweed and apple peel.

Alamy Stock Photo/Stephen Chung

The *Stan Smith Mylo*™, a concept shoe by Adidas, uses Mylo vegan leather.

9780170477499

Positives of vegan leather

Plant-based or mycelium-based leather has a smaller impact on the environment during the sourcing of raw materials and processing. The raw materials are often plant waste.

Weblinks
All You Need to Know About Vegan Leather

Vegan IP

What Types of Plant-based Leather Exist Today?

Negatives of vegan leather

Vegan leather is not as durable or as comfortable as traditional leather. It is usually mixed with polymers, which means it cannot be composted (won't biodegrade) nor is it recyclable. It takes hundreds of years to break down and will end up in landfill or be incinerated.

ACTIVITY

Vegan vs animal leather

1 Watch a video about (or research) a vegan leather made with one of the materials listed above and explain how it is made and what it is used for.
2 Name a product that is often made from leather. Outline two negatives for each leather (animal and plant-based) and suggest a completely different material that is more environmentally friendly and could be suitable for this product.

Worksheet
6.7.5

Bamboo instead of hardwoods?

Most hardwoods come from trees that have taken more than 100 years to grow, and they are valued for their density and durability. Extracting hardwoods from forests can be harmful to the environment, leading to soil erosion, habitat destruction and other negative impacts. However, products made from hardwood tend to last a long time and can easily be reused once disposed of. Hardwood also biodegrades, which means it should not go to landfill where it will create greenhouse gases as it breaks down.

Bamboo is a type of grass that grows very fast (as much as a metre in one day) and regenerates so easily that in some locations it becomes a pest. There are many species, and their stems are of various lengths and diameters. The stems (known as culms) are hollow but they are super strong and lightweight. If used straight from the plant, bamboo is very sustainable, but the uses for unprocessed bamboo are limited due to the tubular shape of the stalks or stems. For many uses, such as floorboards, bamboo needs to be processed. After drying, the bamboo is cut into thin strips or shredded into fibres. These are glued using urea-formaldehyde, phenol-formaldehyde or soy-based adhesives, then cut and shaped to size.

Formaldehyde is a known carcinogen, and some bamboo manufacturers use melamine, which is also toxic. As with any dust, the dust produced by cutting or shaping bamboo 'boards' is a safety risk for people who are working with it.

Bamboo grows mostly in tropical climates. The distance the bamboo has to travel also needs to be taken into account.

Bamboo flooring

Bamboo as fibre?

Global Organic Textile Standards don't give certification to bamboo textiles due to the chemicals involved. Turning rough bamboo stems into a usable fabric requires an intensive and chemical-heavy process, and it is then treated with dyes, bleaches or formaldehyde. The end product is a soft rayon-like fabric. There is a less environmentally damaging method of processing bamboo into fibre, which involves mechanical methods such as shredding, mashing and combing, but it is less common.

Weblink
Truth or Trend: Is bamboo sustainable?

Read more about bamboo and the problems with monoculture crops in the article 'Truth or Trend: Is bamboo sustainable?' on the Eco and Beyond website.

Positives of bamboo

Hardwood and bamboo are both renewable and absorb carbon dioxide when growing. Both are durable, stable and long-lasting in their natural form. Bamboo as a raw material grows very quickly and continues to grow even after harvesting. Hardwood trees take at least 60 years, and in some cases hundreds of years, to mature. Harvesting of hardwood contributes to deforestation and loss of habitat, and it requires energy, which is mostly produced from fossil fuels.

Negatives of bamboo

Transporting bamboo over long distances produces atmospheric carbon dioxide. Processing bamboo into flooring or furniture requires significant amounts of energy. Bamboo can be susceptible to scratches, dents, cracking and warping over time and doesn't offer the same diversity of colour and grain patterns as hardwood. The quality of bamboo varies greatly. Some processed bamboo boards release harmful **volatile organic compounds (VOCs)**. In some cases, people grow bamboo on land that was once covered by native hardwood forest, replacing wildlife habitat with a 'bamboo forest' where wildlife cannot survive.

The creation of bamboo fibres for clothing and textiles involves intensive use of harsh chemicals.

Volatile organic compounds (VOCs)
Harmful gases released into the air from some materials and products. They are considered one of the contributors to sick building syndrome.

ACTIVITY

Bamboo vs hardwood

Think of a product that could be made with bamboo or hardwood and explain, with three reasons, which material would be the most suitable. Use annotated visual methods if it helps.

Worksheets
6.7.6

6.7.7

ACTIVITY

Impacts for producers and consumers of alternative materials

Read all the negatives and positives in this chapter regarding the use of alternative materials. Use your own reactions to decide what are the benefits or drawbacks of each for both consumers and producers. Draw up a table with examples to show your decisions.

Alternatively, find a product made with either bamboo (as a timber) or vegan leather and outline its impacts on consumers and producers.

9780170477499

Summing up

Impacts of technologies

The use of advancing and evolving technologies in production processes can have a number of impacts on both consumers and producers.

Positive impacts for consumers

These can include:

- **lower prices** – technology can help to automate production processes including with the use of robots (increasing productivity and efficiency in production), which can lead to lower costs for businesses and can then be passed on to consumers in the form of lower prices
- **increased variety/more choice** – new materials, new products and manufacturing processes can lead to increased variety for consumers. For example, 3D printing has made it possible to create customised products that were not possible before
- **improved quality** – technology can help to improve the quality of products by reducing defects and ensuring consistency. This can be seen in the use of robots in manufacturing, which can perform tasks more precisely and consistently than humans. The use of sensors in manufacturing can also help to ensure that products are produced to a high standard.

Positive impacts for producers

These can include:

- **increased productivity** – technology can help to automate tasks, which can lead to increased productivity for businesses. This means that businesses can produce more products with the same amount of labour, which can lead to higher profits. For example, the use of automation in manufacturing has helped to reduce the amount of human labour required to produce goods
- **reduced costs** – technology can help to reduce costs by automating tasks and improving efficiency. This can free up resources that can be used to invest in other areas of the business, such as research and development
- **new markets** – technology has also helped producers to reach new markets. For example, online retailers can sell their products to customers all over the world
- **improved safety** – technology can help to improve safety in the workplace by automating dangerous tasks and reducing the risk of human error. This can lead to a healthier and more productive workforce
- **improved innovation** – technology has also helped producers to innovate and develop new products. For example, the use of 3D printing can help producers to create custom products that meet the specific needs of their customers.

Possible negative impacts of technology

Of course, there are also some potential downsides to the use of technology in production processes. This topic includes negative impacts for both consumers and producers.

An important and very large negative impact is the greenhouse gas emissions created by digital products, 'connected things' and other software-related activities. This was estimated to account for 5 per cent of global emissions in 2023 and is predicted to rise to 14 per cent by 2040. Another negative impact is the large amount of energy required to run massive data storage centres (or hyperscale server farms) and the large volume of water required to cool them, especially on hot days. Data storage, and access to it, is an essential aspect of artificial intelligence. This 'online traffic' is expected to increase exponentially as more users adopt AI to complete tasks.

Benefits of the industrial use of technologies include:

- lower prices – often enjoyed by wealthy consumers
- new markets – often enjoyed by large producers, not small businesses.

It is important to note that the distribution of benefits might be very uneven. There is an argument that any development or adoption of technology in the world creates and contributes to inequality, i.e. a bigger gap between rich and poor. This is due to:

- unequal access to technology
- lack of digital literacy and skills
- concentration of technological resources in certain regions of the world.

Other examples of negative impacts are:

- **job losses** – as technology automates more tasks, it can lead to a loss of jobs for low-skilled workers in some industries, especially if displaced workers struggle to find alternative employment opportunities or face reduced wages in new roles. Automation is already having a significant impact on employment in manufacturing
- **increased pollution** – the production of some technologies can lead to pollution, which can have a negative impact on the environment. For example, the production of solar panels requires the use of hazardous chemicals, which can pollute the air and water
- **increased use of resources** – more efficiency means more products and therefore more materials are required
- **possible oversupply** – through being so efficient, or because they base production levels on sales forecasts that turn out to be inaccurate, factories might produce much more than the market needs
- **increased waste** – more products manufactured and then consumed usually means more products that end up in landfill
- **large space required** – large areas of land are taken up for factories and distribution centres
- **increased need for transportation** – more products need to be transported all over the globe
- **unfair distribution of benefits** – the benefits of technology are not always evenly distributed. For example, large businesses may be able to afford to invest in new technologies, while small businesses may not. This can lead to an 'uneven playing field' and make it difficult for small businesses to compete
- **possible decrease in quality** – companies focus on efficiency at the expense of quality (but note that accuracy can also contribute to consistently high quality if set up for this)
- **initial change-over costs** – changing to new technologies can involve expensive set-up costs for producers/manufacturers if it requires updating equipment and production systems
- **the escalating use of technologies in armaments and weaponry for war** – these consume resources (which are lost if they are destroyed or have accidents), create carbon emissions and generally cause destruction and sometimes irreversible environmental damage.

Overall, the use of technology in production processes can have both positive and negative impacts on consumers and producers. It is important to weigh the benefits and drawbacks of each technology before deciding whether or not to adopt it.

9780170477499

Impacts of experimental and alternative materials

Experimental and alternative materials are often used either with the aim of reducing environmental impact and improving sustainability or simply because they are more suitable for their purpose than traditional materials.

Positive impacts for consumers

These materials can have a number of advantages over traditional materials for consumers (and society), such as:

- **lower environmental impact** – some experimental and alternative materials are produced using less resources than traditional materials. This can help to reduce pollution and conserve natural resources
- **possible increased durability** – some experimental and alternative materials are more durable than traditional materials, which can lead to longer product lifespans and reduced waste
- **possible improved performance** – some experimental and alternative materials offer better performance than traditional materials in terms of strength, lightness or flexibility. This can lead to the development of new and innovative products
- **wider choices** – consumers can choose products that they believe to be more sustainable, while some consumers may prefer, for example, to choose products that don't cause any harm to animals.

Positive impacts for producers

These can include:

- **new markets** – experimental or alternative materials can help producers to expand and reach new markets
- **increased innovation** – some materials are very innovative and can 'take over' the use of traditional materials.

Possible negative impacts

Experimental and alternative materials also have some potential drawbacks for all involved, such as:

- **creation of waste** – although many might come from natural resources, they become a waste issue once mixed with chemicals or polymers to make them suitable for a certain purpose if this means they do not break down and may end up in landfill
- **higher cost** – experimental and alternative materials can be more expensive than traditional materials. This is because they are often made using new and untested technologies
- **limited availability** – experimental and alternative materials may not be as widely available as traditional materials. This can make it difficult for businesses to adopt them
- **unproven performance** – experimental and alternative materials may not have a proven track record of performance. This can make businesses hesitant to adopt them
- **unknown consequences of use** – some new materials may cause health issues that take time to become evident.

Overall, experimental and alternative materials offer a number of potential benefits over traditional materials. However, they also have some potential drawbacks. It is important to weigh the pros and cons of each material before deciding whether or not to use it.

Worksheet
6.8.1

Comparing technologies

1. Choose one or more technologies in this chapter that interest you and compare their impacts (both positive and negative) for both consumers and producers.
2. Choose one experimental or alternative material from pages 161–70 and discuss its impacts (both positive and negative) for both consumers and producers.

CASE STUDY

Fungi Solutions

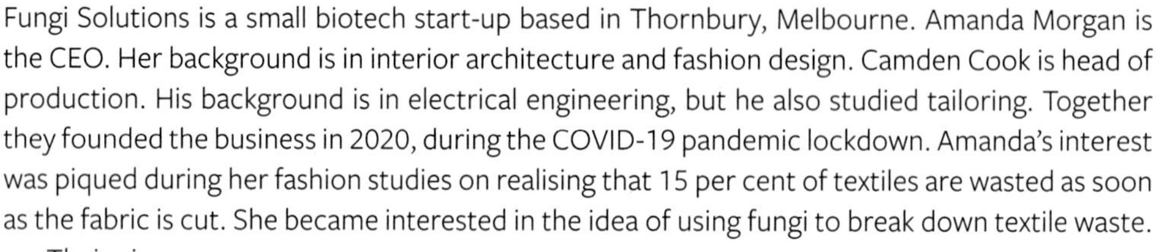

Fungi Solutions is a small biotech start-up based in Thornbury, Melbourne. Amanda Morgan is the CEO. Her background is in interior architecture and fashion design. Camden Cook is head of production. His background is in electrical engineering, but he also studied tailoring. Together they founded the business in 2020, during the COVID-19 pandemic lockdown. Amanda's interest was piqued during her fashion studies on realising that 15 per cent of textiles are wasted as soon as the fabric is cut. She became interested in the idea of using fungi to break down textile waste.

Their aim:

- addressing waste management for small businesses; to reduce what goes into landfill by collaborating with fungi
- product development – prototyping to replace single-use plastics for packaging and other purposes.

In 2022, Fungi Solutions still had a team of dedicated volunteers to do the milling and packing by hand. Their pilot facility had 200 square metres and the plan was to automate processes and move into a larger, commercial facility.

9780170477499

See the explanation of mycelium and mycology ('myco') on pages 161–2.

To grow a usable material from mycelium you need to mimic its normal underground growing conditions. A substrate is needed, which can be made from many waste materials that are shredded or ground down. The substrate is put into a mould or container. Fungal spores are injected into the substrate and then hydrated or temperature controlled as needed. The mycelium grows through the mould material and bonds the cells by its root network, which fuses the particles together.

The resulting material is like an off-white polystyrene or cork, but has a soft velvety surface. It is suitable for packaging as it is lightweight with compressive strength, which means it can be pressed without breaking or can be twisted to break. It is also durable and has good thermal and acoustic performance. It is fire-retardant up to 300 degrees Celsius and is easily extinguished if it does catch fire.

Amanda believes it will be a good replacement for single-use plastics, which Australia is set to phase out by 2025 (this book was printed in 2023). Plastics are cheap and convenient, but the negatives are that they are not good on skin, not good in households (they leach hormone-like molecules) and they never fully break down.

For different purposes, other materials can be added to the substrate depending on the properties required. For example:

- it can be lacquered to achieve a hard surface
- algae and kelp/seaweed can be added to improve the waterproofness and flexibility.

Possible future uses for mycelium-based materials are: interior design, building construction and insulation, agricultural produce and as protective foams for the packaging and storage of technical equipment.

Mycelium has the potential to clean toxic pollutants and to kill microbes in water systems. It does this by digesting the materials into simpler forms in what are known as remediation programs. Fungi Solutions will continue to work with CSIRO on using mycelium for building applications. Other projects include developing packaging materials and using the mycelium of oyster mushrooms to break down cigarette butts.

Positives of myco products

Myco products use 12 per cent of the energy that plastics need for production and produce 90 per cent less carbon emissions. To 'grow', mycelium only requires a dark area with a stable temperature that mimics underground conditions. Myco products are classified as regenerative products because at the end of their useful life, they can be mixed with water and composted within 12 weeks or added to soil as mulch. They are a natural material that reduces waste and minimises negative impacts on the environment.

Weblinks
Fungi Solutions home page

Watch the Fantastic Fungi Film

Negatives of myco products

The ability to scale up production, i.e. to cultivate enough mycelium from fungi, is dependent on investment and adoption.

Go to 'FAQ' and 'Media articles' on the Fungi Solutions website to learn more.

Learn more about fungi in the film *Fantastic Fungi* (also available on Netflix).

Assessment Task for Outcome 1

This task is scored out of 50 and makes up 10 per cent of your total score. It can be any one or a combination of:

- case study analysis
- oral presentation using multimedia: face-to-face or recorded as a video or podcast
- research inquiry.

Weblink
VCAA Product Design and Technologies

Your teacher will decide on the format. You can find the Performance Descriptors (assessment rubrics) on the VCAA VCE Product Design and Technologies web page.

Before you start your task, you need a strong understanding of the topics and terminology in Outcome 1. The learning activities throughout the text and on Nelson MindTap are there to help you with that.

Aiding your long-term memory

Long-term memory, which is important for the end-of-year exam, is assisted by deep understanding of a topic.

- Discussions with your classmates or oral presentations are beneficial for this.
- Recalling what you just learnt, or learnt the week before, on a topic by explaining it to others is also a proven method to assist long-term memory.

A note on using artificial intelligence

It may be tempting to use AI chatbots to complete your work. Be aware that:

- Your teacher can use the same AI and look for similarities in your response.
- Your teacher may be able to use detection software or may choose to ask you to do an oral presentation without referencing your chatbot-generated preparations.
- You may be asked to write sections of the SAC by hand, so you need to know your topics.
- Chatbots are not always correct; they can't fact-check and can present false information or give fake references – you need to double-check. To check properly, you need to know your topics.
- Chatbots do not state the source of their information so you don't know whose viewpoint it might be repeating, and whether the original writer or website has a bias or vested interest in pushing a viewpoint. To understand vested interests, you will need to do further critical research.
- Not knowing or not understanding the topic will be of no use to you in the end-of-year exam. To do well, you need to know your topics!

In 2023, several Australian educational institutions labelled 'generating content' or the 'submission of work' generated by an algorithm, computer generator or other artificial intelligence as a form of cheating. Cheating is anything that is considered to be dishonest in the presentation of your work, which misrepresents what you understand.

Methods you may use to prove your work is your own are to:

- annotate diagrams or drawings in handwriting
- create a list of questions that you answer in conversation without referencing any notes or texts
- create a presentation with photos/images only and talk to it without referencing any notes or texts
- create an infographic with supportive text.

9780170477499

ACTIVITY

Revision activities

1 Each class member chooses a topic, creates a multiple-choice question and provides three incorrect responses and one correct answer. Class shares, edits and saves for exam revision. Methods to help you learn: use Flippity to create quizzes or cards for yourself and your classmates.

Weblink
Flippity home page

2 Think about these possible tasks and how you could develop or combine them to score in relation to the VCAA Performance Descriptors for SAC 1.
   a **Case study analysis:** Prepare a case study on a product from a manufacturing company (or brand) and outline its scale of manufacturing and all the technologies used from this chapter, including their impacts. Choose a product that you are interested in or that is similar to your intended SAT product.
   b **Oral presentation:** Create a presentation on a product made from composite materials and compare it with similar products made of natural materials. Outline the positives and negatives of using the composite material for this product.
   c **Research inquiry:** Research a bioproduct such as a mycelium-based material and write a report explaining how it is made and what it can be used for.
   d **Case study analysis:** Choose a product that you believe will be subject to obsolescence within a time frame and compare it with a high-quality product.
   e **Oral presentation:** Create a presentation that explains how the sustainability strategies in this chapter are different and how they overlap Make a comment on technologies involved and impacts they may have on social sustainability.

Summary
Chapter 6

Quiz
Chapter 6 revision

3 Create a concept map for each of the main topic areas within the outcome. Be as detailed in your links as you can (using a few layers of connections, and phrases or sentences rather than just words). You can also create annotated diagrams or flow charts for certain concepts, such as the sustainability frameworks.

4 Create cards with the titles of each of the topic areas. Then, working in pairs, select a card and:
   a brainstorm together as much information as possible for each area
   b take turns to explain a topic/concept to the other student, who notes this down and then adds anything that might have been missed.

You will be scored according to the VCAA Performance Descriptors, so be sure to check them before submitting your work. Check the task outline from your teacher. See the link on page 176.

Steve Martinuzzo from Cobalt Design

Cobalt Design is a product development consulting business. It employs 28 staff, who work with clients to develop products in the areas of transportation, health and science, consumer and sporting products, and products for industry and public spaces. Steve Martinuzzo is a founder and Cobalt's Managing Director and has a background in Industrial Design from RMIT. Steve shared some insights about his work and the focus of his company.

Teamwork is important at Cobalt. It's good to have a mix in a team – some specialists who are truly gifted in their 'thing', and some all-rounders who are good at many things. It is very similar to an elite sports team, in that you need specialists for a few specific roles, but otherwise flexibility, fitness and teamwork bring the greatest value to the group.

The process of design: using design briefs

In practice, most clients rely on Cobalt developing a brief as part of our proposal. Even where a client has created a brief, it often needs more work (for example, simplifying it where it's too

complex; prioritising objectives when they ask for everything; or making it more strategic). We end up being good at brief writing because we do them so often, whereas most clients only create briefs occasionally.

Research to discover the user's needs

We do this research whenever we can. As people are not especially good at telling you what they want, the ideal is to observe people using current products in as real a context as possible. We call what we learn 'user-driven insights' and they can reveal opportunities to uniquely improve products. In these cases, users will often say, 'Thank goodness someone has fixed that!' It's also important to remember that there is always more than one 'user' to consider. For example, the most obvious end user of an EFTPOS terminal might be a shop customer. But what about the shop assistant, the shop owner, the IT company, the assembler, the client and the banking system? All of these groups of people (or stakeholders) have some interaction with the EFTPOS product or system, so all should be considered to some level.

We do most of our research in-house. Where specialised or market research is needed, our clients usually contract this themselves.

Developing design ideas

Designers at Cobalt use all of the following methods: hand-drawing, digital (tablet) drawing, CAD and prototyping (both physical and virtual).

Focusing in on the best option

Selecting the 'right' design is always an inexact science. In terms of process, we have some common principles but there is no 'one way' to select the best option. Each client, project and start/end point is different and we have to be flexible and adapt the process for each client. Effective communication is really important, and our approach and the amount of detail we give to a client at the selection stage depends on their knowledge and their need to be guided through the process.

Cobalt Design, www.cobaltdesign.co

Concave *Halo* football boot range

9780170477499

The Bombardier tram operating in central Melbourne

Handing over for production

A common point of handover is 'tool release', when we provide fully resolved 3D CAD files and engineering drawings/specifications ready for production tooling. But, as stated earlier, each client is unique and some clients don't need or want us to go that far, whereas other clients want us to be involved through tooling and into early production.

KeepCup *Brew*: A design development example

Cobalt had already worked with KeepCup to develop earlier versions of its sustainable, non-disposable coffee cup. The new KeepCup *Brew* is the result of Cobalt's design thinking process. Our process for the *Brew* included:

- **questioning the questions.** During the early phases of the project, we re-evaluated every assumption and aspect of the original KeepCup and how people use a reusable coffee cup. This involved questioning the way in which users interact with the product, listening hard to user feedback, and rigorously looking for product weaknesses and opportunities for improvements, or new features, materials and manufacturing technologies. User insights revealed that people liked to personalise their cups, and that they often like to drink from the rim (without the cap). We also looked at how far users carried the cups, and how long the coffee needed to be kept hot. These insights and observations were factored into the *Brew* design development.
- **divergent thinking.** Divergent thinking considers a wide variety of ways in which a problem may be solved. Through several rounds of concept development and working closely with the client, we generated a large number of possible product concept ideas. Some of these ideas were radical, but others represented more subtle improvements to the already successful KeepCup product. The key element of this phase was the volume and breadth of ideas created.
- **iterative prototyping.** Iterative processes are those that use repetition. By building mock-ups, prototypes and trial tool parts, we were able to review and test functional design ideas, thermal and performance characteristics and material properties. Lots of rough prototypes were created early in the process. We wanted to explore as many 'big picture' questions as quickly and efficiently as possible, investigating performance and feasibility. The prototypes were then refined in terms of quality and effectiveness. This process allowed physical interaction, testing, evaluation and discussion.

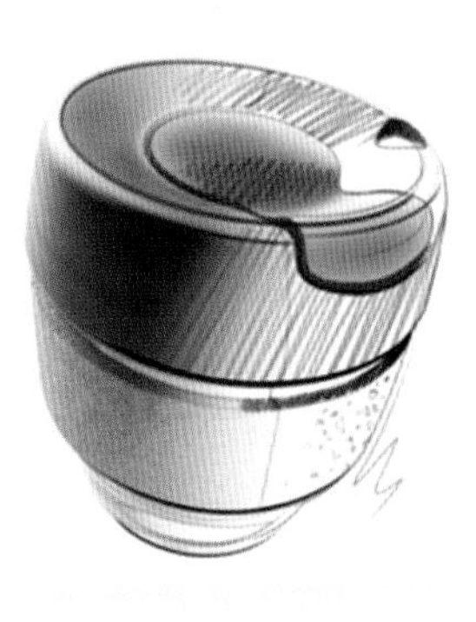

The KeepCup *Brew* at different stages of design

Case studies
Schamburg + Alvisse

Marc Pascal

Alessi

Jim Hannon-Tan

Chapter 6 case study report activity

Choice of materials

The *Brew*'s cup is made from glass, and it is available with a cork band. These materials are functional and environmentally sustainable (both materials are 100 per cent recyclable). The *Brew*'s glass cup is the same tempered soda-lime glass found in caffè latte glasses, making for a robust, break-resistant product – and one that will not shard into dangerous slivers of glass in the unlikely event of a breakage.

Cork is a wonderfully tactile and sustainable material that we have loved for some time. Cork is stripped from the bark of the cork oak tree, so the tree doesn't need to be felled for harvesting. Cork is biodegradable and fully recyclable, and it also possesses excellent chemical and thermal resistance, making it an ideal insulating material for the KeepCup band.

Issues that impact on Cobalt's design work

- Sustainability: We try to address this in every product we design.
- Standards, patents and design registration: These areas can have a big impact when the intent is to be comparable to but not to infringe on someone else's IP.
- Economic pressure: All commercial designs need to make a sustainable profit to justify the risk and investment in design and production. Our designers want to design the best products possible, and planned obsolescence isn't something we consider. However, sometimes the budget constraints for a product mean that some compromises in materials or production processes might need to be made.

What is good design?

At Cobalt, we believe that good design:

- fulfils people's needs. Well-designed products work as expected. They must meet users' current and future needs
- emotionally connects. Good design recognises that people are not exclusively rational
- defines itself globally. Well-designed products are among the world's best of their type
- finds clear space. Good design creates clearly differentiated products

- is socially responsible. Good design reduces waste, hardship and resources. It is accessible and democratic, not exclusive to those with high incomes
- happens through teamwork. Good design is the result of multiple professionals working as a well-managed team
- is 100 per cent commercial. Good design generates success and profit far beyond the development investment
- needs clients with vision and technology. Design is the union of creativity and commerce. Good design only happens with good clients.

The company started a blog highlighting socially responsible design, sharing stories focusing on 'less waste, less use of resources, and creating less hardship' – hence the name: lessbydesign.org.

Why do you like designing?

I like working as part of a team, and seeing our staff grow and be their best. I get real satisfaction in finding a new approach that makes an everyday product or task a little easier or more enjoyable to do.

The Cobalt Design website has more information and many great product case studies. The KeepCup website has information about KeepCup's sustainability. There is a link on Nelson MindTap to a video explaining the design process of the *MoyoAssist Extra-VAD* cardiac life support system.

Weblink
MoyoAssist Extra-VAD

CHAPTER 7

Investigating opportunities for ethical design and production (SAT)

UNIT 3, AREA OF STUDY 2 (SAT): OUTCOME 2

In this outcome, you will:

- investigate a need or opportunity that relates to ethics and formulate a design brief, conduct research to analyse current market needs or opportunities and propose, evaluate and critique graphical product concepts.

Key knowledge

- activities and their purposes within the Double Diamond design approach to investigate and define, generate and design graphical product concepts, evaluate and plan and manage and the relationships between critical, creative and speculative thinking
- design thinking to generate, refine and critique graphical product concepts
- methods used by designers to formulate a design brief, including investigating and conducting research, and recording, collating and forming information about an identified need or opportunity of an end user/s
- elements of a design brief: need or opportunity, profile of end user/s, function, project scope (constraints and considerations)
- relationships between the need or opportunity, designer and end user/s
- purpose of, and methods to develop, evaluation criteria
- relationships between the design brief, evaluation criteria, research and product design development
- processes to develop and communicate graphical product concepts
- characteristics of drawing systems for visualisations, design options and working drawings: level of detail and types of annotations relevant to the design specialisation
- Factors that influence ethical design of a product
- ethical considerations when gathering qualitative and quantitative research, and methods to investigate current market needs or opportunities that include using digital technologies
- methods to conduct ethical research and to appropriately acknowledge sources and intellectual property of others
- methods to examine and test ethical products.

Source: *VCE Product Design and Technologies Study Design*, pp. 30–1

Key skills

- explain activities and their purposes within the first diamond of the Double Diamond design approach
- formulate a design brief that relates to an ethical consideration
- explain the scope of a design project, including considerations and constraints, and identify aspects that require research and testing
- use Factors that influence product design to examine, analyse and critique existing ethical products
- propose and explain methods of ethical research to explore current market needs or opportunities for ethical products
- use ethical research methods to gather, present and interpret research
- explain relationships between design brief, evaluation criteria, research and product design development activities
- develop and use criteria to:
  - inform and evaluate product concepts
  - evaluate processes to design and make the product
  - evaluate the finished product
- demonstrate a range of appropriate drawing systems, using manual and digital technologies
- work technologically to use research and design thinking techniques to generate, evaluate and critique graphical product concepts related to ethical design.

9780170477499

INVESTIGATING OPPORTUNITIES FOR ETHICAL DESIGN AND PRODUCTION

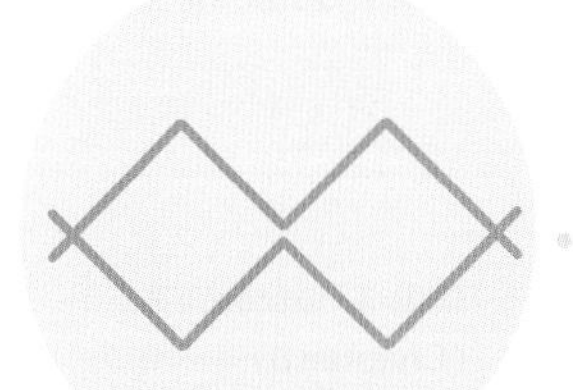

Double Diamond design process

Factors that influence design

How to write a design brief and evaluation criteria

Ethical research on needs and opportunities

Creative, critical and speculative thinking

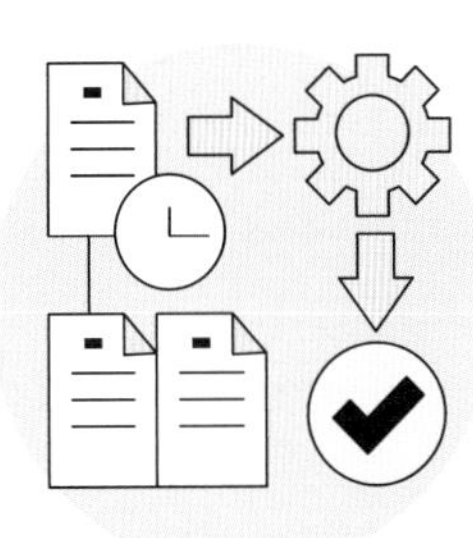

Drawing systems to create graphical product concepts

Worksheets:

- 7.5.1 Course essentials Units 3 & 4 **(p.193)**
- 7.6.1 SAT starter activity **(p.196)**
- 7.7.1 SAT end user activity **(p.198)**
- 7.9.1 Quantitative and qualitative information **(p.206)**
- 7.9.2 Q&Q data methods activity **(p.206)**
- 7.11.1 SAT research on ethical products **(p.209)**
- 7.12.1 SAT end user profile **(p.214)**
- 7.12.2 SAT design brief **(p.214)**
- 7.13.1 SAT evaluation criteria **(p.216)**
- 7.14.1 SAT research plan **(p.216)**
- 7.15.1 SAT evaluating graphical product concepts **(p.219)**

Bonus case studies:

- Etiko **(p.195)**

Templates:

- Design Folio Template

Summaries:

- Chapter 7 **(p.226)**

Quizzes:

- Chapter 7 revision **(p.226)**

Nelson MindTap

To access resources above, visit **cengage.com.au/nelsonmindtap**

School-assessed Task (SAT)

The SAT consists of three outcomes across Units 3 and 4.

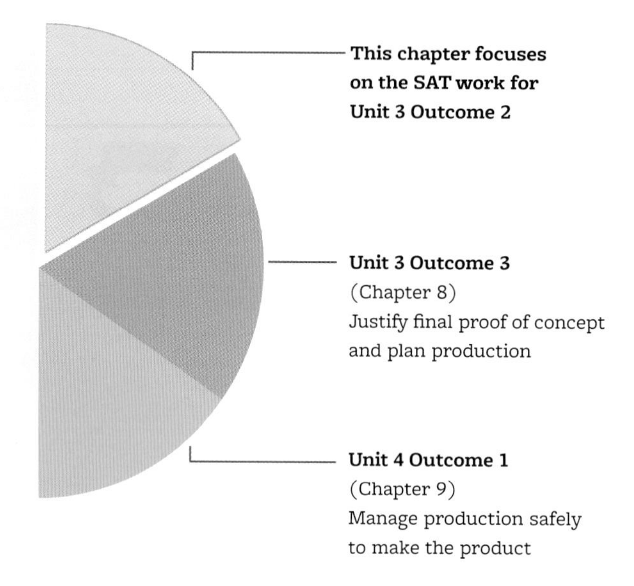

The Double Diamond design process

Refer back to page 6 to read about the Double Diamond design process.

Your SAT and the Double Diamond design process

Your SAT work will be framed by the Double Diamond design process.

You are not expected to cover every activity in the diamonds, and different activities may receive more emphasis or repetition than others. Nor do the activities have to be completed in the order that follows in this chapter. The most important thing is that you keep all your research, ideas, tests or trials and, later on, assemble them in a kind of order that shows your design thinking. Your design thinking can be explained through linking statements or annotations. You will use your 'multimodal record of progress' (or your design folio) to provide evidence of your design thinking.

Your Double Diamond design process might look like any of the following. You may finish your product and see so much possibility for improvement, or get negative but constructive feedback from your end users, that you decide to start again (last row of diamonds) – assuming you have the time.

9780170477499

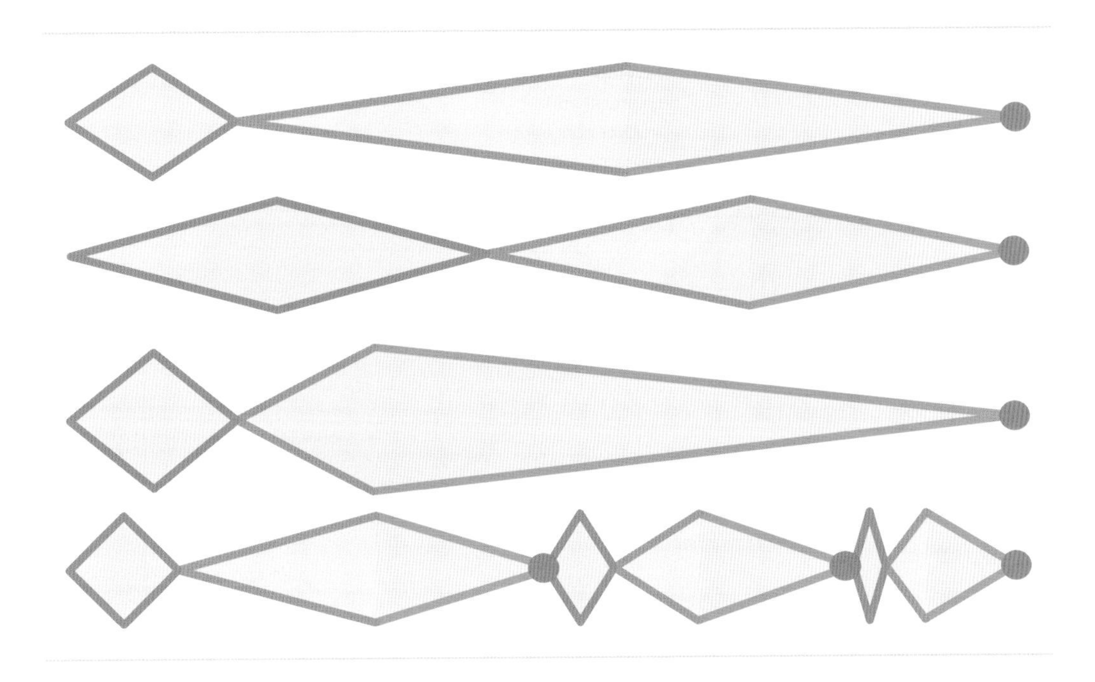

Watch the video, 'What Is the Double Diamond Design Process?' – an animation on how the Double Diamond is applied, from Design og arkitektur Norge (Design and Architecture Norway).

Weblink
What Is the Double Diamond Design Process?

The Double Diamond process usually starts with much research on end user needs completed in the beginning. You might use speculative thinking to consider a forward thinking project. A design brief is finalised once the project and its need becomes clear. Creative and critical thinking are applied throughout to do relevant research, to create many design ideas and to make decisions around them. Once ideas have been refined and decided on, working drawings are developed to use as a basis for making a series of prototypes. Each prototype is evaluated for all aspects of its design and construction and improved in the next prototype. The most successful prototype will then provide the final working drawing and planning information necessary to complete the finished product.

Activities in the Double Diamond design process

The Double Diamond design process is essentially a process of Discover, Define, Develop and Deliver, which is repeated until a successful outcome is reached. Consider that the first diamond is the 'problem space' and the second diamond is the 'solution space'. Outcome 2 in this chapter focuses on the first diamond and part of the second diamond.

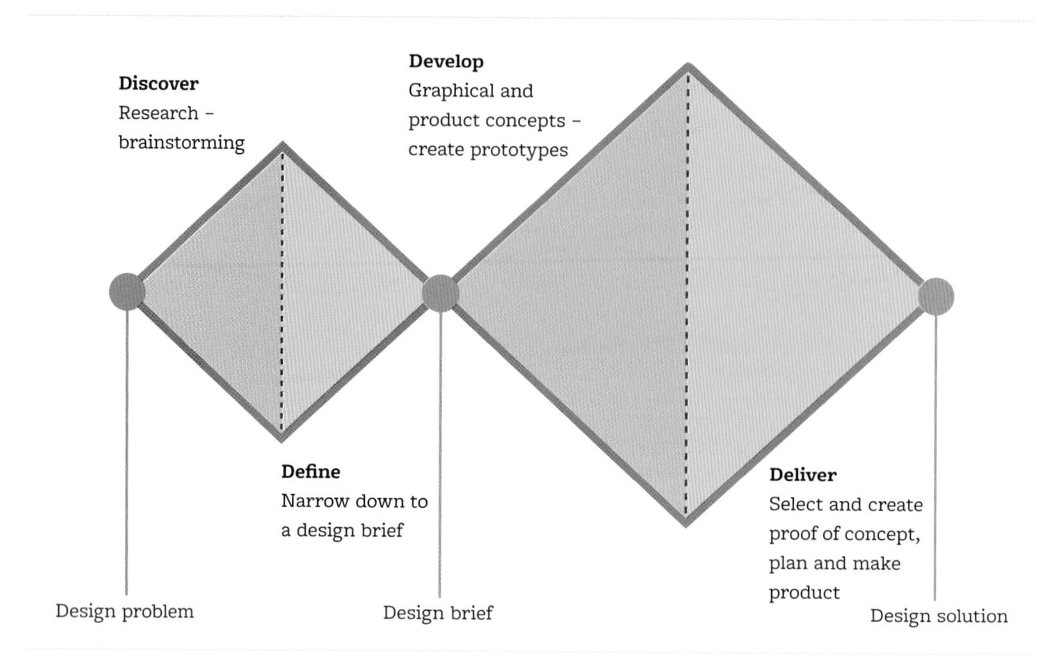

First diamond

Discover

The start of a project is marked by an initial idea or inspiration expressed as a **design problem**. From here comes divergent thought, where the exploration of a broad range of possibilities, ideas and influences creates the foundation for decisions made later in the process. In this stage you are identifying a problem by looking at trends, interviewing end users, looking into their behaviours, needs and perceptions and referring to other relevant information. This initial research helps to identify, clarify and define the problem, opportunity or user need that can be addressed.

Define

With the information you have found, it is now time to start thinking convergently and narrowing down the precise needs of your end users and what exactly your design project will be aiming to achieve. You will look at the research you have done and discard the ideas or directions that you consider won't help the end user or solve their need, or those ideas that are beyond your capabilities, those that may be too expensive or take too long, or those where you don't have enough information to move forward. Here you will make decisions on what your project will be and define this in a **design brief**.

Second diamond

Develop

At this point you once again think divergently. Your research in this stage will be practical and problem-solving, and will assist you in coming up with many different ideas that could become a solution to your design brief and possible ways of implementing them. Not only will you develop graphical concepts, you will also develop physical concepts in the form of models, maquettes and prototypes. You will explore many possibilities for making your

9780170477499

product, the materials to use and how to construct it as well as obtaining end user feedback on all versions. This is called iteration – repeating these activities, refining and improving them each time.

Deliver

Now it's time to think convergently again and narrow the whole process down to what will help you make the most suitable solution to your design brief. At this point, you choose the most appropriate and effective methods you have explored so far, plan your schedule and construct your finished product – the **design solution**. Through all segments of the diamonds, you will be involved in evaluating your ideas and your solution/s.

Critical, creative and speculative thinking

Creative and critical thinking

You will use design thinking all the way through the Double Diamond design process. Good-quality thinking involves using both creative (divergent) and critical (convergent) thinking at the same time. Refer back to the explanations on creative and critical thinking in Unit 1 on pages 13 and 14.

- Watch the video 'Convergent Thinking Versus Divergent Thinking', which explains the difference between the two types of thinking and how they work together.
- This website of the Creative Education Foundation also explains how creative thinking and critical thinking work together (like the fuel and the brakes in a car).

Weblinks
Convergent Thinking Versus Divergent Thinking

What is CPS?

Speculative thinking

Speculative thinking involves questioning current accepted practices of design and consumer culture and focusing on possibilities for an imagined future. It requires a **synthesis** of what is already known with new information, and using this to speculate on the impacts of science and technology on humanity and the planet. The focus is on satisfying future environmental, social and economic needs of humanity in a beneficial way.

Questions to ask of common solutions:

- How appropriate is this?
- How useful is it?
- How suitable is it?
- What impact does it have on the environment in the making of it, its use and its disposal?
- If it uses recycled materials, can these materials be reused or recycled again? Or is it just delaying their final disposal in landfill or incineration?
- What needs to change for it to solve problems in new ways?

Speculative design practice poses questions in order to create alternatives for our world both today and tomorrow, between technology and people (society). The biggest question that is asked is, 'What if ...?' When thinking about our possible futures, speculative design looks at how things could be, and considers any negative implications of bringing newly designed products into our life.

Speculate
To form a theory, ideas or opinions about something without all the facts or information; to invest in the future in the hope of gain; to think and make guesses about what might happen or how something might work; to make a prediction based on evidence that is followed by actions to find the outcome

Synthesise
To research, collect and make sense of information in a way that has meaning to you, connecting it with information you already have to create something new

Read further under 'Regenerative design as a subset of speculative design' on pages 189–92 to see questions that can be applied in speculative design thinking.

Great Wrap

Speculative and Critical Design

There is a subcategory of design, labelled Speculative and Critical Design, that seeks to challenge 'normal' ways of doing things. It lives in the world of art, research and/or education and encourages exaggeration and offence to provoke debate. According to Professor Cameron Tonkinwise of the University of Technology Sydney, it is a form of 'Design-focused Science Fiction or design philosophy'.

Designers and future thinkers in this field develop products, services or environments that are not (yet) wanted, not yet considered possible or that are for worlds that do not yet exist but may one day, for better or worse. Its designers use design prototypes, models or concepts rather than fully finished products to imagine alternative futures and explore the potential consequences of technological and cultural developments. They may not know the intended result of their work, but the research and the process becomes very informative.

The goal of Speculative and Critical Design is not to produce commercially viable products, but rather to stimulate critical thinking and encourage reflection on the role that design plays in shaping our future world. It is a form of design activism that seeks to challenge dominant narratives and promote alternative visions for the future. Practitioners believe it's not about making up an alternative future, present or past. They believe that by taking human behaviour into account and blending the knowledge, skills and technology from the past, present and future, they can develop radical and ambitious solutions that are thought-provoking.

An example scenario for speculative design

Melbourne Design Week is an annual event and in 2023, the theme was 'Design the world you want'. This theme was meant to provoke designers into thinking about current and future challenges such as loneliness, climate disaster, social inequity and the cost of living. This theme also demanded that designers:

- be honest and responsible with the materials they use, from extraction to their afterlife – that they show **transparency**
- think about value being other than financial, and give value to social, environmental, cultural, natural and political concerns – to show **currency**
- think about the impact of design long after we are gone – or **legacy**.

9780170477499

AAP Image/JOEL CARRETT

Objects on display during Melbourne Design Week 2023

Gensler is a US-based global architecture, design and planning firm, which is always looking ahead and speculating about what humans will need in the future and what market opportunities the business can fulfil. They look at the future of cities, the future of work, the future of lifestyle and the future of healthcare. Read about Gensler's predictions of trends at the firm's website.

Weblink
Product Development

Regenerative design as a subset of speculative design

Regenerative or 'restorative' design can be considered a method that not only looks into the future but seeks to 'fix' or reduce damage already done. It is a way of designing something positive that will 'generate' better outcomes for humanity and the planet. It seeks to restore and renew the energy and materials used in design. It is a step further than 'sustainable' design, which is interpreted so differently by so many. Designers working in this field will think beyond getting the product manufactured and having a successful sales account, to how it impacts the communities where it is manufactured and whether it contributes to their hardship or helps to improve their quality of life. Other key factors such as social unrest, demographic shifts and climate change are also considered.

Some companies use their profits in a regenerative model to fund environmental programs. The interesting thing about regenerative design is that it doesn't always put customers or end users first; it puts communities as the focus, and aims to impact the wellbeing of many as opposed to that of the individual. It is always thinking about the future – for example, how to turn the anxiety people have about the world into optimism and excitement.

Asking questions of the future

Regenerative design asks 'big' questions of the future such as:

- How will the planet cope with more than 8 billion people?
- How will technology influence our daily lives? How can it do this positively for all people, not just those in the developed world?
- How will people live, work and relax?
- How do we house all people while others live in more space than in the past?
- What will the climate be like? What will be the temperature extremes and what might help humanity cope with them?
- How will people get around – what transport will be used?
- How will families relate? Will families only consist of one child? What about the large number of people who will never have children? What does ever-increasing pet ownership mean for cities? What does this mean for housing and furniture?
- How will people cope/live after the destruction caused by war, floods, fires or droughts?
- What will we do to deal with or change the prediction that clothing waste will use 25 per cent of all carbon by 2050?
- How could the clothing industry cut its environmental impact and also take into account impacts on developing countries where it provides a lot of employment, resulting in social and economic benefits?
- How do we manage solar energy and other renewable energy sources, and change habits, activities and use of products to fit in with that?
- How will society cope with increased life expectancy? What will be needed as more people withdraw from the labour market and live longer?
- How will society cope with possible food and water shortages, and the need for products to preserve food more effectively?
- How do countries deal with increased migration of people fleeing war or climate change, and the resulting need for more housing and other objects, without penalising those people?
- How can we design solutions that can adapt to future situations? Do they need to be of a solid block of material that exists after its disposal? Or is there another way? Can modular design help? How can a product's materials go back into the system?

Shutterstock.com/Holli

School Strike 4 Climate protest in Australia, 2019. Regenerative design responds to issues such as the threat of climate change.

9780170477499

Regenerative design is not always about being new. Things can be adapted/adaptive and/or renewed/renewable and include social and environmental considerations.

Regenerative design not only requires designers to tap into empathy and 'human-centred design', but also the circular economy and the triple bottom line, where companies and manufacturers measure success beyond purely financial targets.

Refer back to 'Circular economy' on page 138 and 'Triple bottom line' on page 142.

Regenerative design requires:

- listening to all stakeholders/communities and considering social outcomes and environmental impact
- considering viewpoints from many different perspectives and from many corners of the world
- rethinking of all the 'usual' methods and materials used.

It requires a fine balance with business needs, but many big companies are moving towards this model and, in doing so, are considering their partnerships more carefully. The use of technology to reach further into communities and share research improved significantly during the pandemic lockdowns of 2020 and 2021. This ease of communication enabled better 'listening' to others. Forward-thinking companies use regenerative ideas and speculative thinking to design for all of humanity, to create new solutions that benefit everyone.

Indigenous practices can inspire regenerative design

Many 'regenerative' designers look to the culture and traditions of ancient and not-so-ancient societies for inspiration and information. Many indigenous cultures continue time-honoured practices that keep them in tune with nature, with the unspoken aim of nurturing all natural resources on hand for the future. They have done this for centuries because, if they didn't, they would be faced with not having their needs met.

Aboriginal and Torres Strait Islander peoples have a deep-rooted philosophy of 'taking only what you need and nothing more', with all activities aimed at providing for the future. Indigenous peoples in northern Russia make offerings to aspects of nature often associated with gods. Balinese culture reveres the beauty of the natural world and its bounty. Balinese people respect that we share the planet with nature, and individuals play a part by offering something beautiful every morning to please their gods.

South American indigenous cultures speak to their ancestors through nature. Through this comes a strong desire to protect the environment in everything they do.

Ubuntu is a South African Zulu word meaning to be understanding, forgiving and to have compassion for humanity. It means that a person is a person through other people, that we are all part of a larger entwined communal, societal and spiritual world.

Alamy Stock Photo/Bill Bachman

Indigenous practices such as extracting material from a tree for canoes or implements always made sure the tree could continue to survive.

Weblinks
Flourish

New Mindsets Will Help the World Flourish

Designing the Circular Economy: Sarah Ichioka on regenerative design

Design for Our Collective Wellbeing

For further research:

- The book *Flourish* offers a plan for designers, clients and change agents alike to build a thriving future, together. Read an extract from the book and a review of it on the RIBAJ website.
- Watch a video featuring Sarah Ichioka talking about her book *Flourish* with co-author Michael Pawlyn at Melbourne Design Week (starting at the 9-minute mark). *Flourish* focuses on circular design for the future. The 40-minute mark shows questions from the audience.
- For a one-hour conversation on regenerative design in 2021 promoted by BMW Designworks (a subset of BMW group since the 1970s), visit the BMW Group Designworks website.

How to use speculative thinking

Start by choosing some of the questions on page 190 and creating/adding your own questions that might be relevant to what you decide to focus on for your SAT. Your research, tests and trials will be important tools to assist in this, along with your critical and creative thinking. From this you will make many ideas and suggestions, followed by informed decisions. This will help inform your design brief and evaluation criteria.

For more ideas on what your SAT project could focus on, see the heading 'Need inspiration?' on page 195. You can also look through some of the suggestions in Chapters 3 and 4 outlining products that create a positive impact.

Your SAT 'folio'

Multimodal
Using multiple methods or modalities of presenting work, e.g. text, drawings and images, videos and audio

The following topics address the work you are required to complete for your SAT folio. Your folio can be a **multimodal** record of your progress through the Double Diamond process.

Your folio can be created using any of the following or a combination:

- digital applications, such as word processing for written sections, presentation apps such as MS PowerPoint and OneNote, Google Slides and Sites, Prezi
- CAD work downloaded in a format that your teacher can view
- hard-copy pages with handwriting and hand-drawn images, as well as printed digital material, inserted into a folio with plastic pockets
- audio recordings explaining your progress or design thinking
- video to show your production processes
- a digital presentation that is a compilation of all your digital work with embedded audio files, video files and scanned drawings
- different forms of information/content can be combined into a (private) website.

When deciding which modality to use, you need to consider ease of access and file size of videos. This is important for your own convenience and for assessment purposes. Your teacher must be able to access your work. If you intend to enter Top Designs, consider creating your folio in a convenient form that can be easily compiled as a PDF.

Advantages of working digitally or keeping digital copies of all your work are that:

- they are safeguarded, particularly if you save them to a cloud service, such as your school server, Google Drive, Microsoft OneDrive, iCloud, Dropbox and many more
- they can easily be retrieved for interviews or applications at the end of the year
- they don't take up physical space in your home
- pages and items can be resized and rearranged easily
- files can be printed out at any stage.

Your teacher may give you direction about the format/s to use at your school for SAT submission.

Factors that influence design

In Chapter 13 you will find more details on the Factors that influence design. They are the typical areas that all designers will address in some way. You will use them throughout your SAT for:

- writing your design brief and evaluation criteria
- organising your research
- proposing and evaluating concepts
- assessing what materials and technologies are available to you.

Revision activities

Choose at least three of these activities to help you revise.

Worksheet
7.5.1

1 Draw a chart or graph to show what percentage of your total score is for the SAT. Annotate the three outcomes that make up the SAT.
2 Once you have received the relevant SAT Assessment Criteria for the year, draw your own pie chart or bar chart to show the proportion of the scores for each outcome. Annotate the items required for each criterion.
3 Draw your own Double Diamond and annotate your understanding of the main activities in each diamond.
4 Draw a diagram that helps you understand the difference between critical and creative thinking. Annotate with thinking techniques that belong to each type of thinking.
5 Explain what speculative thinking focuses on. Brainstorm three ideas on how it could apply in your SAT project (or do this later as part of your SAT).
6 Read about the possible multimodal methods for recording progress for 'Your SAT'. Use the table below to make your own list for those you know how to use, those you would like to learn how to use and how you could or will use all of them. Consider in your SAT that you will have written items, diagrams, drawings, printed photos/images, samples and possibly audio and video and all of it needs to be stored safely. You should also refer to the SAT Assessment Criteria for the current year to see the scope of the task.

Multimodal method	*My experience with using this is:*	*I can learn about this by:*	*I will use this for:*

7 Read Chapter 13, 'Factors that influence product design'. Create a mind map with the first layer of shapes listing the Factors, and the second layer as annotations of your understanding of their meaning; *or* make a list of the Factors for your own reference throughout Units 3 and 4.
8 Extension exercise on the Factors: annotate how the Factors are addressed in a commercial product; *or* complete this as part of your SAT research on ethical products related to your intended project. See 'Analysing existing products: are they ethical?' on pages 206–9.

An ethical design problem

You will start your SAT work in the Double Diamond with a **design problem**. This is a brief statement that outlines the situation you are designing for, the type of product you are thinking of designing, for whom and why. Your design problem can be a real personal, local or global design need or opportunity.

In other words, you will design and create:

- **personal** – a one-off piece for one end user. If designing for yourself, you would see yourself as a member of a target group or 'tribe', i.e. as part of a group in a similar age bracket or with the same tastes, lifestyle, budget, etc.
- **local** – a product for typical end users of a particular target market in your circle of neighbours, family, friends, town or city
- **global** – for a broad group of end users outside your own world, i.e. for a particular group of people you know little about and need to research.

In the first diamond, and after completing research on your design problem, you are expected to formulate (create) a design brief for an ethical product. Read the topic 'Ethical design of a product' on pages 206–9 to help you. You should **not** 'know exactly what you are going to make' at the start of your SAT. However, you do need to have identified a design problem, need or opportunity that allows you to complete research and explore *lots* of ideas on what you could design and make. Your design brief must allow for your own creative input. You can use speculative thinking or regenerative design for ideas, or read the table 'Need inspiration?' on the next page. Refer back to pages 189–92.

These are examples of design problems:

- At present it is difficult for people to wash, dry and store plastic bags so that they can be used repeatedly for other purposes.
- Large quantities of clothing are emptied into landfill each year, while people keep buying the new fast fashions.
- Single-use sachets for food and detergents, which are multilayered (plastic and aluminium) and have minimal value for recycling, litter the waterways of Indonesia and the Philippines, yet sales are predicted to keep increasing.
- All available cycling gear looks similar and is conservatively styled, plus it makes you hot once you have been riding for a while, and it is very hard to change out of while cycling.
- Autistic children find it difficult to sit quietly or peacefully to read or play games in the family home.
- Cats need to be entertained and fed while being contained and/or transported.
- Children in some developing countries walk to schools where there are no desks or equipment, and they need to carry everything required to and from their classes.

Households struggle to deal with the quantity and variety of plastic bags used to carry products home.

9780170477499

Need inspiration?

17 goals of Sustainable Development	Click on 'More information' on each goal. Your project might be loosely related but inspired by one of these goals, e.g. Goal 12. However, remember that for this subject you are required to construct some type of useful product, not a service or activity. See more on page 70 in Chapter 3.
The Index Project (previously named INDEX: Design to Improve Life, Denmark)	A non-profit organisation promoting designs to improve life. The biennial Index Award offers a generous cash prize. Many winning designs are technological systems, but you can find ideas. Go to 'Award', then 'Winners and Finalists', and wait. Each case starts with an explanation of the 'design problem'.
Wool4School	The Wool4School competition. For your SAT you could use its design brief and resources from past years – or the current year, and enter the competition, too!
United Nations Environment Program	Read about the international treaty to end plastic pollution by 2024. Take a look at the information (real-time data) published on the UNEP website in relation to its environment program.
The *Guardian* newspaper	Subscribe to the *Guardian* for free to read one of their hundreds of articles on the environment for inspiration.
Break Free From Plastic	Take a look at what this organisation does, see if its campaigns give you inspiration for creating solutions or use its resources for reports or data.
ClientEarth	ClientEarth is an environmental charity that uses law to protect life on Earth. Some of the issues it tackles might give you ideas.
Only One	A non-profit focused on ocean health. Be inspired by some of the short films on Tide Turners to help you think of ideas.
Victoria and Albert Museum	Design challenges for students 14–16 years old in the UK, provided by the Victoria and Albert Museum.
Pew Research Center	A non-profit and research fair fact tank that informs people about issues, trends and worldviews, and uses polls to do so.
One Army projects	This group creates many projects (and educational videos) to tackle global problems such as Precious Plastics, eWaste and Fixing Fashion. Story Hopper videos are a One Army project.
DW Planet A	Deutsche Welle is Germany's international broadcaster, with many videos in English outlining environmental issues.

Weblinks
The 17 Sustainable Development Goals

The Index Project home page

Wool4School Design Competition

United Nations Environment Program

The Guardian

#BreakFreeFromPlastic

ClientEarth

Only One

V&A Innovate

Pew Research Center home page

One Army home page

Research Videos from One Army

Story Hopper

DW Planet A

Market opportunities

A market opportunity is a need for which there are currently few solutions or products available to purchase. This is a business term that indicates there is a 'gap' that a new product can fill, benefiting both the business and the customer. Sometimes this process is called 'finding the pain points' for consumers and then designing something that appeals to reduce those pain points.

Refer to the design problem examples listed on page 194 and read the section 'Deciding on an end user' in this chapter to help you think about a market opportunity that could be the focus of your SAT.

Case study
Etiko

Using existing market research

- Statista is a market research site that has paid subscriptions but allows you to access basic information for free. It lists the main topics on the home page and for each topic there are many graphs and reports available.
- Google Trends lets you see specific phrases that people are searching for. You can brainstorm words around the type of product you are intending to design for your SAT and see if it suggests any extra ideas for you, or channels to pursue.
- Online tools require subscriptions but can be used freely for short periods. They provide keywords searched by people in relation to a business.

Weblink
Statista home page

Weblinks
9 Free Market Research Tools You Should be Using in 2021

The Index Project: Winners and finalists

- Watch the video '9 Free Market Research Tools You Should be Using in 2021' for an overview of some key tools.
- Need some guidance with writing your design problem? Go to the Index Project website to see the list of award winners. Each case study starts with a design problem statement.
- Check out the James Dyson Awards, where UK university students tackle a wide range of design problems and opportunities.

Worksheet
7.6.1

Template
Design Folio Template

SAT ACTIVITY

1 Look at previous student work from Top Designs and make a note of the design problems it tackles. In the previous Study Design (2018–2023) you will find the design problem written in the end user profile.
2 Brainstorm with your classmates or on your own, and write down as many ideas as you can for:
   a design problems you could 'solve'
   b market opportunities
   c speculative design ideas
   d regenerative design ideas
   e ethical design ideas.
3 Circle and annotate those ideas that have more appeal to you for your SAT project.

Deciding on an end user

Human or non-human

Your end user can be human or non-human. However, you will be looking at statistics largely through the human lens. For example, if you choose to design a product for a pet, you may decide to interview pet owners or look at sales of similar pet products. It is humans who would purchase this type of product, use it in conjunction with the animal, care for and maintain the product and then dispose of it at the end of its life. You will, however, be able to watch the animals at play or while feeding, walking or sleeping and record your observations. You can also get feedback from experts 'in the field'.

Throughout this text, it will be assumed that your end user is a person, but if you are designing for a non-human, you will need to extend and apply the same instructions or refer to the 'owner' or manager of the non-human or an expert in that field.

End user must be real and accessible

Your end user must be a real person from whom you are able to get feedback. For example, it isn't feasible for you to choose a famous film star, model or sports star as you will not be able to get their feedback. Choose family, friends or school/work colleagues that you know you can communicate with, either in person or virtually. Or you could choose one or more people who represent the type of people your product is for.

You may, however, find that your end user likes the particular style of a famous person, and you might therefore be able to use the 'star' as inspiration.

You may be designing for end users in another country and be unable to communicate with them. In this case you could ask interested classmates for feedback or you could gain feedback from someone who has lived or worked in that country or setting.

9780170477499

The end users are the people who give you information for the design brief and feedback on all your ideas and prototypes and, ultimately, on the finished product. Regardless of your choice of end user, aim to see them as belonging to a target group. You can design for yourself but you need to also see yourself as a member of a target group (or tribe) and get feedback from others in the same group. Read more on demographics in Chapter 13, 'Factors that influence product design'.

Shutterstock.com/Rawpixel.com

Examples of end users

Typical end users in a target group	Broader end user groups
Male teenage skaters/surfers	Arthritis sufferers
20- to 30-year-old female city office workers	Slum dwellers in Asia
Labourers who carry tools	People with impaired vision
People needing gym-to-office wear	Bedridden patients
Board members or families requiring furniture	People without access to clean water
Children in a childcare centre	City workers who drive a car
Teenagers who move from one parent's house to another	People who live in cities with toxic air
Young adults who prefer to wear gender-neutral clothing and accessories	Consumers interested in sustainability

If designing for one end user, you need to think of them as part of an end user group or target market.

Defining the end user

Clearly defining your end users in a profile will come later, when you write the design brief. This will happen after you have completed some research. The research will help you 'look around' for ideas and possibilities and learn more about the end users and their needs that you are considering for your SAT project. This could be in the form of interviews or by making your own observations and taking notes. Your research will be enhanced by putting yourself in the shoes of your end users to understand their point of view. This lets you deeply understand the need and create the best solution. See page 210 for more on an end user profile.

Watch the videos 'Think Like an Innovator' and 'The Innovate Journey' from the Victoria and Albert Museum in London. Both will give you an idea of the journey you will take with your SAT.

Weblinks
Think Like an Innovator

The Innovate Journey

Worksheet
7.7.1

Template
Design Folio Template

SAT ACTIVITY

Think about possible end users for your SAT project. At this stage it might be a broad group; you will define your end user more clearly in their profile for your design brief.

1 Start contacting people as your end users to check if they are willing to give you feedback for the whole of your SAT.
2 Write a brief note/message/email to send to them for their response.
3 Start collecting images to create a visual profile to convey who your end users are and what their interests and tastes are.
4 Review the Factors that influence product design and start creating questions to ask your end users. Think about research that will give you both qualitative and quantitative information.

Research

Research is your next step, helping you to:

- build up knowledge of your end users and their needs
- investigate a market opportunity or a need.

You may want to start from the point of an end user and their needs. If so, read about suitable end users for your SAT under the heading 'Deciding on an end user' on page 196.

Primary and secondary research

Primary sources
Sources of information that is obtained first-hand

Secondary sources
Published accounts of information supplied by others

In your folio, aim to include research from both **primary sources** and **secondary sources**. Primary research is what you do yourself, i.e. tests, trials, photos, interviews, observations, etc. Secondary research is information already published by others in any form of print or online. In your folio, you need to make it clear whether the information came from something you did, or something someone else did. It is acceptable to insert the secondary research, but you should explain why you have included it, how it is helpful and which bits are of the most help by highlighting and annotating them. You should also acknowledge the source, making it clear and obvious to the viewer. See the heading 'Acknowledging IP in your folio' on page 199.

Watch a short video explaining the difference between primary and secondary research.

Ethical research

Weblinks
Market Research: The difference between primary and secondary sources

Throughout your SAT, all your research should be done ethically. There are strict 'rules' in Australia for conducting medical research that requires funding and is intended for publication. We can borrow some of those guidelines, as follows.

- Respect people's **privacy** – only include information that is relevant to the project and do not share it with others who are unrelated to the project.
- Be **honest** about your intentions and present any information truthfully and accurately, especially if there is a possibility it may become public.
- Treat people **fairly** and with respect:
  - Consider any special needs of your end users.
  - Reference and acknowledge any work by others in your folio. See more on this on pages 199–201.
- **Respect** and look after the environment and public resources, and take care of animals if any are involved – particularly if your own primary research involves tests or chemicals that could have a negative environmental or health impact.
- **Be responsible** – comply with any relevant legislation, regulations or guidelines.
- If involved or engaged with any Aboriginal and Torres Strait Islander peoples, recognise, value and respect their heritage, knowledge, cultural property and connection to land; recognise their right to be engaged in and see any outcomes of your research.

9780170477499

Acknowledging IP in your folio

Intellectual property (IP) is the property of someone's mind, something that a person creates, and is protected by laws. Read more about IP and the various forms it takes in Chapter 13, 'Factors that influence product design'.

Whose work is it?

All information and images – photographs, drawings, text and any other medium – that you have not created must be acknowledged.

Any time you use the work of others – i.e. something already published on the internet or in hard copy – you need to acknowledge the intellectual property (IP) of the creator or quote the source.

In your folio it must be clear which work was completed by you and which work has come from others.

Shutterstock.com/IOIO IMAGES

Clear acknowledging of IP owners or sources is best done on or next to the work in your folio.

Acknowledging the intellectual property (IP) of products/images

Any image (such as a photo, illustration or diagram) you find, and that you haven't created yourself, has been created with great effort by someone else. As a moral right (part of the Copyright Act) the creator of the work has the right to be attributed. This means it must be clear and obvious to the viewer who the creator is. Read more on the 'moral rights' page of the Copyright Agency Australia website.

In the example at right it is clear and obvious that the designers are Luke Sales and Anna Plunkett from the Australian brand 'Romance Was Born'.

Weblink
Moral Rights

Note that online retail sites or 'libraries' such as Google Images, Etsy, eBay, Pinterest, AliBaba, Farfetch, Shein, Supre, Gumtree and Amazon are *not* the creators of works; they simply showcase or sell the products that appear on them. You must dig deeper to find the creator's name. A better way to search for inspiration is to select a brand or designer you like and search directly for images of their work. See more hints on page 201.

See the example of LyndyLouDesigns page on Etsy.

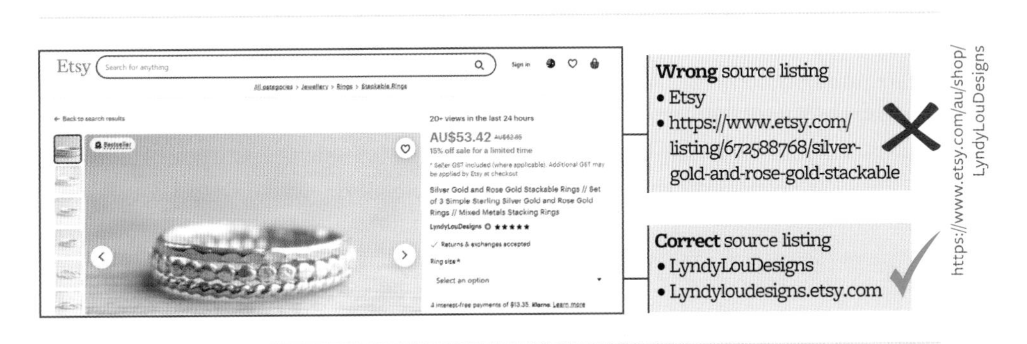

https://www.etsy.com/au/shop/LyndyLouDesigns

Acknowledging the source of information

When you insert text from a publication (text that you haven't created yourself), you need to include the source. If it's from a website, then find out the name of the organisation.

Weblinks
Cite This For Me

EndNote

For example, instead of pasting a long URL, simply state the name of the website, as shown in the example below. To justify why it is in your folio, circle and annotate the text to explain why it is useful to you or highlight areas that are important to your project.

Some websites will help you keep track of written information sources. They include:

- Cite This For Me, which will keep track of sources if you lose the URL but can remember the name of the website. You need to register with an email
- EndNote, which is commercial software for keeping track of research that allows users a free 30-day trial.

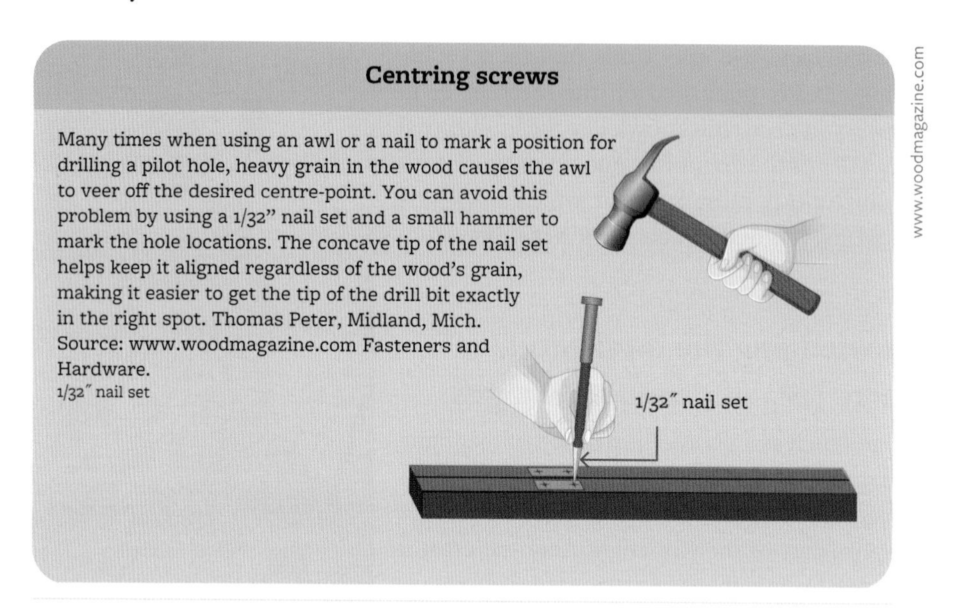

Information such as preparing to attach hinges can be inserted in your folio, but you must include the source (as in this example) and annotate why it is useful to your project.

Use of a bibliography in a design folio

A bibliography is not required in your Product Design folio: it is far more effective to acknowledge sources of other people's 'work', or the creator's name, on the page where you have inserted that work. That way, you are following the moral rights code and making the creator's name clear and obvious to the viewer.

If you do choose to use a bibliography, it needs to be completed properly. A bibliography is an important part of essays and research work and, if you intend to continue down that path at a tertiary level, it might be a good practice to start now.

Weblink
Easy Cite referencing guide

There are many protocols and 'rules' for creating bibliographies, and different methods or 'styles', including RMIT, Harvard, AGLC4 and Vancouver. All styles, however, require you to note any part of your work that is not your own. To assist with this, the pages of your folio need to be numbered and all images given a number – e.g. 'Figure 1' – and these numbers must correspond with how they are listed in the bibliography. You can look at several different styles and their conventions on the Royal Melbourne Institute of Technology's (RMIT) Easy Cite page. You should use the same style all the way through your folio.

9780170477499

Note that adding a list of long URLs at the end of your folio is not a valid method of acknowledging the IP of the creators of the work you have used. The person looking at your work will have no way to decipher which URL belongs to which image or to what information. Refer back to the section 'Acknowledging the intellectual property (IP) of products/images' on page 199.

When acknowledging IP in your folio:

- Give each image a caption that includes the name of the creator of the work, e.g. '*Comb of the Wind* sculpture by Eduardo Chillida, Spain'.
- Place the caption so that it is clear which image it refers to.
- Include where you copied the image from, e.g the name of the book, website or gallery.
- If the source is digital, add the name of the website, the date you accessed the information, then the URL.
- Avoid long URLs that quickly go out of date; search further to find the creator's own website (or the brand website).
- If you are adding your own photos, you can add the copyright symbol followed by your name and the year, e.g. ©[Name] 2023.
- Make it very clear which work is yours and which was created by others.
- Include the name of any software you used to create your drawings or designs (visualisations, models, design options, working drawings and/or prototypes).

End user research

End user research can identify:

- how users deal with current products (including their issues and problems)
- why they need the product or what it needs to do
- what could be improved in these products
- how a completely new product could address their need.

A significant proportion of user research is conducted with potential consumers (typical end users or target market), ranging from focus groups and in-depth interviews with target market groups, to more focused and detailed **ethnographic** and observation-based techniques.

Ethnographic
Related to a branch of anthropology that studies peoples, cultures and their customs, habits and mutual differences

Qualitative and quantitative research methods

Market research is done through interviews, focus groups, user trials, questionnaires and surveys. It is used to gather users' feedback on existing products, and to identify any desires for new products. It aims to discover some 'truths' about what customers really want, and what they are happy or unhappy about. Depending on the nature of what is being asked, the resulting information can be quantitative (numerical) or qualitative (literal).

To find out more about market research, go to page 195.

Qualitative and quantitative research

Information from research can be either qualitative or quantitative:

- Qualitative research gives us descriptive content (it describes something or gives opinions). It is gained through open-ended questions or observation of users.
- Quantitative research gives us numerical content (numbers as data).

Weblink
Quantitative and Qualitative Marketing Research

For a detailed explanation watch the video 'Quantitative and Qualitative Marketing Research' on YouTube.

Focus or discussion groups often give qualitative data.

Qualitative research

Qualitative research seeks descriptions or suggestions that are related to how people think or feel about something. They describe a quality or characteristic and might include observations of human behaviour and the reasons for it, such as why and how people make decisions. It is sometimes referred to as research based on reality. This type of data can be collected through interviews, observations (notes taken by the observer) and focus group discussions.

Because this type of information can be quite involved (or complex) and harder to analyse than numbers, it is often done with smaller numbers of people.

Methods of collecting qualitative data

Qualitative research does not happen in a lab or while sitting at a desk. Collecting data for qualitative responses may require organising a group, and it will involve asking people for descriptions and/or observing what they do as well as how they perceive and interpret things. Questions need to be open-ended to allow participants to provide detailed responses from their own viewpoint, rather than choose from a fixed set of possible expected answers.

It is a time-consuming method and therefore expensive, making it suitable only for working with a small group. However, the information is richer, deeper, gives more insight and is often more useful than quantitative data as it can present unexpected information.

The main methods for collecting qualitative data are:

- individual interviews – these need to be conducted like a conversation or discussion
- focus groups – useful for discussion of circumstances, behaviour or opinions relevant to a target market or about a particular product
- observations – the researcher needs to watch, listen and take notes on behaviour around the product's use, or use the product themselves.

9780170477499

Qualitative questions usually begin with 'How', 'Why' or 'What'. To be of good quality, this type of research requires a lot of thought and preparation, so that the questions elicit useful responses.

Examples of qualitative questions are:

- What is it like using this product in the rain?
- How do you feel when using this product when you are under stress?
- Why do you need to have quick access to the external pockets of this bag?
- What are the things you find frustrating about this product?
- How would you describe your ease at performing tasks with this product?

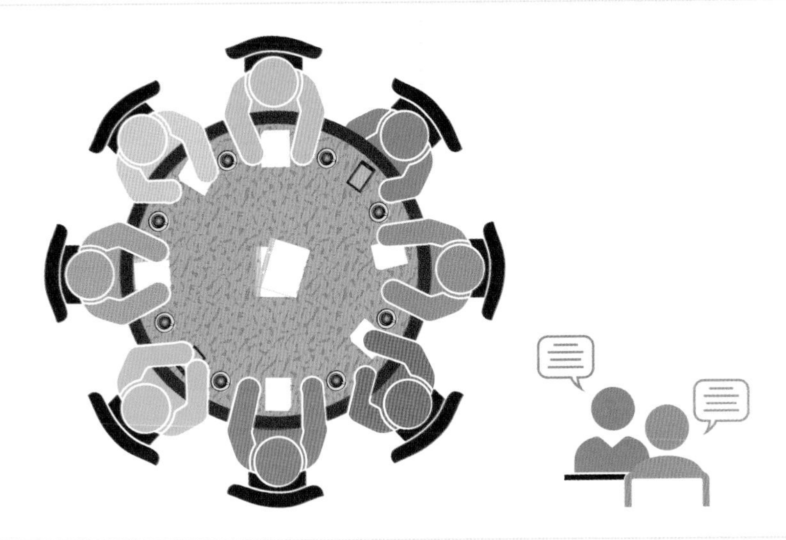

Quantitative research

Quantitative research seeks information that is measurable, that is based on numerical data, and that can be categorised or put into a rank order. We can remember its meaning by thinking of the word 'quantity', which denotes 'amount'. This type of information is most useful when taken from large groups of people.

Examples of quantitative data might be:

- 63 per cent of 300 people surveyed scored the product's function 3 out of 5 for ease of use
- 80 per cent of people chose the blue handle over the yellow handle.

Height in centimetres, age in years and weight in kilograms are also examples of quantitative data. However, information about the user themselves, or the number of times they use a product, may or may not be useful to you. It may be more helpful to get 'feedback' on using a particular product or actions related to using the product you intend to design and make. You need to word your questions carefully to get suitable and useful information.

Note that if you need feedback to help evaluate a product on the market, then questions about the end user's habits will not be helpful at all. Think about why the following question would not help you evaluate a product:

- How many times a day do you use this product?

This question does not focus the response on the product itself. This type of question may be helpful, however, if you are looking for feedback on a product that is still in the designing stage and need to make decisions related to durability, for example.

iStock.com/studiocasper

Quantitative data is numerical and is easily translated into graphs and charts with the use of software.

Methods of collecting quantitative data

As with qualitative data, collecting data for quantitative responses will involve asking people questions. However, these questions are usually 'closed' questions, with a limited range of responses. Often, the respondents can fill in this information online or in a digital format, rather than respond face to face.

The questions are asked with choices for the response, such as:

- a numerical answer (score or rating)
- a choice from a set of options
- yes or no.

Suitable questions might include:

- How would you rate the handle placement on the case out of 5 (where 1 is worst and 5 is best)?
  O 1 O 2 O 3 O 4 O 5
- Do you like the available colours?
  O Yes O No (This would give a numerical result.)
- Do you prefer zippers or buttons?
  O Zippers O Buttons (This would give a numerical result.)
- How important is having two handles to you?
  O Very important O Moderately important O Not important

Quantitative data is easily analysed by software that can quickly generate graphs or charts and make the information visual. It is suitable for large groups of people.

Representing data visually can also be used with smaller groups. The bar graph on page 205 represents the height variation between six people. Three are between 100 and 150 centimetres tall, two are between 150 and 200 centimetres and one is more than 200 centimetres. The pictograph represents the same group of people.

9780170477499

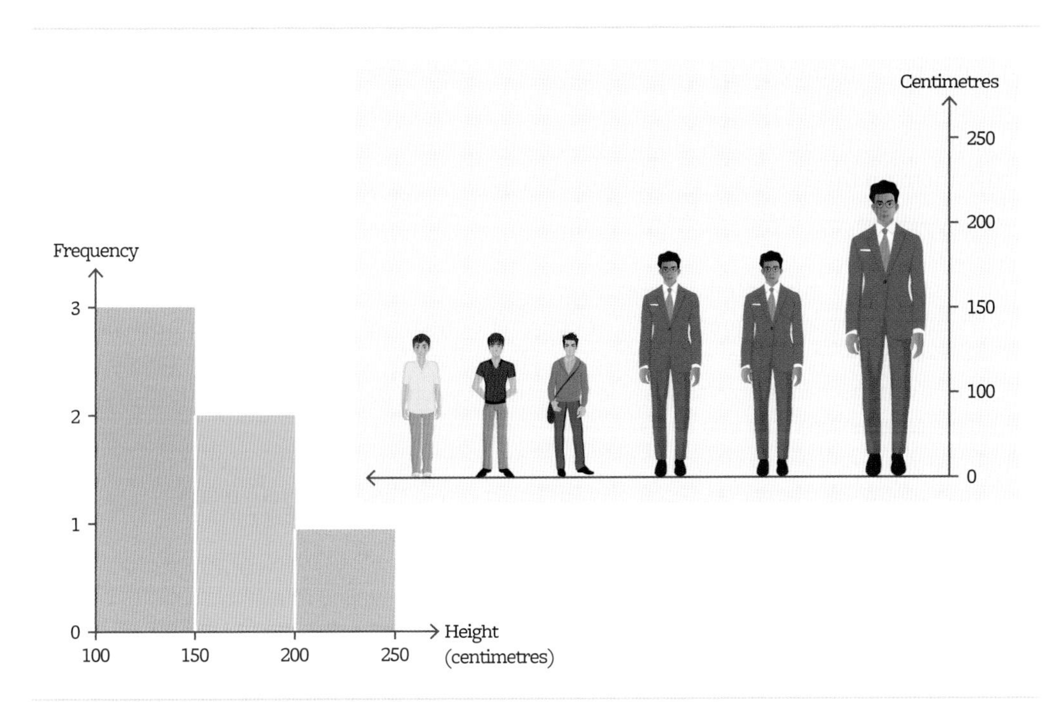

Quantitative data in business

Manufacturers use quantitative data to assess sales of a product to see how it is performing with consumers. This has certain advantages:

- Sales can be tracked through barcodes and stock inventories.
- Product returns can be tracked, particularly through warranty systems. If a specific type of product is returned frequently, this indicates that there is a problem somewhere in the design, quality and/or manufacture of the product. If the product needs to be replaced or serviced frequently, this increases the cost of that product to the manufacturer.
- Other quantitative data could be sizes, prices and energy ratings to allow consumers to compare products to suit their need and budget. This is common on consumer advocate sites such as the Choice website.

In a business, access to sales and tracking data is typically limited to specific employees. It is important to keep in mind that this information may not be useful for evaluating a particular product. Evaluating products requires carefully considered key questions to generate helpful feedback. These will vary, depending on whether the purpose of the information is to help:

- design a new version of a product
- design a product to outdo competitors
- design a completely new product to meet a market opportunity
- evaluate and compare similar products for consumer information.

Online sites that use both qualitative and quantitative information to compare products for consumers are:

- Choice Australia
- 'Indy Best' from the *Independent* (a UK newspaper).

Weblinks
Choice

Indy Best

Qualitative and quantitative research in your SAT

You will be using quantitative and qualitative methods to collect more details about the end users of your product. Note that, to be useful, quantitative data needs to be collected from several typical end users.

Look at the examples of suitable questions on page 204 (for qualitative information) and consider how they could give feedback on an existing product or a new product in development. Consider the type of questions that will help you gather information about the end users or communities that will help you with decisions for your SAT project. Remember to base them on the Factors that influence design.

Worksheets
7.9.1

7.9.2

ACTIVITY

Quantitative and qualitative information

1 Create one cartoon image each for quantitative and qualitative information (e.g. a pen writing sentences for qualitative, and a graph or pie diagram for quantitative) and beside each image write as many things as possible that relate to each type to aid your recollection of the difference between the two types.
2 Swap your work with a classmate and check to see if you are both clear about quantitative and qualitative information. Discuss any differences and check with your teacher.

Template
Design Folio Template

SAT ACTIVITY

1 Decide how you will collect qualitative data from your end users, and plan how to do it. For example, will you ask open-ended questions? If so, create at least four. Will you make observations? If so, explain at least four observations you will make, and what you will be looking for. How will you report on what you find (e.g. in text and/or photos)?
2 Create at least four questions to ask your end users for quantitative information.
3 Start applying both methods of data collection and recording the data you collect.

Recording, collating and forming information

Keep all the relevant information and images you find and always record the source and acknowledge the IP (see page 199). At some stage, and with the aid of your convergent (critical) thinking, you will discard anything that is completely irrelevant.

Here are some tips:

- Make notes on why you kept things, what ideas they sparked or what you like about them.
- Make notes on further questions you have or other information you will need to research later.
- Use headings to help the viewer understand what your folio page is addressing.
- Make annotations to indicate how the research is assisting your ideas.
- Consider a clear and uniform size for images that appear on the same page.
- If your folio is multimodal (see page 192), consider how you will record, collate and present research information.

Ethical design of a product

Analysing existing products: are they ethical?

What is an ethical product? This may depend on a person's values. Many people consider that being sustainable is an ethical priority. Yet how can we determine the sustainability of a product? The information that companies provide about sustainability is often opaque (hard to understand) or misleading. Often, the first important point that people want to

9780170477499

know about for a product to be ethical is the use of labour to produce it. For example, do workers in the countries where the product is made work in safe conditions in factories and have safe methods of transport? Do they receive reasonable rates of pay – a living wage? Are they under undue pressure to work long hours, or do they receive threats of punishment or dismissal for speaking out about poor conditions? You may choose to analyse products that don't seem very ethical. This will give you the opportunity to create a 'better' product through your design development process.

Some types of information can, however, be very hard to find. Some manufacturing companies don't even know how their own contractors treat their workers. A good example is the technology company Apple, which espouses workers' health and wellbeing, yet was apparently unaware (or didn't believe) that workers in Foxconn's various iPhone factories in China were so badly treated that they rioted at various times over the past two decades.

Examples of ethical issues

We have to depend on our own knowledge, or on educated guesses, about the issues we see as ethically important. Examples might include:

- materials – how they are sourced or extracted, how much energy is required to process them, are they safe for workers to process and use and what usually happens to them at the end of the product's life
- impact on the environment
  - how purchases of this product impact the environment (think of fast fashion and minimal uses before disposal)
  - whether making the product creates a lot of waste, such as scrap material or toxic chemicals
  - whether the country in which it is made has lax environmental laws
- quality of the product – how long it lasts, whether it functions as expected or has aesthetic appeal
- whether it is designed and made locally – in some situations this is an ethical choice, as local manufacture supports a community and does not require long-distance transport
- Fairtrade certification – no matter which country a product is made in, provides financial and social safeguards for those involved in materials sourcing
- safety – is the product safe under normal conditions? Are the materials safe for humans if used in households? Is it safe for those who are more vulnerable (children, the elderly, those with disabilities)?
- value for money – whether it satisfies the purchaser's need
- whether the promotion matches the actual product
- whether a product, particularly if it is to be used in public spaces, is designed so that all can use it (universal design and accessibility)
- whether a product is culturally appropriate.

Research for your SAT

Researching existing products that fulfil the same function is a good way to start your research for your SAT, as this will start the process of opening up your mind to different approaches and possibilities and to ways in which your design could be ethical. Alternatively, you could search for and present research on products that are very different from your own SAT proposal, and consider ideas that you could 'transfer'. For example, transfer the ideas or approaches incorporated into a car's seat belt into the straps of a bag design, or the sustainable practices in building into furniture design.

Examples of research projects

See the following two examples of research on ethical products. Each includes statements on why they were chosen as 'ethical' and how this could influence ideas for a SAT project. To improve your knowledge of the Factors that influence product design, annotate products in your research to indicate how several Factors are addressed.

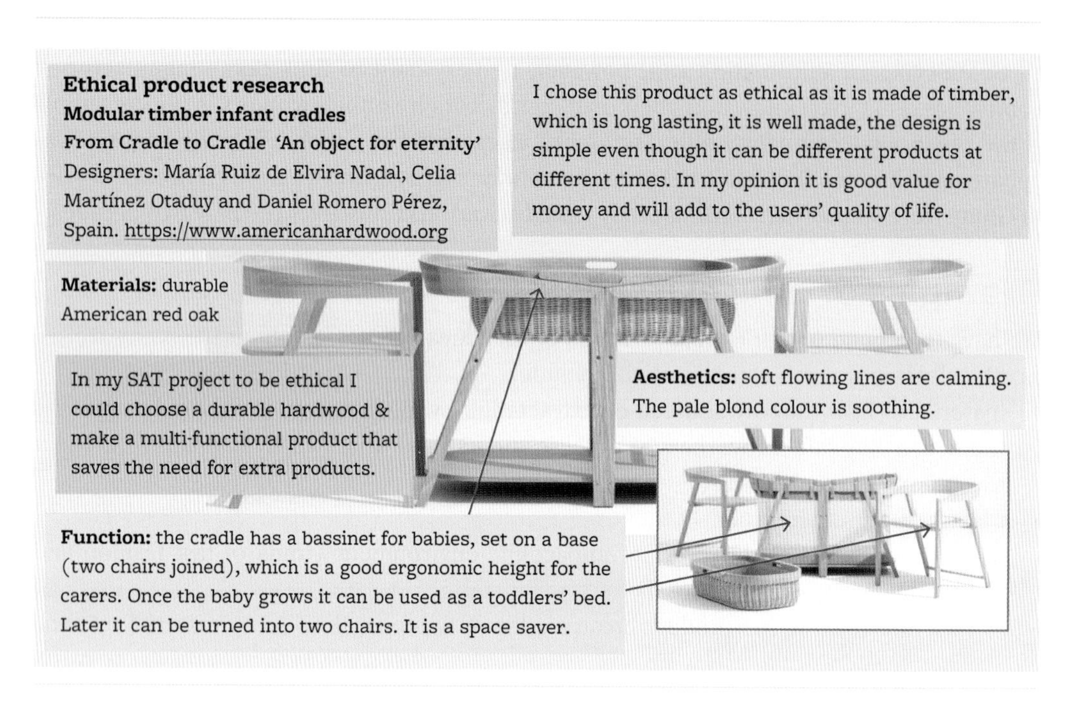

Ethical product research

Australian Animal Felt Play Set by Poshom

https://thefairtraderstore.com.au/collections/kids-toys-and-gifts/products/australian-animal-play-set

I chose this product as ethical because:

- it is designed in Australia (employs Australians)
- it is handmade by women artisans in Nepal who are paid above award wages
- it is education for children, helping them to learn the names of Australian animals and recognise their appearance
- it is 100 per cent New Zealand wool, which is biodegradable, soft and non-toxic for children
- Pashom donates a percentage of profit to the continuing education of the artisans' children and uses Fairtrade principles.

In my SAT project to be ethical I could choose a biodegradable fabric such as pure wool and design a product that is educational for children.

Possible ethical products to research include:

- the Gravity Light on the Index Project website
- Bliss Bean Bags custom made Australia
- Flow Hive Australia on the Honey Flow website
- wooden coding toys such as Cubetto on the Primo Toys website
- BetterLiving Sensory Cushion for dementia sufferers
- Mr Stacky vertical gardens
- portable garden beds on the Vegepod website
- Memobottle
- sensory/orthopaedic floor mats on the Muffik website.

Weblinks
Gravity Light

Bliss Bean Bags

Flow Hive Australia

Primo Toys home page

BetterLiving Sensory Cushion

Mr Stacky Vertical Gardens

Vegepod home page

Memobottle

Orthopaedic Floor: Snake set

You can refer back to case studies on products that make a positive difference in chapters 3–5 (Unit 2).

Researching ethical products

1 Complete and present research on one or more ethical products. Ideally, focus on products similar to what you intend to design and make, to help with ideas. Choose products where you can find enough information that is relevant to this area of study.
2 Include the name of the creator or brand – remember to dig deeper if you found the item on an online retail site.
3 Explain why you or your end users think these products are, or are not, ethical and make a suggestion on how this could influence your SAT project.
4 Annotate the Factors that influence product design and make a brief comment on how they are addressed.

Worksheet
7.11.1

Template
Design Folio Template

Unethical products

You may find it helpful to present research on products that you think are unethical, explaining why and including ideas on how you could design a similar product that is more ethical for your SAT project.

The design brief

All your research and analysis so far has been to help you finalise your design brief. At this stage you may still be working out your end users, or you may have decided on them while doing your initial research. Now is the time to make a final decision and profile them in your design brief.

Elements of a design brief

A design brief is a written statement outlining the design problem, need or opportunity and its requirements. It sets the direction for the design process. It should not be overly detailed (one A4 page is sufficient) and it should not be a description of a product, but rather an explanation of the problem, need or opportunity. There must be room for your design input later in the process. Therefore, if your end users ask for a specific, well-defined product that you can almost visualise immediately, you will need to explain to them that that is not a suitable scenario for your Year 12 SAT work.

Your design brief is expected to include:

- profile of end user/s
- identification of the intended function
- the project scope, which includes the constraints and considerations with reference to the Factors that influence product design.

See the example in Unit 1 on pages 21–2.

Profile of end user

This needs to give an impression of who your typical end user is. It can be a written paragraph that summarises your initial research, and can include material created visually with a few 'slogans'. The examples below are from Virtual School Victoria 2022 students Anneleise Van Den Broek, Victoria Cannon and Halide Karatas.

End user profile

My end user is an 18-year-old female student living on a farm in rural Victoria. She loves being outdoors in the bush with nature and animals.

She's creative! Her hobbies include crochet, gardening, cooking, looking after her chickens, painting, and finding new ways to live both naturally and sustainably.

Her wardrobe is heavily inspired by 60s and 70s fashion.
She loves bright, earthy colours.

Design problem: finding an outfit that she can wear gardening and caring for her animals that is comfortable, stylish and sustainable.

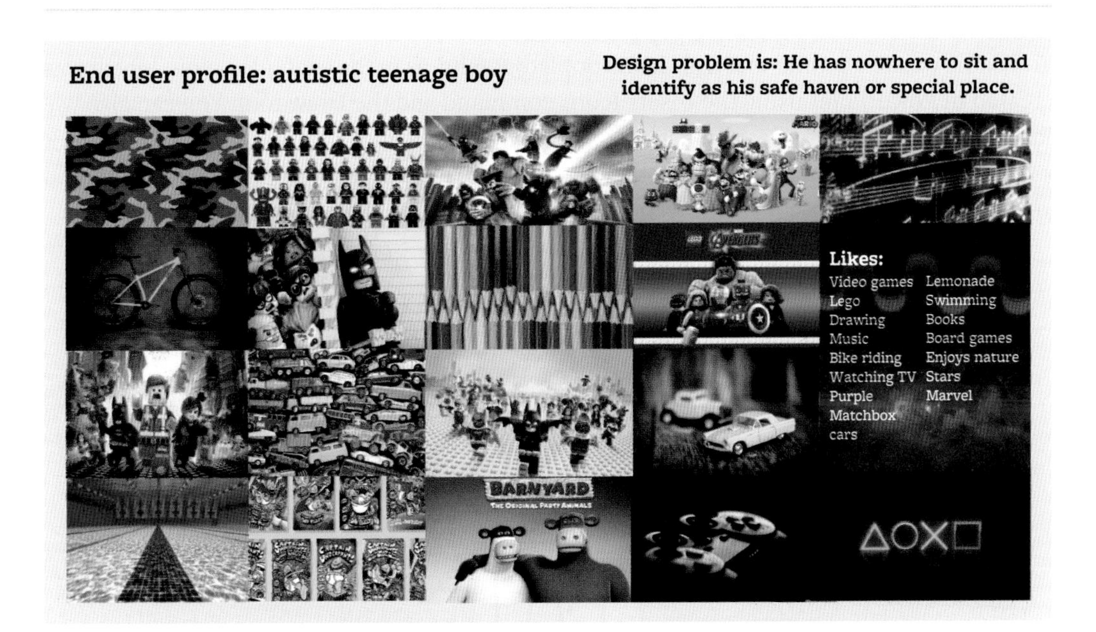
End user profile: autistic teenage boy
Design problem is: He has nowhere to sit and identify as his safe haven or special place.
Likes:
Video games
Lego
Drawing
Music
Bike riding
Watching TV
Purple
Matchbox cars
Lemonade
Swimming
Books
Board games
Enjoys nature
Stars
Marvel
BARNYARD

End user profile
Female Muslim teenager, lives in inner city Melbourne, likes to dress well for formal events.
Loves: Earthy tones.
Pearls and silk: her favourites.
Design problem: There is not any modest formal dresses in the market.
Favourite flowers: daisies and sunflowers.
She enjoys going to the gym and working on herself to eat well and be healthy.

Weblinks
Pictogrammers

Where to Find Beautiful and Useful Free Icons

Make My Persona

Creating an end user profile using icons

You could create a profile that uses icons or pictographs, as in the example below, which was created in PowerPoint using the Insert Icons command and the Draw tool. You can also use Pictogrammers or go to Canva for a list of sites with free icons.

Make My Persona is a free website that allows you to make a 'persona'. It is very limited and you can't download without sharing, but it does ask some questions that may give you ideas.

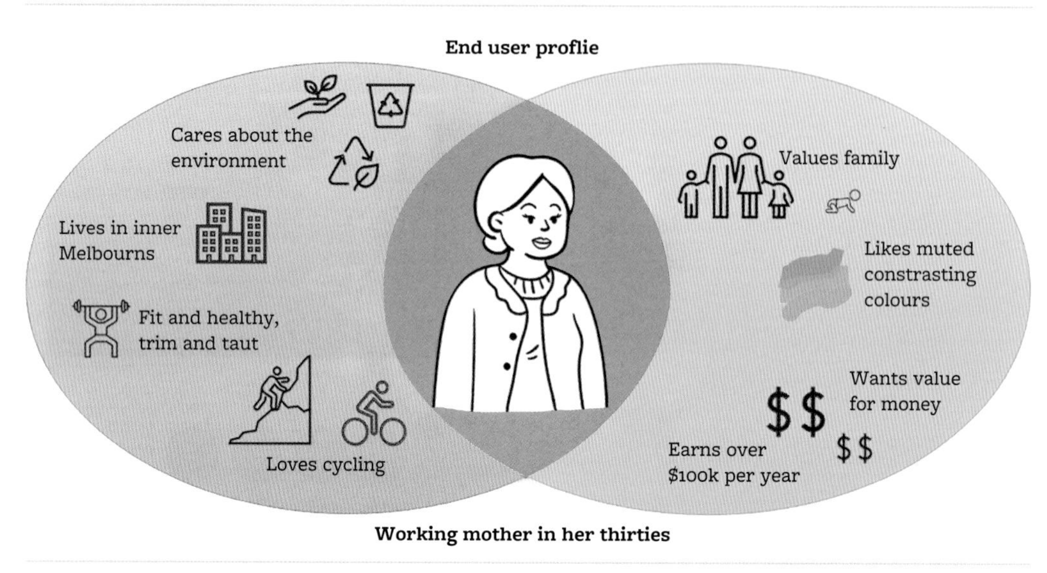

An end user profile, created in PowerPoint, using Insert Icons for images

Intended function

This brief paragraph needs to give an overall explanation of what the product is expected to do, the situation or context, relevant background information, how the product might be used and why it is needed. You may also borrow your statement from the design problem to insert here if it is still relevant. See 'An ethical design problem' on page 194.

Project scope, including constraints and considerations

The project scope states the date for completion and the budget. For all students of Units 3 and 4, the finish date for the completed product will be sometime in late August. The scope also includes the constraints and considerations, giving more detailed requirements. They cover the relevant Factors that influence design. For clarity of communication, it helps if these are written in dot points and each dot point only relates to one Factor. The table shows some examples. See more examples in Chapter 13, which focuses on the Factors that influence design.

Constraint and consideration examples	Design Factor
It must have compartments to store [name particular items and their size and number/amount if relevant]	Function
It must sit stably on the base	Function
The materials must be durable and fire-retardant	Technologies (materials)
It needs to be suitable for a tall person	End user
It needs to reflect the work of Frank Gehry (architect) with silvery organic aspects	Aesthetics
Must be well made and long-lasting	Ethical (sustainable) Technologies (construction processes)

Formulate a design brief

Your SAT design brief must relate to an ethical consideration for a real personal, local or global design need (or opportunity). Your research on an ethical product will help with your ideas. The design brief governs the remaining SAT work. It sets out the needs and requirements clearly.

Sample design brief

Design brief

Profile of end user

The end users of this product are typically aged in their late 60s and showing signs of dementia. Dementia sufferers lose their mental capacity in different ways (losing speech, mobility, memory, etc.) but all feel anxious at various times. They get stressed easily if they don't know what is going on around them. Many enjoy being in nature and feel relaxed when surrounded by colours that remind them of nature. End users also respond to connections to their past and to things that are familiar to them. Many end users lose fine motor skills, so they cannot operate objects with small parts, e.g. buttons or controls. Some enjoy simple, repetitive physical motions that they find calming and that distract from their anxiety.

Intended function

My end users need something in their hands to entertain, comfort and distract them from rising anxiety due to dementia.

Project scope

The object is to be used to relieve anxiety and give the dementia sufferers a sense of calm. It should allow caregivers, family members and healthcare professionals the ability to redirect and soothe anxious behaviours commonly exhibited by dementia sufferers. It needs be engaging and provide sensory stimulation, while also being easy to interact with and hold. It should be safe and durable (withstand being dropped and cleaned).

Constraints and considerations	Factors
• Needs to engage through a range of sensory elements - may include textures, colours, motion, smells, sounds	• Function, Aesthetics
• Should be mostly made from natural and biodegradable materials, but can have some plastics - all must be food safe (non-toxic)	• Technologies: materials • End user: safety
• Must have items/aspects that can be fiddled with easily (without requiring fine motor skills)	• Need, Function
• Any attachments must be secure	• End user: safety
• Must be portable - needs to be easy for them to lift and move, and comfortable to hold	• End user, Function
• Needs to be durable and easy to wash/sanitise	• Technologies: materials
• It must be finished by the end of August	
• The budget is $120	

Aspects that require research and testing

Your design brief holds the secrets for your research, testing, trialling and exploration of ideas. Once it is completed, use a highlighter to go over phrases in your design brief that:

- you will need to research for further information
- will require design exploration or material experimentation
- will require learning and practice of any skills (for trials).

Annotate these highlighted phrases with suggestions for relevant activities you could do. Or create a graphic organiser (visual method such as a table, concept map, etc.) to assist you.

Some examples of activities are:

- Research similar products and their functional aspects.
- Research different types of products to borrow ideas from.
- Measure up and calculate any relevant dimensions, including body size if relevant.
- Research and practise/trial some joining techniques or attaching and using fasteners.
- Draw many ideas for parts, and many ideas for the whole product.
- Photograph the environment where the product will be used.
- Interview end user/s to find out styles that appeal to them or what they find aesthetic (beautiful).
- Collect images of particular styles/aesthetics that appeal to the end user/s.
- Explore ideas by drawing and combining elements of this style into your own ideas.
- Test some functional ideas with miniature models or prototypes.
- Investigate availability of environmentally friendly materials, their suitability and cost.

You will then undertake this work during the design process for your SAT.

SAT ACTIVITY

1 Create an end user profile that is a written paragraph or communicates visually. Remember to read about ethical research and apply the same guidelines here. If your end user profile is going to be seen in public, then you need to explain that to your end user and omit any private personal information.

2 Write a design brief for an ethical product that takes into account the research you have completed on your end user and their needs, the market opportunity that you researched or speculative design thinking that you have considered. Make sure you include sections that cover the end user profile (from Question 1, above), the intended function and the product scope (including the constraints and considerations).

3 Add a column to label which of the Factors that influence design each requirement (constraint or consideration) belongs to. There may be more than one.

4 Highlight phrases that will require some research or design explorations and annotate some suggestions.

Worksheets
7.12.1

7.12.2

Template
Design Folio Template

Evaluation criteria

Evaluation criteria are a set of expectations that help you evaluate something. In VCE Product Design and Technologies, they come directly from the design brief. They explicitly repeat the design brief requirements, particularly the constraints and considerations. They should not refer to a product requirement that has not somehow been referred to or mentioned in the brief. If, however, you think of something important, then simply go back to your design brief and add it in.

See the examples in Unit 1 on pages 22–3 for how to write evaluation criteria.

Questions are preferable to statements as they can be expressed in specific and relevant ways that demand or allow some (brief) discussion. Vague questions with a simple yes or no as the answer are not so helpful. Criteria for the budget and due date will have simple answers, close to a yes or no, but if they are not met, your evaluation will include why.

Evaluation criteria are to help you evaluate:

- your design work (product concepts, both graphical and physical). They also assist with decisions and refinement around this work
- the process you followed to design, prototype and make the product
- the finished product.

9780170477499

See 'Product concepts' on page 217.

Unsuitable criteria for graphical concepts

Criteria that will be difficult or impossible to address in your graphical product concepts (drawings in the design development stage) are those related to the fit of the product, whether the budget or due date was adhered to, the construction quality of the finished product and the materials used in the finished product (since they are not used yet; they are only suggested or proposed in your designing stage). These factors relate to aspects assessed *after* the product has been made, and therefore cannot be used to evaluate your drawings.

See the table below for examples of how to write clear and specific evaluation criteria from your design brief requirements and how to differentiate questions to be used when evaluating graphical product concepts (drawings) from those used to evaluate the finished product.

Requirements from the design brief	Possible wording of questions for evaluating graphical concepts	Possible wording of questions for evaluating the finished product
The design needs to have a bright colour scheme	Is the colour scheme in this idea (or design option) sufficiently bright?	Are the bright colours of the finished product appealing to, or suitable for, the end user?
Need to include storage for the end user's fragile items	Does this design include storage that could be suitable for fragile items?	Does the container protect the end user's fragile items as required?
Materials must be waterproof but also flexible	Are the suggested materials in this design going to be suitable?	Are the materials used in the finished product suitable?
Needs to have suitable sensory activities to occupy my end user	Has the design included sufficient and suitable ideas for sensory activities?	Do the sensory activities on the finished (name of product) occupy the end user (include a brief type of end user) and help them relax?
Needs to fit (name of end user)'s size 10 body (or available space).	Cannot be answered from design option drawings but could be applied to working drawings if all measurements are provided.	Does my finished product fit (end user) nicely? Or fit easily into the required space (include a brief description of the space)?
Must be finished by mid-August	Can't be 'answered' from a drawing, but an estimate could be a response to a question such as, 'Can this design option be designed and constructed by mid-August?' Or 'Is it within my capabilities and time frame?'	Was the (name of product) finished by mid-August?
Budget is $230	Can't be evaluated from a drawing, but could be applied later in the process when the exact amount of materials are costed and/or purchased.	Was the finished (name of product) made within the $230 budget?
Construction must be of high quality	Can't be evaluated in any drawings.	Was the finished (name of product) constructed to a high quality to be long-lasting after constant use?

Read further to see how you will use evaluation criteria and some suggestions under the heading 'Evaluate and critique graphical product concepts' on page 225.

Worksheet
7.13.1

Template
Design Folio Template

SAT ACTIVITY

1 Develop criteria questions directly from your design brief to be used at the end of the SAT when your product is finished. Aim to cover most of the Factors that influence product design, but make sure they are relevant (at least 6–8 criteria questions in total).
2 From the questions, extract and reword a minimum of four evaluation criteria that will help to evaluate your product concepts with your end user/s. Remember these are for design brief requirements that *can* be judged in drawings, diagrams, models or prototypes, and need to be worded accordingly. Suggestions for Factors to focus on are: functional aspects (all catered for in the design?), suitability of suggested materials, aesthetics to suit the end user, complexity of design to manage within budget and time frame.

Note: these criteria can be determined later with your end user when seeking feedback on your visualisations, to help refine or evaluate your design options, or when evaluating your physical product concepts.

Further research

Inspiration for concepts can come from anywhere and at any time. Many of us turn to the internet at first to gain inspiration, unaware that a set of algorithms have some control over what we find. It's important to remember there are other physical ways in which designers find inspiration, such as:
- browsing in shops, travelling, walking short distances, communicating with nature, museum and art gallery visits, etc.
- taking photographs of nature or objects
- any form of visual, mental and physical break from a computer screen to stimulate and refresh the mind
- thinking of questions to create an initial idea.

Note that any of the research presented in this chapter may be completed before or after you write your design brief. If it is completed beforehand to help you finalise your design brief and end user, then it should be placed in that order in your folio. If it is completed after your finalised design brief, then place it after the design brief in your folio.

Possible reasons for you to conduct further research include:
- to gain more information from the end user
- to get aesthetic ideas from other product types or from nature
- to get ideas and information on how to use materials
- to get functional ideas, construction methods or ideas for fastening and joining
- to check the availability of particular colours in suitable materials.

Worksheet
7.14.1

Template
Design Folio Template

SAT ACTIVITY

1 Refer back to your design brief and the aspects you identified for research or testing.
2 Optional: Create a plan or mind map to identify specific things or activities you could or will do to complete this research or testing.
3 Present any further research in your mode of choice or on the template provided on Nelson MindTap.
4 Follow the tips under the heading 'Recording, collating and forming information' on page 206.

9780170477499

Product concepts

In this study you will be creating:

- graphical product **concepts** – visualisations, design options and working drawings and/or virtual prototypes on CAD (see the infographic below)
- physical or virtual product concepts:
  - physical product concepts – something you can pick up and hold that is also still in idea form; could be a model, a mock-up or a prototype. This could be for parts of the product as well as the whole. See page 233 in Outcome 3
  - virtual product concepts – a realistic digital version of how a finished product could look
- final proof of concept – the chosen product concept (most likely, the last prototype) where all the design decisions have been made, checked and agreed on and that contains all the information to create a working drawing and plans for your final product. See pages 239–40.

Graphical product concepts: visualisations, design options and working drawings

Refer back to the information on these types of drawings in Unit 1 on pages 23–5. You can also read more in Chapter 12.

The three drawing types are outlined in the following infographic.

Visualisations could be seen as divergent thinking as they should represent many different ideas – here the quantity is important

Design options are still using divergent thinking but are moving towards convergent as they are starting to narrow down the choices for further refinement – here the quality is important

Working drawings are using convergent thinking as precise measurements are decided on – here the quality is also important

It is important to understand the three drawing types and the differences between them. Refer to the table below and relate them back to the Double Diamond and the different thinking types. For example, it would be considered that visualisations use divergent thinking (because you are exploring creative possibilities), design options use both divergent and convergent (as you are still being creative but are narrowing down decisions) and final working drawings would demonstrate convergent thinking, as most decisions would by now have been made and you are being analytical about manufacturing details.

Don't be concerned about the order in which you do these drawings or any models or prototypes. You can complete them in the order that suits your thinking. Be sure to keep all of them (even those on scrap paper or rough models) or take photographs.

Here are two approaches you can use:

- Create pages of visualisations for one design option, showing a clear link between these ideas and your research and a clear link from these initial ideas (for parts and the whole product) into one design option. Create a further 'set of visualisations pages' for a second design option that also shows the links between ideas – and so on for consequent options. This approach should show how each set of rough ideas moves or morphs into a detailed design option for the whole product and, from there, moves into physical product concepts.
- Create pages with a huge variety of visualisations, showing clear links with your research, and choose the best ideas or combinations to develop into design options.

Types of graphical product concepts	Purposes of communication	Techniques/methods to use
Visualisations	• To explore a multitude of ideas • To flesh out how parts could look or function • To develop ideas for colour, texture, shape, pattern, line and proportions (Design Elements and Principles) • To show ideas gained and influenced from research • To explore placement of parts and functional aspects of a design or how it will be constructed • To explore rough ideas for the whole of the product • To get end user feedback	• Quick rough sketches, some in pencil or black pen and some coloured • Material scraps used to 'build' 2D images • Diagrams • Explorations of Design Elements and Principles with pencils, paints, etc. • Use of any software All visualisations should be annotated to connect your ideas, research and decisions and should include end user feedback.
Mood board: in between and not mandatory (may be done earlier)	• To define/finalise the overall 'feeling/style' or what appeals visually to your end user/s and to use as a reference to guide design options	• Mood board – a purely visual exploratory method (digital or hard copy)
Design options (often known as 'presentation drawings')	• To give an overall realistic (3D) image of the whole product; to show details; to indicate true proportions; to represent surface, materials or texture and to indicate scale • To indicate (by annotations) the materials and construction methods or processes to be used; and the design brief requirements • To get end user feedback for refinement and selection	• Fashion illustration • Crating, airbrushing • One- or Two-point perspective • Isometric drawing • Use of CAD All design options should be rendered, annotated and include end user feedback. Exploded pictorial drawings to show details can be included.

9780170477499

Types of graphical product concepts	Purposes of communication	Techniques/methods to use
Working drawings (accurate, with measurements and to scale)	To give accurate details for prototyping and production such as: • all shapes, sizes and dimensions • position/placement of components, fasteners and parts • materials for all parts • any internal details and an indication of how parts fit together Working drawings are mostly 2D and 'line' drawings – i.e. they do not need to be rendered in colour. However, 3D CAD drawings are also acceptable.	Non-resistant materials Textiles: • Flats, trade sketch or technical drawings (include extra construction drawings or diagrams of parts and joining methods as needed) Resistant materials: • Orthogonal drawing including relevant sectional views, exploded views, hidden or enlarged detail • CAD drawings in either 2D or 3D, as suitable Consider using CAD for accuracy – Adobe Illustrator for textiles and any suitable CAD program for resistant materials. Exploded details of small or hidden parts can be included.

SAT ACTIVITY

1 Look at your researched images for inspiration and start quick drawings.
2 Insert your visualisations near the research that inspired you.
3 Annotate to explain your design thinking.
4 Present at least two annotated design options using your evaluation criteria to guide you.

Worksheet
7.15.1

Template
Design Folio Template

Need help visualising ideas?

At first you may be daunted by coming up with ideas. Perhaps the use of artificial intelligence software might help you? Check that your school allows access to this type of software. If not, you will need to resort to your own brainstorming.

Using AI if permitted

Craiyon is AI software for drawing images from text prompts. Type in a command such as 'mobile phone carrier in a belt'. The software will come up with nine variations. You may find, however, that none of the nine suggestions are suitable. What you are likely to find is that several of them are *almost* suitable. You can then start your own visualisation drawings based on improving one, or several, of the ideas generated by Craiyon. You need to ensure that you are adding your own creative and critical input and making significant changes to the AI suggestion. You can take a screenshot, make it clear which software you used and how it inspired you, and annotate which ideas you will take forward.

Weblink
Craiyon

Visualisations in your folio: presenting multiple ideas

Regardless of the quality of your drawings, it's important that you show a lot of ideas in your folio in the form of visualisations. It doesn't matter how rough they are, so long as they communicate many ideas for parts of your SAT product or the whole product. They can be drawn by hand in 2B or 4B pencil or pen, or you can use digital methods.

Many students start the subject with one idea in their head of something to make. It's important to note that one of the criteria that you will be marked on is your ability to present many ideas and to indicate which part of your research inspired them. To stretch your idea into multiple ideas you can use one or more of the many creative thinking techniques.

Creative thinking techniques

Consider using a trigger word activity such as SCAMPER and brainstorm different things you could change on an existing drawing or product; or elements you could 'steal' from a completely different product or from nature. Using SCAMPER, you may write some ideas, but always move to drawing them so that your design development communicates visually.

Do an internet search on the acronym SCAMPER to find out the list of prompts that might help you be creative.

An area that can definitely be explored in drawings is mixing and changing around combinations of the Design Elements and Principles.

Read more on the use of the Design Elements and Principles in Chapter 12 on page 346.

Tips for better drawings

If you lack confidence in drawing, there are certain steps you can take to make your drawings look better. Try the following:

- Use a black pen on blank paper (good for scanning into software and editing, colouring, repeating, etc.)
- Draw a basic idea and use a light box to copy the outline, then create different versions of the idea by changing features, shapes, colours, proportions, etc.
- Divide your page into a grid (in your head or using light lines) and in each 'cell' draw a small squiggle or a shape copied from one of your researched images. Go back and turn each squiggle into an idea for a part of your product.
- Bring the page alive by adding splashes of colour to one or more of your ideas to heighten some of the images. Use watercolour as a background or highlight, or any suitable media.
- Scan your drawings into any software to manipulate them further.

Weblinks
Inkscape
ibisPaint
Procreate for iPad
Procreate Pocket
Procreate Education
Tayasui
The Best Drawing Apps and Art Apps for 2023

Software you could use to create visualisations includes:

- Adobe Creative Cloud products such as Illustrator or Photoshop Sketch – check whether your school has a subscription
- Inkscape Draw Freely
- ibisPaint X – visit the website to download the version suitable for your device (free or subscription versions) and for tutorials
- Procreate by Savage for use on iPads, or Procreate Pocket, for iPhones. Tutorials can be found at Procreate Education. This is an Apple product and requires a subscription
- Tayasui – many different apps for sketching, colours and painting.

Or search at the Format website for more free drawing apps.

Critical thinking techniques

To evaluate and critique your visualisations, apply your evaluation criteria and get end user feedback. Make the end user feedback visually different from your own annotations. See more under 'Evaluate and critique graphical product concepts' on page 225.

At this stage, you will be selecting a combination of the best ideas to create your design options.

9780170477499

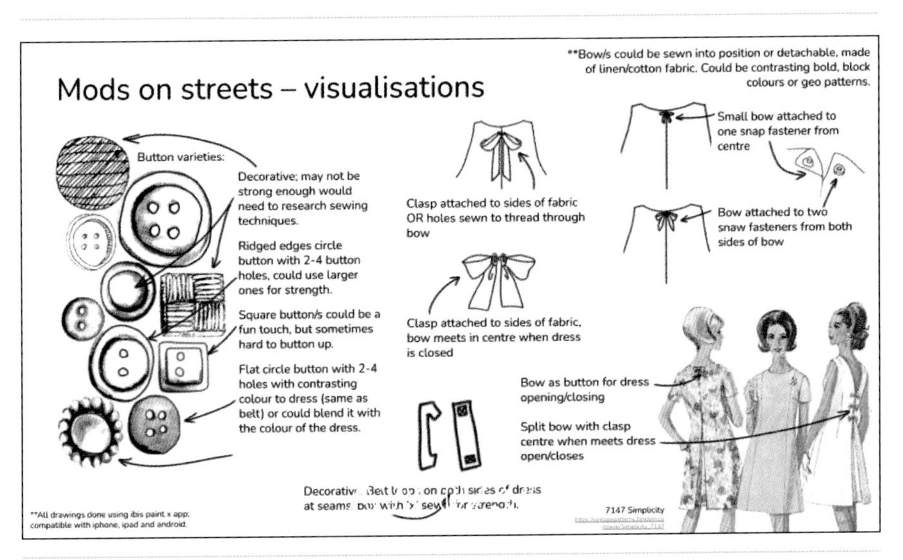

Visualisations by student Esma Corlu using ibisPaint X, as acknowledged in the bottom left of her page. IP of research images also acknowledged: Simplicity pattern 7147.

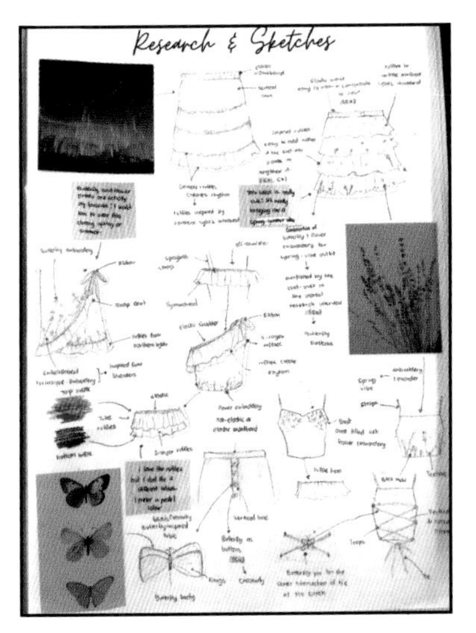

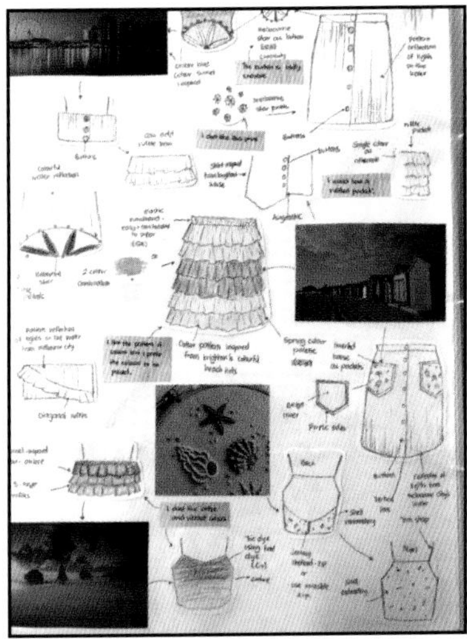

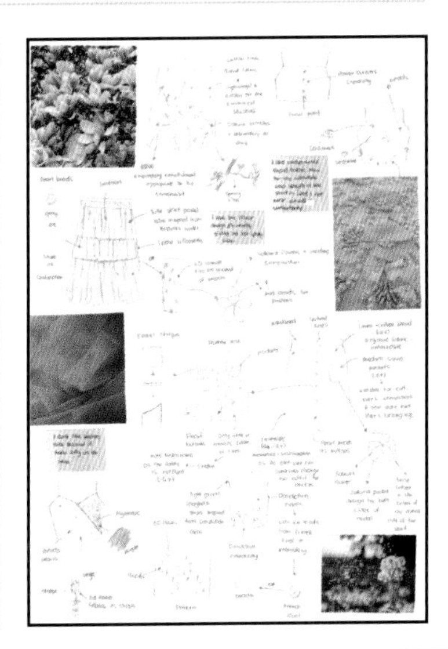

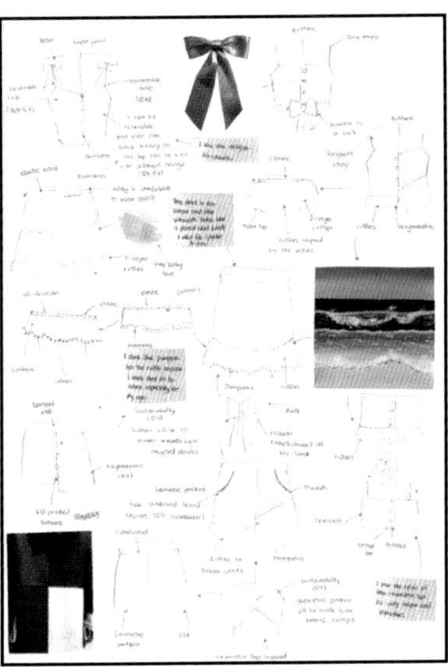

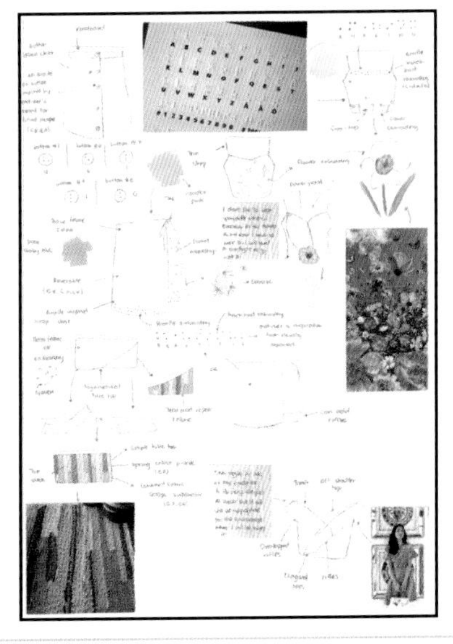

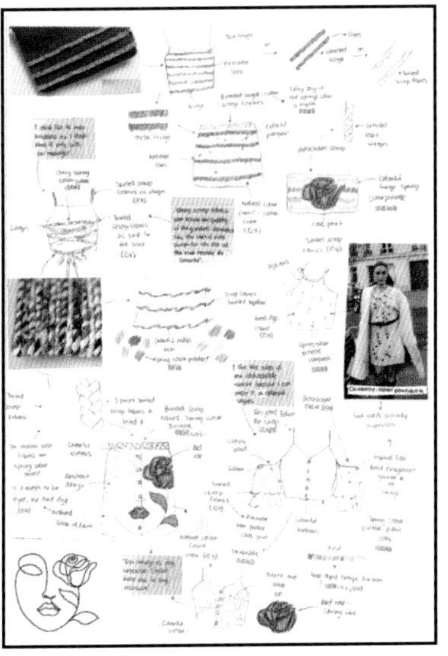

Several folio pages of visualisations, clearly showing inspiration from nature, by student Janna Porras

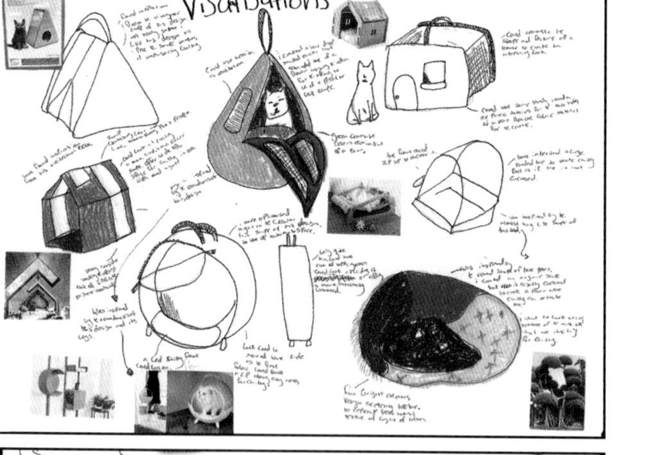
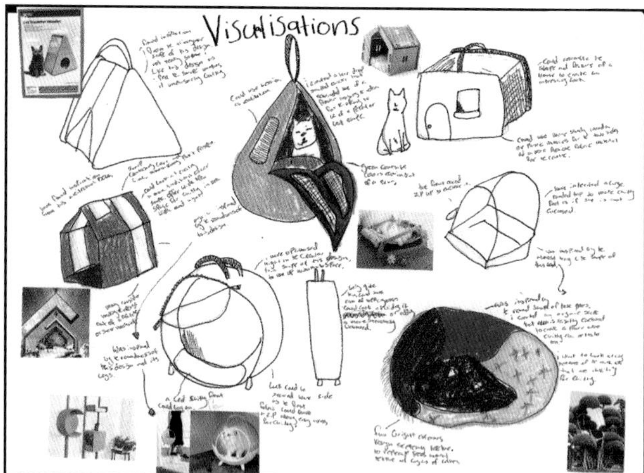

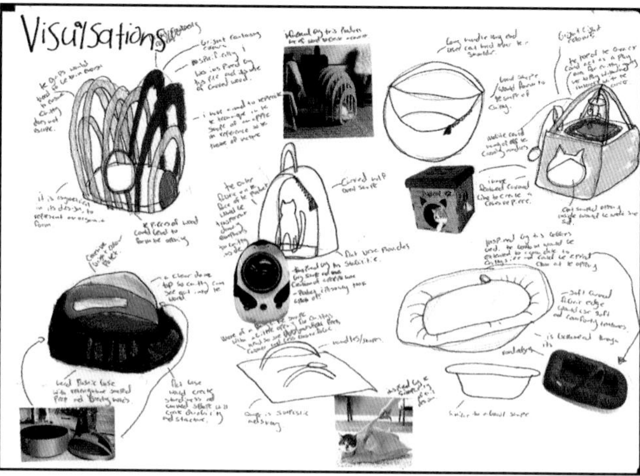

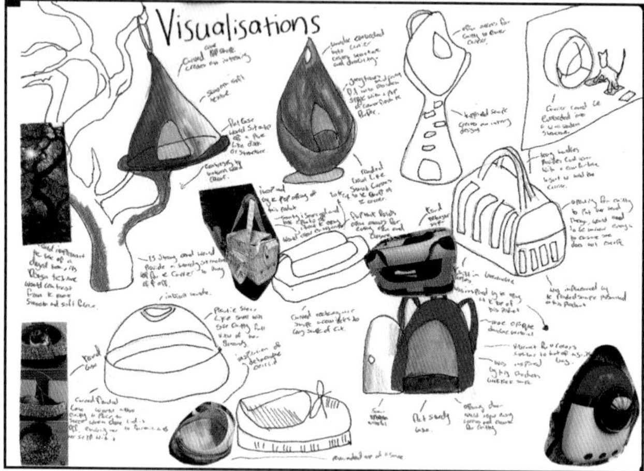

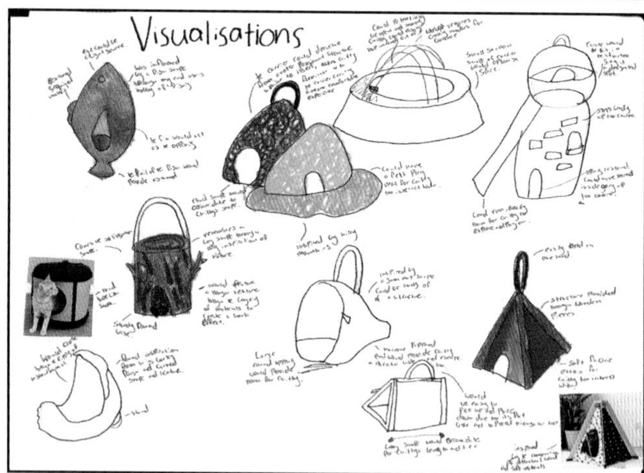

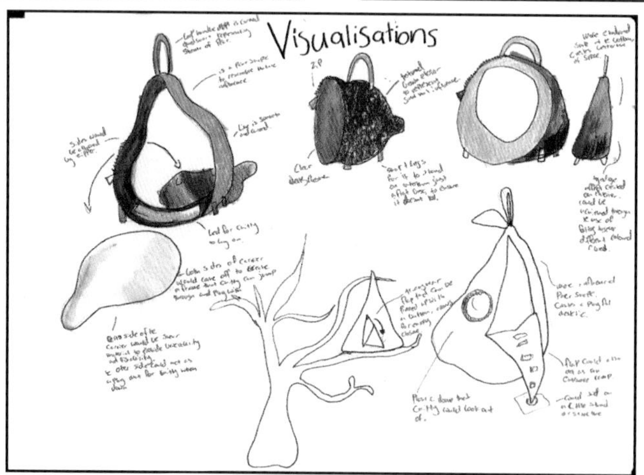

Several folio pages of visualisations, clearly showing inspiration from similar commercial cat carrier products, by student Amelie Thompson. Amelie had previous pages with IP acknowledged on all researched images and annotations as to why they were in her folio.

Design options in your folio: presenting a choice of refined ideas

Design option drawings need to be better quality than your visualisations. Make sure they are visibly different from each other, not just variations of colour on the one design. They need to be detailed, realistic depictions to show an end user how the product could look. They should be rendered (coloured) and annotated.

9780170477499

Annotations communicate to the viewer what is not apparent or obvious in the drawing. You are not required to label parts or colours that are obvious. However, annotations should communicate the materials that might be used, the possible construction processes (including any **fasteners or functional components**) and how the design option drawing meets the needs of the design brief.

Exploded pictorial drawings to show smaller details can also be included if necessary.

Fasteners or functional components
Zippers, buttons, velcro, hinges, latches, belt buckles, loops, press studs, snap/lock fit, handles

Design option 1

End user feedback

I like the shape and different textures of the play mat. It would keep Chitty happy and busy. The added wood features keep it sturdy + upright. The handle + zipper system are a great idea, providing ease of use.

Would have a zip coming from both the left and right side of the base. For both the front and the back, having a total of 4 zips, the zips would be attached with a fabric handle that reaches the widest part of the carrier creating one swift easy motion of pulling the handle up or down to open or close it.

Short singular strap intended to be held with one hand resembles a pear steme.

x2

the handles of the zipper come up at the top to create two longer handles giving the end user the choise to hold the carrier over their shoulder.

felt triangular pockets filled with cat treats

inside of back

felt

Chitty can play and interact with the back when inside the carrier, or when it is laying flat on the ground.

outer of back would be green colour, to create the full look of the pear

each piece would be made of a different coloured and textured material, the shapes are inspired by the segments of a pear.

hand feted felt to create texture

varying colour and cotton to resemble a natural pear.

Woven piece of long felt strips.

Pin tucks created close together in blue fabric to create ruffled texture

Cat not to scale just demonstrating were she would sit.

Calico fabric interior.

wooden legs – possible from a heavy duty dowl.

Play mat and front are attached at the base of the carrier, carrier would sit on the ground when standing upright, relieving presure of off the structure and providing easy access for Chitty.

would be conected at the base of the carrier and when open would lay flat on the floor in front.

front door would be a clear woven material to provide visability and breathability.

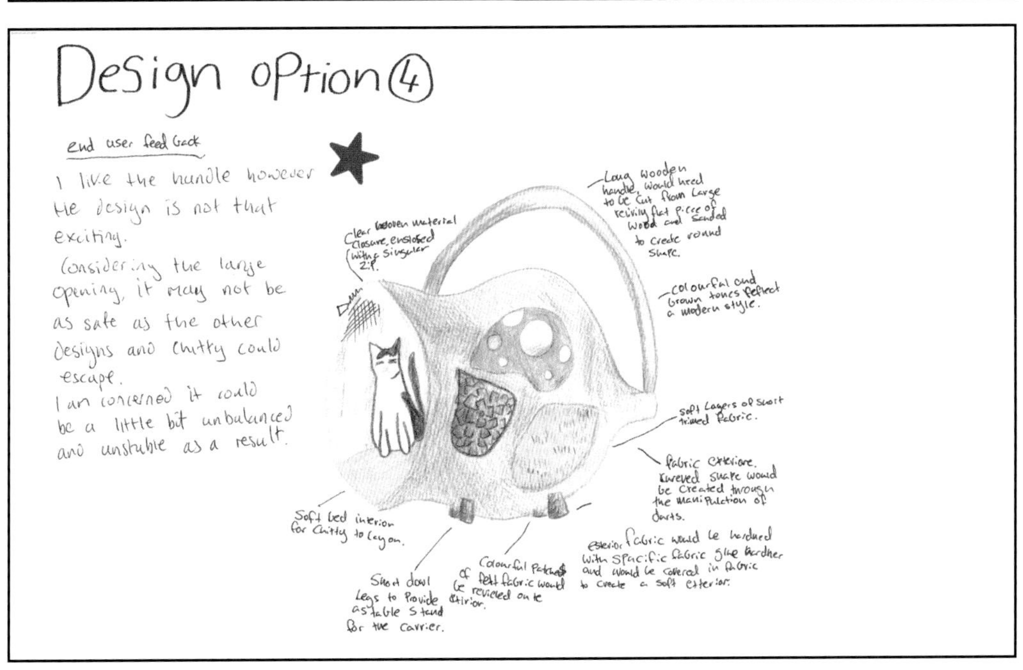

Design option drawings for a cat carrier by student Amelie Thompson (note the detailed annotations)

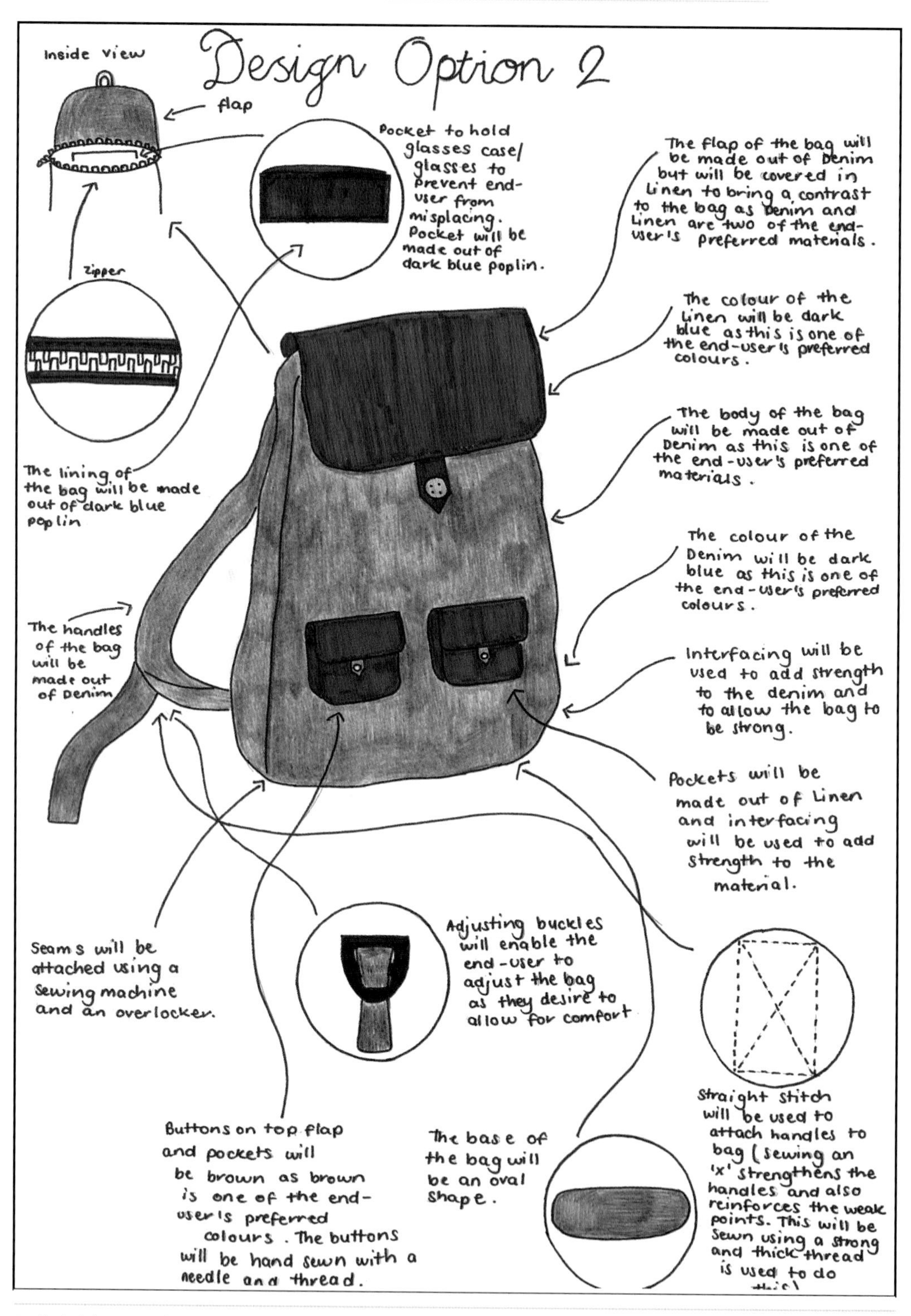

Exploded drawing of a backpack design by student Sarah Hampton

9780170477499

Evaluate and critique graphical product concepts

Once your design options are complete, it's time to get feedback to help you select the best idea. Now is the time to use your evaluation criteria to critique your design work. This is considered convergent thinking (critical and analytical). If you have access to your end users, you need to get their feedback; if not, consider using your classmates, teachers, relevant 'experts' (people with experience in the area) or family. To critique your work you need to seek mostly qualitative information rather than quantitative – see pages 201–6.

Almost all design drawings can be evaluated on these five basic factors:

- functional aspects (all catered for in the design?)
- suitability of suggested materials
- suitability of suggested construction methods
- aesthetics to suit the end user
- complexity of the design to manage within budget and time frame.

If you have not yet worded some of your evaluation criteria suitable for applying to drawings, use the five factors listed above to elaborate on (adding your own relevant details) to create your own criteria to judge your drawings, particularly your design option drawings. Consider adding a small table on the same page as each design option drawing, with room for feedback and suggestions from end users. Differentiate the end user feedback from your own annotations visually, e.g. in callouts, highlighted text or coloured text.

Based on this evaluating and critiquing you may decide to create a working drawing to use for your prototypes from only one design option or several working drawings from two or three design options for different prototypes to compare. You may also decide to leave your working drawing until after you have your final proof of concept.

SAT ACTIVITY

1 **Evaluate and critique:** Show your visualisations to your end users and discuss your idea formation and how you have attempted to meet the design brief requirements. Add any feedback in coloured boxes or a coloured pen to differentiate it from your own annotations.

2 **Evaluate and critique:** Ask for feedback on your design options (ideally, not just a score out of 5) using the evaluation criteria to guide you. You can ask for any positives or negatives visible in the design or for anything that has been neglected or any possible improvements. Add the feedback as you did for the visualisations.

3 **Justify:** With your end users, choose the most suitable design option. Use the feedback for each criterion applied to all the design options to explain and support your choice.

Template
Design Folio Template

Summary
Chapter 7

Quiz
Chapter 7 revision

Working drawings in your folio: to specify dimensions

Working drawings are accurate, dimensioned drawings. Mostly they are 2D, but some 3D drawings can be included to indicate the way the product will look when the pieces are assembled together. They could be considered outline drawings in that they communicate the exact flat shapes of the whole product and its pieces. They are not pattern pieces, nor are they rendered. Garment 'flat' drawings do not include the human body.

Working drawings communicate the specifications – shapes, sizes/measurements, components and construction methods – that will be followed to construct the product. They indicate the placement of components, known as attachments, fixtures, fasteners or (in textiles) notions. Refer to the table under the heading 'Graphical product concepts' to see how working drawings differ from the other drawing types.

In this area of study for your SAT, you will develop a working drawing for the design option you feel you are most likely to take into the prototyping stage or a working drawing for each of the design options that you intend to prototype. Note that, once you have selected the chosen product concept in Area of Study 3, after prototyping, you will most likely need to adjust these drawings or create a new and final working drawing. The thinking and detailed planning involved in producing these drawings reinforces your knowledge of your design. It becomes important for calculating the steps in production. All iterations (versions and repetitions) of your working drawings will go into your folio.

See more examples of working drawings on pages 335–42.

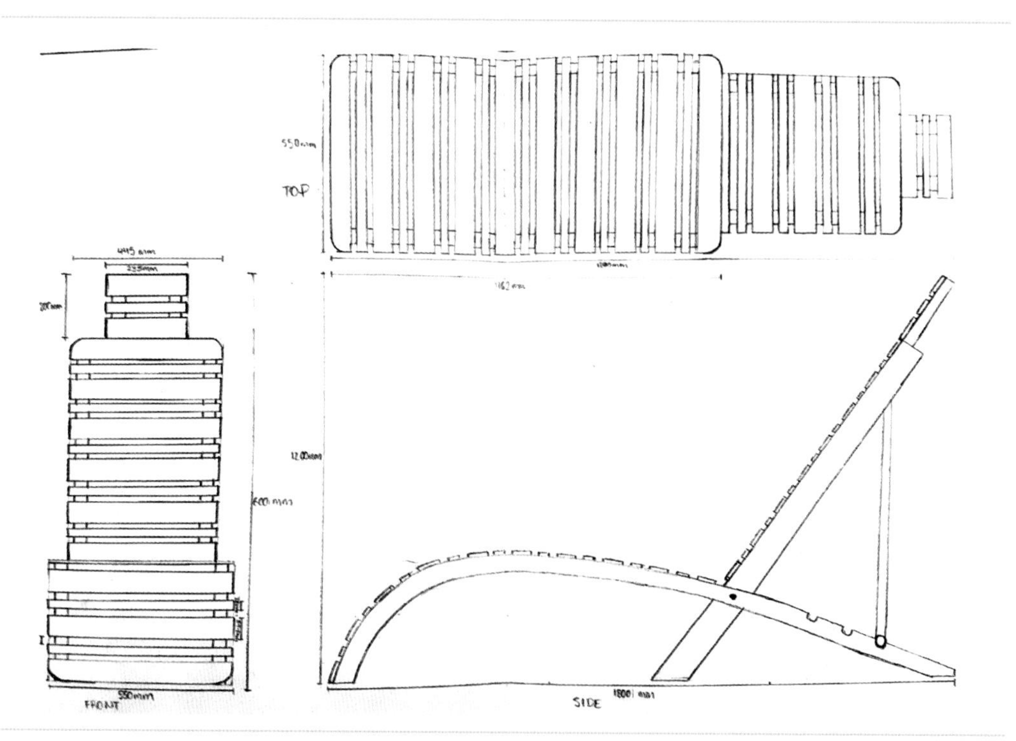

Working drawing for an outdoor lounge chair created by student Charlee Turner

9780170477499

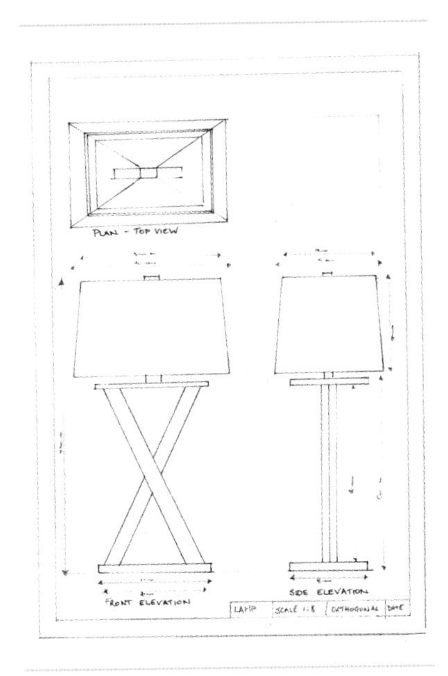

An orthogonal drawing of a lamp, showing front, side and top views

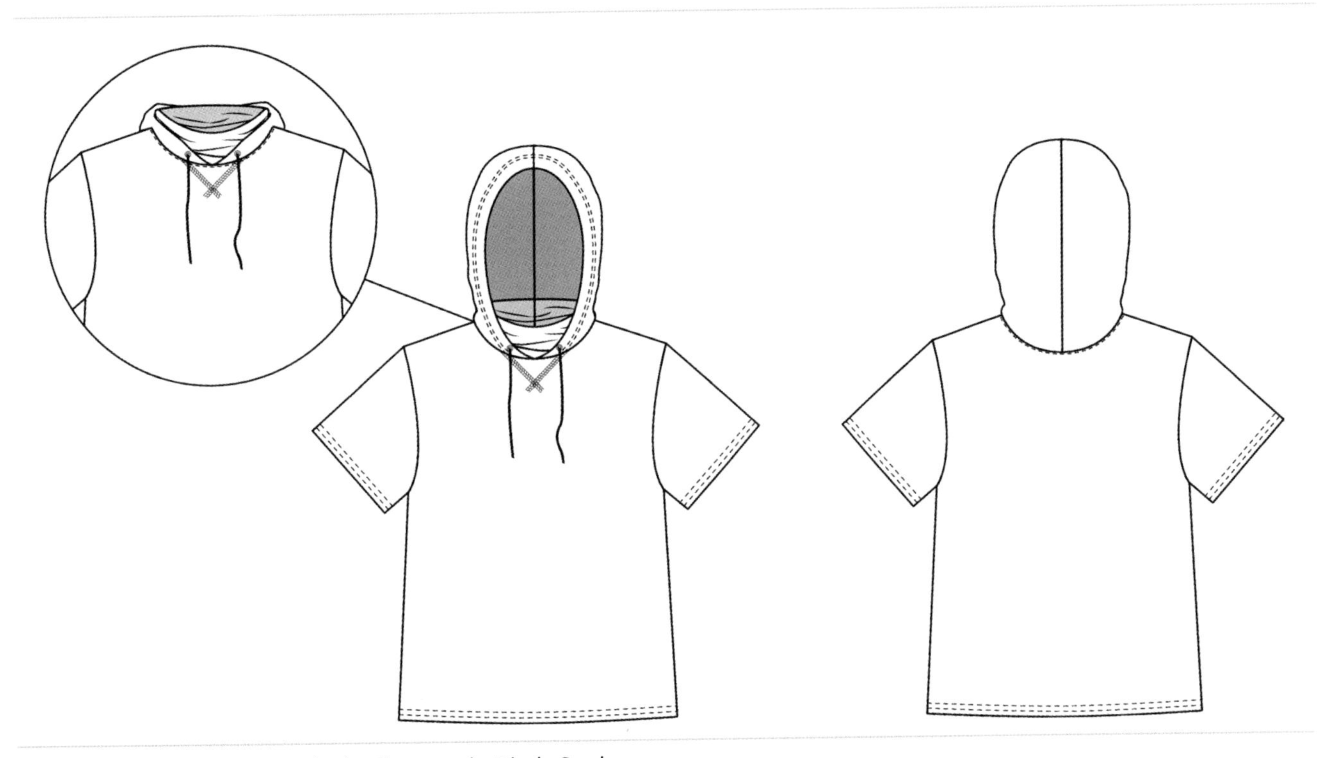

A flat drawing with detail, created using Illustrator by Nicole Crozier

CHAPTER 8

Developing a final proof of concept for ethical production

UNIT 3, AREA OF STUDY 3 (SAT): OUTCOME 3

In this outcome, you will:

- evaluate product concepts related to ethical design, synthesise and apply feedback to justify a final proof of concept, and plan to make the product safely.

Key knowledge

- activities and their purposes within the second diamond of the Double Diamond design approach to generate and design physical product concepts, produce and implement, evaluate and plan and manage the relationships between critical, creative and speculative thinking
- design thinking to generate, refine to generate and design physical product concepts, produce and implement, evaluate and plan and manage critique the physicality of product concepts
- ethical research methods to gather quantitative and qualitative data including end user feedback
- social, environmental, economic and worldview considerations related to selecting materials, tools and processes
- role and components of scheduled production plans: timeline that includes production steps, estimated times and quality measures; materials and costings list, tools and processes; and risk assessments and safety control measures
- methods used to record progress in scheduled production plans and reasons for modifications to the design, planning and timing
- methods to research characteristics and properties of materials, including experimentation techniques and trial processes
- purpose of prototypes and a final proof of concept
- risk assessment associated with selecting and using materials, including chemicals and other substances, tools and processes.

Key skills

- explain the activities and their purposes within the second diamond of the Double Diamond design approach
- work technologically to research, test and use experimentation techniques and/or trial processes to design and evaluate physical product concepts and develop prototypes to select and justify chosen product concept, and develop final proof of concept
- use design thinking techniques to evaluate selection of materials, tools and processes in relation to ethical considerations, and discuss impacts on individuals, society, the economy and the environment
- justify selection involving different degrees of difficulty associated with the manufacture of a product in terms of materials, tools and processes used to make the product safely
- synthesise research and feedback from end user(s) in order to justify a final proof of concept
- develop a scheduled production plan, including assessing risks in production and recording implementation of safety control measures.

Source: *VCE Product Design and Technologies Study Design*, pp. 31–2

9780170477499

DEVELOPING A FINAL PROOF OF CONCEPT FOR ETHICAL PRODUCTION

Refining and evaluating product concepts into prototypes (virtual or physical)

Materials characteristics and properties, experimentation

Construction processes: degrees of difficulty and trials, justifying tools, equipment and machinery

Roles and components of a scheduled production plan and how to record modifications

Selecting a final proof of concept

Gathering user feedback with both quantitative and qualitative research

Templates:
- Design Folio Template

Summaries:
- Chapter 8 **(p.251)**

Quizzes:
- Chapter 8 revision **(p.251)**

Case studies:
- Charlwood Designs **(p.233)**

Nelson MindTap

To access resources above, visit **cengage.com.au/nelsonmindtap**

School-assessed Task (SAT)

The SAT consists of three outcomes across Units 3 and 4.

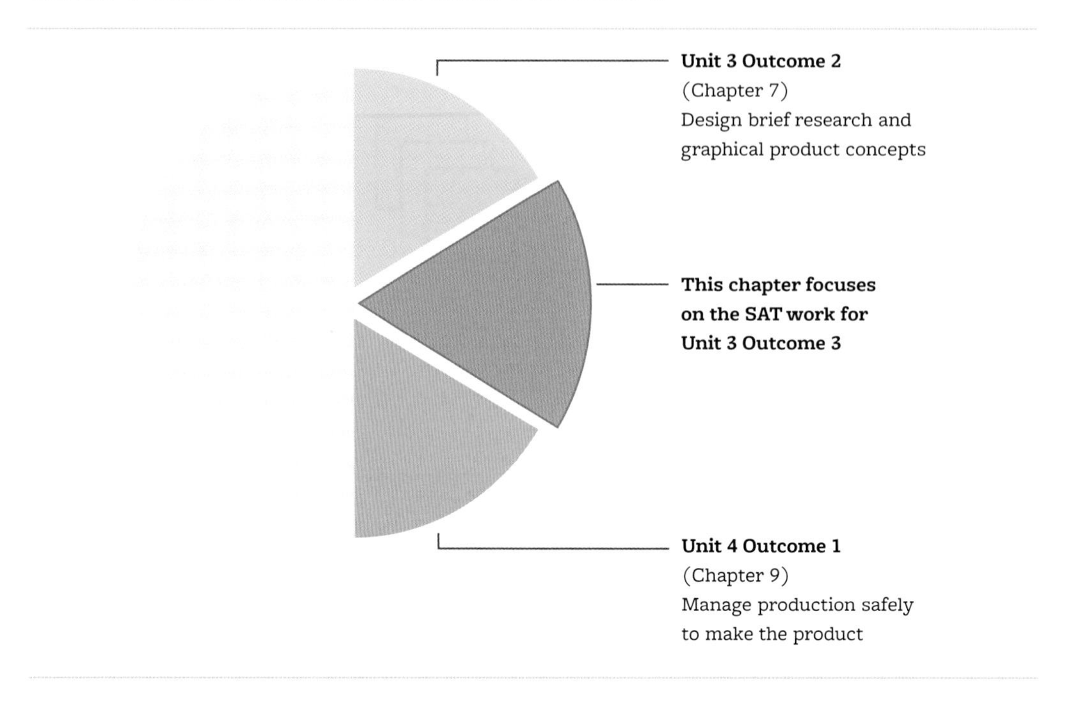

Design thinking

You are already on your design thinking journey, and now you will continue by working on prototypes to 'perfect' your design details.

Again you will be using your creative, critical and speculative thinking skills to conduct thorough research, tests and trials along with models and prototypes to create the most suitable solution for your design brief.

As a reminder, the thinking skills are used together to gather, generate, explore and evaluate design ideas and their suitability.

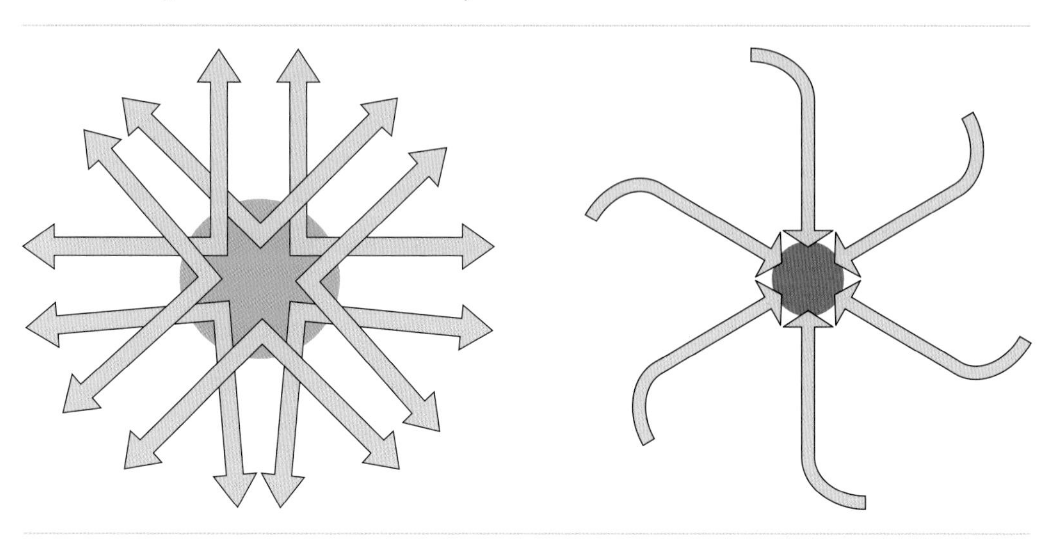

9780170477499

- Creative thinking is also called divergent thinking – e.g. generation of broad ideas, use of imagination, unusual combinations.

Read more in Chapter 1 on page 13.

- Critical thinking is also called convergent thinking – e.g. evaluation, reasoning, analysing, comparison, processing and refining of ideas.

Read more in Chapter 1 on page 14.

- Speculative thinking could be called 'thinking about the future', where forward-thinking questions are posed, information is synthesised and ideas considered for their appropriateness and usefulness – e.g. asking deep questions, forming hypotheses, proposing and recommending ideas and suggestions that are based on research requiring critical and creative thinking. It includes future environmental, social, economic and worldview considerations and is sometimes called 'science fiction design'. Read more in Chapter 7 on pages 187–92.

Ethical considerations

When learning about materials and processes for your SAT task, you will continue to think and work ethically. This requires you to:

- respect the privacy of your end users
- research and select materials that have minimal negative environmental impacts
- check and select materials that are safe, and have not been sourced in a way that harms people or the environment
- work in a way that values the 'worth' of materials and creates minimal waste
- make decisions and choices about your product that are in the best interests of the end users
- treat your end users with respect during the design and development stages of the process (e.g. respecting privacy, no animal cruelty, etc.)
- work safely and collaboratively while respecting your teachers and classmates.

End user feedback (quantitative and qualitative)

You are expected to use methods to gather quantitative and qualitative data from your end users to provide information for and to refine your work when testing, trialling and developing prototypes.

Re-read the information on quantitative and qualitative information on pages 201–6.

Consider using your evaluation criteria to help you. You can reword the criteria to give you both types of information.

SAT ACTIVITY

Plan and prepare your methods of collecting quantitative and qualitative information on your work about materials, processes and prototypes. You can refer back to your previous methods for gathering research and tweak them to apply to your practical work on tests, trials and prototypes.

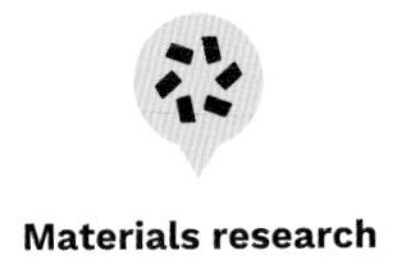

Materials research

Testing materials, tools and processes

At any stage in the design process, it is really useful to experiment with and test materials or trial your skills, as this increases your applied knowledge, skills and experience. This will form part of your research and inform your design and production decisions. Evidence of this should be placed in distinct pages in your folio (at the stages of the process when you carried out the testing and trialling). The purpose is to understand the characteristics and properties

Resistant materials
Materials that are hard when knocked, such as wood, metal and plastics

Non-resistant materials
Materials that are soft and that move when touched, in particular textiles (fabrics, fibres and yarns)

Worksheets
Materials worksheets

Weblinks
Sustainable Product Design

Sustainable Fashion Design

Precious Plastic Universe

How to Fuse Plastic Bags Together

of materials you are using, to be experimental and to learn more about suitable technologies to use. This information helps you to make the best, most informed choices for your end product.

To see suitable tests you could perform on **resistant materials** and **non-resistant materials**, and also for ideas on how to plan and structure your own tests for any creative use of materials, read all of Chapter 14, 'Materials and testing' and complete the Materials worksheets on Nelson MindTap.

Sustainable materials and processes

When choosing materials for your product, make sure you research and consider their impacts. Think about:

- the social, environmental and economic issues related to specific materials
- how the sustainability frameworks you learned about in Unit 3 Outcome 1 might help you to choose more sustainable materials and production processes.

For example, if you want the Design for Disassembly framework to guide your design decisions, you may:

- choose only a small number of different materials
- if you are using plastics, research and select ones that can be recycled, and ensure they are labelled for end-of-life disposal
- choose joining processes that use screwed or snap joints that can be easily pulled apart, rather than permanently glued joints (particularly important when joining different materials).

Research as much as you can about where your materials come from:

- Have they travelled long distances from their source, using lots of fossil fuels in transportation?
- Does the country where the material comes from have strong OH&S protections for workers?
- Have the materials been certified as sustainable (independent labelling/accreditation)?
- Are the materials safe (non-toxic) for workers when sourced and processed, for you when making the product, for your end user while using the product, and for the environment when being disposed of?

When choosing production processes, research the amount of water and energy used for each process, and the waste they create.

For more detailed information about the environmental impacts of each material category, read the relevant sections in Chapter 14. The Business Victoria website has really useful guides for Sustainable Product Design and Sustainable Fashion Design.

Precious plastics

You may want to do some tests involving making, reusing or creating your own plastic materials. This will require a lot of research. A good resource is the Precious Plastic Universe website by One Army, a global movement dedicated to serving and protecting planet Earth using open-source methods online to share information. Another useful resource is the video 'How to Fuse Plastic Bags Together' on YouTube.

9780170477499

SAT ACTIVITY

1. Select the tests or trials you will do, and make a plan to carry out those tests. Make sure that any tests you choose to carry out are safe.
2. Take photographs of your technical exploration – annotate to explain what you have learned.
3. Write up your work and present clearly in your folio with headings and explanations to communicate with your 'viewer' the purpose of your tests and trials, and how the results have influenced your decisions.

Template
Design Folio Template

Prototyping

Developing and refining physical product concepts

In industry, designers and manufacturers design, construct and test a series of prototypes until the design is considered totally ready for production. In your SAT you are expected to develop and refine the physical aspects of your product concepts (i.e. their appearance, functionality and construction). Physical product concepts are refined through models or prototypes, but product concepts can also be explored and improved using CAD and drawing software. In design, the process of completing and refining a concept through a series of variations is called 'iteration'. To iterate means to test, adjust, repeat, adjust, try again, test, repeat and so on. Each iteration leads the way to the final suitable solution. Before selecting your final proof of concept, you would complete several prototypes and each would be numbered.

It is up to you whether you include virtual product concepts (see page 234) or actual ones (see page 235), or both.

You will need to provide evidence of this process in your folio. Once you select the **chosen product concept** from your iterations – the one that is most suitable for your design brief – it will become the **final proof of concept**. This is the 'final' prototype, from which you will calculate and plan your production. See 'Finalising a proof of concept' on page 239.

Case study
Charlwood Designs

The Kami project

The Kami project by James Chapman was a 12-month research project at Swinburne University of Technology into the issues of compact living environments. It required a commercial product solution.

Kami is a completely flat-folding stool/side table that takes up minimal space when flat and turns into a 3D structure by lifting from a single central point. Heavily inspired by and designed through origami, the piece folds along its fabric 'crease-lines', which run between rigid internal **PET** panels. Through carefully planned fold-geometry, the piece is fully self-supporting and self-locking.

The materials used in Kami are exclusively variations of PET, and thus the piece is fully recyclable, without any need for disassembly.

James did extensive prototyping and countless refinements until it folded up exactly as he envisioned.

See the flow of ideas, from graphical concepts to physical concepts and finished product, at the More Than Boxes website.

PET
Polyethylene terephthalate, a recyclable plastic widely used in drink bottles and other products

Weblink
Kami

Virtual product concepts

These are digital images and can be done using computer-aided design (CAD) software such as Adobe Illustrator, Trimble SketchUp, TinkerCAD, AutoCAD, Fusion 360, Microsoft 3D Builder (free), Solid Works and many others. (Be aware that most digital prototype generators found online are only suitable for digital applications such as phone apps.)

As your virtual product concepts evolve, they should include:

- suggestions for which materials to use for all parts
- thickness of material (relevant for resistant materials)
- the joining methods to be used
- placement of components or parts.

Once you have a rendered virtual prototype, follow these steps:

- Save the image and number it as V1 (Version 1) or take a screenshot and save as V1.
- Arrange to get feedback from your end users, including any questions they may have.
- Present this version with feedback and/or annotations on its own page or slide.
- Make your own suggestions under a small heading, e.g. 'What I will change/try'.
- Continue with more versions and feedback, numbering the versions clearly and presenting each one on its own page with feedback.

Using AI for virtual product concepts

You might be tempted to use artificial intelligence to help you generate concepts or develop some concepts further. This may or may not be successful as it depends a lot on what you type into the software. The more detail there is in your command, the better. If you use AI, you need to make it clear on your folio page which method you used, just as you would with other software. Be sure to check with your school whether you are permitted to use AI on any school network. Online AI software may charge you to generate ideas.

Any prototypes generated from AI will not have the accuracy of dimensions that you will get from a CAD program. A lot more work is required for you to check and finalise your working drawings. AI image generators are going to be more helpful for visualising ideas rather than nutting out all the construction methods.

Weblink
OpenAI

Here are some examples of the command 'paper model for a wedding dress' typed into DALL-E 2 at OpenAI.

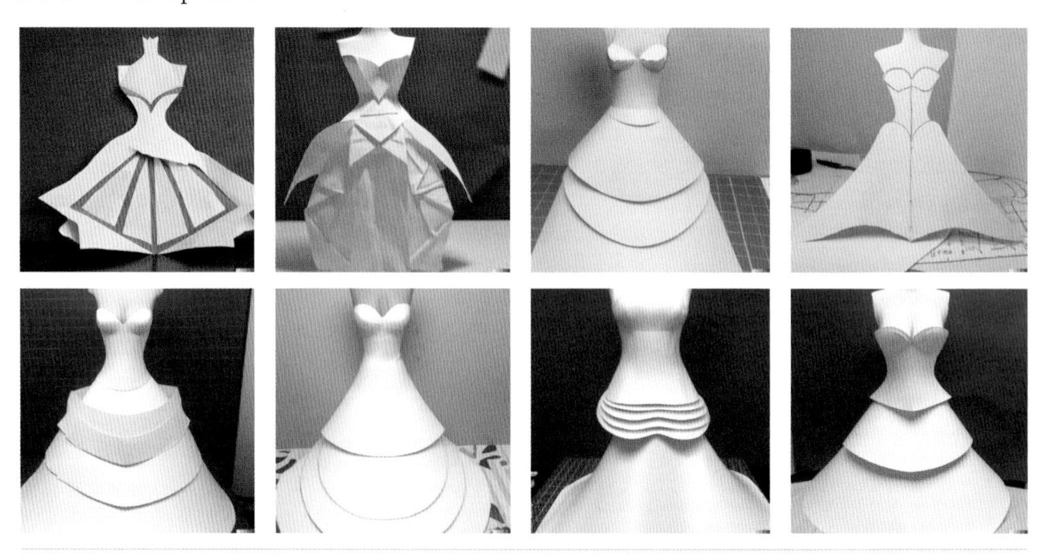

The top four images were generated using DALL-E 2 with the command 'paper model for a wedding dress'. The bottom four came from Image 4 with the command 'generate variations'.

9780170477499

After using AI-generated paper models, you would need to check that the ideas are 'doable' – in other words, that you can actually make them. You can choose several of the images to:

- inspire your own physical paper models
- create more visualisations
- change your design option drawing
- develop a working drawing
- create a **toile** or prototype in more stable materials.

For a discussion on the use of AI in fashion and the importance of creativity in choosing the generating instructions, read the article 'Fashion's New Fakes: How AI will change what you wear' on *The Sydney Morning Herald* website (posted July 2023).

Toile
A toile (pronounced twahl) is a prototype or trial version of a garment. A toile is made from cheaper, plain fabric for testing and fitting. Toile is also a translucent linen or cotton fabric originally used for fittings. Calico is a common fabric choice for toiles as it is inexpensive, is a stable, woven fabric and you can draw lines and write notes on it. If you intend to make a garment in a floaty silk or jersey knit, choose a fabric for your toile that has similar drape.

Weblink
Fashion's New Fakes

Physical product concepts

Physical product concepts are prototypes that are actual, that you can pick up and feel. Many materials can be used. You may construct the prototype using:

- recycled or pre-used materials
- materials considered to be low-fidelity – such as paper or card
- less expensive materials or substitutes for costly, rare or precious ones
- materials that are easier to manipulate without sophisticated machinery.

Suitable materials for physical product concepts

For products to be made from resistant materials	For products to be made from non-resistant materials
• Folded paper, card or cardboard • Used plastic containers, cut or melted • Tin • Thin plywood, custom wood or medium-density fibreboard (MDF) • Any waste materials that could be suitable • Lego bricks or characters – quick and easy to change, but not suitable for all products	• Calico • Checked cotton • Inexpensive draping fabrics • Any scraps of fabric • Any of the materials listed at left for resistant materials could also be suitable

Student Nicole Crozier's 'prototype' demonstrates the complicated patchwork to be incorporated in a fully completed garment.

Weblink
Autodesk Instructables

Join Autodesk Instructables to download instructions for creating this cardboard sneaker to help you create your own physical concept made from cardboard.

Further drawings and graphical product concepts

During the process of developing and refining prototypes, you may need to work things out with more drawings. Perhaps you can see after the first prototype (or later versions) that things won't work out as you hoped.

In this case, you may need to create some further:

- visualisations – to come up with some more ideas either for parts or the whole product
- design options – to present a realistic detailed drawing of a new design
- working drawings – to finalise dimensions, placement of parts, joins, etc.

Present your drawings to show your progress after each prototype version that they are relevant to. Be sure to get end user feedback and use it to state what you will do in the next prototype.

Low-fidelity prototype
A basic prototype that has just enough detail to communicate the essence of an idea

Don't be too concerned about whether your work here is physical, virtual, a **low-fidelity prototype** or a true prototype – the most important thing is to show the progression of finalising the design of your product with end user feedback. Read more on low-fidelity prototyping at the Victorian Government's website.

Weblink
Low-fidelity Prototyping

Chosen product concept

At some stage you will decide on the chosen product concept. A physical prototype of your chosen idea will become your final proof of concept. Read more under the heading 'Finalising a proof of concept' on page 239.

Prototyping examples

Some examples of prototypes are (or could include):

- a fully constructed garment (a toile) made to size from low-quality materials
- a piece of furniture made to scale (not necessarily to true size) from pine or plywood – stained or painted to represent other materials
- a small area sample of high-intensity handiwork such as beading, embroidery, appliqué, carving or embossing (photographed, with an indication of where it would be found on the whole product)
- joins (both traditional and experimental) and an indication of where they would be used in the whole piece
- the use of modelling foam, covered in plaster and painted to represent plastic or other materials
- parts of your product 3D printed, or a scaled 3D print of the whole product idea
- a piece of jewellery made in silver or copper instead of gold.

Maquette
A scale model, made from substitute materials or using less material than the final product

In general, prototypes are true to size, but check this with your teacher.

Small models or **maquettes**, which are smaller in scale, are often used in the designing stage to work out proportions and how pattern pieces fit. Compared with full-sized models, small models also reduce the amount of materials used.

9780170477499

It is important that you:

- check the SAT Assessment Criteria from VCAA (new every year and published in January on the VCE Product Design and Technologies home page)
- photograph and number each iteration (version) of your physical product concepts (prototypes) or, if using software, save and number each version
- collect end user feedback guided by your evaluation criteria
- number each version clearly
- include annotations explaining how the design brief has been satisfied and/or what will be checked/tried on the next version
- add relevant comments about which parts or aspects don't work, don't look right or might be difficult to join, etc., and why.

Weblinks
VCE Product Design and Technologies Study Design

10 Types of Prototypes (With Explanations and Tips)

A prototype can be scaled, but is usually full-scale, and is somewhat functional. You can read about prototyping at indeed.com for ideas, but note that some of the explanations and tips apply to software development.

Silver paper, fabric and net pinned together to judge placement for a hat design, from students in the So You Think You Can Design competition, run by DATTA Vic. (top); and a rose folded out of paper to check the amount of fabric needed (bottom), by student Riona Finn

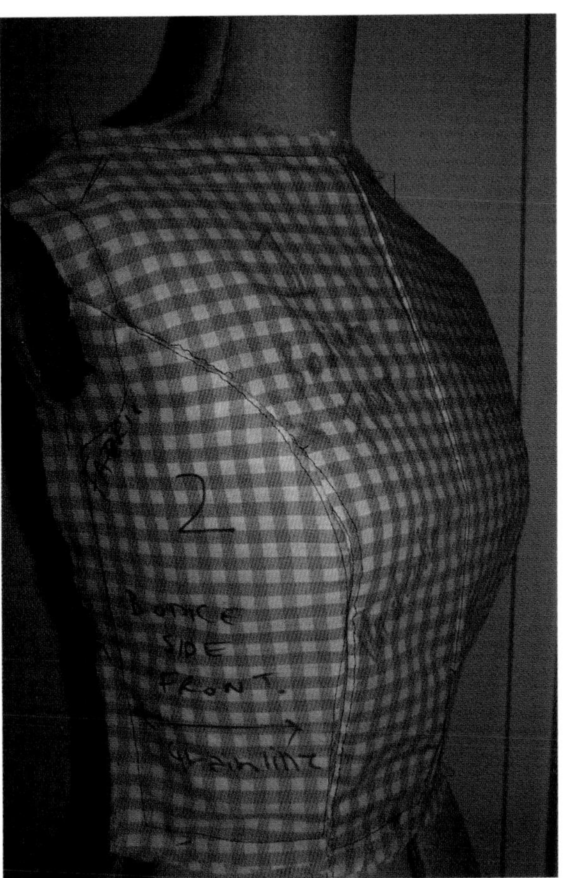

Constructing a toile from check gingham, by VCE student Tamara Bottomly

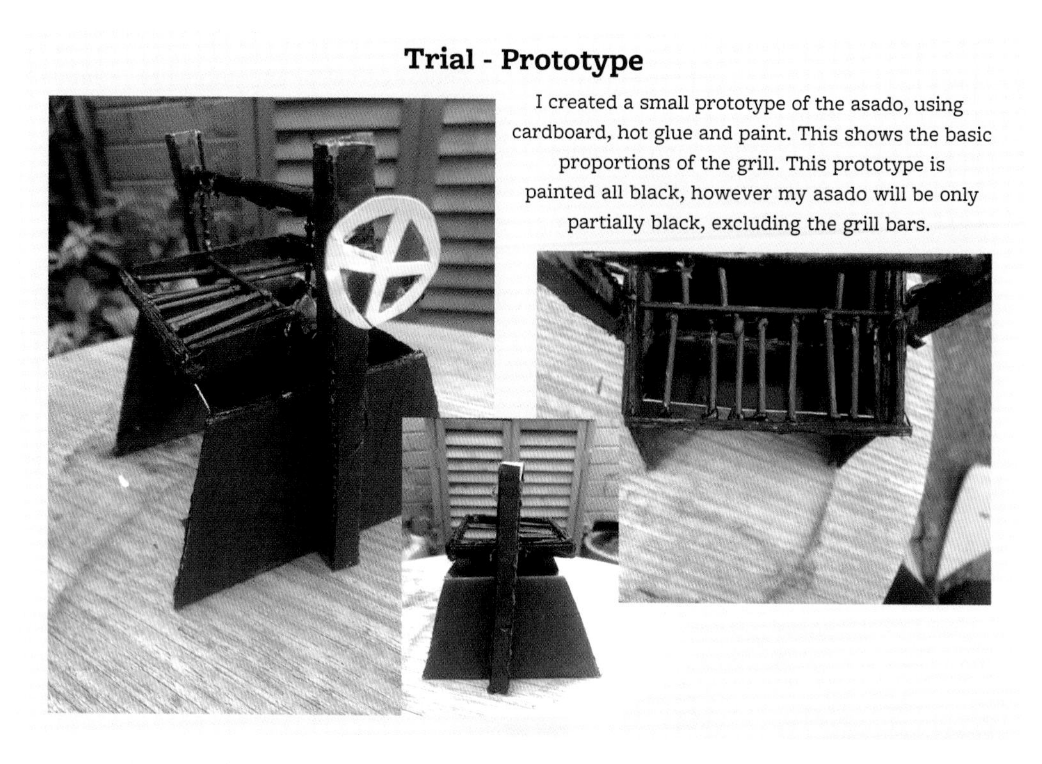

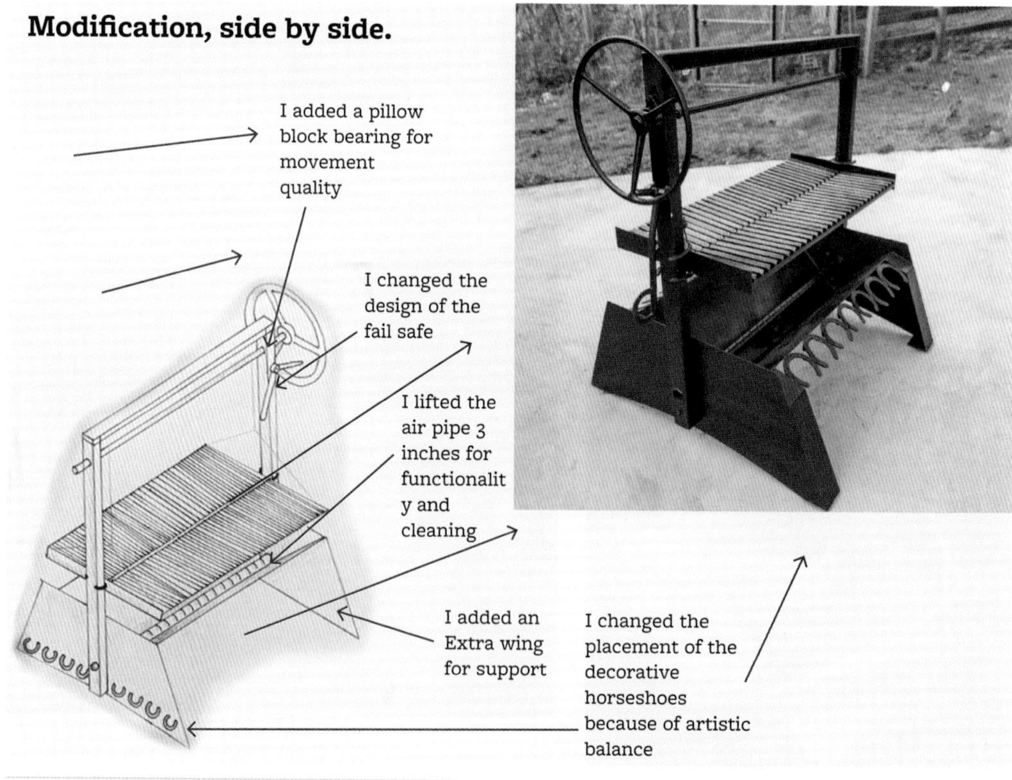

A small-scale low-fidelity model (top) using cardboard, hot-melt glue and paint to test the proportions of an asado BBQ design, by student Bregje Van Vark, and the final product (bottom)

9780170477499

Finalising a proof of concept

A suggested timeline or sequence for finalising a proof of concept might be:

1 Use your most suitable design option and its working drawing to construct small, scaled-down paper or card models (low-fidelity prototypes) or a physical or virtual prototype.
2 Seek end user feedback and suggestions for improvement.
3 Complete drawings if required, either for new ideas or to finalise parts.
4 Complete your next model or prototype.
5 Seek end user feedback and suggestions for improvement.
6 Make a decision on your chosen concept, justifying it with a clear statement in your folio.
7 Use it to develop your final proof of concept.
8 Adjust the design option drawing or working drawing to suit.

A final proof of concept is your final design wrapped up! It is to scale – i.e. all measurements are correct in relation to each other – and gives you all the information needed to move into the production of your final product. You may need to create a completely new working drawing or adjust your previous working drawing to be accurate. Make sure you show each version of your working drawing in your folio, explaining what has been changed in each version.

Note that your chosen concept may come from your drawings (graphical product concepts), but would need to be further developed into a final proof of concept through prototyping. The next step is to complete planning to make your product, based on your final proof of concept and its working drawing.

A low-fidelity maquette made with cardboard, foam and glue, by student Sophie Glover, sitting on the open adjustable table of the half-scale plywood model of a window seater

Scale model made with plywood, hinges, foam cushioning and glue by student Sophie Glover. From the model Sophie could see the need for additional strengthening to withstand up to 200 kilograms. Spline mitre joints, which would also add visual interest, and supporting beams in the substructure made from hardwood, would be included in the final product.

Template
Design Folio Template

SAT ACTIVITY

1 Complete a number of virtual or physical product concepts and record all your work and decisions with photos and text.
2 Get end user feedback, both qualitative and quantitative, on all your 'prototyping' work.
3 Write a statement to justify your final proof of concept.
4 Adjust your previous working drawing or create a new one.

9780170477499

Developing a scheduled production plan

Once you have your final proof of concept and an accurate working drawing, you are ready to develop your scheduled production plan. It is important that this planning, even if it is brief, is done before any production starts. Any adjustments can be documented during production and in your 'record of progress', commonly called a journal.

Components of the scheduled production plan

Component	Briefly	Role
Production steps	How?	How to construct your product – a list of steps to follow
A timeline	When?	To give milestones and dates for the steps
Materials list and costing	What with?	To be prepared with all materials and bits required
Risk assessment	For safety	To look ahead and avoid accidents and incidents
Quality measures	Good job	To create the best quality work possible

It helps to start with a (predicted) list of **production steps** required for construction. You may feel that you don't have the skills yet, but forcing yourself to think ahead will help you to be more independent from your teacher during production. It will reinforce your existing knowledge and indicate areas where you will need help or where you will need to do some research or practice. If production areas are less familiar, you may need to seek advice.

Production steps

Production steps are a sequenced plan of construction. They list and describe the construction processes (steps) in an order that you expect to follow. They include details of how each process will be carried out, equipment and material needs, particular safety precautions and the estimated time needed. You must identify any teacher or outside assistance required, as there may be some processes that rely on skills or machinery that you do not possess or have access to. The people providing assistance need to be located, and you will need to make arrangements with them.

Shutterstock.com/byswat

Plans can be handwritten or digital.

See the following page for an example of sequenced production steps and note how the student has communicated teacher assistance.

You can find many excellent examples of production steps in the Top Designs catalogues produced every year by Melbourne Museum. You can change the date at the end of the URL to see other years.

Weblink
Top Designs

Planning your production steps

Think about the major processes needed to construct your project, and then add more detail. Remember, a plan is a prediction, so it doesn't need to be precisely accurate – it's how you expect to make the product. The actual steps taken will be documented in your journal throughout production. It's helpful if the steps are numbered at this stage.

Major construction processes are as follows:

- preparing the material, measuring and marking out
- cutting, shaping, forming or 3D printing the various pieces
- joining, which will include the name of various joining methods you will use and how different pieces will be joined
- assembling (putting together all the pieces – some of which are already joined)
- embellishing, decorating and finishing.

An example of production steps by a student for an outdoor table

Step	Process and materials	Tools/machines/equipment	Safety measures	Est. time	Date
1	Mark out and measure dressed cypress timber	Tape measure, pencil, tri-square, workbench		45 min	8 June
2	Cut all legs, rails and planks for table top to length	Docking saw (electric)	To be cut by Mr Ahmed	50 min	15 June
3	Mark out the mortise and tenon joints on the top of the legs and ends of the rails	Tri-square, pencil, marking gauge, ruler		50 min	19 June
4	Drill holes for the mortises in the legs	Bench/pedestal drill	Make sure wood is clamped to drill bench, wear safety glasses	40 min	22 June
5	Pare back mortises	Chisel, bench vice	Cut away from hands	2 h	26 June
6	Cut out tenons on the rails	Handsaw, chisel	Use bench hook or clamp wood to workbench	2 h	29 June
7	Test and ensure mortise and tenon joints fit, fixing	Gentle tapping with mallet, fine work with chisel	Make sure work is secure	?	29 June
8	Use template to mark out gentle tapering shape for legs	Template, pencil, workbench		20 min	3 July
9	Shape legs and smooth off	Bandsaw, rasp, file, sandpaper, sander	Bandsaw to be used by Mr Ahmed	3 h	3 July
10	Join rails to legs, check squareness	Sash clamps, large square, tape measure to check diagonals		1 h	27 July
11	Mark out dowel joints for joining table top pieces	Square, ruler, pencil		30 min	31 July
12	Drilling holes for dowel joints, checking drill bit size, making sure top side of planks goes face down	Dowelling machine	Clear all debris Tie back hair and loose clothing Make sure drill bit is in correctly	30 min	31 July
13	Cut dowels for joints, making them a little shorter than the depth of the hole	Vice, handsaw	Make sure work is secure	30 min	31 July
14	Join planks for table top with suitable outdoor glue	Sash clamps	Make sure sash clamps are stable and safely placed on work surface	1 h	7 Aug
15	Assemble table (not sure how to do this yet)	Metal support brackets, portable electric drill and screwdriver	Manual handling – ask for help to move table	2 h	14 Aug
16	Sanding in gradations and oiling	Sander, sandpaper, brushes, oil	Dust mask, move table outside	?	21 Aug
Total hours expected				18+	
Due date					24 Aug

Remember to choose an order of steps that logically works for your product. You may need to swap, repeat or move around some of the steps listed opposite.

Production steps for textiles

Sometimes the safety precautions in textile projects become repetitive. To avoid this, delete the 'Safety measures' column and add a row underneath the table that lists safety precautions to be taken throughout production, or refer to the **safety controls** in your risk assessment.

Safety control
A precaution you will take or put in place to remove or reduce a risk. You must demonstrate that you know exactly what to do to remain safe. List the exact PPE or set-up required.

Timeline

A timeline sets out the dates at which you expect to have different steps of your construction completed and the final due date for completion. You could do this in a column of your production steps plan, as can be seen in the example above, inserting a date for each step – but this may become crowded. You can see this in the example for an outdoor table.

If you prefer to create a separate timeline, use Microsoft Excel or a table in Microsoft Word, like the example shown, or use free online software or templates. Check that the number of steps in the timeline matches the number of steps in the sequence plan (as in the outdoor table example). In this type of timeline, the steps can be written in an abbreviated form.

A separate timeline (instead of a column added to production steps) allows you to mark it up later (in a different colour) according to the actual date steps were completed. This can be part of your modification evidence.

An example of a timeline for the outdoor table

	Week 1		Week 2		Week 3		Week 4		Week 5		Week 6		Week 7		Week 8	
	8 June	12 June	15 June	19 June	22 June	26 June	29 June	3 July	27 July	31 July	3 Aug	7 Aug	10 Aug	14 Aug	17 Aug	21 Aug
1 Mark/measure	■															
2 Cut pieces		■	■													
3 Mark joints				■												
4 Drill holes					■											
5 Pare mortises						■										
6 Cut tenons							■									
7 Check joints							■									
8 Mark out legs								■								
9 Shape legs								■								
10 Join/check									■							
11 Mark joints										■						
12 Drill holes										■						
13 Cut dowels										■						
14 Join planks											■	■				
15 Assemble table													■	■		
16 Sand/finish													■	■	■	■

Template
Design Folio Template

SAT ACTIVITY

Start your scheduled production plan

1 Create a table with your production steps to show your prediction/planning of how and when they will occur, with estimated times.
2 Add the dates by which you expect to finish each step in the last column, or create a separate timeline.

Risk assessment and management for safety

Safety is not only vital during production; it is integral to all aspects of Product Design and Technologies. Many processes involve a level of risk, such as using equipment, materials and chemicals that may be harmful if used inappropriately. Products made can also be dangerous.

Risk management

Risk management is required throughout your production in Outcome 1 of Unit 4. It is a process used to recognise **hazards** and assess, manage and control **risks**. Before you can manage any risk, it needs to be assessed. Hence a risk assessment must be done before any production starts. Your risk assessment planning document covers the first three steps of risk management, as can be seen in the table below.

See more on risk assessment on page 247.

Hazard
A dangerous thing, action or behaviour

Risk
The severity or degree of harm arising from an accident, ranging from minimal (requiring first aid) to serious (requiring hospitalisation), and the likelihood of an incident happening

Likelihood
The probability that a given event will happen; often rated low, medium or high

Risk management has four steps	Concepts to be understood
Step 1: Identify hazards and potential injuries	**Hazard** – a dangerous thing, situation or action; something in the work environment that has the potential to cause harm to people; e.g. oil spills, a cord across a walkway, sharp spinning blades
Step 2: Assess risks. Refer back to Unit 1, Outcome 2, on pages 52–3 to read more and see the risk assessment matrix.	**Risk** – the degree of harm and the chance (or **likelihood**) of something happening, i.e. that a hazard such as a cord across a walkway might cause tripping and a head injury, or that fingers could touch the sharp spinning blade and be badly cut **Harm** – death, injury, illness (physical or mental) or disease
Step 3: Control hazards and risks	**Controls or precautions** – things, methods or actions that eliminate, prevent or reduce the risk so far as reasonably practicable; e.g. use another method to cut, or use the guard to limit chance of cuts; never have cords across a walkway unless you tape them down
Step 4: Review (check) controls	**Check** – don't assume things are under control; inspect at regular intervals to see whether they are working as expected and are effective; communicate with others exposed to the same risk; e.g. look to see where cords are placed and fix if necessary; check guard is in correct position

Textiles

The most common **injuries** or illnesses in textile production result from:

- poor manual handling practices
- slips, trips and falls
- musculoskeletal problems due to poor ergonomic set-up, bad posture and repetitive movements over prolonged periods
- eye strain due to poor lighting
- inhaling or absorbing hazardous chemicals in fabric, thread and dye
- incidents with sewing machine needles.

Injury
The actual harm or damage that could occur to a person, ranging from scratches and bruises to broken limbs, head injuries, entanglement injuries, electrocution or death, etc.

Wood, metal and plastics

The most common injuries in wood, metal and plastics workshops result from:

- incorrect use of equipment or lack of training
- failing to use safety guards
- machinery that is not maintained to a safe standard
- not wearing the right **personal protective equipment (PPE)**
- an unsafe working environment
- poor electrical safety measures
- poor manual handling practices.

Longer-term harm might come from dust inhalation or exposure to loud noise.

Personal protective equipment (PPE) Equipment such as safety glasses, shields, masks and gloves, needed for protection when using a machine or equipment

Shutterstock.com/Juice Verve

General safety controls

The guidelines in the tables that follow are a general reminder of the safety aspects you need to keep in mind when producing your product. Please note that the safety controls in your risk assessment need to be more specific and relevant to each hazard.

Guidelines to avoid injuries around any workshop	
Manual handling – lifting and moving materials	• Ask for help when lifting more than 10 kg (such as the product you have made); a two-person lift is safer. • Use suitable lifting equipment or a trolley for carrying materials whenever possible. • Learn, use and state the correct method for lifting.
Slips, trips and falls	• Always wear suitable footwear, non-slip if required. • Clean up any liquid spills, use detergent to clean greasy floors. • Keep floor clear of obstacles, especially those hard to spot, such as uneven edges in flooring, loose mats, open drawers, untidy tools or electrical cables. • Remove rubbish such as material scraps to avoid trip/slip hazards. • Return tools and other items to their storage areas after use.

Guidelines for wood, metal and plastics workrooms	
Safe working	• Only use equipment once you have been taught how to use it correctly. • Know the location of, and how to use, the emergency stop buttons on all machinery. • Report all accidents and faulty equipment to the teacher. • Keep your work area tidy and return tools to their storage area. • Use tools for their correct purpose and always carry sharp tools (such as chisels) with the point facing downwards.
Personal safety	• Understand why different PPE is used and take responsibility for wearing it when required, e.g. always wear eye protection when operating machinery, hearing protection for noisy operations and fully enclosed shoes at all times. Wear suitable face masks to protect from dust/fumes, and have ventilation.
Machine safety	• Never use a machine without the correct training from your teacher or a qualified instructor. If you are unsure about operating procedures, check with your teacher. • There should be no more than one student operating a machine or within the safety zone at one time. • Always check that machine guards are in place. • Be aware of the hazardous parts of a machine – sharp blades, trapping or crushing points, etc. – and keep body and clothes away from these.
Portable power tools	• Check the lead, plug and machine for damage and report any damage to your teacher; do not use a damaged portable power tool. • Make sure the machine is turned off at the power point when checking settings and fittings or changing parts, such as router bits. • Work must always be firmly secured in a vice or with clamps.
Substances	• Read the safety data sheet (SDS) and understand all the hazards associated with the chemicals/materials you are using; follow the safety precautions on the SDS.

Guidelines for textiles	
Using a sewing machine or overlocker (machine-powered needle, poor posture, poor lighting, long periods of sewing)	• Have a good ergonomic set-up and posture to avoid musculoskeletal injuries/strains of the neck, back and arms. • Take breaks and stretch. • Have good lighting to avoid eye strain. • Always turn machine off when changing parts or setting up. • Keep fingers away from machine-powered needle (and blade on overlocker).
Scissors (sharp pointy blades and weight)	• Keep blade closed when not using and when transporting, and do not place on table edges to avoid dropping onto feet. • Cut fabrics on a flat surface, not while holding them in your hand.
Pins and needles (sharp ends)	• Always pick up and store in a pin container.
Iron (hot surface) and steam press	• Use an ironing board and keep fingers away from hot iron. • Keep hands away from any steam being released. • Check electrical cords for damage. • Turn off when not in use.

The information following is adapted from *Summary of the Occupational Health and Safety Act 2004: A handbook for workplaces* by WorkSafe Victoria.

Weblink
Summary of the Occupational Health and Safety Act 2004

In the workshop, you must take reasonable care for your own safety and the safety of others who may be affected by your actions or omissions. You must cooperate with any actions taken by your teacher to comply with the OH&S Act and regulations. You must not intentionally or recklessly interfere with or misuse anything provided at the workplace in the interests of health, safety and welfare.

Duty not to recklessly endanger people at workplaces (S32)
It is an offence, without lawful excuse, to recklessly engage in conduct that exposes, or may expose, a person at a workplace to the risk of serious injury.

Risk assessment

A risk assessment occurs before production to enable risk management during production. It aims to predict possible **incidents**, accidents and injuries and to eliminate them or lessen their severity through safety controls. Risk assessment is an important **occupational health and safety (OH&S)** tool used to prepare and follow safe practices in every aspect of design and construction.

In VCE, you need to apply risk assessment not only to machinery and equipment, but also to:

- materials
- hazardous substances/chemicals you might use
- manual handling
- your product.

Incident
An event that comes close to causing injury and, in a workplace, should usually be reported to management

Occupational health and safety (OH&S)
Programs that aim to ensure workplaces are safe for all who enter them

Set out your risk assessment in columns, as in the chart that follows. Use the definitions of the column headings and the safety information on these pages to help you. Your teacher will also tell you what is expected.

Risk assessment chart for one piece of machinery

Process or activity	Hazard	Possible harm/injury	Risk/Likelihood	Risk controls
e.g. Using the router	Cord across walkway	Bruising from being tripped	Low/Medium	Tape cord down or run above workbench
	Blade cutting through electrical cord	Electrocution	Very high/Low	Do not start until cord is secured as above
	Sharp spinning router bit	Finger injuries	Medium/Low	Check router bit is set up correctly and tightly
	Dust and flying debris	Irritation of or damage to eyes, nose, throat and lungs	High/Low	Wear safety glasses, tie long hair back, wear sturdy shoes
	Work falling off table	Bruising or cuts	Low/medium	Make sure work is secure with clamps, follow teacher instructions or those in the machine's manual

Hints for risk control

During the **design stages**, reduce risks by:

- researching and selecting the safest machinery, tools and processes (also chemicals and materials) to use
- considering design features of your product that are least likely to cause injury.

During **production**, reduce risks by:

- taking responsibility for your safety by researching the risks, understanding them and using safe work practices
- reading and following the safe operating procedures for any machinery you use, such as wearing appropriate PPE and cleaning up dust and spills
- being correctly trained in processes and the use of machinery and equipment
- substituting safer processes for those that have a high level of risk, or arranging for an expert to complete them for you
- seeking permission if required
- being supervised by a qualified person.

Weblinks
How to Conduct a Risk Assessment

Plant Hazard Checklist

Guide to Safety in Wood Products Manufacturing Industry

Some resources

The following are related to industry but have information relevant to the classroom:

- You can read more on degrees of severity and likelihood (in relation to risk assessment in the mining industry) at the Victorian Government's WorkSafe website.
- WorkSafe also produces a **plant** hazard checklist.
- Go to the WorkSafe 'Guide to Safety in Wood Products Manufacturing Industry' and see page 13 for hand tools, page 17 for manual handling and much more.
- For machinery used on wood, metal and plastics, ask your teacher to help you locate Safe Work Procedures documents on the Department of Education website. Safe Work Procedures (SWP) are a useful guide for risk assessments and to clarify how to work safely when using machinery. A SWP should be displayed next to any piece of machinery.

Plant
Any machinery, equipment, appliance, implement or tool; any component of any of those things; anything fitted, connected or related to any of those things

Weblinks
Plant and Equipment Risk Management

Hazardous Manual Handling Solutions in the Textile Industry

For manual handling in the textiles industry, you can check the WorkSafe website and see what is relevant to your classroom or your own facilities.

Safety data sheets (SDS)

It is essential that you know about the dangers of the chemicals and materials you are using, particularly as the harmful effects of hazardous **substances** are often not seen for a very long time. An example of this is the many people who were exposed to asbestos dust in workplaces and later developed mesothelioma, a rare and aggressive cancer. Prolonged exposure to any type of dust is harmful: even bakers who for years are exposed to wheat/cereal dust can develop serious lung issues.

Substance
Any natural or artificial material, which might be in the form of solid granules, liquid, gas or vapour

Safety data sheets, or SDS (previously material safety data sheets, MSDS), provide information about substances and their associated hazards, and are written in reasonably straightforward language. They inform you whether a substance is classified as hazardous to health or not and what safety procedures should be followed, as well as how to dispose of the material after use.

Companies that make or distribute materials and/or chemicals are required to provide an SDS when asked. By law, all schools/workplaces must have an SDS on all of the materials/chemicals they use that are classified as hazardous. Your teacher should have SDS readily available for you to read, but you can also use a search engine and find an SDS online for most finishes and chemicals used in schools. These can be downloaded and put into your folio, and you can highlight the sections related to hazards, health effects and safety precautions.

9780170477499

 See more on safety data sheets on the Safe Work Australia website.

Weblink
Safety Data Sheets

It is important that your risk assessment covers *all* the hazards relevant to your project and *only* those hazards that are relevant.

Processes: some suggestions for risk control at different stages of production

Stage of production or use	Areas of hazard – examples	Possible injuries – examples	Risk controls – examples
Preparing materials, measuring, marking, cutting, sewing or overlocking	• Machine-powered – moving parts, blades, needles • Waste materials and flying objects, etc. • Electrical cords and use of some electrical equipment • Dust, vapours, noise	• Striking, crushing, entanglement, cuts, bruising, lacerations and abrasions, amputation, degloving, scalping, stab wounds • Bruising, lacerations, head injury, electrocution • Breathing problems, nose, throat or lung irritations, eye and hearing damage	• Correct training, use appropriate PPE (be specific) • Check electrical cords • Use the safest materials and chemicals • Ensure good ventilation • Use machine guards
Joining and assembling	• Toxic glues	• Lung, eye, skin and internal organ damage	• Outline SDS safe use procedures
Sewing or overlocking	• Poor posture, lighting or ergonomic set-up • Prolonged periods of sewing	• Musculoskeletal problems: neck, back or eye strain	• Have good ergonomic set-up with good posture • Take breaks and stretch
Finishing	• Vapours and fumes from finishes (paints, oils, etc.) • Heat and steam from iron or steam press • Dust from sanding	• Inhaling toxic chemicals – damage to lungs, eyes, skin and internal organs • Burns • Inhaling dust – damage to lungs, nose and throat	• Outline SDS safe use procedures • Use appropriate mechanical ventilation • Outline what you learned from training • Use a dust extraction system and wear a particle mask

Materials and costing list

A materials and costing list needs to be developed for construction of your final product. In your materials list, include all the components and extras that will be needed. If it is helpful, you can put in an estimated cost and then, once you buy the materials, add the actual cost.

- A materials cutting list is for planning and costing wood and metal projects (this is sometimes called a bill of materials); see the example below.
- A materials and notions list is for planning and costing fabric projects (notions are zippers, buttons, etc.).

Shutterstock.com/New Africa

An accurate materials list helps you to minimise waste and your costs. It also helps you to think more responsibly about your material use and reduce your environmental impact.

Coffee table: materials and costing list

Part of product	Amount	Length	Width	Thickness	Type of material	Cost per unit	Total cost per part
Legs	4	400 mm	45 mm	45 mm	Blackwood	$	$
Rails – short, etc.	2	450 mm	60 mm	19 mm	Blackwood	$	$

SAT ACTIVITY

Continue with your scheduled production plan

1 Create a table for your risk assessment, including everything that has an element of danger.
2 Create a list of materials with costs.

Template
Design Folio Template

Quality measures

Quality measures and quality checks will help you to avoid mistakes and ensure that your product turns out to be the best possible quality. These measures help you to have quality control. Preventing errors also prevents waste of time and materials.

Quality measures can be written as a standard you expect to achieve. They can be accompanied by 'tips' ('how to achieve') to help you complete processes accurately and expertly.

Some general quality measures are set out in the following table.

Examples of quality measures

Aspect	Quality measure	How to achieve (optional column)
When selecting material	• No flaws or holes • Straight grain • Not warped	• When purchasing, look for flaws: tiny holes that might affect strength or appearance, variations in colour or even thickness, or skewed weave (fabric) • Look for warping, check the grain suits your construction, check for knots where strength may be required (wood)
When marking and cutting out pattern pieces from the material	• Aligned with the grain as required • Correct size and accurate to 1 mm • No jagged edges	• Check that pieces are laid out accurately, along or with the grain as required for the material • Lay pieces as close together as possible to minimise waste (all materials), but include room for saw cuts when marking out (wood, metal and plastics) • Double-check all measurements and re-measure if necessary • Check for squareness on face and edge (wood) • Make sure angles are accurate, especially any right angles, by using a set square (wood) • Mark lightly or use an instrument for marking that will not show up on final product (or can be easily removed) • Concentrate and cut as accurately as possible using conventional guiding methods
Sewing seams	• Straight stitching • No missed bits • Even stitches and correct tension	• Check machine set-up and try on scraps first to check • Ensure the machine has the correct needle for the fabric
Timber joints	• No gaps larger than 1 mm • Joins are square	• Do a test join with a scrap of similar timber
Joins with welds or solder	• Close fitting and firm strong contact • No blobs of solder or weld	• Do a test run

9780170477499

The table on page 250 identifies a few quality measures in the selection, measuring, marking out and cutting of materials and some joining methods. Write up your own quality measures that are relevant to, and important for, your particular materials and processes. You can use textbooks or online videos related to your materials to assist you.

The last column in the table is optional, but can be extremely useful for explaining how you will achieve this quality, and as a place to include the tools, equipment or technique you will use to help you achieve it. One excellent way to achieve a high level of quality is to practise! Do as many trials as you need to improve your work.

After the process is finished, do a quality check by, for example, looking at and inspecting your work, to decide whether it needs redoing or is of satisfactory quality. You can add to your quality measures list throughout production as you learn better ways of doing things.

SAT ACTIVITY

Continue with your scheduled production plan and complete:

- a quality measures list.

Making a plan for recording your progress: a journal

Template
Design Folio Template

Summary
Chapter 8

Quiz
Chapter 8 revision

Most students record their production work in a journal. Many use Google Slides, Microsoft PowerPoint or a simple table in a Word document for this – all are simple to use with images and text. Any software that allows you to easily upload photos and add dates, the time taken to complete a stage and captions is suitable. You might also include short videos, showing critical stages or describing problems that you have solved.

One class member or the teacher might take photos of each class member's work every session and upload it to a server for access.

In this journal you can record:

- the date and time taken in each practical session
- one or more photos of what you achieved in the session, each with a caption explaining it
- end user feedback, in a highlighted or colour-coded box, on aspects that were modified
- your own reasons for changing (modifying) the design, the production steps or your timeline
- important safety control measures and precautions
- your reflections on what you have learned about materials and processes through your experiences.

Additional ways to record progress in your folio (after your journal) include:

- a copy of your production steps with notes showing changes
- a copy of your timeline to mark the actual date of processes
- new design option drawings or working drawings indicating/annotating changes.

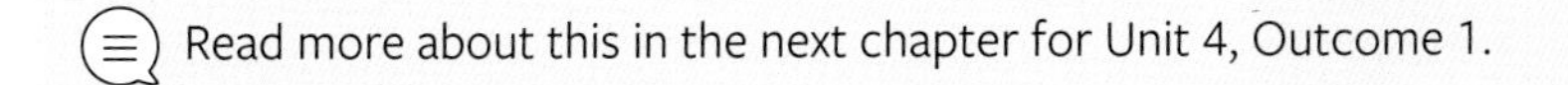

Read more about this in the next chapter for Unit 4, Outcome 1.

UNIT 4

ETHICAL PRODUCT DESIGN AND DEVELOPMENT

Areas of Study 1 and 2

Chapter 9: Area of Study 1: Managing production for ethical designs (SAT)

You learn about:

- implementing a scheduled production plan including:
  - » risk management
  - » time management and techniques for goal setting
  - » making a quality product and degrees of difficulty in processes.

You complete a journal for recording progress and documenting decisions and modifications.

Chapter 10: Area of Study 2: Evaluation and speculative design

You learn about:

- research and development (R&D)
- speculative design thinking and innovation and entrepreneurial activity
- product development process for products that integrate new and emerging technologies
- ethical considerations associated with products using new and emerging technologies
- relationships between market research and the product development process
- qualitative and quantitative methods of evaluating products
- using data to evaluate the success of a range of products and to justify enhancements and /or improvements
- key factors that determine the success of a product.

The SAT

Unit 3, Areas of Study 2 and 3: In response to a design brief, you created graphical product concepts and refined them through prototypes and experimentation. A final proof of concept allowed you to plan your product's production.

Unit 4, Area of Study 1: You will follow the scheduled production plan to make the product designed in Unit 3.

9780170477499

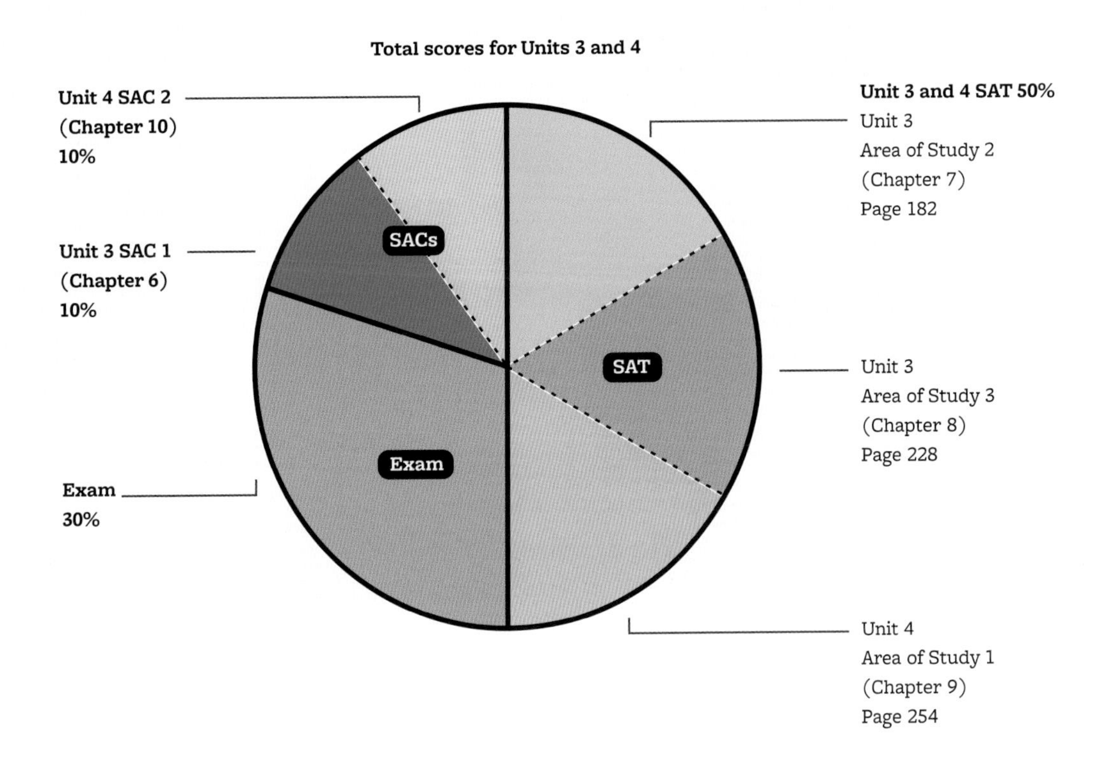

Assessment in Unit 4

- **SAT:** Unit 4 Outcome 1 – after the proof of concept has been finalised, you will manufacture the product safely using a range of complex tools and processes, keeping a record of progress.
- **SAC:** Unit 4 Outcome 2 – the finished product is evaluated and compared with other similar products, and questions are posed about future developments.
- **Exam:** In the end-of-year exam, you will respond to exam questions to summarise your knowledge of, and skills in, the subject content.

Remember to check the VCAA website for SAT assessment criteria, and for sample and past exams.

CHAPTER
9
Managing production for ethical designs

UNIT 4, AREA OF STUDY 1 (SAT): OUTCOME 1

In this outcome, you will:

- implement a scheduled production plan, using a range of materials, tools and processes and managing time and other resources effectively and efficiently to safely make the product designed in Unit 3.

Key knowledge

- risk management associated with selecting and using materials, including chemicals and other substances, tools and processes
- techniques for goal-setting when implementing a scheduled production plan, and methods used to record and report progress that include monitoring efficiency and effectiveness of production activities, and management of time and other resources and the overall project
- methods for documenting decisions and modifications made when following the scheduled production plan.

Key skills

- apply risk management throughout production
- use materials, tools and processes safely, demonstrating different degrees of difficulty
- devise and use methods to manage time and other resources effectively and efficiently to make a quality product
- record and report progress, and justify decisions and modifications when implementing a scheduled production plan.

Source: *VCE Product Design and Technologies Study Design*, pp. 34–5

9780170477499

MANAGING PRODUCTION FOR ETHICAL DESIGNS

Implement a scheduled production plan

Record the process

Risk management

Time management and techniques for goal setting

Making a quality product and degrees of difficulty in processes

Documenting decisions and modifications in a journal

Summaries:

- Chapter 9 **(p.259)**

Quizzes:

- Chapter 9 revision **(p.259)**

To access resources above, visit **cengage.com.au/nelsonmindtap**

Nelson MindTap

SCHOOL-ASSESSED TASK (SAT)

The SAT consists of three outcomes across Units 3 and 4.

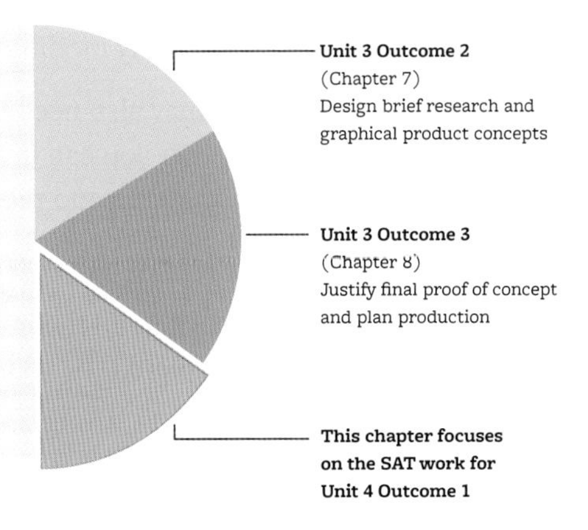

Production

In this stage of the design process, you will implement your scheduled production plan to safely and competently construct your product. You may find you need to 'revisit' other steps in the Double Diamond design process from time to time. You will continue to use your creative, critical and speculative thinking throughout.

For example, many students think that evaluation happens at the end of the process, when the product is finished, but the evaluation of choices and procedures, using critical thinking, will be happening constantly. This continual evaluation may mean that you need to do more research, create some more design ideas (creative thinking), carry out some more trials, experiment with different materials, refine the working drawings, do another prototype or model of a small part or change some components. This is key to the iterative design process, and is normal practice!

Quality and degrees of difficulty

In Unit 4, a high-quality 'end product' is expected. You are also expected to demonstrate different degrees of difficulty in the work. This means using construction processes that require some form of judgement or skill. Some schools will have technologies to do this for you and only require your CAD and set-up skills. Alternatively, it may be more difficult but you may need to complete something by hand if you don't have access to technology that would make it simple and easy. What is important is that your journal (or record of progress) makes clear the part you played.

3D printing could be used to replace parts that may not be possible to create using normal classroom facilities. Electronic or digital parts may be non-functioning, but their placement and position can be indicated on your prototype. All 'end products' should be full scale.

Implementing your scheduled production plan

- Refer to your **production steps** and start preparing the space you need.
- Gather your **material/s** and the necessary equipment, tools or machinery (technologies).
- Take note of the date, and compare it with your **timeline**.
- Take photos of the raw materials as you process them and make a note of dates for your journal.
- Identify when work is outsourced.
- Check your **risk assessment** and organise any relevant **controls** or precautions.
- Check your **quality measures** and any tips you have to achieve them.
- Record any modifications or changes made during production as you go.

You may also need to carry out extra equipment training or practise some techniques to ensure that you can implement your production plan safely and to a high level of quality.

- Photograph and record what you learn when practising skills.

Working safely: risk management

Refer back to all the information in the previous chapter on risk management and the work you completed for your risk assessment. Make sure you know exactly how to keep yourself and others in the room safe from any harm. Do this for each workshop session.

Time management

Don't spend time waiting for your teacher: there are many things you can do instead of waiting. Look at the next step to see what you can do, such as a test run or a trial. Spend time updating your journal.

Techniques for goal setting

As the weeks go by, refer to your timeline and, if you are behind, look at ways to maximise your output in class, ask for help or simplify something in your design. You may have other duties, an illness, teacher absence, a school camp or family issues that interfere with your workshop time. You will need to mark this on a copy of your timeline. In this case, you will need to consider how you might make up time or how your design can be modified to take this into account.

Showing modifications to your timeline

An example of a modified timeline

	Week 1		Week 2		Week 3		Week 4		Week 5		Week 6		Week 7		Week 8	
	22 May	24 May	29 May	31 May	5 June	7 June	12 June	14 June	19 June	21 June	10 July	12 July	17 July	19 July	24 July	31 July
1 Mark/ measure	×															
2 Cut pieces		×	×													
3 Mark joints				×												

An example of a timeline comparing predicted completion of steps (in pink and crossed) with actual completion (in green). In this case, extra columns may be needed to add extra dates.

A journal: recording your production progess

In this study, you are expected to record your progress during production with photos and words (you can also use videos). In this text, we will be calling that record a 'journal' regardless of the method you use to create it. Remember, it can be multimodal.

A journal is part of your SAT work, like a section in your folio. It helps others to see your progress and understand why certain decisions were made and the context in which you made them, and identifies what was done by your own hand, by machine or by an 'expert'. It also cements your own knowledge and terminology. It helps you to evaluate your progress through production and provides evidence for authentication.

Your journal documents the production steps as you complete them. It should be supported by photographs, so aim to have access to a digital camera for all sessions (you may need to organise this with your teacher). Don't forget to check the quality of the photos, too. Upload your photos onto a computer straight away, so that they don't become 'lost'. Keeping your journal on a school server or a cloud server (such as Google Slides) is a good backup.

Make sure you record:

- clear photos – not blurry, and without any shadows; it's best to put the 'work' between you and an open window
- dates for work completed and the amount of time the work took
- a description of production work carried out, including the tools, materials and processes used
- observations of things you have learned and experienced.

Also record if relevant:

- assistance given or processes outsourced
- quality measures you implemented
- the safety procedures followed
- aspects that could be improved.

It is also very helpful to include a statement at the end of each session entry that tells you what you need to do in the next session. This saves you relying on your memory, and can save time and provide the answer to the question, 'Now, where was I?' at the start of your next workshop session. This can be done even if you haven't yet inserted the photos.

For example, you may describe your day's progress like this:

Session 1 **Date: 29 May** **Time: 1.5 hours**

Today I marked and cut the front panels for my cabinet drawers. I expected it to take 90 minutes. Unfortunately, I measured incorrectly and will need to remake one of the panels.

Next session I need to: get all the tools and templates out again to measure that panel correctly, then cut it out carefully again.

Template
Design Folio Template

See examples of journals in Chapter 11, on the SAT visual checklist.

Date:		Time taken:	
Summary of production, any outsourcing or assistance; safety precautions; errors etc.			
Quality measures used			
Modifications (with reasons)			
Next session I need to ...			

Possible layout for your journal

Modifications to designs or plans during production

If you make changes to your final proof of concept during production, check with an end user to get their feedback. Modifications to the design or plans can be noted in your journal. Modifications can be communicated visually:

- with diagrams – this will help when preparing for exams, where you may be required to draw diagrams of processes
- using a copy each of the timeline and the sequence plan, with modifications scribbled onto them in a different colour or highlighted and annotated
- with photos of mistakes or errors and how they were fixed
- with annotated photos as normal journal entries
- by copying relevant sections of the working drawing that have changed and redoing them.

Alternatively, you could use a modification sheet to keep track of any changes.

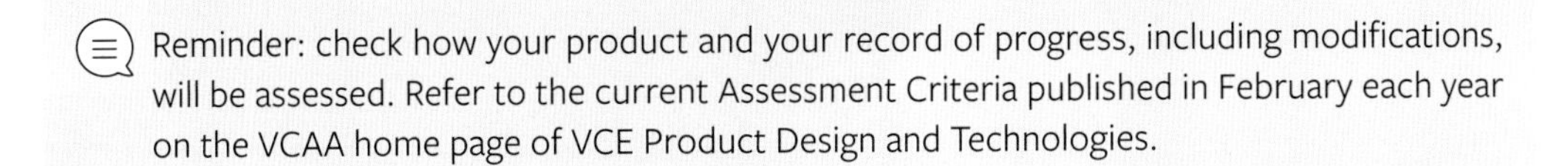
Reminder: check how your product and your record of progress, including modifications, will be assessed. Refer to the current Assessment Criteria published in February each year on the VCAA home page of VCE Product Design and Technologies.

Summary
Chapter 9

Quiz
Chapter 9 revision

Weblink
VCAA VCE Product Design & Technologies home page

Chapter 10

Evaluation and speculative design

UNIT 4, AREA OF STUDY 2: OUTCOME 2

In this outcome, you will:

- synthesise data to evaluate a range of products, including making judgements about the success of each product
- discuss product designs in regard to entrepreneurial activity, innovation and sustainability and/or other ethical considerations.

Key knowledge

- the role of research and development (R&D) and its importance to entrepreneurial activity and innovation
- speculative design thinking and innovation and their importance to entrepreneurial activity and the success or failure of products
- product developments that integrate new and emerging technologies
- environmental, economic, social and worldview issues associated with new and emerging technologies
- relationships between market research and the product development process
- qualitative and quantitative methods of evaluating a range of products, including their own product, that are evidence-based, including criteria and end user feedback
- methods to interpret and present data
- key factors that determine the success of a product.

Key skills

- discuss the importance of research and development
- describe the product development process in industry through the analysis of products that integrate new and emerging technologies
- use speculative design thinking to discuss and analyse strategies that encourage innovation and entrepreneurial activities
- discuss sustainability and other ethical considerations for products that use new and emerging technologies
- construct qualitative and quantitative research to collect data on a range of products, applying ethical considerations and using digital technologies where appropriate
- collate, interpret and synthesise data to evaluate the success of a range of products
- use data to justify enhancements and/or improvements to a range of products.

Source: *VCE Product Design and Technologies Study Design*, pp. 35–6

9780170477499

EVALUATION AND SPECULATIVE DESIGN

Research and development (R&D)

Speculative design thinking and innovation and entrepreneurial activity

Product development process for products that integrate new and emerging technologies

Ethical considerations associated with products using new and emerging technologies

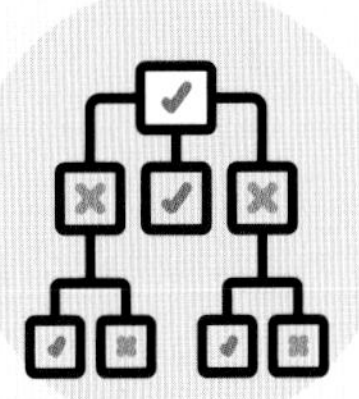

Key factors that determine the success of a product

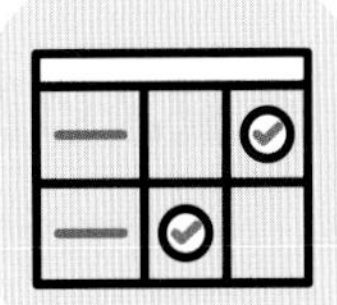

Using data to evaluate the success of a range of products and to justify enhancements and/or improvements

Qualitative and quantitative methods of evaluating products

Relationships between market research and the product development process

Worksheets:

- 10.2.1 R&D (Research and development) **(p.266)**
- 10.6.1 Researching technologies **(p.273)**
- 10.8.1 Key factors for success **(p.277)**
- 10.9.1 Evaluation criteria **(p.281)**
- 10.9.2 Presenting your data **(p.283)**

Summaries:

- Chapter 10 **(p.284)**

Quizzes:

- Chapter 10 revision **(p.284)**

To access resources above, visit **cengage.com.au/nelsonmindtap**

Nelson MindTap

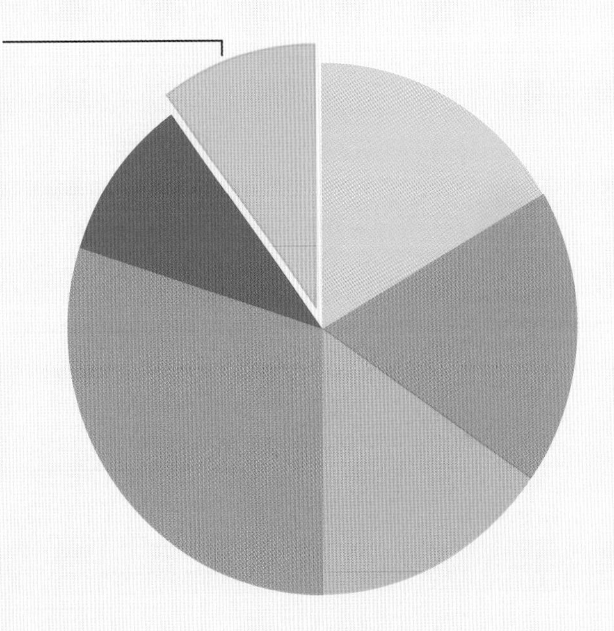

Evaluate a range of products

In this Area of Study/Outcome, you will evaluate a range of similar products, including the one you created for the SAT. Choose two or more commercial products that are similar to your SAT product. Identify the particular brand and a particular model of each product. It may be helpful to purchase the products so that you and the end users can assess their performance (how well they work), quality of materials and construction. Of course, this will depend on their size, cost and availability.

Make sure you choose products:

- for which you can find plenty of information relevant to this Area of Study, e.g. the material they are made of, the country they are made in, other sustainability issues and some background about how they were developed
- that use new and emerging technologies
- that may have used R&D.

Examine all the key knowledge before choosing the products to evaluate, so that you have clarity on what information is needed.

Read more on pages 279–80 under the heading 'What products can be evaluated?'.

What is R&D?

Research and development (R&D) is scientific/technical and experimental in nature. It is focused on creating new knowledge, materials, technologies, processes and systems. In manufacturing, it can also be about products, what they can be made of or how they can be made.

The Australian Bureau of Statistics (ABS) website defines R&D activity as:

> systematic investigation or experimentation involving innovation or technical risk, the outcome of which is new knowledge, with or without a specific practical application, or new or improved products, processes, materials, devices or services ...

The *Frascati Manual 2015*, published by the Organisation for Economic Co-operation and Development (OECD), defines R&D as:

> creative and systematic work undertaken in order to increase the stock of knowledge – including knowledge of humankind, culture and society – and to devise new applications of available knowledge.

According to the Frascati Manual, activities must be: novel (new), creative (based on original concepts), uncertain (with an unknown outcome or unknown results), systematic (planned and budgeted, with evidence) and transferable (leading to results that are reproducible).

The Frascati Manual also says that 'R&D activities in the mechanical engineering industry often have a close connection with design'. In small businesses, R&D completed for new prototypes or new manufacturing processes might include calculations, designs, working drawings and operating instructions.

Weblinks
Research and Development Tax Incentive

Enviroloo Hypothetical Case Study

Windwake Hypothetical Case Study

Biofnatics Hypothetical Case Study

Read about the Research and Development Tax Incentive on the Australian Government's Business website.

9780170477499

Forms of R&D

According to the Australian Tax Office, R&D activities must start with an idea, or hypothesis, and involve experiments that will have unknown results; these can take place in a laboratory or on a factory floor and are for the purpose of generating new knowledge (knowledge that is not already available).

The activities that are classified as R&D differ from business to business. R&D may not result in immediate financial gain, so is often considered a risk on investment. It can be difficult to manage, as the researchers do not know in advance exactly how to accomplish the desired result. See more under the heading 'Investment and risk'.

You can watch case studies in relation to the R&D Tax Incentive on the BusinessGovAu channel on YouTube.

Shutterstock.com/Gorodenkoff

Inventing and designing new components can be a form of R&D.

Some of the forms that R&D can take are:

- inventing or developing new materials (e.g. lighter, stronger, warmer, more elastic, photovoltaic for use in solar cells, suitable for 3D printing, using advanced composites, having improved sustainability) or new uses for materials
- looking at materials developed in other fields (e.g. food and medicine) and determining their suitability for various uses
- developing new equipment and machinery, technologies and/or systems
- developing new ICT (writing and testing code) for use in design and product manufacture, and/or distribution.

It is important to note that the Australian Tax Office does not consider market research and market testing as R&D for the purpose of tax incentives because the methods they use are not normally experiments.

The automotive industry is an example of a sector that invests a lot of money in R&D. Car manufacturers use R&D to research and improve:

- the use of alternative fuels/energy
- engine efficiency/performance and fuel economy
- body shape and aerodynamics
- the durability and tactile qualities of interior materials
- the ability of body materials to withstand impact or compress in a road accident
- the use of sensors and computer-controlled safety systems
- the quality and performance of outsourced components
- the development and use of new construction machinery, processes and systems
- the use of AI in the manufacturing process.

Role of R&D in manufacturing

A company's goal when investing in R&D is to innovate and introduce new or improved products and services to market and to increase profits. R&D helps companies to:

- stay competitive in a global market
- improve productivity, which can also increase margins and outpace competitors
- stay ahead of customer demands or trends.

R&D benefits businesses, communities and the broader Australian economy and is important for helping companies to grow and succeed. A healthy and viable manufacturing sector that exports products to other countries helps to balance national trade figures, creating employment, wealth and opportunities for others. Strong economic growth helps to raise living standards by stimulating enterprise, encouraging innovation and providing a positive environment for technological progress.

R&D's importance for innovation and entrepreneurial activities

A business is being entrepreneurial when it is prepared to work hard and take risks to develop a new product or service for profits. Companies want to 'own' new knowledge to help in their entrepreneurial activities, to increase customer satisfaction and to stay ahead of competitors.

- To be competitive, innovation is required – not just in products, but in processes and systems. You can understand why innovation is important by asking simple questions: 'What if a company wasn't innovative in the way it did things? What if it produced the same design in the same way, year after year?' Of course, there are some products produced in this way, but they are usually of very high quality and with a long and proven reputation.
- To be innovative, designers can use **speculative thinking** along with R&D to consider what will suit customers' needs in the future. Read more about speculative thinking on pages 266–7.
- To be **entrepreneurial**, something new, useful and innovative is needed.

Speculative thinking
Thinking ahead about new ways of doing things to benefit all of humanity

Entrepreneur
A person or group that starts their own business and takes on financial risks to bring a product to market in the hope of profits

Consumers now have worldwide access to an unlimited selection of products in a variety of price ranges. When similar products are advertised at equal or similar prices, it is design and innovation that gives one product the 'competitive edge'. Some consumers may want something special or personal and be prepared to pay a lot for it. This item may be special because of the type of material used or because of something clever, beautiful, sophisticated or quirky about the design. It may be more useful, lighter, faster, easier to use than the alternatives. Perhaps the product is more sustainable.

9780170477499

Designers work to catch the eye of consumers, with style and a carefully thought-out product that offers something 'new'. When combined with innovation in materials and processes, this can set a product apart. The following examples can be seen as innovative:

- Munjoi *All Dai Shoe* – it can be worn in four different ways and is made completely from plant-based materials, with less than half the CO_2 emissions of normal sneakers
- *Safee* – a small image projector that sits behind a cyclist on the bicycle seat and projects the relevant indicator signs on the rider's back. It enhances the safety of bike riders at night
- *Pure Spout* – an add-on filter to drinking fountains that removes lead and copper water contamination from old pipes.

Knog

The innovative Knog *Oi* bike bell

For a company to stay in the market and be competitive, it needs to bring innovation into the product development process. This innovation could be in:

- the product itself
- the systems for organising the way the product is manufactured
- the way the product is distributed
- the manner in which the product is made available to customers
- the way the product deals with sustainability issues
- the servicing of the product that is provided to customers and post-sales customer support.

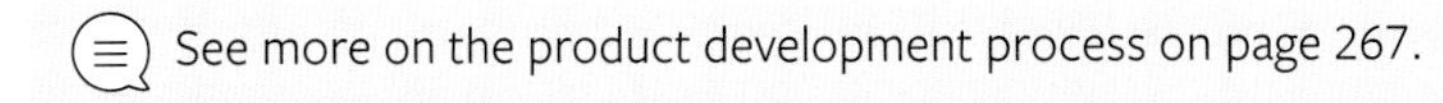
See more on the product development process on page 267.

Investment and risk

R&D brings innovation but it requires significant financial investment in education, training and suitable technology, and the results may not be known or guaranteed in advance. It is therefore a long-term approach, but it may lead to new intellectual property (IP), i.e. patents, copyrights or trademarks, which provide some protection of ownership for a certain number of years and therefore financial benefit.

Refer back to Chapter 7 on page 199 to review intellectual property.

As well as there being uncertainty about whether R&D will be worth the cost to a company, another risk is that there might also be a 'spillover' of knowledge to competitors. However, it is considered so important to our economy and global competitiveness that the Australian Government offers significant tax incentives to reduce the associated financial risks.

The Australian Government also supports R&D by funding university grants and keeping a register of Research Service Providers (RSP) such as CSIRO (Commonwealth Scientific and Industrial Research Organisation) and ANSTO (Australian Nuclear Science and Technology Organisation). These agencies provide opportunities for companies to collaborate with them as research partners. A list of RSPs can be found on the Australian Government Business website.

Small companies will outsource R&D for a variety of reasons, including their limited size, the cost of high-quality R&D facilities and the difficulty of gaining access to experts. CSIRO works with R&D projects that have budgets in the range of $10,000 to $50,000.

Weblinks
CSIRO Kick-Start

Charlwood Design

- See more details and a list of case studies (from the 'CSIRO Kick-Start Success Stories' menu) such as Ceres Tag (pet tracking collar), Conflux Technology (additive manufactured heat exchangers), Molten Labs (robots to detect and deactivate landmines), ULUU (seaweed to plastic), Worn Up (fabric recycling), and watch a short video, on the CSIRO Kick-Start page on the CSIRO website.
- One company that is on the list of Research Service Providers in the field of Product Design is Charlwood Design, Melbourne. Visit the Charlwood Design website for more information.

Worksheet
10.2.1

ACTIVITY

1 Choose five questions from the following to answer. Choose questions that are very dissimilar from each other.
   a Give two reasons why R&D is important for businesses and organisations.
   b How can R&D help a company to stay competitive in a fast-changing market?
   c What are some examples of successful products or services that were developed through extensive R&D?
   d How can R&D help a company to improve its existing products or services and create new ones that better meet customer needs?
   e What are some potential challenges or risks associated with investing in R&D?
   f What are some of the ethical considerations that companies should take into account when conducting R&D?
   g How can R&D contribute to broader social and environmental goals, such as sustainability and social responsibility?
   h How can governments and other organisations support R&D efforts in the private sector, and what are some potential benefits of such support?
   i What role do collaboration and cross-disciplinary approaches play in R&D, and how can companies foster a culture of innovation and creativity?
2 Research one form of R&D (you could use the CSIRO Kick-Start website) that could be relevant to the type of product you are designing and making for your SAT. Present the information to your class and include ideas for how it could be incorporated into your SAT product. Ask the class for their ideas, too.

Speculative design thinking

Speculative design thinking poses the 'big' question, 'What if ...?' Speculative design thinkers look to future possibilities to solve current problems and to create a better society, environment and world. They then use creative and innovative thinking and cutting-edge

9780170477499

R&D to make some of those possibilities a reality. Speculative design thinking can help companies to be innovative and take entrepreneurial risks by encouraging them to think deeply about how the world is changing and what impact their current decisions and practices will have on humanity and the planet. They can look at how current practices need to be transformed, what is on the horizon in technology, how R&D can help and the best decisions they can make to improve life for all of us.

Strategies to encourage innovation and entrepreneurial activities

If a company wants to be innovative and entrepreneurial when using a speculative design thinking approach, it needs to:

- provide funding and resources to create and foster a culture in which creativity, experimentation, risk-taking and failure are seen as learning opportunities
- have a supportive environment for training and upskilling staff, and promote diversity and inclusivity in the workplace
- embrace emerging technologies, look to R&D
- use a range of perspectives to guide decisions, including market research, futures research, trend analysis, research related to sustainability and ethics, etc.
- encourage collaboration among their team and in partnerships with other businesses – as shown on page 274 by the example of the Havaianas Market *Zip Top Slides*, in which Havaianas collaborated with Market – or with academia, industry or government bodies
- apply for tax incentives or grants as a 'start-up' (new business)
- know what appeals to consumers and how to get them 'on board' or attract them, i.e. what is new and exciting, up-to-date, cutting-edge or trendsetting.

When evaluating products for this outcome, you will be assessing whether companies have used methods or approaches that are innovative and/or entrepreneurial.

The product development process

It is important to remember that product development is a process for commercial products in mass or low-volume production; it is not relevant to one-off products. All commercial products go through a version of the product development process – from abstract concept to design, production and distribution, and then on to retail.

Market research and feedback from retail is an important part of the process and feeds into the next round of design and development, which could lead to an improved version or a completely new product. See more on market research and the marketing elements on page 269.

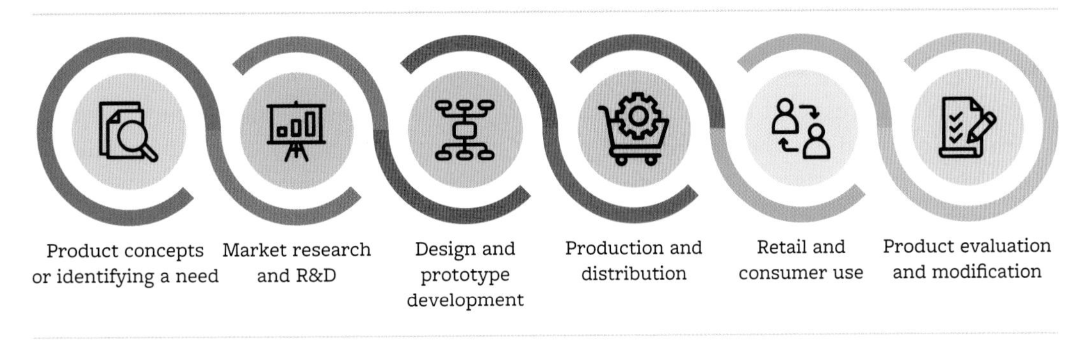

Starting points: product concepts or identifying a need

New products may start from a range of possible triggers – for example:

- market research identifying a consumer or end user need that is not properly met by existing products
- the need to replace or modify an old product that has become obsolete or unpopular
- access to a patent for a product concept or idea
- technological innovation that allows new products to be developed or existing products to be improved.

Research

When developing or modifying products, it is important to investigate relevant information and conduct trialling or testing. Areas of research include:

- market research – into the needs, interests and behaviours of end users, the products offered by competitors, etc.
- R&D that is available, such as the innovative uses of materials and production processes (including the application of new technologies).

Research may be used throughout the product development process for specific purposes that are relevant to each stage.

Graphic representation of the product development process by student Paige Anderson

Design and prototype development

Designers may respond to market research or create something new. They come up with designs and get them approved by the company (in-house or by the company that hired them). Issues are sorted out through computer modelling and physical prototypes (in the clothing industry these are often called samples) until approved again. Prototypes are also checked by the manufacturers to sort out any construction or quality issues. Once the final prototype has been tested (both technically and through user feedback) and approved, it is ready for production (of multiples).

Production and distribution

Once the prototype has been finalised and production processes have been planned and organised, the product is put into production at low or high volume, depending on the number of units that are expected to sell (based on research of the market size). A company will choose a manufacturer that has suitable production technologies for quality and speed and that charges suitable rates (costs). Components are outsourced if necessary (especially in vehicle manufacture) and suppliers are organised. The production process is carefully monitored, quality is checked at all stages, and modifications to the process are made if they are needed.

Distribution and transportation move the product from the factory to wholesalers or retailers for sale.

Retail and consumer use

The product is sold through appropriate outlets or directly to the consumer. Increasingly, products are sold online. Consumers use the product until it breaks down, goes out of fashion or becomes obsolete. They may replace the product, buy a modified version, or seek a new product to meet their needs. Market research is used again to see what can be improved and what is going well.

Product evaluation and modification

Market research in many forms, such as sales and returns data from retailers and consumer responses or reports from focus groups, is used to refine or modify the product. Information is also gained from the sales of products by competitors. All this information is used to evaluate the success of the product. If minor aspects of the product need changing, these changes are implemented with minimal alterations to production. Major product changes or redesigns require the company to work through the cycle again.

Market research in the product development process

Market research gathers information about people and their needs: who wants which products, where they live, how much they are prepared to pay and how best to make the product known to them. It occurs throughout the product development process (mainly performed by the marketing team) and is vital for the product's success. The information is sometimes grouped into five areas known as the marketing elements.

The marketing elements

In marketing there are five elements, known as the '5 Ps': people, product, price, place, promotion. By defining each of these, a company can develop effective marketing strategies.

Note that you do not need to know the 5 Ps for VCE Product Design and Technologies, but they will help you understand marketing and the difference between R&D and market research.

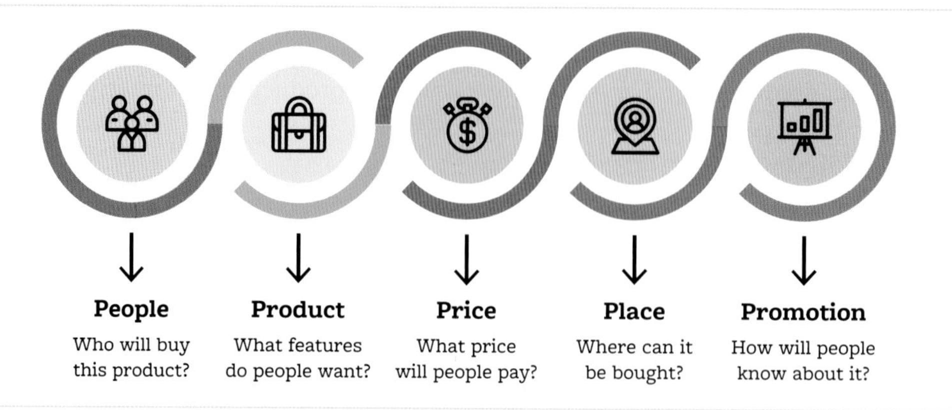

1 People: who will buy this product?

For marketing purposes, 'potential customers' are divided into target groups, and information about the tastes and needs of people within a group is pinpointed. People are broadly placed into target groups according to personal factors, age brackets, interests, their spending habits and where they live.

2 Product: what features do people want?

What is special about the product that makes it appealing to the specific target group, such as how it functions, its status, its price, its sustainability, its new technology or its appearance?

Market research is used to identify the particular features of a product's design that appeal to the target group, or improvements that could be made.

3 Price: what price will people pay?

Price is determined by what a potential customer might be willing to pay. Pricing strategy considers the cost of competing products. The cost of making and distributing the product also has to be taken into account to ensure that it eventually creates profit. Market research also identifies what potential consumers are looking for in terms of status, quality and value for money. The retail price can be high for new and innovative products, as consumers are willing to pay for them.

Weblink
Manolo Blahnik

4 Place: where can it be bought?

The product needs to be placed where the target market does their shopping. For example, shoes with broad appeal – such as sports shoes – are available in department stores and smaller stores, often in large retail centres. Other types of shoes are sold in specific locations, such as seaside resorts, country towns, outer suburbs or inner-city suburbs. Expensive handmade designer shoes such as Manolo Blahniks are sold with other luxury brands in exclusive shopping strips around the world.

Not all places for distributing or selling products are physical stores. Many products are sold through catalogues, direct ordering and, increasingly, through the internet, particularly since the lockdowns during the COVID-19 pandemic of 2020 and 2021.

5 Promotion: how will people know about it?

Promotion communicates information about the product, and alerts people (the target market) that it is available (where and how much). This is done by targeted advertising in print and online, sponsorship, in-store demonstrations, product placement in entertainment,

9780170477499

direct mailing, emailing, phone calls and social media. A whole new sector of people known as 'influencers' have developed since the advent of Instagram in 2010, and many of them promote products to their followers (who sometimes number in the millions) in exchange for a fee or for gifts or free travel. Instagram, with text and images, quickly became a favoured method of product promotion thanks to its ease of use, immediacy and reach to the target market.

Brands need to be careful when choosing influencers to promote their products or people with whom they showcase a strong relationship. If at some future time an influencer or famous person makes public statements that don't align with the brand's values, it can be a costly problem to deal with in terms of reputation and unsold stock. An example is what happened to the Adidas Yeezy line of runners in 2022. You can read more on the Sneaker News website.

Weblink
Sneaker News: Adidas Yeezy

Market research is used to identify what sort of promotion strategy is economically feasible and is likely to have the greatest impact on the target group.

A graphic representation of the product development process, including market research, by student Ollie Nwobu

Considerations in the use of new and emerging technologies

You will also be looking at how effectively new and emerging technologies are used or integrated in the manufacture of the products you choose to analyse. Refer to Chapter 6 and Unit 3, Outcome 1, for new and emerging technologies listed in this study.

The following is a list of ethical considerations, including sustainability issues, in relation to the use of new and emerging technologies. Keep in mind that all these considerations are **inextricably** linked.

Inextricably
In a way that is impossible or difficult to disentangle or separate

Environmental, economic and social issues

Some of the issues associated with new and emerging technologies are:

Environmental issues

- increased need for energy to power technology, especially for the hyperscale data server 'farms' needed for AI and 'interconnected' devices, and the impacts of different sources of energy
- increased use of raw materials, leading to depletion and environmental degradation
- faster rate of production, meaning more products are produced
- more transportation required and negative impacts of vehicle emissions and waste such as bilge water from cargo ships
- generation of waste and pollution from production, usage and disposal of products
- more products going to landfill
- problems in disposal of technologies that are out of date or will go out of date
- **detrimental** impacts on air, soil and water
- climate change caused by emissions from production

Detrimental
Negative and destructive

Economic issues

- creation of new job opportunities, but also job displacement as automation increases
- **disparities** in access to technology and the benefits it brings (disparity between wealthy and poor people, developed and developing nations), leading to economic inequality or making existing inequality greater
- changes to traditional business models and market structures, leading to economic disruption

Disparities
Great differences

Social issues

- changes to the way people interact, shop and communicate, including increased screen time and decreased face-to-face interaction, less physical activity
- disparities in access to technology, leading to a digital divide and social inequality
- disparities in quality of life due to low employment opportunities
- impact on privacy, security and personal data, leading to concerns about surveillance and data misuse
- the impacts on the health and wellbeing of low-paid workers in less regulated countries

Worldview issues

- shifts in values and beliefs about what is important, desirable and acceptable
- increase in production, its consequent waste and where it ends up
- emergence of new ethical, legal and moral questions related to technology and its impact on the nature of labour and therefore society
- changes to cultural norms and traditional ways of living (due to war, pandemics, floods, fires, deforestation, reduced birth rate, aging populations and smaller families, etc.)
- challenges to our economic and social order (e.g. how goods and services are distributed, the increasing gap between the rich and poor, rising crime rates associated with lack of opportunities and poverty, how to treat the employment shifts and job losses created by automation, how to deal with and finance education and healthcare)
- challenges to commonly agreed principles and values that are acceptable in different societies and determine how a society's members interact with one another and with the broader world (such as individual rights, justice, equality, freedom and responsibility); often shaped by factors such as government policies, cultural norms and historical events

9780170477499

All of the points above shape people's experiences of work, education, healthcare and other aspects of daily life, and influence the overall wellbeing of individuals and the society as a whole. They are just some of the many issues associated with new and emerging technologies, and it is important to constantly consider them afresh as technology continues to shape our world. You will consider some of these issues when evaluating products for this outcome.

You can also refer back to the topic 'Impacts of technologies' on pages 171–2 of Chapter 6.

ACTIVITY

1. Draw a diagram to explain the differences between market research and research and development (R&D). Annotate the focus of each.
2. Draw a diagram to represent the product development process, using cartoon images to represent the type of product you are designing and making in your SAT.
3. Make a list of the technologies covered in Chapter 6 from pages 125–33 that have been used in the making of the products you are analysing. Alternatively, in a diagram, elaborate the designing or manufacturing stage with annotations (or cartoons) to include the emerging technologies used in these products.
4. Read the text under the headings 'Environmental, economic and social issues' and 'Worldview issues' on page 272. Combined with your own research and knowledge, make a statement on how the companies that make the products you are analysing use emerging technologies and how they deal with several of these issues.

Worksheet
10.6.1

Success or failure of products

What is a successful product? For a company, a successful product would be a 'hot' item that sells well for years. For consumers, it might be a product that works well and lasts for years. Sometimes, a product is extremely successful when it first hits the market but this initial success does not last.

Some products are successful due to their aesthetic appeal, e.g. the *Juicy Salif*, a lemon juicer designed for Alessi in 1988 by Philippe Starck. Its limited functionality (it was unstable, unsafe and messy to use) has not dented its continued market success. Other products are successful because they offer affordable solutions to everyday needs, e.g. Ikea researches customer needs to create household items that are reasonably priced.

A product's success can change for many reasons. Read more about the success that Adidas originally had with the Yeezy range of runners, as mentioned on page 271. In 2022, however, the company's brand ambassador and collaborator Ye, formerly Kanye West, made public comments that were not aligned with their values. Adidas was left with the massive problem of working out what to do with Yeezy products worth more than $500 million. The choices were to dispose of them by incineration or landfill (with negative environmental impacts), donate them, or repurpose or rebrand them. Whichever method they chose would cause them losses of close to $1 billion. To sum up, the product was originally very successful, but to continue selling it would impact on the reputation of the company, and perhaps cost them more in lost sales of all products. In this case, it was something outside the product itself that impacted on its success.

Take a look at the 'flip-flops to zip-top shoes' by Havaianas on the next page. They would appear to be destined for success for the company – but do consumers really want two shoes in one?

CASE STUDY

Havaianas

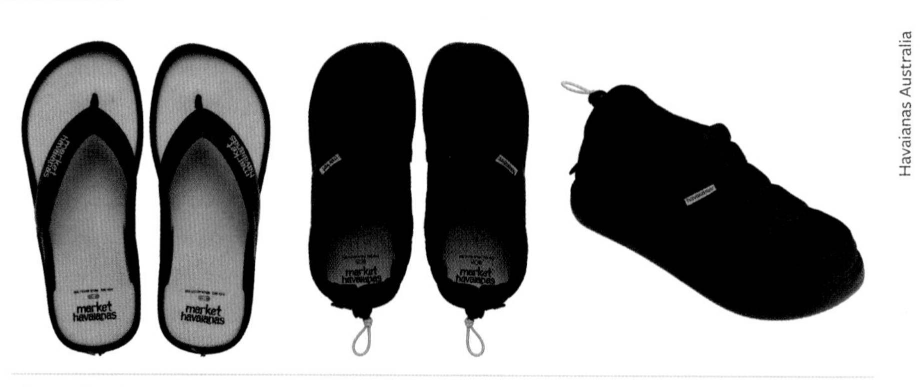

Havaianas Australia

All-weather Market Havaianas, made in Brazil – from flip-flops to zip-top shoes, all in one

Havaianas is a Brazilian brand of flip-flop sandals (known in Australia as thongs) created and patented in 1962. According to Havaianas, the design was 'inspired by the Japanese Zori sandal; a sandal with cloth straps and woven rice straw soles. Havaianas replaced the straw with premium Brazilian rubber and updated the design to look fashionable'. They soon became a global brand.

The two-in-one Havaianas 'shoe' was a collaboration with a Los Angeles–based brand, Market, in 2022. Their successful first collaboration earlier that year saw the creation of thermo-reactive flip-flops, which change colour from black to red when left exposed to heat as a warning to wearers. You can read about this collaboration on the Highsnobiety website.

Market is a streetwear brand from Los Angeles that was founded in 2016. It is known for its collaborations and partnerships with sporting champions, famous singers and rappers and its link with culture and music.

Market Havaianas Zip Tops are sandals that turn into sneakers – or the other way around! The zipper on the side opens or closes the upper, and the sandal from the beach turns into a shoe for the street. The top is warm like a puffer jacket. The footbed is cushioned and has a textured rice pattern.

Weblinks
Market x Havaianas is Too Hot to Handle

Havaianas: About us

Zip Tops are made from:

- sole: 100 per cent rubber
- insole: 100 per cent EVA
- upper: 100 per cent polyester
- zipper: nylon.

CASE STUDY

Rothy's: sustainable shoes and accessories

Rothy's shoes and accessories are from San Francisco, USA, manufactured in Dongguan, China, and owned by Alpargatas, Brazil.

Rothy's goal was to reach circular production by 2023 – that is, a continuous loop that renews itself, from material and manufacturing to product and recycling. The company's vision is to use twice-recycled materials in new products – to close the loop, like nature does. Founders Roth and Stephen set out to turn plastic bottles into shoes, endeavouring to create a timelessly stylish shoe that perfected comfort and design – while drastically reducing the excess produced by traditional shoemaking. Highly efficient manufacturing is made possible by their wholly owned factory, which is unusual in the industry. This includes knit-to-shape technology and skilled craftspeople who finish each piece by hand, with great care and attention. It also allows them to achieve a big reduction in waste by limiting overproduction and using a cleaner supply chain.

Bags and shoe uppers are made exactly to shape by precision knitting machines, which reduces by 30 per cent the normal waste in a 'cut and sew' production. Skilled artisans polish the hardware (buckles, etc.) and hand embroider the Rothy's logo.

9780170477499

Happy conditions for its workers mean an exceptionally high staff retention rate. Rothy's has a diverse group of employees and they are committed to furthering diversity, equity and inclusion efforts – from strengthening channels for employee feedback to improving hiring practices and mental wellness initiatives.

Weblinks
The Art of Shoemaking

Behind the Design: Comfort and Sustainability – Men's by Rothy's

What Goes Into Our Thread?

Recycling Shoes Gives Old Rothy's New Life

Imagine What a Shoe Can Do

Meet the Monty

The Future Is Circular

Meet Our Materials

See how the shoes are made on the page 'The Art of Shoemaking' on the Rothy's website.

Rothy's bags and shoes are made from innovative materials – natural, renewable, bio-based and recycled. They work ardently to source materials that are easier on the planet. In 2020 they made the list of Fast Company's Most Innovative Companies, and in 2021 *Time* named them as one of the most influential companies for innovation in recycled materials.

The ways in which Rothy's uses innovative materials include:

- Rothy's thread, which is made from plastic bottles and ocean-bound plastic
- footbeds made of algae-based foam
- outsoles made with 35 per cent natural and renewable materials (rubber) and minerals (sand)
- insoles containing 30 per cent bio-based material and recycled rubber, which are the result of expert exploration on repurposing materials. Every shoe and bag is machine washable!

- Watch the video 'Behind the Design: Comfort and Sustainability – Men's by Rothy's' on Rothy's channel on YouTube.
- See how the thread is made on the page 'What Goes Into Our Thread?' on the Rothy's website.

To achieve circular production, the aim is to keep old Rothy's shoes out of landfill. As of September 2022, well-loved (but clean) Rothy's can be taken to any of their retail stores to be recycled. Customers receive a discount code to use on a future purchase.

The returned shoes are passed on to disassembly and recycling partners, where they are separated and sorted into upper, insole and outsole (according to the materials). They are then shredded or baled and sent to the recyclers, who will repurpose the materials into something new, such as thread to be reused, carpet backing, insulation or athletic flooring. As the shoes are long-lasting, customers are encouraged to resell, gift or donate products that are still in good shape when it's time to dispose of them.

Rothy's goes to great lengths to appeal to its target market by making many YouTube clips.

- Read about Rothy's sustainability and recycling programs on their website.
- See examples of marketing clips on Rothy's YouTube channel.
- See a list of materials on the Rothy's website.
- See photos and descriptions of materials on the Rothy's website.

Packaging is strong and made from recycled material and is able to be recycled again. Output matches demand at every step in the supply chain (a lean manufacturing principle), which also reduces waste. Carbon emissions are offset by partnering with Carbonfund and Pachama – two organisations that fund forestry conservation, protection and restoration projects. Rothy's uses 100 per cent renewable energy sources in all its offices, workshops and retail spaces.

It took 3.5 years of **research and development (R&D)** to perfect the knit-to-shape technique and to launch the first product. Knit-to-shape is the foundation of Rothy's sustainable practice, for which they have received many environmental accreditations.

Alamy Stock Photo/Patti McConville

Alamy Stock Photo/Imaginechina Limited

Rothy shoes made with 'knit-to-shape' technology, perfected after three and a half years of R&D

Key factors that determine the success of a product

There are several key factors that determine the success of a product through the eyes of consumers and from the company viewpoint.

Key factors that influence the success of a product

For customers	Strategies of successful companies
The product meets the needs and demands of customers, e.g. it functions as required, is durable and meets a specific need or matches a current trend.	Understanding the target market and their needs is crucial to solving the need.

9780170477499

For customers	Strategies of successful companies
Customers want a good 'user' experience with the product, e.g. they want it to work well and be comfortable and safe.	Designing an intuitive and enjoyable experience for the user.
Some customers will look for a product that is innovative and offers unique features and benefits	A new approach to solving a problem is important and led by strategies within the company (see page 267 for strategies).
Many customers want high quality and are prepared to pay for it.	Developing a product that is reliable, efficient and performs as promised is likely to bring success. For highly priced products, good quality is critical in building a good reputation and retaining customers.
Many customers are looking for the cheapest price or what they see as value for money.	It's important to set the right price for a product in relation to competitors, for attracting customers and for generating revenue for the company.
Other customers are swayed by marketing and promotion.	Implementing a well-planned marketing and promotion strategy with a strong brand image can help a company to build brand awareness, attract the target market and build brand loyalty.
Customers need to be able to purchase the product when they want it and without too much difficulty.	The product must be readily available and convenient for customers to purchase through a well-established distribution network.
Many customers want support/help or questions answered about using the product or how to fix the product.	Providing good customer support and after-sales services can help to build customer loyalty and increase repeat sales.
Customers want products that are up to date and they want to purchase them when they are 'in fashion'.	Creating a product that is adaptable and flexible to changing market conditions and customer needs is more likely to be successful. Launching the product at the right time in the market is important.
Many customers want products that are sustainable and ethically sourced and made, and that have minimal negative impacts when used and disposed of.	The ability to prove its sustainability and ethical credibility through transparent supply chains and efforts to support disadvantaged communities. Putting in place systems that continue the company's responsibility for the product throughout its life cycle.

These are just some of the key factors that determine the success of a product. The relative importance of each factor can vary depending on the specific market and product.

ACTIVITY

1 Read the key factors that influence the success of a product and identify those that apply to the products you have chosen to analyse that are similar to your SAT product. Group them into the 'Factors that influence design' (i.e. need or opportunity, function, end users, aesthetics, market needs and opportunities, product life cycle, technologies, ethical considerations).
2 Use these key factors in the table above to assist you in analysing the products you have chosen and explain how successful you think each product is or will be.

Worksheet
10.8.1

Evaluating a range of products and data

Using qualitative and quantitative methods

Read about qualitative and quantitative information and how to gather it on page 201. Remember:
- Qualitative research methods could be interviews, focus groups or observations.
- Quantitative research methods could be surveys, measurable experiments or information from websites on sizes, prices, weight or other measurable/countable features.

iStock.com/FG Trade

A variety of data, both quantitative and qualitative, needs to be analysed.

Interpreting data

Your evaluation of products should include both qualitative and quantitative information. Any information you get from secondary sources (such as websites) will need to be sourced (acknowledged) and combined with your own interpretation.

Digital methods to collect and present data

Where appropriate, use digital methods such as an online survey, recordings or video of interviews or focus groups, but check with your teacher first.

Online sites that use both qualitative and quantitative information to compare products for consumers, and that you may find helpful, are:
- Choice Australia, a not-for-profit consumer advocacy group that conducts scientific testing of products – check if your school or local library has a subscription to give you full access to Choice's product comparisons
- Indy Best, from the *Independent* newspaper (UK edition), which reviews products by category, e.g. 'home and garden', 'kids' or 'sports and fitness' – you can use the search facility to find products
- Which?, a not-for-profit consumer 'champion' from the UK. Similar products are listed in a table, with direct links to online retail sites, and compared. Products are rated (e.g. out of five stars) through surveys of thousands of consumers and sometimes from lab testing
- Good Housekeeping UK, which compares many services and consumables as well as products. Be sure to choose a range of relevant products – ones that have a purpose very similar to your product, and where you can find enough information to complete your research and evaluation task.

Weblinks
Choice

Indy Best

Which?

Good Housekeeping

What products can be evaluated?

Suggestions for a range of products to evaluate include:

- products similar to your SAT product that have an innovative aspect
- products that are related to your SAT product, such as accessories, to which your evaluation criteria can be applied and are relevant.

Products similar to your SAT product

Choose products that are similar in type to the product you design and make in your SAT. This allows you to use the criteria you developed in Unit 3, Outcome 2, for evaluating your finished product, to apply them to similar commercial products and perhaps gain extra ideas before completing your production.

Examples of products that are similar to your SAT product

Your SAT work is:	Consider selecting a range (at least two) of:
• a piece of clothing/wearable product	• similar pieces of clothing or wearable products
• a piece of furniture or a household or outdoor item	• pieces of a similar type of furniture, or household or outdoor item
• a set, or piece, of jewellery	• similar sets, or pieces, of jewellery
• a health-related product or assistive or labour device	• similar products that are assistive or labour devices or health-related
• a musical instrument	• similar types of instruments, e.g. if you made a guitar, any type of guitar
• a sport- or travel-related product	• similar types of sports/travel products
• a toy	• similar types of toy

How to evaluate?

You should choose a minimum of two similar commercial products as well as your own SAT product. As well as referring to your evaluation criteria from Unit 3, Outcome 2, you need to look at entrepreneurial activity, innovation and sustainability and/or other ethical considerations.

Suggested questions are below. Choose only those that are relevant to guide your evaluations.

Entrepreneurial activities

When evaluating products, you will be thinking about the entrepreneurial activities that each company carried out and risks they took to get their product onto the market:

- How risky do you think it was for this product in terms of market success?
- What methods has the company used to make this product known to you and other potential consumers? For example, what promotional methods have they used?
- Do you think this product will do well with consumers for years, or decades?
- Is there a lot of competition on the market for this type of product?

Innovation of products

Innovation can be defined as creating products (or processes) that are new, different and generally accepted as an improvement on what exists. They may also provide increased value in addressing needs in a new way. Choose several suitable questions to apply from the following:

- What is innovative about this product?
- Can you find out whether the company has any intellectual property protections, such as patents or design registration, for this product?

- Has the design of the product applied new knowledge or technology?
- How is it new and/or different from usual products of this nature?
- How is it an improvement on previous products of this type?
- Has it addressed a problem or need that did not previously have a solution?
- Has it created a new or different market through new features or a change in style direction?
- Is it an incremental improvement on other products?
- Is it a radical breakthrough for this type of product that has disrupted existing markets?
- Has it applied disruptive technologies or new business models?
- Has it challenged established practices?

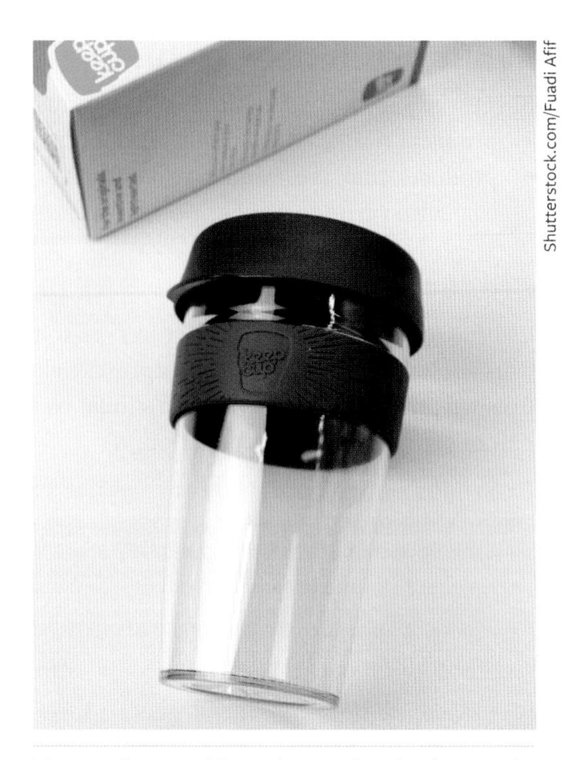

Shutterstock.com/Fuadi Afif

The KeepCup reusable cup has continued to innovate by adding new features, addressing customer concerns and changing needs, changing materials and collaborating with other businesses, all with the aim of reducing the use of disposable cups.

Sustainability and/or other ethical considerations

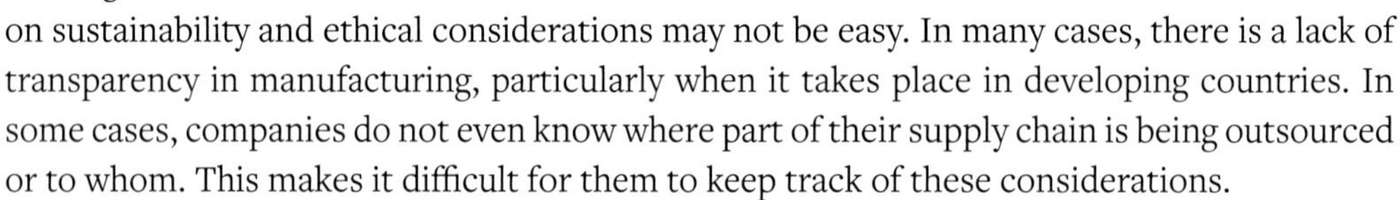

Finding accurate and clear information on sustainability and ethical considerations may not be easy. In many cases, there is a lack of transparency in manufacturing, particularly when it takes place in developing countries. In some cases, companies do not even know where part of their supply chain is being outsourced or to whom. This makes it difficult for them to keep track of these considerations.

You will need to use your own educated guesses, basing your conclusions on what you know about:

- materials
- the quality of products (given that higher quality products can be more sustainable because they are more reliable and long-lasting, reducing the need for a replacement product)
- the country they are made in and:
  - its environmental standards and regulations, i.e. for waste, pollution and environmental degradation. Manufacturing processes can release harmful chemicals, pollutants and greenhouse gases into the air, water and soil, leading to health hazards and environmental damage
  - labour management – is poverty typically exploited to cut costs, with poor working conditions, low wages, long working hours, unsafe working conditions, forced labour or child labour, which violates human rights and ethical standards and is often referred to as modern-day slavery?
  - accepted practices of corruption in that country
- how this product or company has approached sustainability
- what materials have been used and what their sustainability issues are. Are the materials reusable over and over? For example, metals that are not mixed with other metals or materials can be recycled without limit (further research will be required)

9780170477499

- whether the materials will biodegrade in a home compost system or whether they need industrial-scale conditions to biodegrade (further research will be required)
- whether the materials will end up in landfill. This is most likely to happen to materials that are a blend or composite, or that are expensive to recycle (further research will be required)
- if the products could be considered 'fast fashion' or if their design builds in planned obsolescence
- whether you think the company uses greenwashing. See more on this in Chapter 6, on page 157.

ACTIVITY

1 Develop a range of new evaluation criteria that cover some of the dot points in each of the areas above (to add to the list of your product's evaluation criteria). Consider the following areas, and select aspects that are relevant to your type of product:
   - R&D
   - speculative design thinking
   - their connection to entrepreneurial activity
   - the use and impact of new and emerging technologies (particularly related to sustainability and ethics)
   - market research and market effectiveness
   - materials and their environmental impact.

2 Identify and plan the research you need to carry out to assess these new criteria. Consider how you will compare your product with the commercially available products in these areas.

Worksheet
10.9.1

Shutterstock.com/doublelee

Shutterstock.com/martinpixel

Shutterstock.com/Vitalii Stock

For the products you choose to evaluate, you will need to conduct research and make educated guesses, based on your knowledge of materials and regulations in the country where the product was made, to make certain judgements.

Evaluating your own product

You will refer back to your evaluation criteria for your finished product and end user feedback. Where applicable, these criteria should also be applied to the similar commercial products you have chosen to evaluate.

Some examples of evaluating are in the following table (the table includes criteria from different projects). You could draw up a table like this **for each product** you have chosen to evaluate as well as your own SAT product. End user feedback will provide data for you to justify **enhancements** and/or improvements to both your own SAT product and the other commercial products you choose to evaluate.

Enhancement
Something that will increase or improve quality, visual appeal, functionality or value

Sample evaluations of different types of products

Evaluation criteria question	How I made a judgement and/or collected data	My conclusion on the success of each criterion with end user feedback	Suggested enhancements or improvements
Is the colour scheme sufficiently bright?	I checked this by showing it to several end users for feedback.	I think the colours are great, they are really bright. The end users said: • 'How gorgeous!' • 'The colours are perfect.' • 'Yes, that will stand out.' • 'Good but too much of the same colour.' • 'The colours are really bright and fun looking.'	Adding some contrasting colours on the rims and edges to brighten it up.
Does the container include storage that could be suitable for fragile items?	I put the fragile items in the container and asked my end user to carry them around and got feedback.	The fragile items looked nice and safe to me. However, my end user said they would also wrap them in cloth for extra protection.	Add extra padded fabric on the inside to increase protection for fragile items.
Do the sensory activities suit the needs of the end user?	I watched several end users interact with the sensory activities.	I think they worked well. Out of five 'end users', they all fidgeted with the secure pocket contents and looked relaxed.	I realised that one of the pieces could have been a lot longer as it kept slipping out of end users' hands.
Does the garment fit the end user nicely?	My end user tried it on.	It fitted my end user very well. However, they mentioned it was a bit loose at the top (provide more detail).	Take more precise measurements before adjusting the pattern and cut out pieces accurately. Also do fittings before final closing of joins.
Does the product fit easily into the required space?	I put the product in its intended position for the end user to see how it fitted.	The product fitted perfectly in the space. My end user said, 'It looks so snug.'	Perhaps make it a tiny bit smaller next time for ease of getting in and out.
Does the storage unit allow for easy assembly and disassembly?	To check this, I asked an end user to disassemble and reassemble the unit and timed them.	I think the disassembling and reassembling is very straightforward and fairly fast. I'm proud of it. It did take close to 40 minutes for each user. They struggled with getting the shelves back in the right place, but all agreed after one try they felt more confident to do this easily.	I think it might help to put instructions, both written and visual, for disassembling/ reassembling under the bottom shelf ... or perhaps online.

9780170477499

Presenting your data

Your data needs to be collated, interpreted by you and presented in a clear and concise manner, and it needs to be relevant. Comb through this chapter and work out what is relevant to your range of products in relation to your SAT product. You should also check the VCAA Performance Descriptors of the Unit 4, Outcome 2 assessment task to check what is expected.

Whether you are presenting an analysis, an inquiry or your research on products, consider the most suitable method to use. It could be, for example:

- setting up an A3 page with three or more columns, as suggested in the following table
- using Word or PowerPoint in landscape view and taking as many pages as you need
- creating one A4 page for each product and for each aspect.

If you are going to talk or record your presentation, consider also how much text to add and how you will remember everything you want to say.

It is advisable to complete your work on analysing commercial products well ahead of evaluating your own SAT product to avoid a large workload at the end of Unit 4 (which most often ends in Term 3). The following table provides a possible layout for your evaluation of products.

The table could include the following sections for adding evaluation criteria questions:

- your SAT evaluation criteria
- the marketing elements
- R&D, Innovation, and the use of new technologies
- entrepreneurial activities
- sustainability, speculative design and worldview issues.

Explain how you researched or tested to compare your products, either in a separate column or in your product-by-product responses as shown below.

	Commercial product 1	Commercial product 2	My SAT product
Marketing elements			
What is the product?	[Photo and source]		
[Add your evaluation criteria (one row each) in this column]	In each column, explain how you checked each product against the criterion, and include your end user's feedback/response and your own response regarding how well the product fulfils the criterion. For example: I observed two end users using the [name of product] and they said: End user 1: 'I found it hard to ...' End user 2: 'I liked how it ...' Personally, I found ...		

Go to Nelson MindTap for a template to record your product evaluation. Choose between one template with three columns, allowing you to compare each product on the same row, or one template page per product with an outline of 'data' to present.

Worksheet
10.9.2

Assessment

Tasks can be any one or a combination of:

- data analysis
- oral presentation using multimedia: face-to-face or recorded as a video or podcast
- product analysis
- research inquiry.

Summary
Chapter 10

Quiz
Chapter 10 revision

A summary of your task

1. Read all of this chapter.
2. Select the products you will evaluate, making sure you have enough information on them to evaluate them using your own SAT criteria. It would be a considerable advantage to have each product in your hands (to feel, use, read the labels, etc.) but this will depend on the cost, size and availability of the product. If this is not possible, using internet sources will suffice.
3. Research aspects of the areas in the list below that are relevant to your product, and those that you have selected. Make note of your sources and explain how you gathered the information – through lab testing or surveys of readers, consumers or subscribers, etc.
4. Collate, analyse and summarise your research data and responses. Apart from using the questions under 'How to evaluate?' and your SAT evaluation criteria to guide you, you also need to cover:
   a. **market research and entrepreneurial activity.** Begin with the marketing elements and by defining the following: Who is the product for? What features of the product are important to the end user? Its retail price? Where is it sold? How is it promoted? Outline the entrepreneurial activity that could have brought each product to the market (i.e. development, promotion and retailing).
   b. **the product development process.** Outline this process for each product, including any 'new and emerging' technologies involved in its creation. You may need to make an educated guess.
   c. **innovation and R&D.** Compare the degree of innovation in each product and explaining why each is more or less innovative than others on the market. Outline the R&D that has been, or could be, incorporated or your prediction of R&D that would improve the product (or make it more innovative).
   d. **sustainability.** Using the information you found on each product, and your knowledge of materials and sustainability strategies from page 138 in Chapter 6, make a statement on the sustainability positives and negatives of each product.
   e. **speculative design thinking and worldview issues.** Sum up the relevance of these types of thinking to each product or how they could be applied to consider the future and the impact of the product in relation to worldview issues.

   Points a–e above could be applied to your chosen products for analysis early in the semester, before you have finished your SAT product.
5. Ask your end users to examine the information you have collated, and give their own feedback on each product. Depending on what stage of production you are up to, they might be able to suggest something from these commercial products for you to include in your own SAT product.
6. When your own product is finished, ask your end users to look at, inspect, check, use or try out your product to get their feedback as a response to your evaluation criteria.
7. Make judgements about how successful each product is (including your SAT product). Use your research data and end user feedback to support your judgements.
8. Ask your end users for suggested improvements.

9780170477499

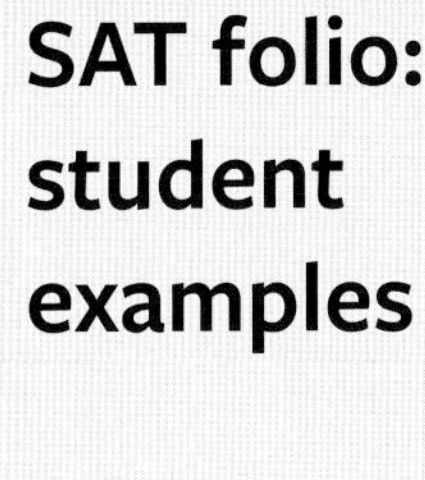

CHAPTER 11

SAT folio: student examples

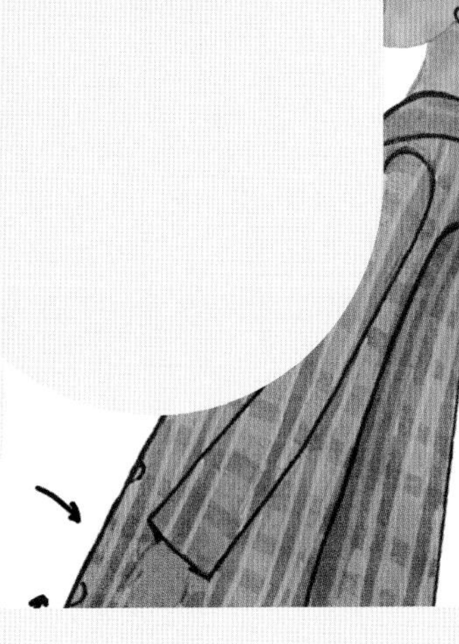

Units 3 and 4 SAT outcomes and assessment requirements

Outcome	Description
Unit 3 Outcome 2	Investigate a need or opportunity that relates to ethics and formulate a design brief, conduct research to analyse current market needs or opportunities and propose and evaluate graphical product concepts.
Unit 3 Outcome 3	Evaluate product concepts related to ethical design, synthesise and apply feedback to justify a final proof of concept, and plan to make the product safely.
Unit 4 Outcome 1	Implement a scheduled production plan, using a range of materials, tools and processes and managing time and other resources effectively and efficiently to safely make the product designed in Unit 3.

Assessment tasks

- multimodal record of evidence that records:
  - » formulation of a design brief and gathering evidence of research that explores market needs or opportunities
  - » generation, design and evaluation of product concepts
  - » justification of final proof of concept
  - » scheduled production plan

and

- practical work that demonstrates:
  - » use of technologies to develop physical product concepts including prototypes and finished product
  - » management of time and other resources
  - » record of progress during the production process and decisions and modifications made to the scheduled production plan

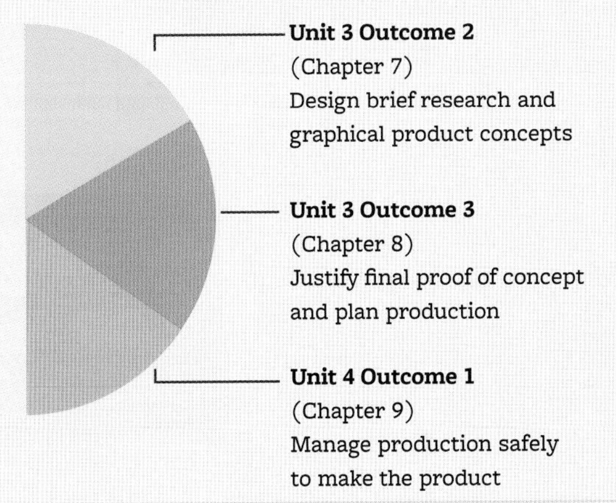

Design Folio Template

Visual checklist for SAT folio

Outcome 2: Investigating opportunities for ethical design and production

UNIT 3

Product Design and Technologies

- SAT folio
- contents

—
—
—

Brainstorm

- possible design problems to focus on

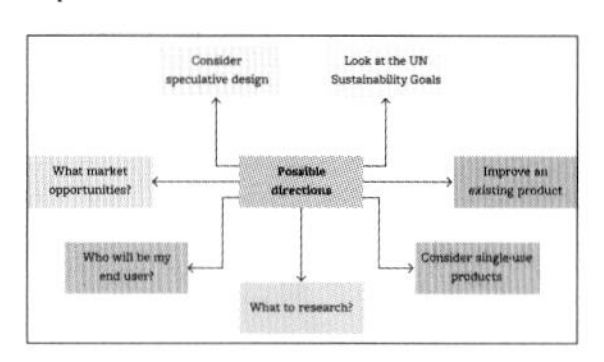

- identify the design problem

Research

- existing ethical products
- quantitative: end user info/statistics/surveys
- qualitative: end user interviews, in-depth

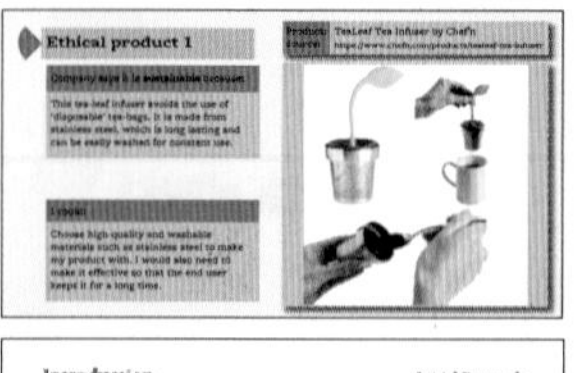

Introduction Initial Research

As the COVID-19 Pandemic continues for the third year, with no sign of stopping anytime soon, lockdowns, PPE and social distancing are becoming the norm. Massive amounts of masks are being consumed and thrown out everyday, leaving a huge toll on our landfill. Due to COVID mandates, mask waste has jumped by nearly 9,000, the two biggest consumes, UK and Australia.

"They can take up to 450 years to break down if they're dropped in the environment."

Design brief

- end user profile

My end user/s is/are ...

- intended function

- scope: constraints and considerations

Must be ...
Needs a ...

Evaluation criteria

- reflecting the design brief requirements and addressing the Factors that influence design
- criteria for graphical and physical concepts and processes
- criteria directed at the finished product

Further research (2–6 pages)

Primary (done personally)

- interview/survey if needed
- measurements
- trials/tests
- own photos of background information or inspiration

Secondary (desk research)

- inspiration (IP acknowledged)
- ergonomics (sources quoted)
- prices/availability
- materials
- sustainability (sources quoted)

Research can be done throughout your SAT whenever it is relevant

Graphical product concepts

Idea development (2–6 pages)

Visualisations mixed with research images and annotated

- end user feedback throughout

Brainstorming/graphic organisers

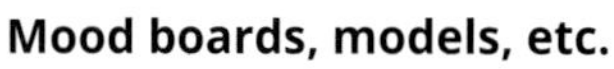

Mood boards, models, etc.

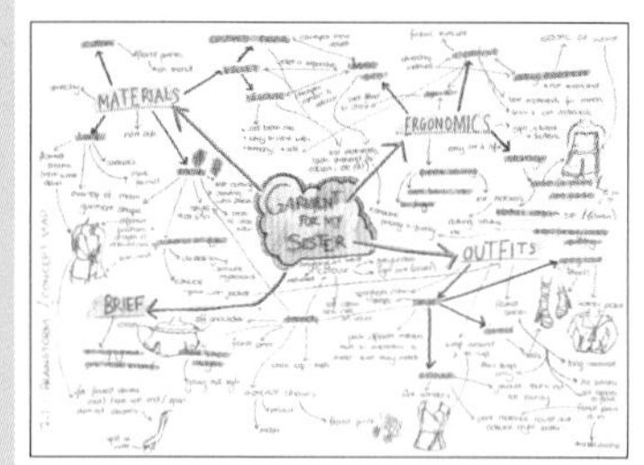

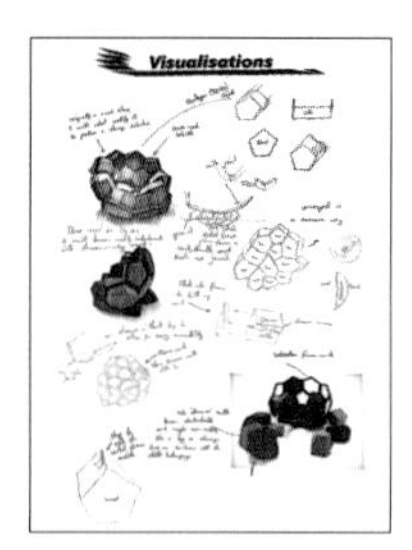

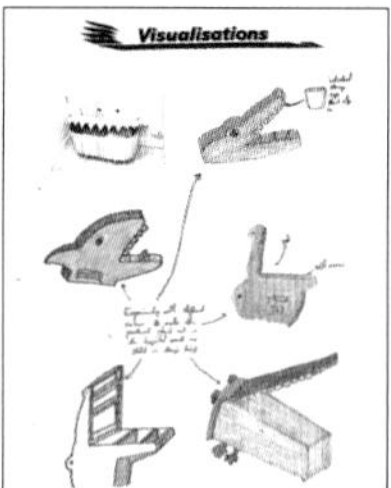

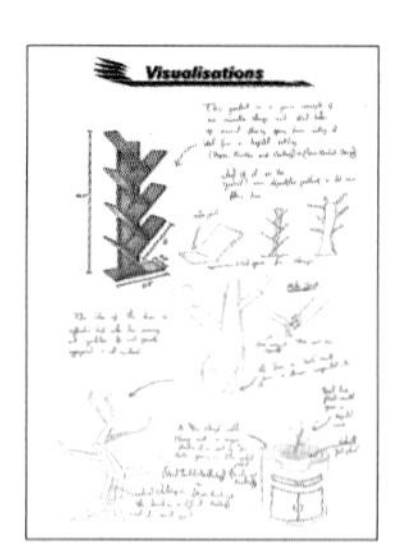

Design options

- annotated to communicate possible materials, construction processes, design brief requirements and with end user feedback

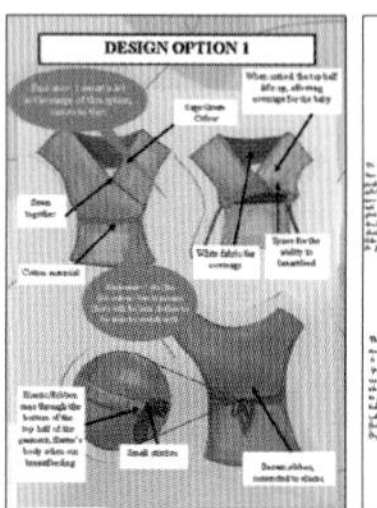

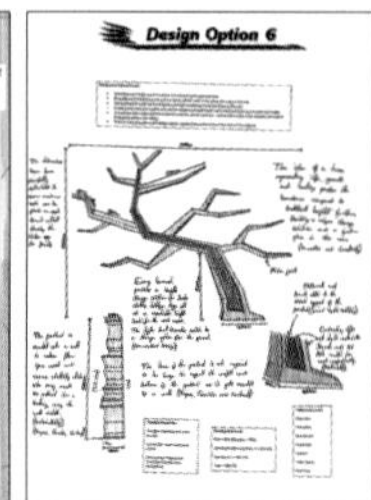

Working drawings 2D (1+ pages)

- resistant materials: orthogonal drawing
- non-resistant: flats
- may include: plans, templates, CAD

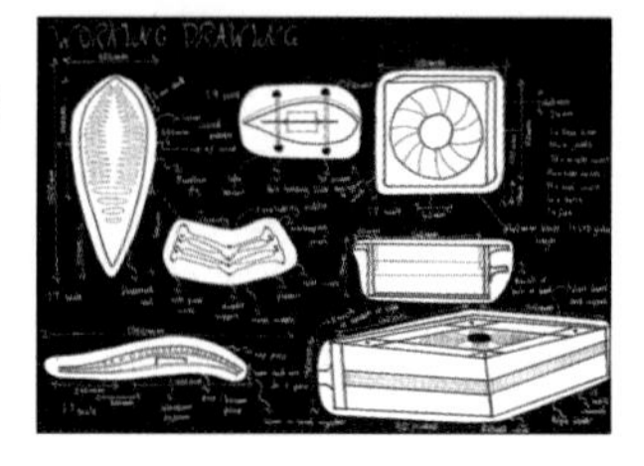

9780170477499

UNIT 3

Outcome 3: Developing a final proof of concept for ethical production

Physical product concepts (2–6 pages)

Decisions

- photos of working through physical product concepts (models and prototypes – actual, or virtual in CAD) including end user feedback

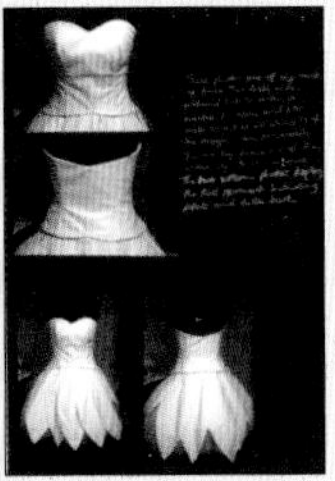

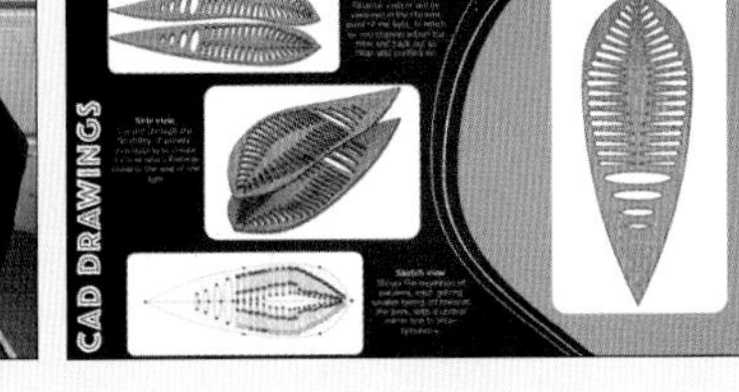

Materials and processes

- tests and/or trials

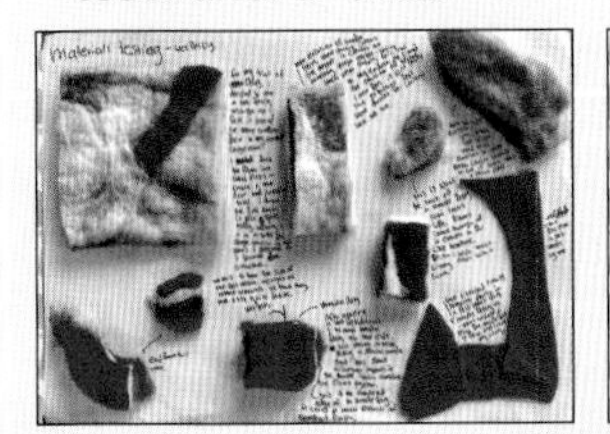

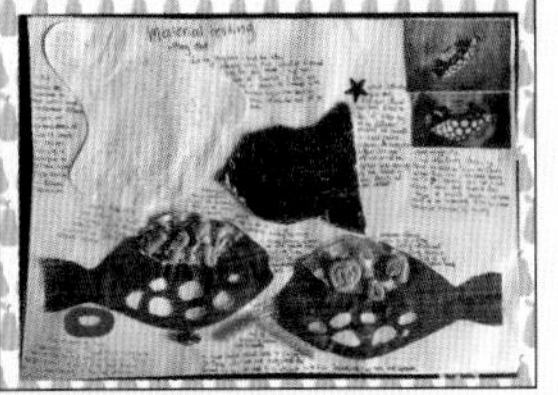

Final proof of concept

- justification with drawings/photos of final prototype to assist planning

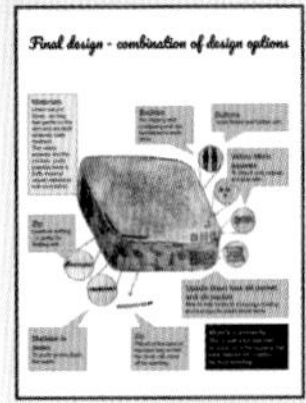

Scheduled production plan

Production steps and timeline (1 page)

	Process & materials	Tools, equipment	Safety	Hours	Dates
1					
2					
3					
4					
			Total time		
			Expected finish date		

Materials cutting and costing list (1 page)

Part of production	Material or part	No.	Cost
			$
			$
			$
			$
		Total costs	$

Risk assessment (1–2 pages)

Main process	Hazards	Injuries	Risk/likelihood	Precautions

Quality measures (1 page)

Process/skill	Quality expected
Marking, measuring, cutting	
Joining, i.e. all seams	
Halving joint	

Images (L–R): Laura Mazzarella, Rachel Krause, Amelie Thompson, Ella Costanzo, Sophie Glover, Victoria Cannon

UNIT 4

Outcome 1: Managing production for ethical designs

Journal – record of progress and modifications (and finished product)

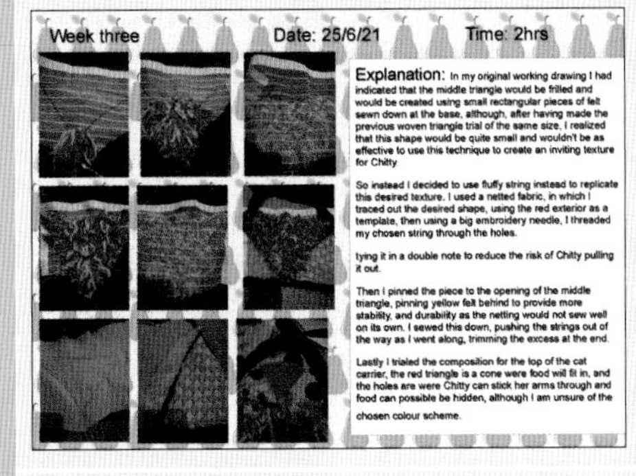

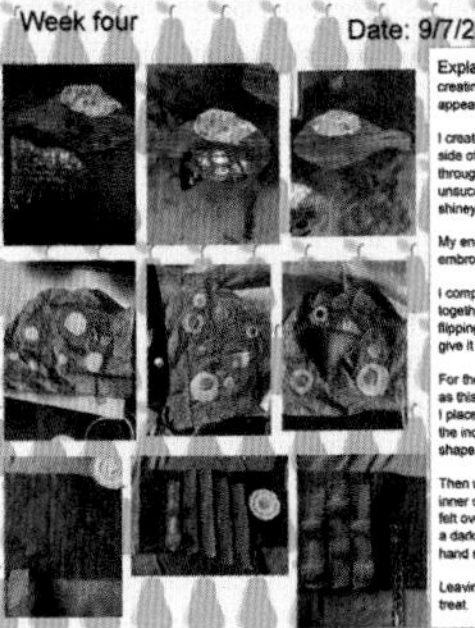

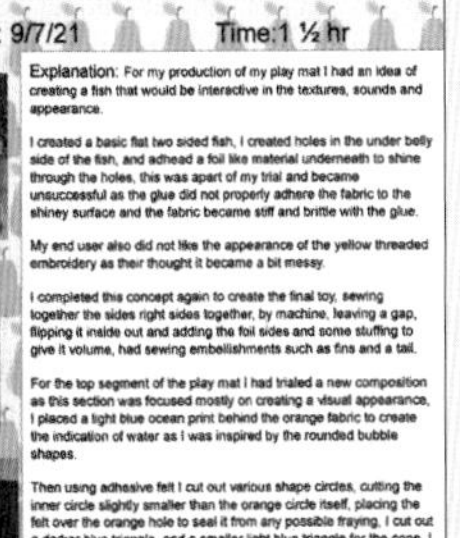

Images (L–R): Amelie Thompson, Janna Porras

NOTE: your folio pages may be in a different order, particularly with your testing and trialling of materials. You should present your pages in the order of design thinking that works for you.

Examples of items for your SAT

The following pages contain examples of high-quality student work to demonstrate different ways of presenting your SAT work and a suggested order in which you could do so. However, this is not the only way to present your work or the only order. Your work should be individual, guided by your personality and your end users' needs. Work is loosely grouped in the requirements with very brief instructions, but you may find your project needs more work in one area of the Double Diamond than another.

Weblinks
What Is the Double Diamond Design Process?

Miro

To revisit the Double Diamond design process, you can watch the video, 'What Is the Double Diamond Design Process?' on the Design og arkitektur Norge's YouTube channel and you can revise the sections at the start of Chapters 1 and 7.

Investigating and defining

At this point you are:

- choosing an end user and looking at their needs and/or market opportunities
- researching around the design problem and synthesising your data.

Brainstorming

The following are examples of how you could do this.

- Brainstorm to suggest several ideas for the design problem and end user you will focus on for your SAT.
- Read through pages 194–8 in Chapter 7 for ideas, how to write a design problem and what makes a suitable end user.
- You can draw by hand or use software such as Miro to capture your brainstorming process and ideas in any type of graphic organiser.

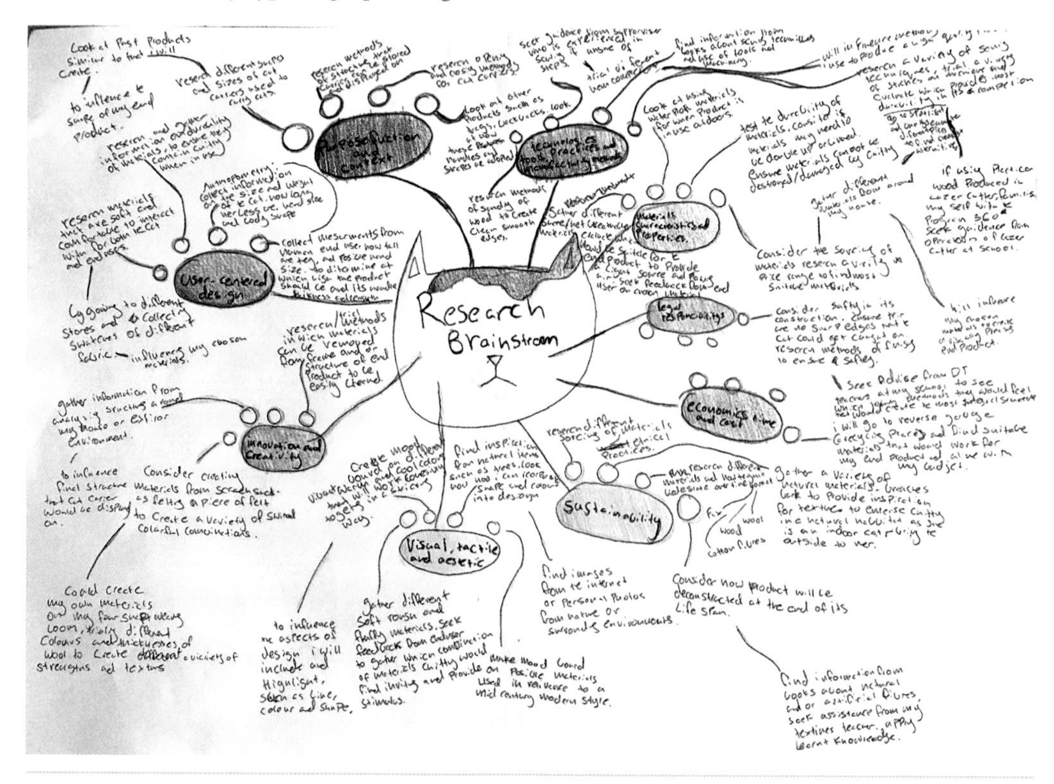

Brainstorming ideas for a product for cat owners, by student Amelie Thompson

9780170477499

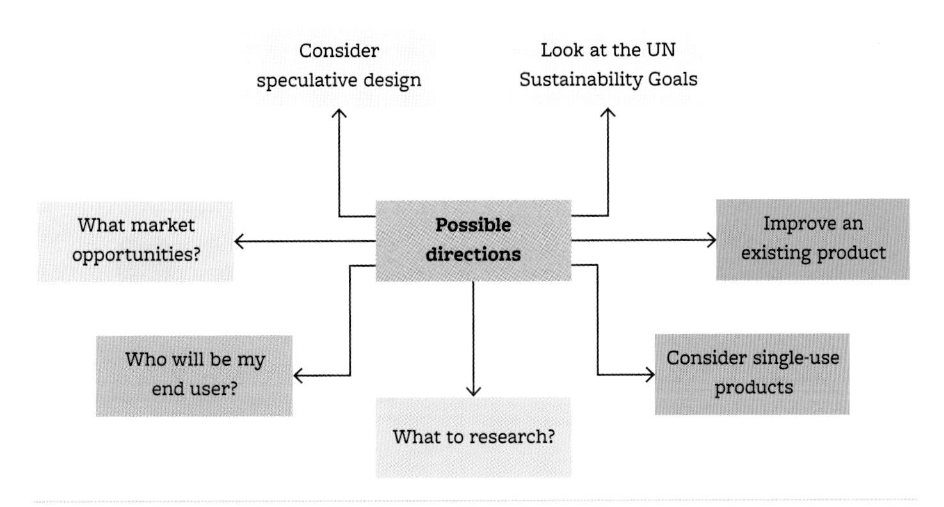

A simple mind map created by the author using Miro

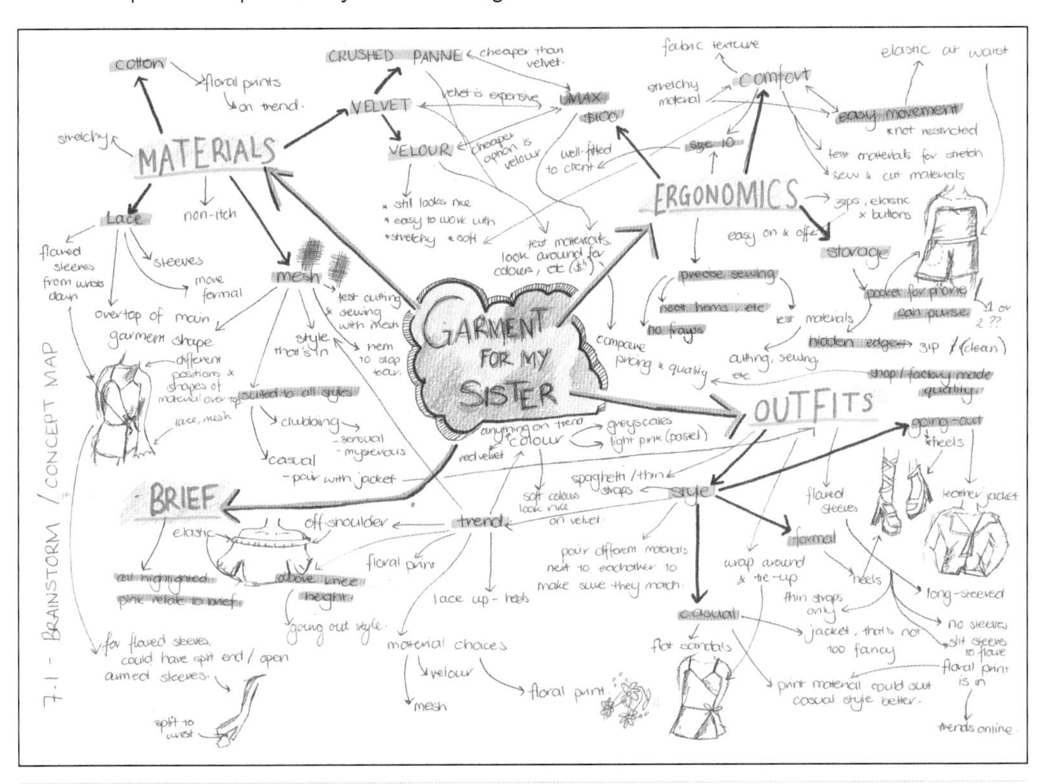

A concept map by student Paige Anderson

Research

Gather information about your end user and their needs by undertaking research. Use the Factors that influence design to help you create interview or survey questions to collect both qualitative and quantitative information. Complete primary and secondary research. All secondary research must acknowledge the IP of the creator, or the source of the information.

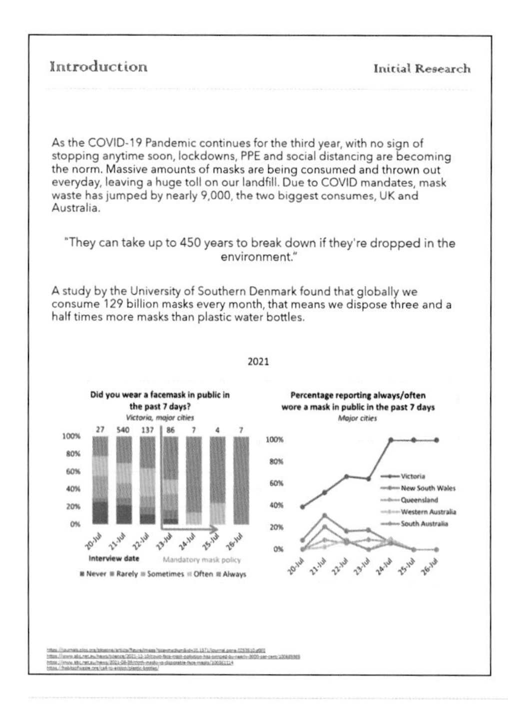

Introduction

Initial Research

As the COVID-19 Pandemic continues for the third year, with no sign of stopping anytime soon, lockdowns, PPE and social distancing are becoming the norm. Massive amounts of masks are being consumed and thrown out everyday, leaving a huge toll on our landfill. Due to COVID mandates, mask waste has jumped by nearly 9,000, the two biggest consumes, UK and Australia.

"They can take up to 450 years to break down if they're dropped in the environment."

A study by the University of Southern Denmark found that globally we consume 129 billion masks every month, that means we dispose three and a half times more masks than plastic water bottles.

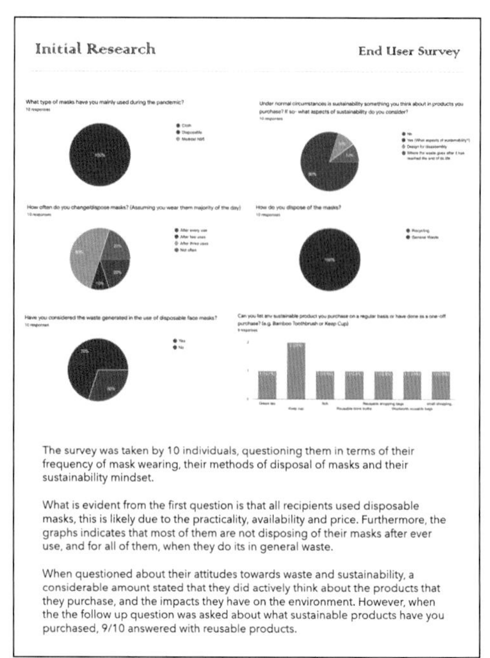

Initial Research

End User Survey

The survey was taken by 10 individuals, questioning them in terms of their frequency of mask wearing, their methods of disposal of masks and their sustainability mindset.

What is evident from the first question is that all recipients used disposable masks, this is likely due to the practicality, availability and price. Furthermore, the graphs indicates that most of them are not disposing of their masks after ever use, and for all of them, when they do its in general waste.

When questioned about their attitudes towards waste and sustainability, a considerable amount stated that they did actively think about the products that they purchase, and the impacts they have on the environment. However, when the the follow up question was asked about what sustainable products have you purchased, 9/10 answered with reusable products.

Two pages of end user research by student Hamish Pittendrigh

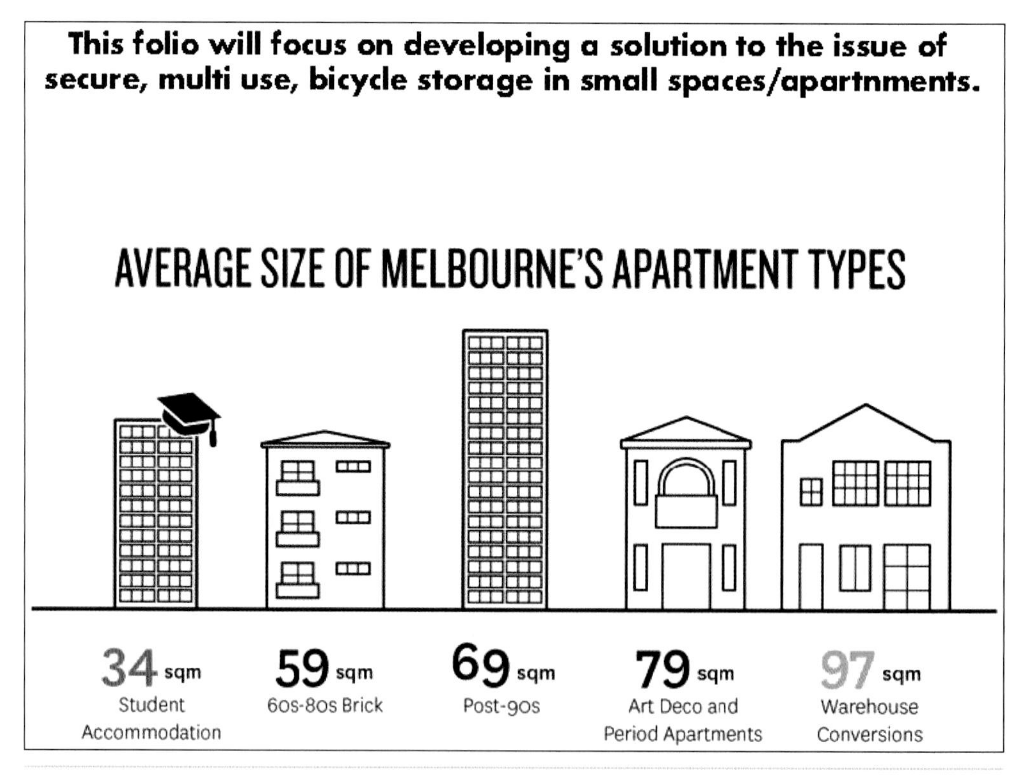

A small section of a folio page by student Kent Pittendrigh Smith. The page included references and years for the information presented in the infographic.

9780170477499

Research into ethical products

Choose and analyse two or more ethical products that relate to your end users' needs, or that are similar to the type of product you will design and make. Refer to examples on pages 208–9 in Chapter 7.

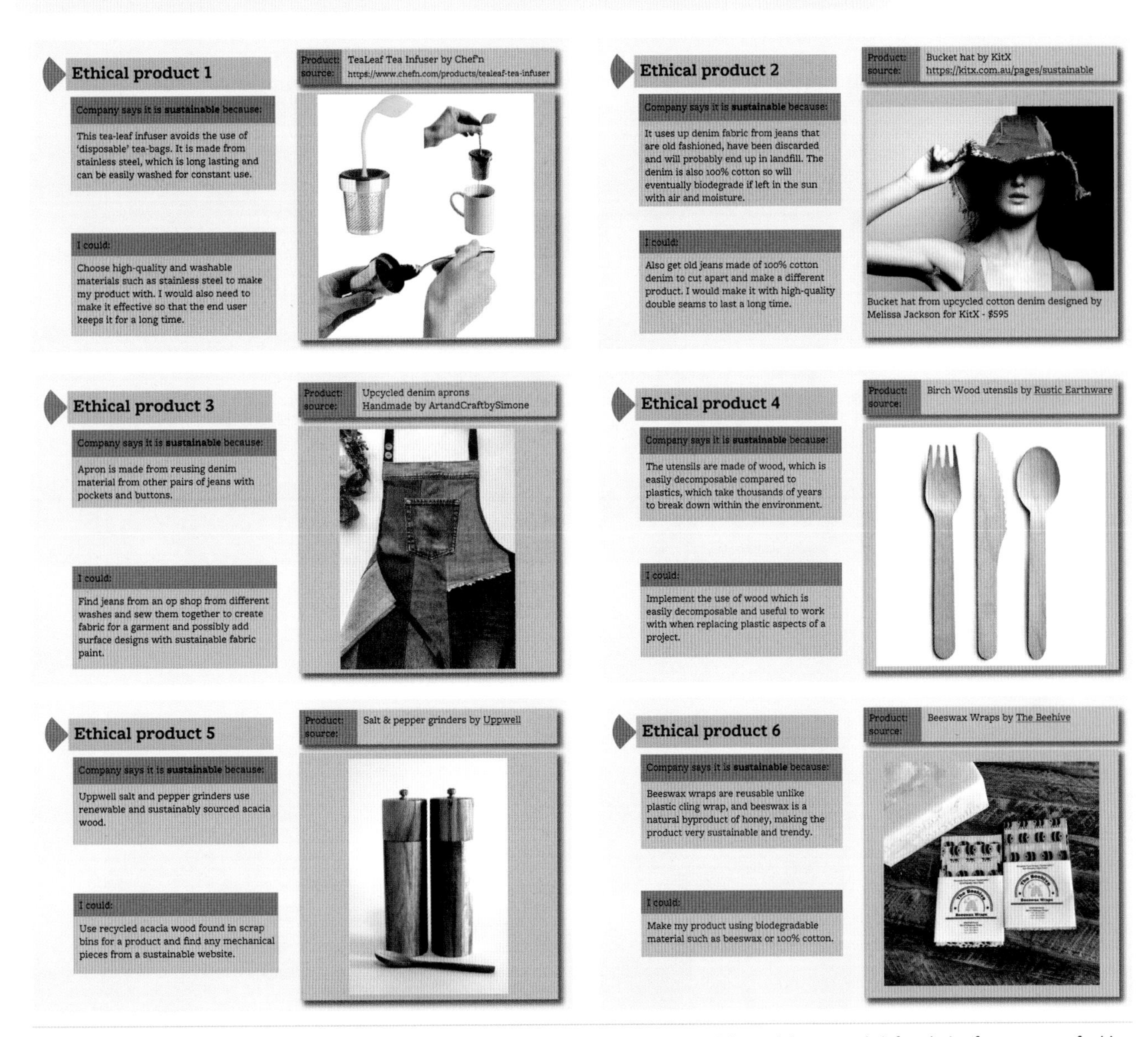

The examples shown are by the author using Google Slides for layout, with one product per slide. Each has a very brief analysis of one aspect of ethics only: sustainability claims of the commercial products.

Writing your design brief and evaluation criteria

Organise your research and its presentation, and add annotations about how useful it is, i.e. why it is in your folio. Check in again with your end user, discuss your research and use this information to create the **design brief** with an ethical consideration. The first section requires you to create an end user profile.

An end user profile

Create an end user profile to add to your design brief, which can be visual or descriptive, as in the design brief box on page 293.

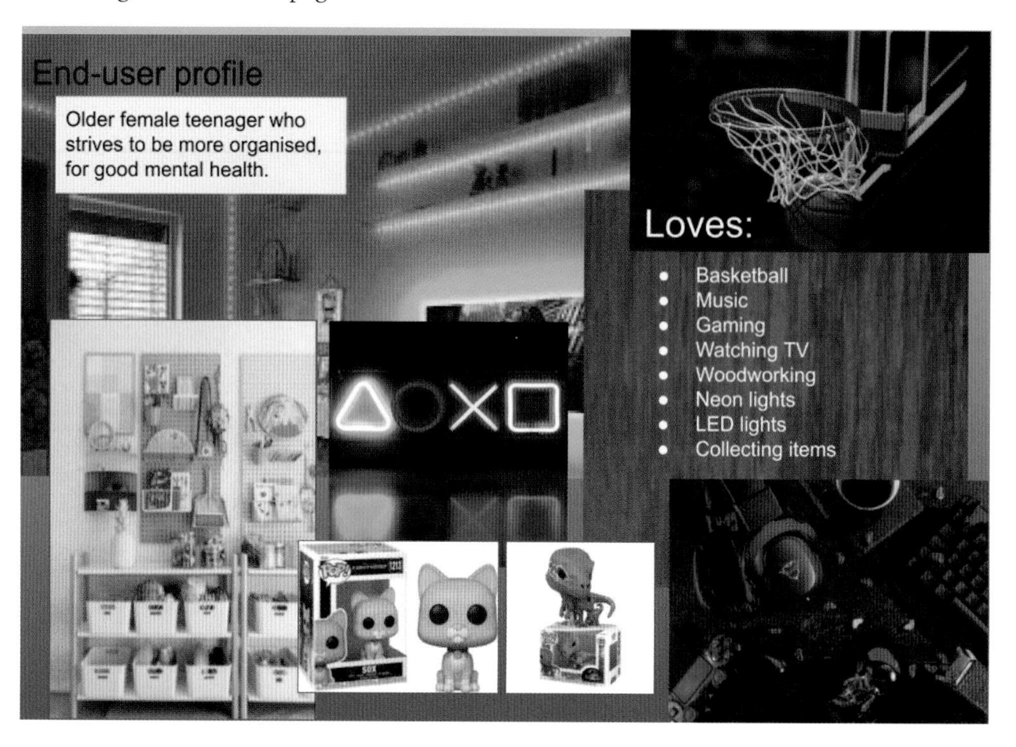

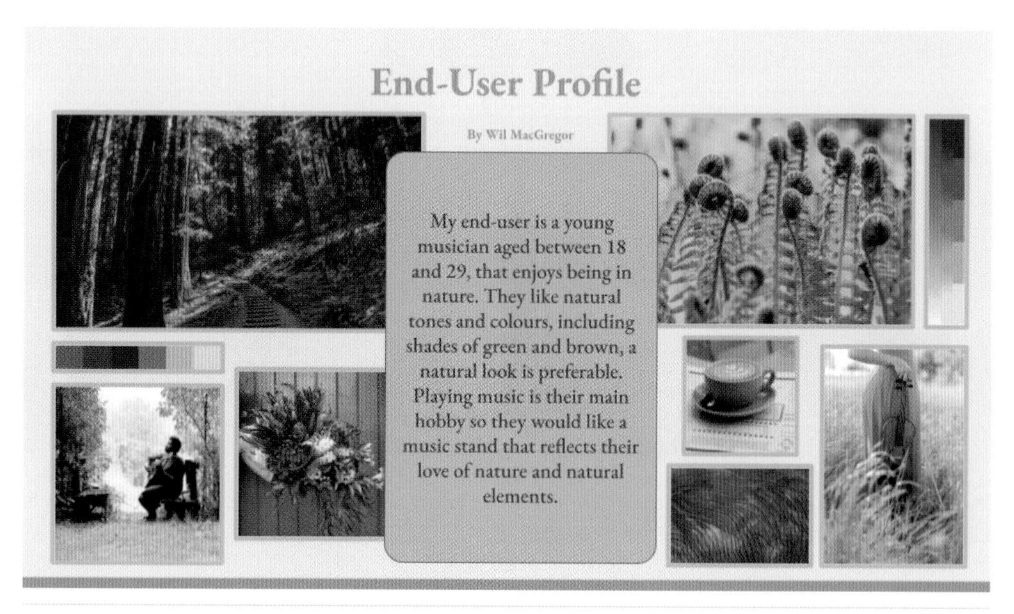

End user profiles by students Rori Barry (top) and Wil MacGregor (bottom), created separately from the design brief

9780170477499

Design brief

Design problem, need or opportunity
Similar to the statement you made in the beginning but finalised.

End user profile
Helpful information and/or images to identify your end user group and their life situation or interests relevant to the design problem.

Intended function
The primary function or purpose of the product.

Scope of task (description), including constraints and considerations (with reference to the Factors using colour coding or adding a column to indicate the relevant Factor – see page 212 in Chapter 7)

1. Date and budget
2. For example, desired aesthetics
3. For example, material characteristics/properties required
4. For example, extra functions, size

Highlight phrases that cover areas for research, design activities or experiments, and link to an annotated phrase outlining what you could do (or use a graphic organiser – see page 288). See page 213–14 in Chapter 7.

Evaluation criteria

The next step is to create the **evaluation criteria** directly from your design brief. These can be written as questions only. You may consider including notes for each evaluation criterion on:

- examples of further research to help achieve it
- ideas/areas related to this criterion to explore when designing
- how to get input and feedback on this criterion from the end user on your graphical and physical concepts (your design drawings and modelling, which can be virtual or actual)
- how to get feedback from the end user when the product is completed.

Write your evaluation criteria as numbered questions.
Refer to the requirements in your design brief (rephrase them as specific/targeted questions).
Aim to cover the relevant Factors (6–8).

The criteria will be used to:

- inform and evaluate graphical and physical product concepts
- evaluate processes to design and make the product. See page 214 in Chapter 7
- evaluate the finished product.

Further research for inspiration and ideas

It helps if you start your research with a plan. Take the highlighted phrases from your design brief and add some details on what research is needed. See an example of a research plan on the next page.

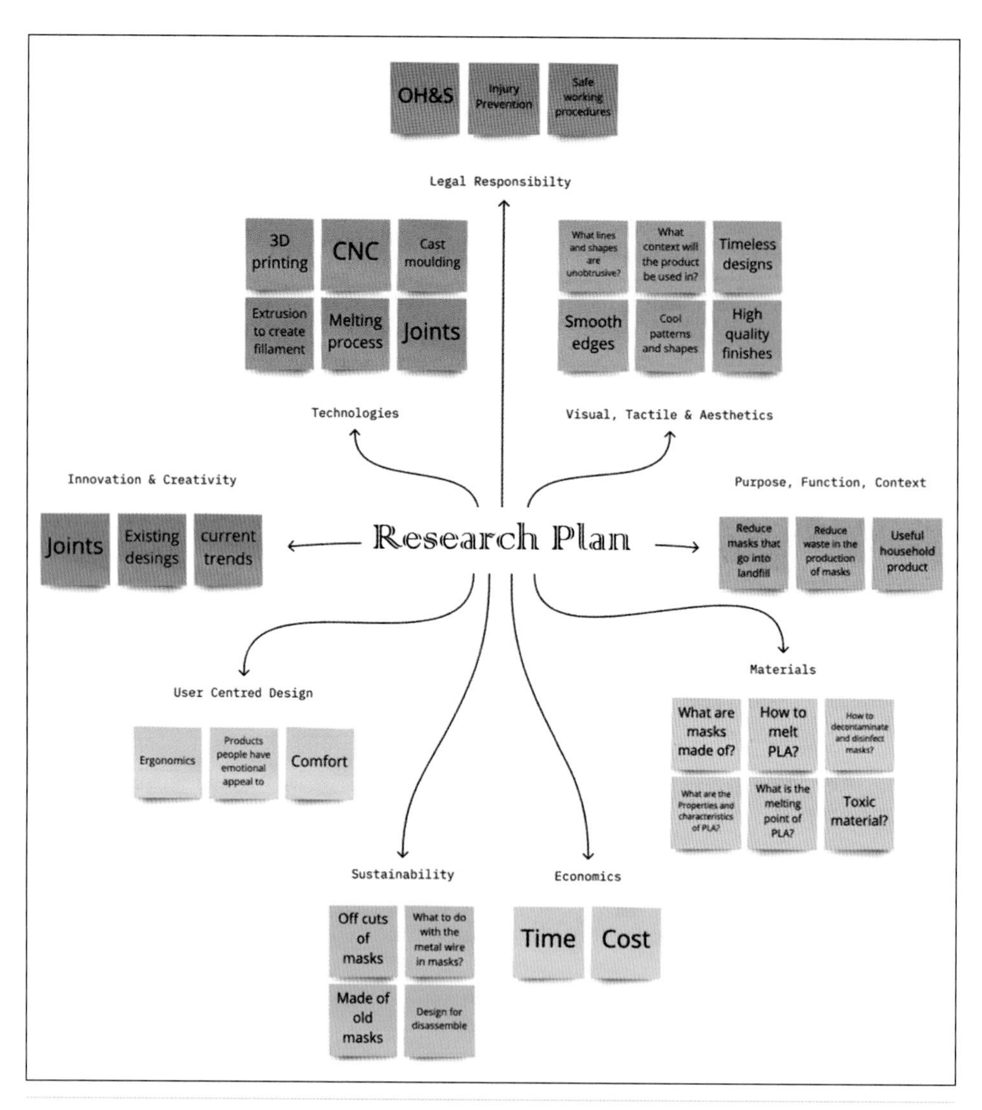

Concept map to plan research, drawn using Miro by student Hamish Pittendrigh

Weblink
Miro

Generating and designing

Gather and collate all your relevant research, making sure you have acknowledged the intellectual property in a clear way, with some annotations on why the research is in your folio or what aspects you are inspired by. The following examples show research and visualisations presented together.

Graphical product concepts

Visualisations showing influence from research

Develop a range of drawings that explore possible solutions and investigate different ways the product could work and look. Be creative, take design risks, speculate about 'what could be'. Evaluate and critique your **visualisations** with your end users to get feedback on what ideas could be taken into the design options. You can do this by applying your evaluation criteria (remember to word them appropriately for drawings) with your end user. Use coloured boxes, a different font or highlighters to visually differentiate their feedback from your comments.

9780170477499

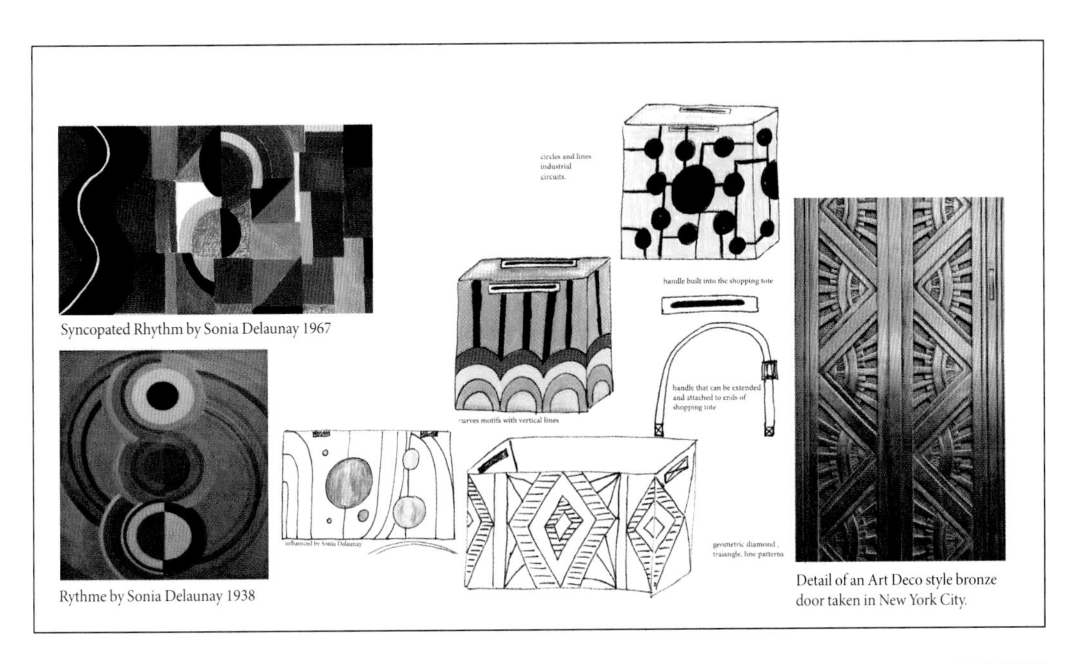

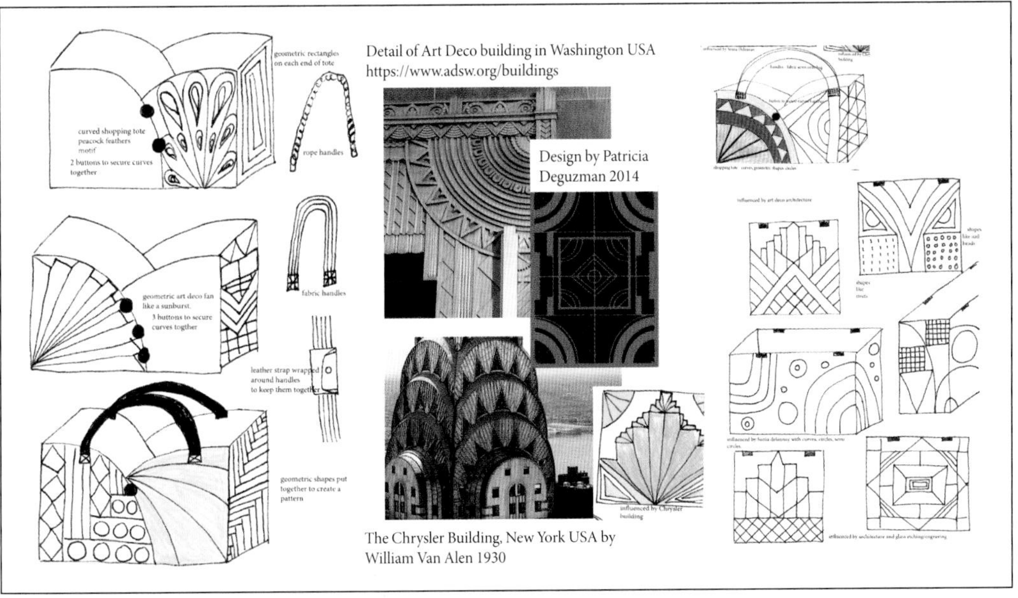

Visualisations from student Connor Nevins for a container inspired by Art Deco design and architecture

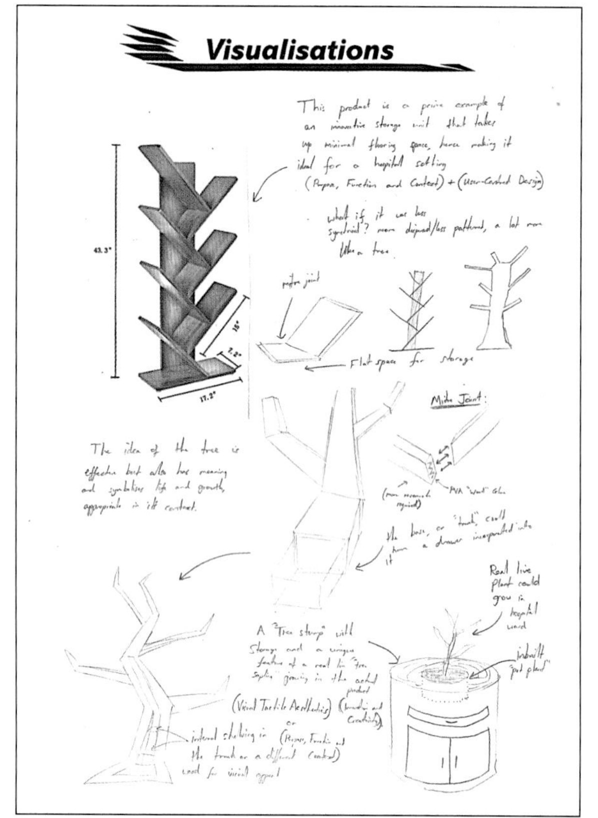

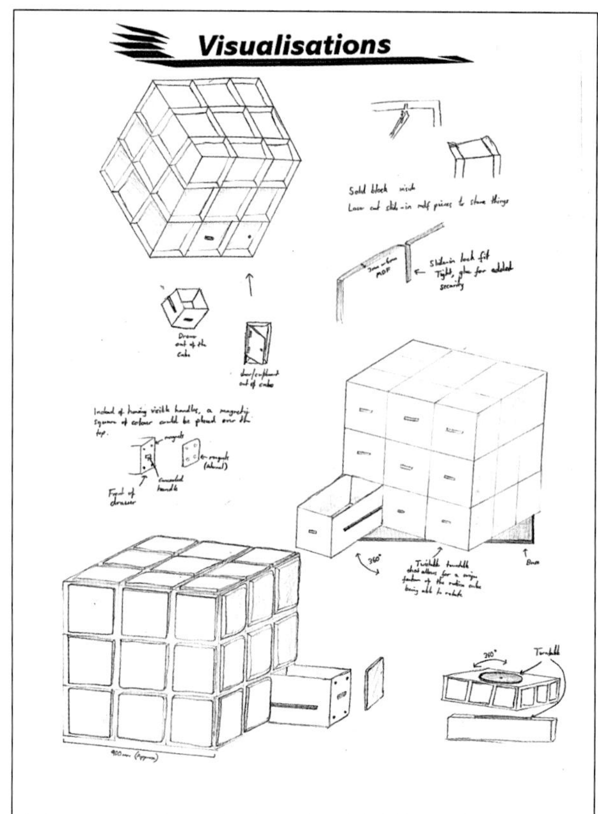

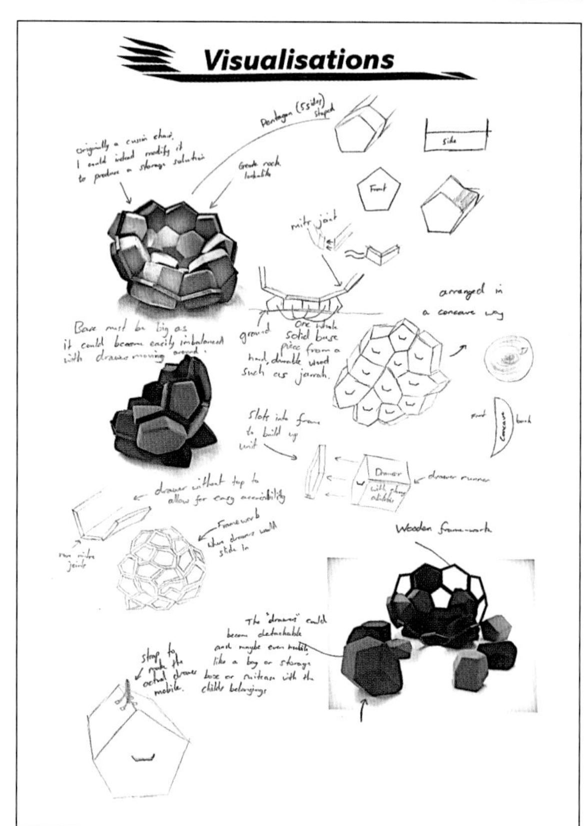

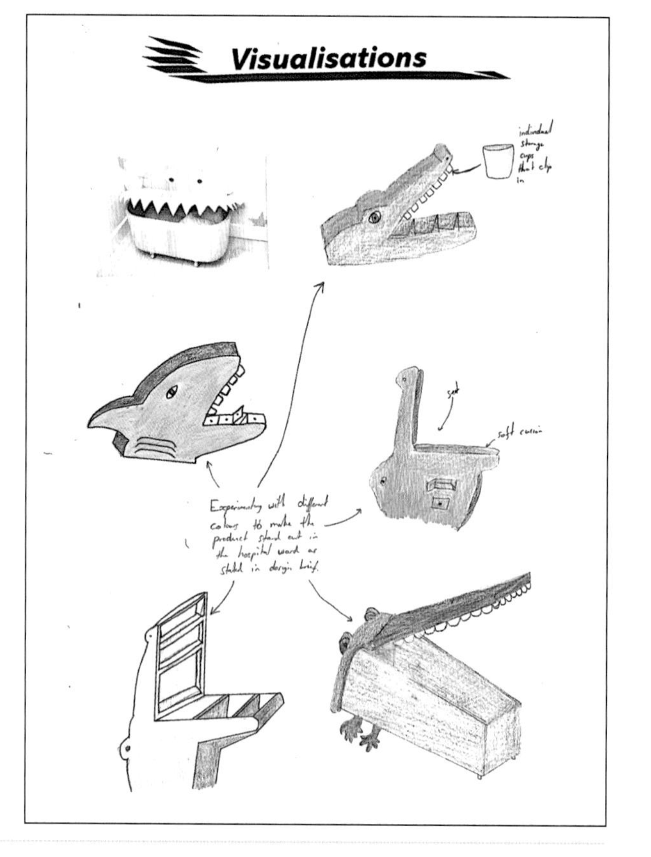

Visualisations from student Harrison Carr for a storage item

9780170477499

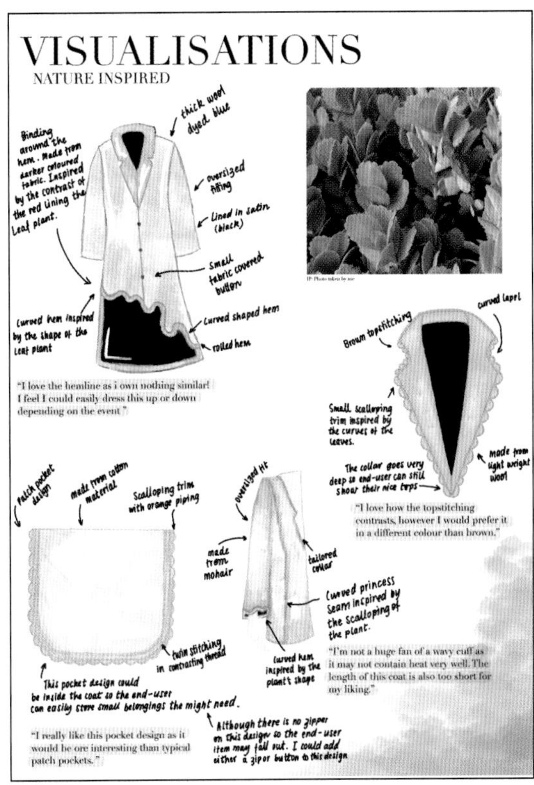

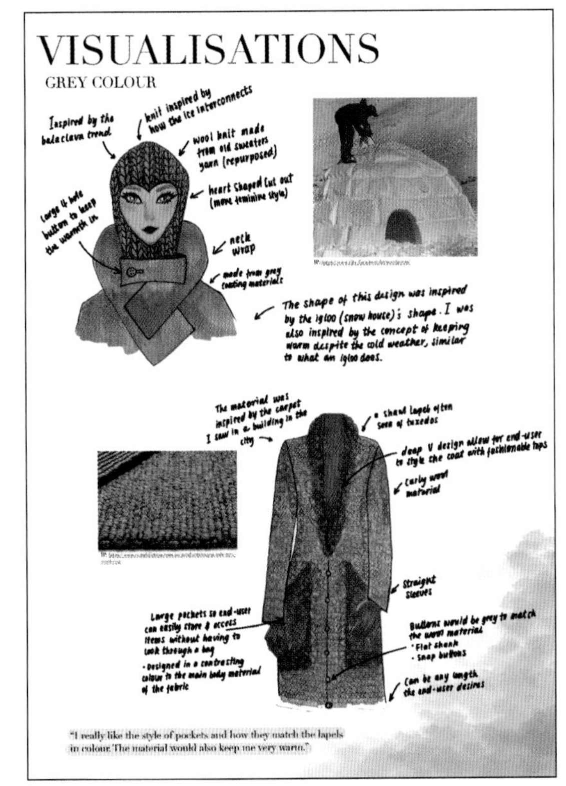

Four pages of visualisations inspired by luxury bags and coats (top left), her own garden photos (top right), igloos and carpet (bottom left) and the 'Great Wave' woodblock print by artist Hokusai and water droplets (bottom right) from student Izzy Donat, finalising ideas before design options. Drawings were done by hand and inserted into Goodnotes for Mac, where they were annotated and end user feedback was added in yellow highlights.

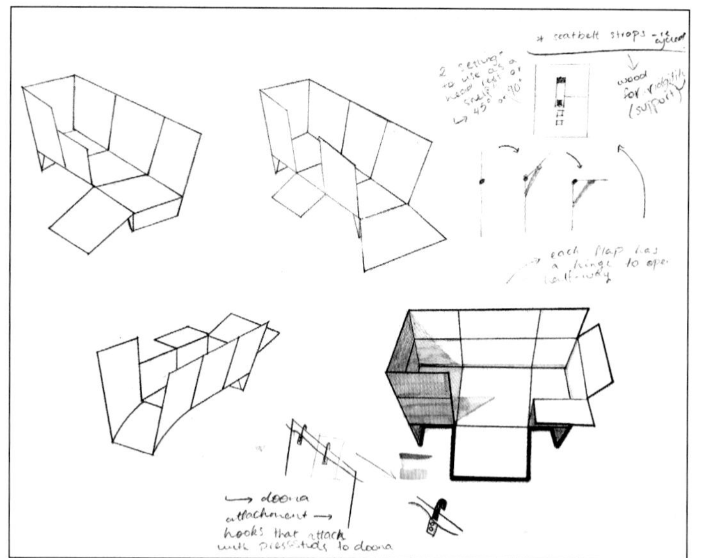

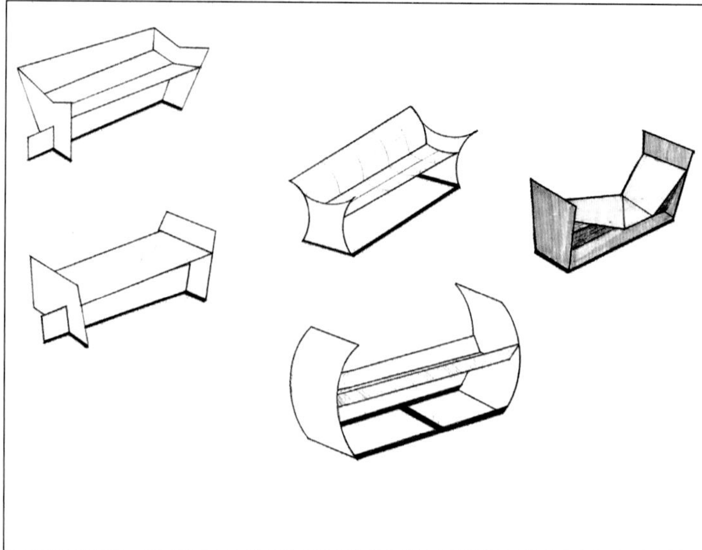

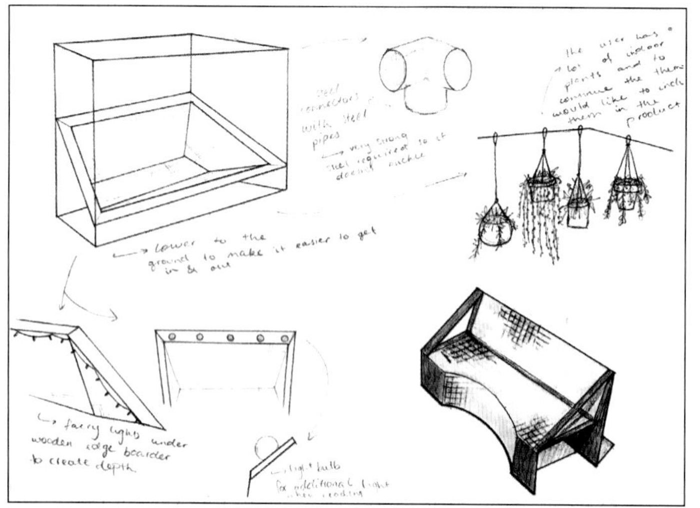

Visualisations by student Sophie Glover for a window seat. See her cardboard and plywood models on pages 239–40 in Chapter 8.

9780170477499

Design options (at least two)

Select the best ideas and concepts from your visualisations and turn them into a range of clear and detailed design solutions (your design options). Evaluate and critique your **design options** with your end users, and get feedback on the most suitable design option/s that could be taken into prototyping. You can do this by applying your evaluation criteria (remember to word them appropriately for drawings) with your end user and adding their feedback in an empty corner of the page. Consider:

- functional aspects (all catered for in the design?)
- suitability of suggested materials and construction methods (from your annotations)
- aesthetics to suit the end user
- complexity of the design to manage within budget and time frame.

Select the design options that you will take into physical product concepts, and create suitable working drawings.

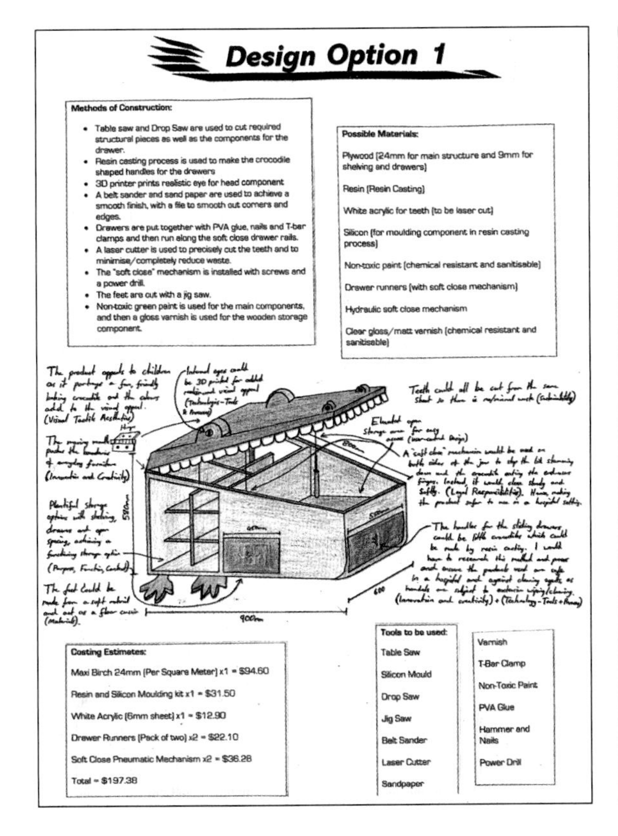

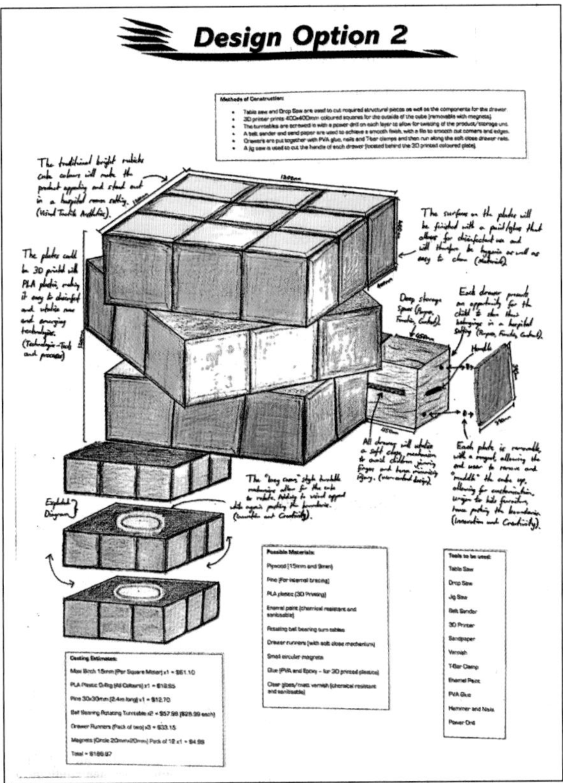

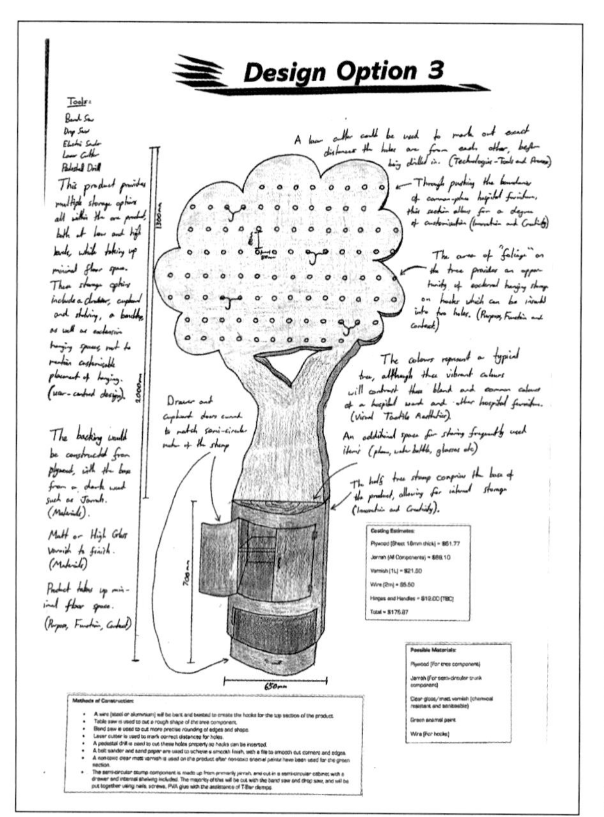

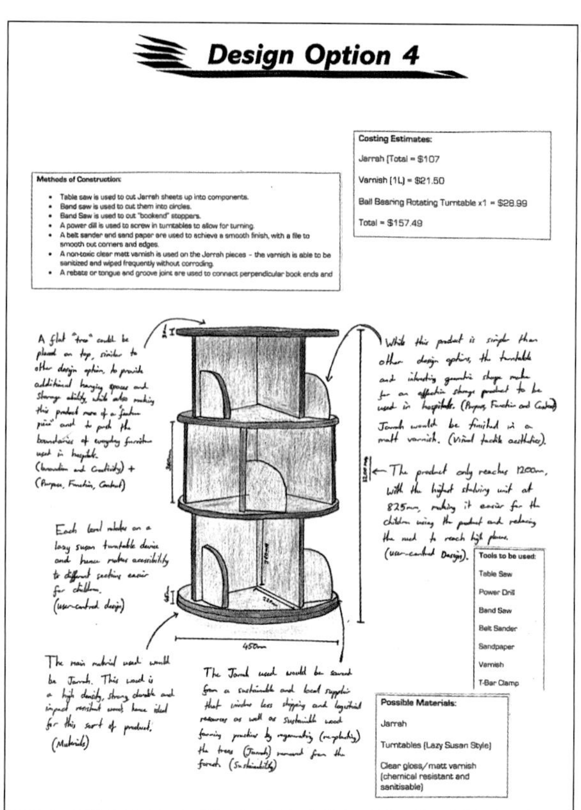

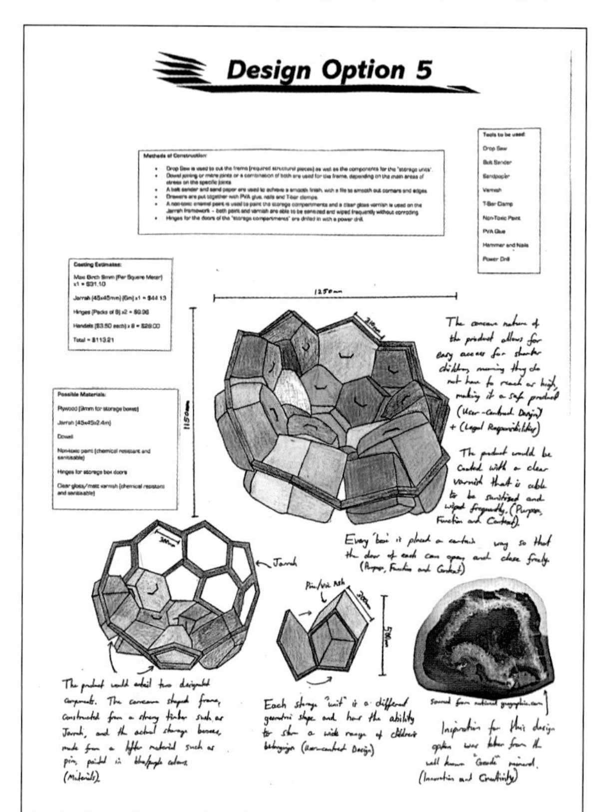

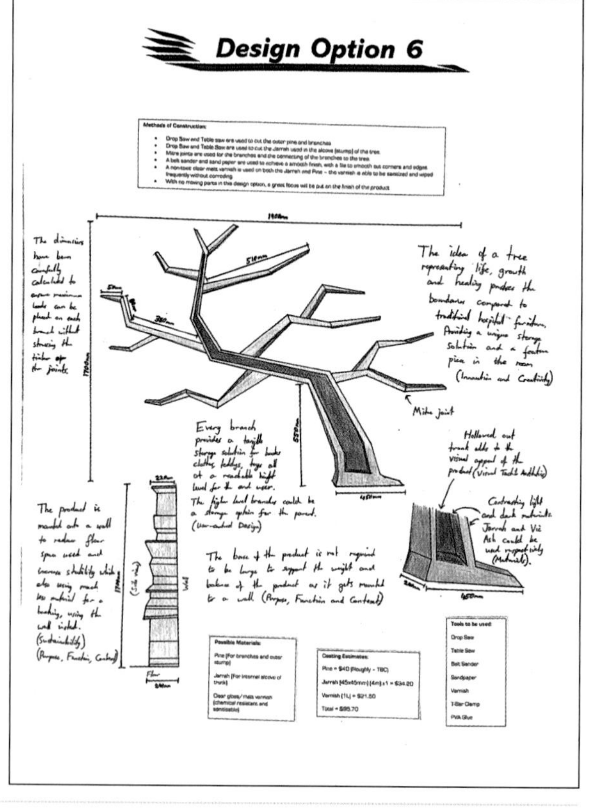

Design options for a storage unit from student Harrison Carr. Note that is not mandatory to complete six options; quality is more important than quantity. Only one view of each design option is required.

9780170477499

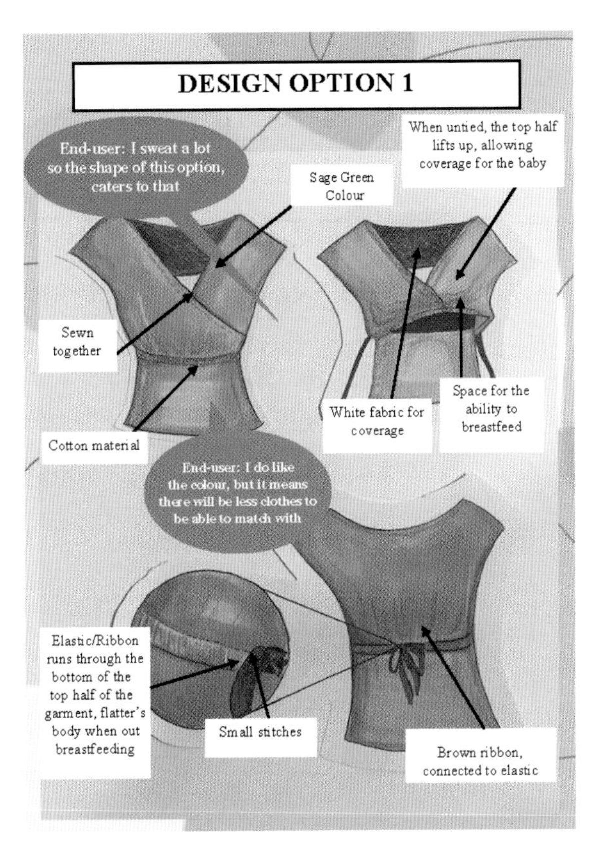

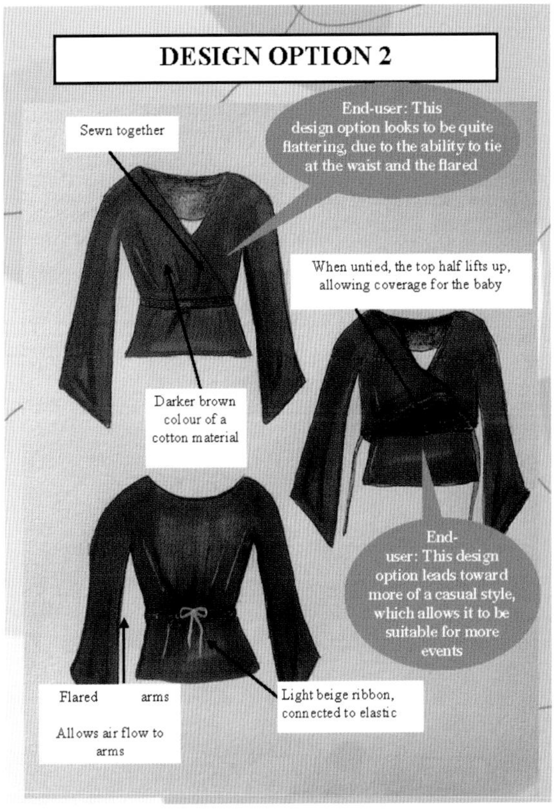

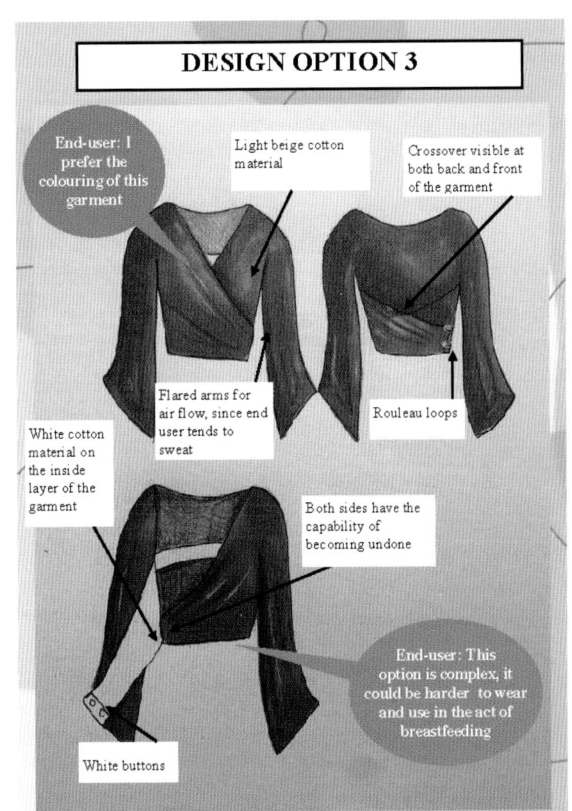

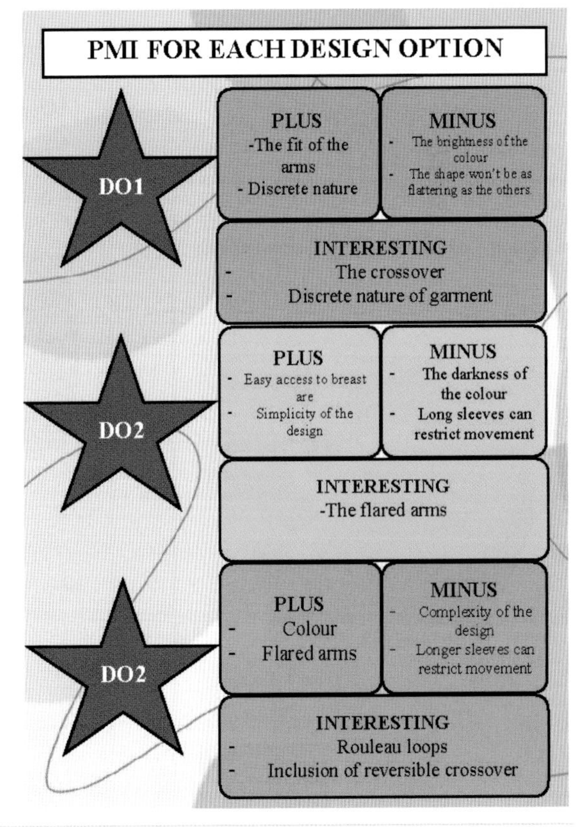

Design options for a top for a future breastfeeding mother by student Mia Rowland, followed by a PMI for the three options

Working drawings

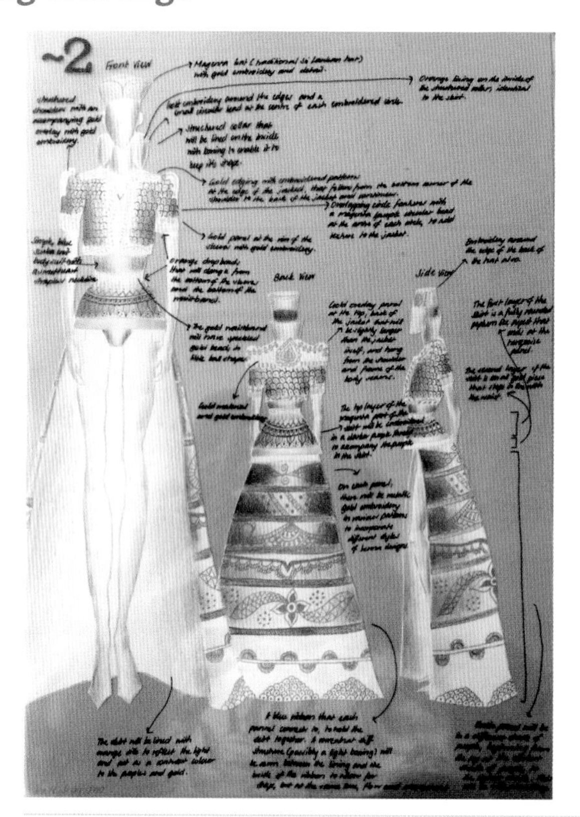

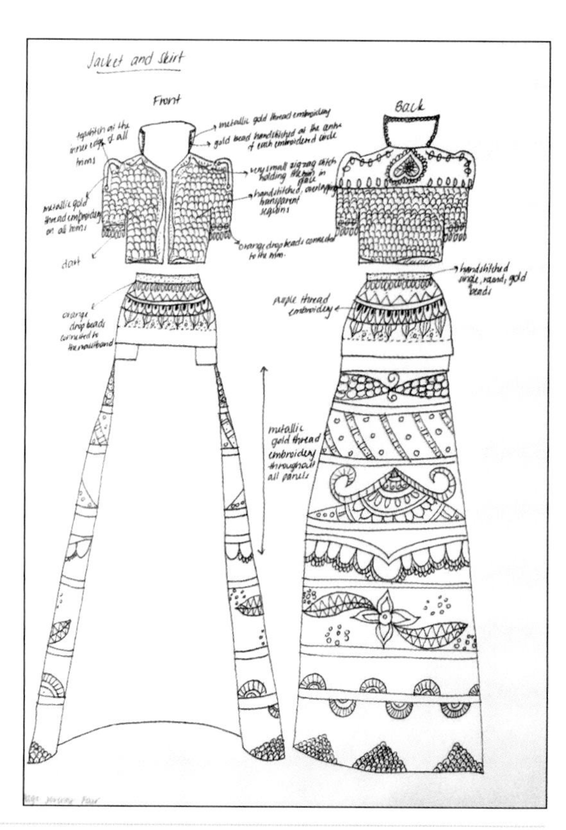

One design option (left), showing detail on the same page, and its working drawing (right) by Isabelle Hanna

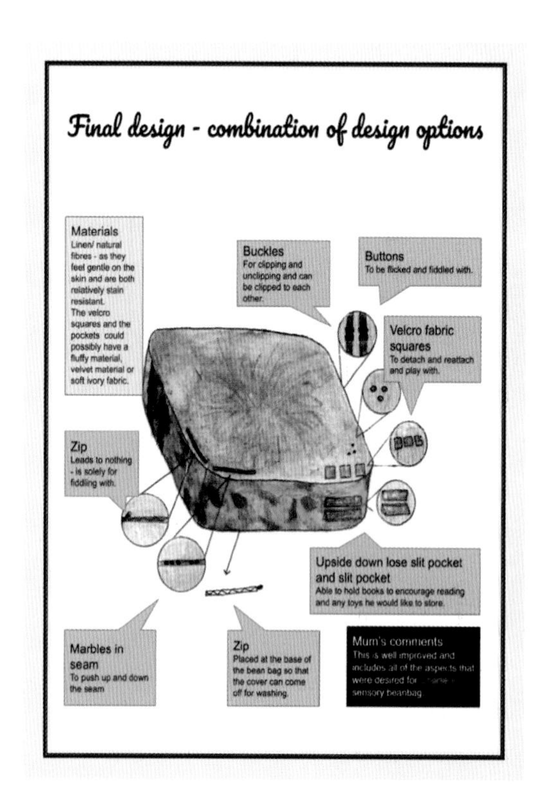

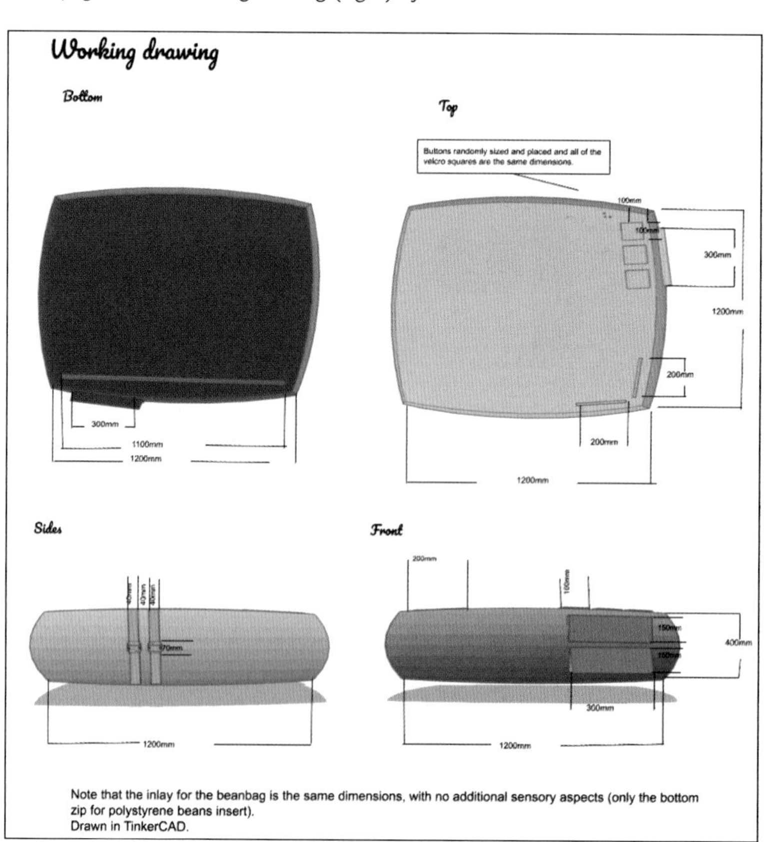

Final design drawing (left) and initial working drawing (right) of a sensory bean bag by student Victoria Cannon

9780170477499

See Chapter 7 on Unit 3, Outcome 2, for more examples of working drawings.

Working drawings and prototyping (actual or virtual product concepts)

Working drawings will need to be finalised when final proof of concept is created. See the examples in Chapter 8.

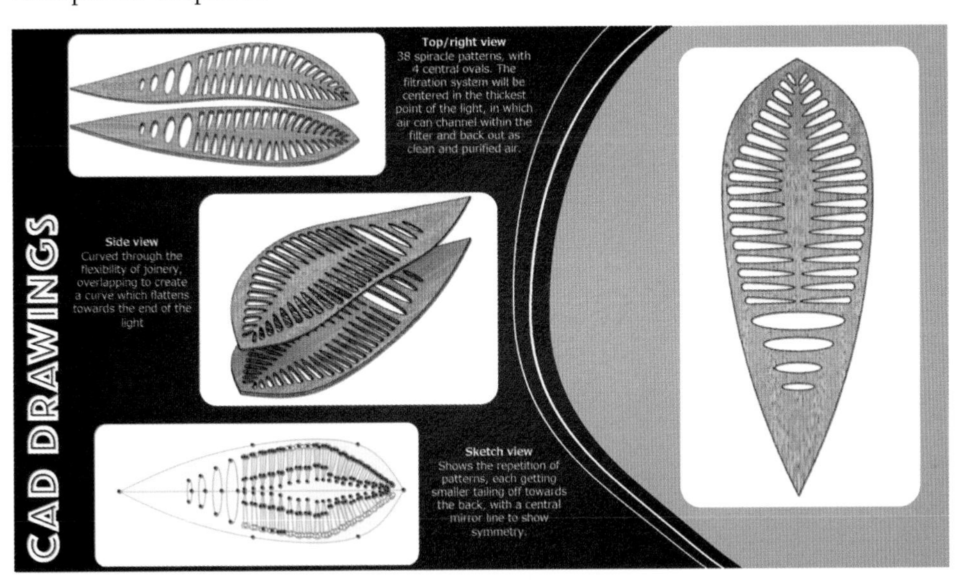

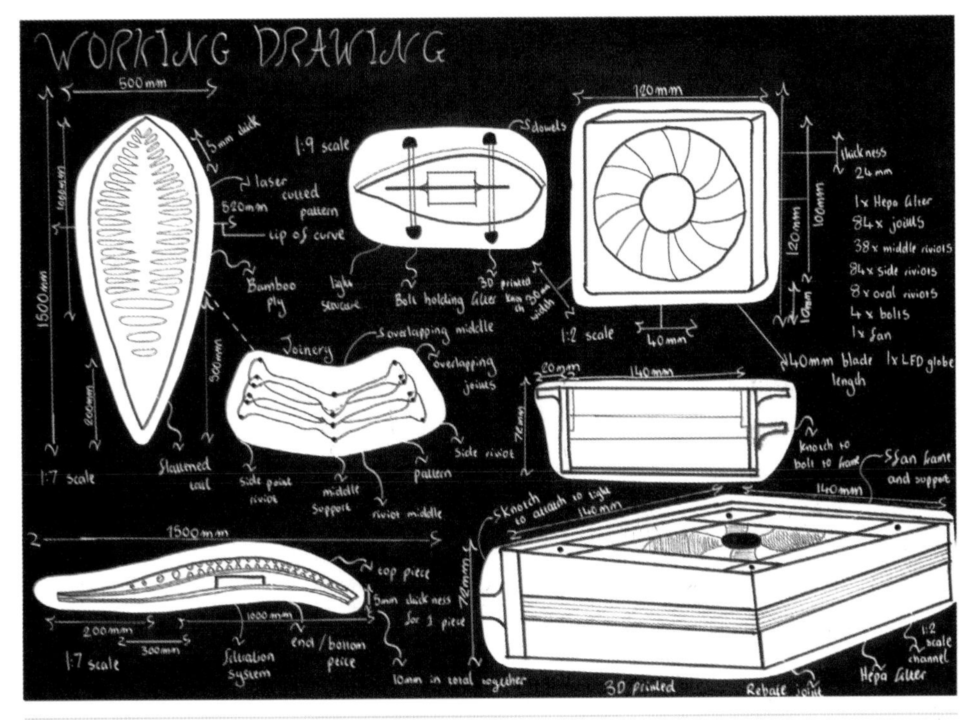

Folio pages by student Ella Costanzo, showing both CAD drawings and hand-drawn working drawings

Product concepts: models, mock-ups, prototypes, toiles

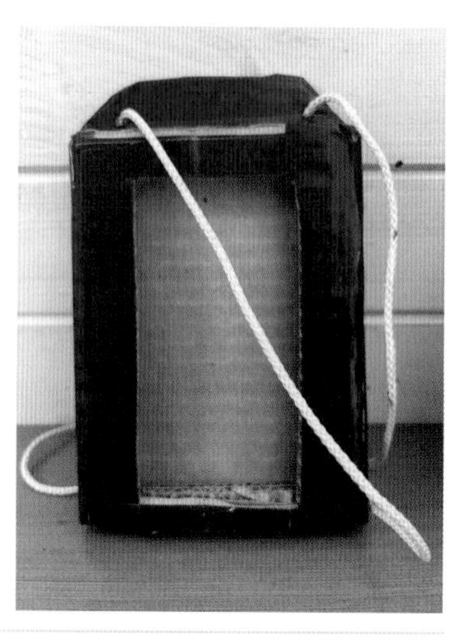

Full-size low-fidelity model/prototype of a mobile phone carrier made of cardboard, coloured propylene tape and 4 mm polypropylene rope, to check fitting of phone, fragility and ease of carrying it over the shoulder

High-quality toile by student Rachel Krause, with explanation. This toile was used to check shapes and fittings and to practise the construction processes.

9780170477499

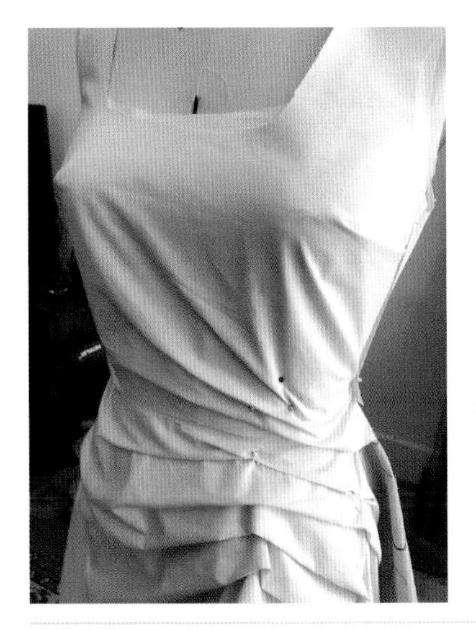

Checking, adjusting and cutting a calico toile on a mannequin for an outfit by student Laura Mazzarella

Evaluating models and prototypes

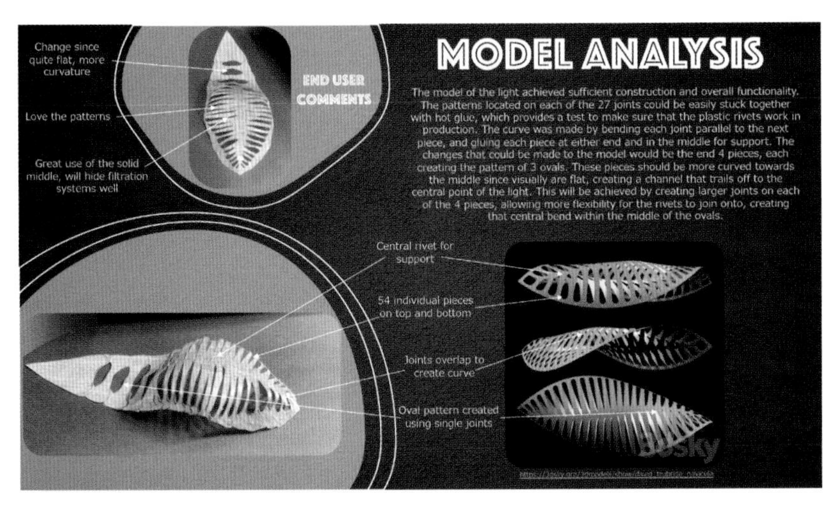

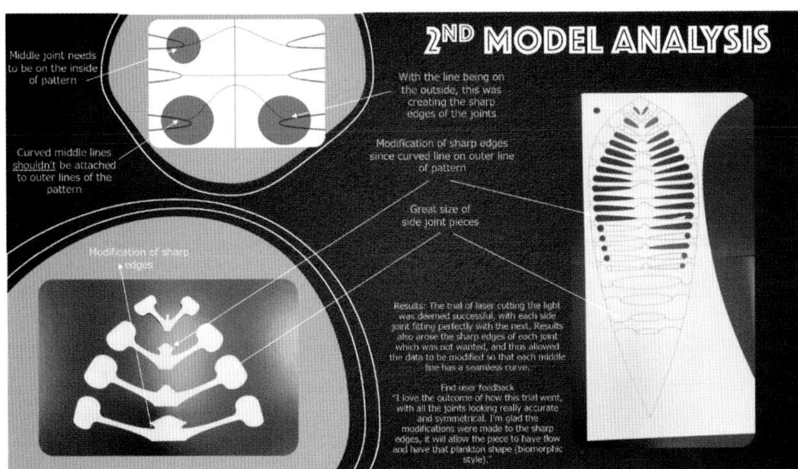

Checking and analysing prototypes for a lamp/air filter unit by student Ella Costanzo, with both physical and virtual product concepts

Materials worksheet
Construction process trials

Materials tests and trials

Materials tests and trials may be done before, during or after prototyping.

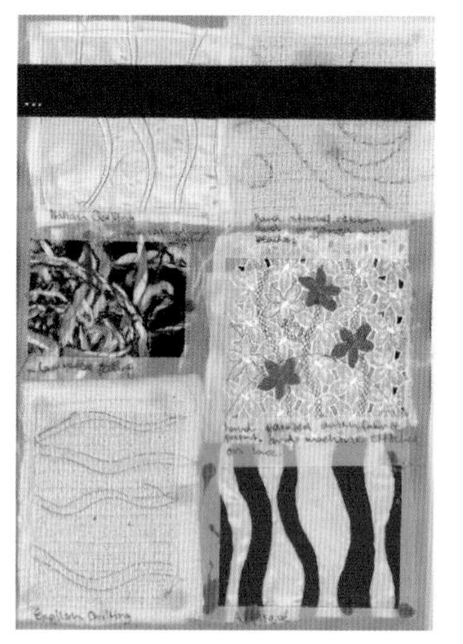

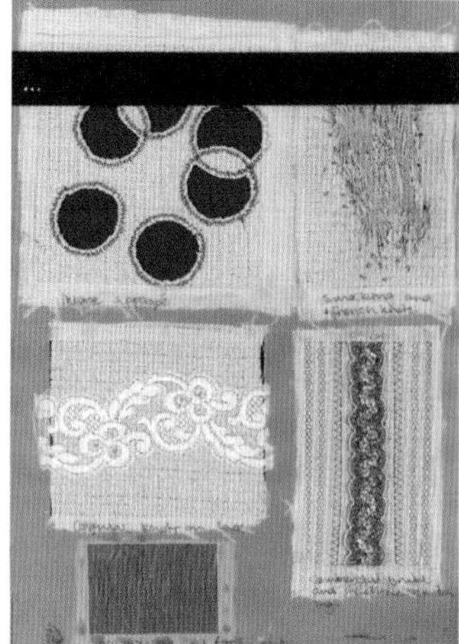

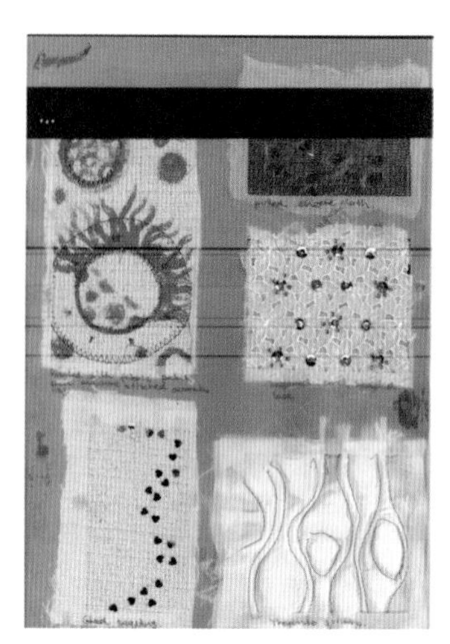

Evidence of process trials, from the folio of student Grace Wallace

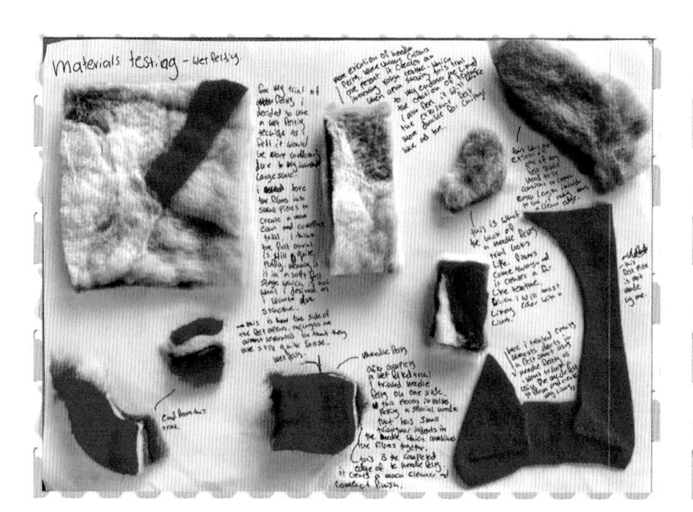

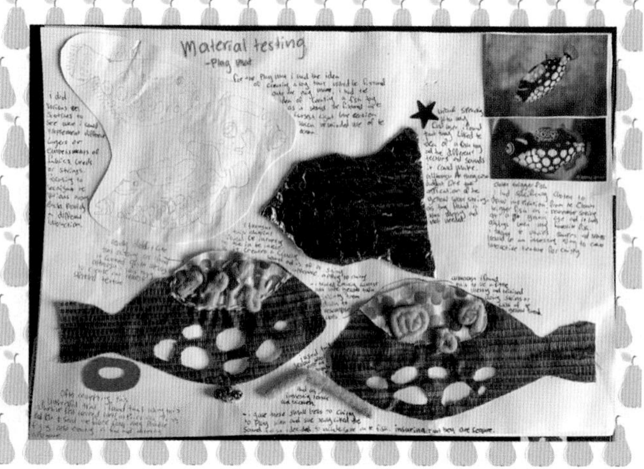

Evidence of materials testing by student Amelie Thompson for a cat carrier with cat entertainment

9780170477499

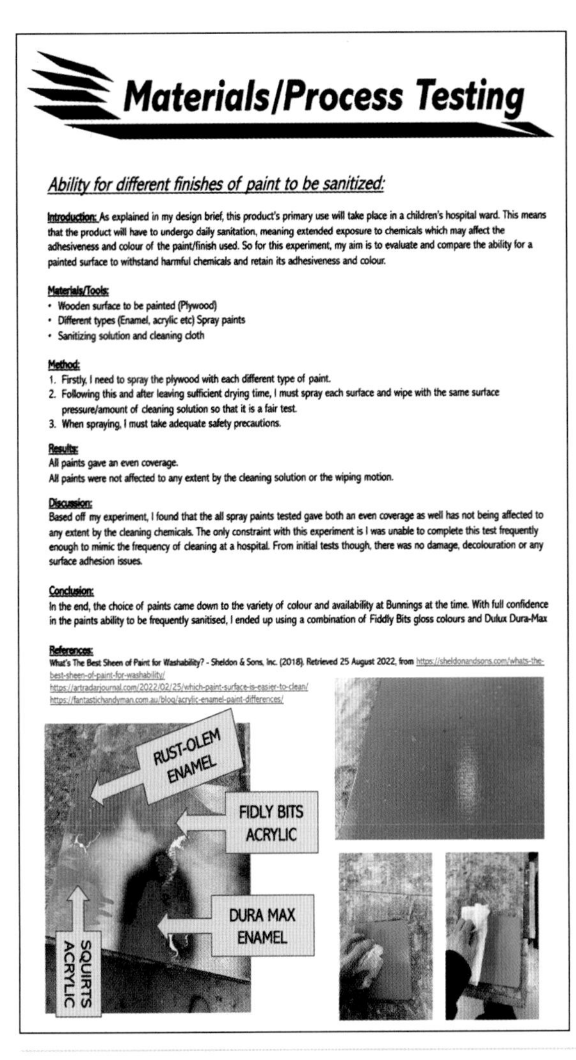

Materials/Process Testing

Ability for different finishes of paint to be sanitized:

Introduction: As explained in my design brief, this product's primary use will take place in a children's hospital ward. This means that the product will have to undergo daily sanitation, meaning extended exposure to chemicals which may affect the adhesiveness and colour of the paint/finish used. So for this experiment, my aim is to evaluate and compare the ability for a painted surface to withstand harmful chemicals and retain its adhesiveness and colour.

Materials/Tools:
- Wooden surface to be painted (Plywood)
- Different types (Enamel, acrylic etc) Spray paints
- Sanitizing solution and cleaning cloth

Method:
1. Firstly, I need to spray the plywood with each different type of paint.
2. Following this and after leaving sufficient drying time, I must spray each surface and wipe with the same surface pressure/amount of cleaning solution so that it is a fair test.
3. When spraying, I must take adequate safety precautions.

Results:
All paints gave an even coverage.
All paints were not affected to any extent by the cleaning solution or the wiping motion.

Discussion:
Based off my experiment, I found that the all spray paints tested gave both an even coverage as well has not being affected to any extent by the cleaning chemicals. The only constraint with this experiment is I was unable to complete this test frequently enough to mimic the frequency of cleaning at a hospital. From initial tests though, there was no damage, decolouration or any surface adhesion issues.

Conclusion:
In the end, the choice of paints came down to the variety of colour and availability at Bunnings at the time. With full confidence in the paints ability to be frequently sanitised, I ended up using a combination of Fiddly Bits gloss colours and Dulux Dura-Max

References:
What's The Best Sheen of Paint for Washability? - Sheldon & Sons, Inc. (2018). Retrieved 25 August 2022, from https://sheldonandsons.com/whats-the-best-sheen-of-paint-for-washability/
https://artradarjournal.com/2022/02/25/which-paint-surface-is-easier-to-clean/
https://fantastichandyman.com.au/blog/acrylic-enamel-paint-differences/

Material Testing **Polypropylene**

Test 3 - Bend Test

For this test I used a 15cm long, 3mm thick piece of polypropylene, to evaluate its bending performance. The test consists of a clamped down workpiece, with a 1kg "C" Clamp dangling from the end. The piece of plastic does not bend very much, due to its high rigidity and density. Ultimately it held up well against the test, further proving its viability as a table tennis bat.

Test 4 - Remelt

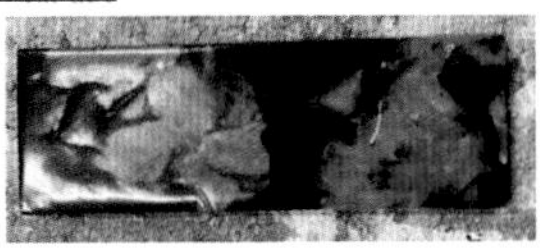

This test identifies the materials ability to be remelted, once in solid plastic form. I created two plastic blocks, one made from blue masks, the other from black masks, then once solid, broke up the two and placed them back in the baking tray. The image on the right shows the final product, it worked out exceptionally well, combining nicely with no degradation of rigidity and strength.

Test 5 - De-bubbling

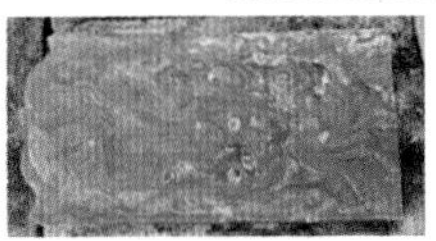

This test investigated the ability to remove bubbles from the surface of the plastic. Due to the melting process, large bubbles and air gaps form on the surface of the plastic which need to be removed. I experimented using a soldering torch to melt the top layer, closing off the air gaps. This partially worked, removing large air pockets, which self filled with the melted plastic surrounding it. However, the more heat that was applied, the more bubbles it generated, so in practice there is an even middle ground you must achieve between filling big holes, and creating new, smaller holes.

Evidence of materials testing by students Harrison Carr (left) and Hamish Pittendrigh (right)

Final proof of concept

Your final proof of concept could be demonstrated by a prototype, either actual or virtual. It could include a refined design option drawing or working drawings annotated to show modifications (from initial graphical concepts) and end user feedback. Explain why it is the best solution for the end user, referring to the design brief and evaluation criteria. Include information about whatever is needed to inform the scheduled production plan to construct your final product (i.e. material requirements, processes, special equipment, aspects that might need to be outsourced, etc.).

Finalising your drawing

Make sure you update all the measurements and components needed to construct your finished product.

Planning and managing: Producing and implementing

Scheduled production plan

Refer to the visual checklist at the start of this chapter and detailed instructions in Chapter 9 and on pages 241–4.

Journal with end user feedback throughout production, including modifications

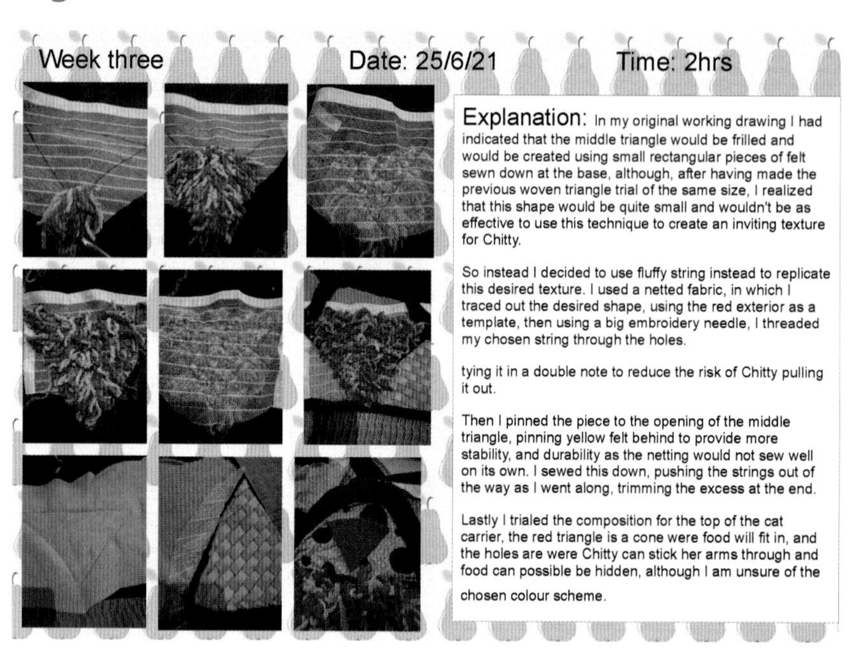

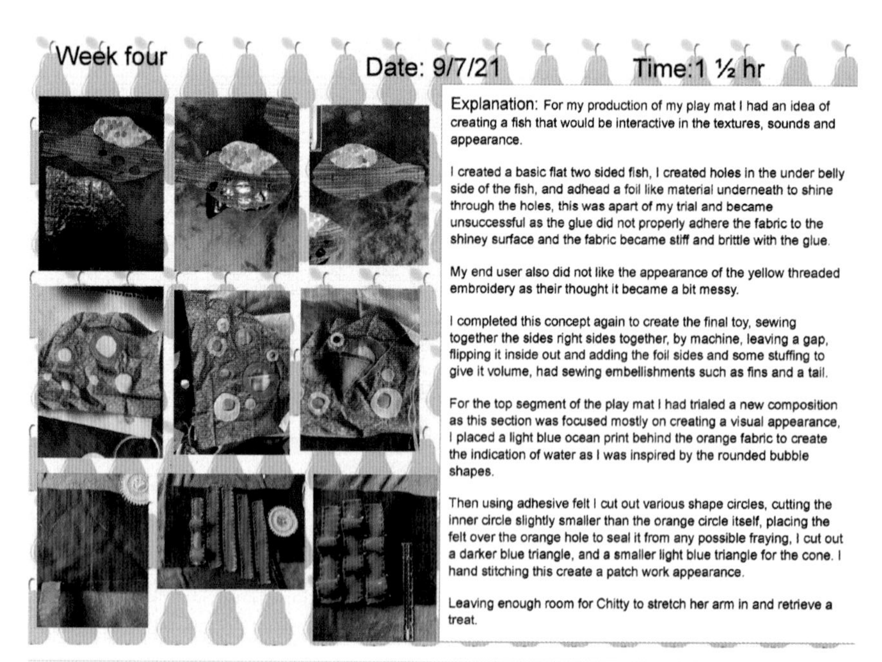

Two journal pages, with date and time, created by student Amelie Thompson when making a cat carrier with cat entertainment

9780170477499

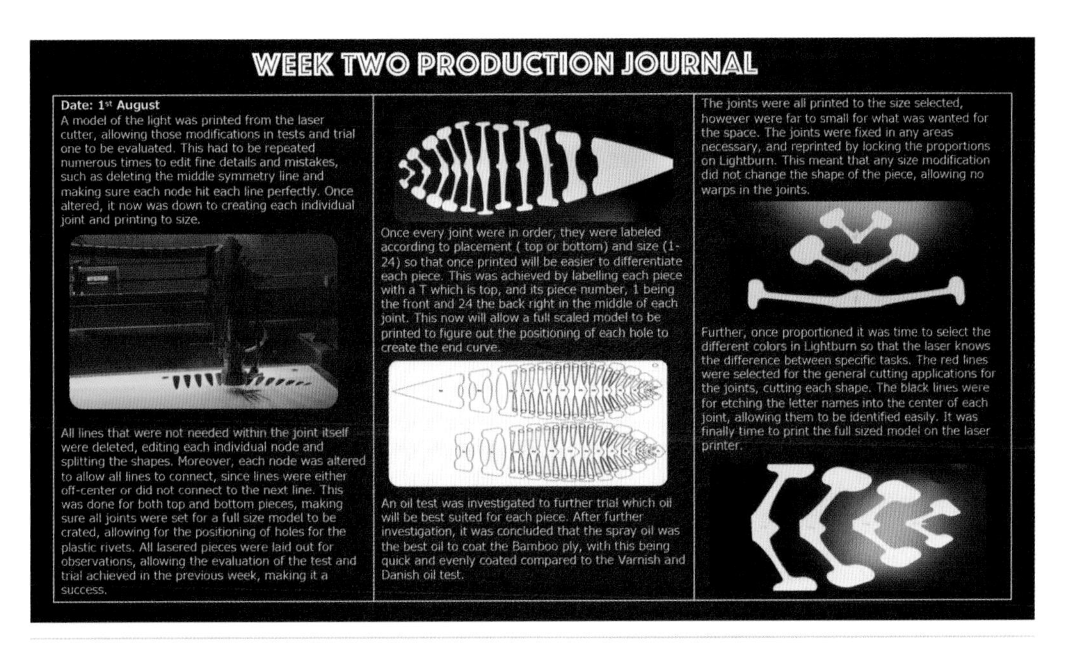

Journal page created by Ella Costanzo when making a suspended lighting/air filter unit

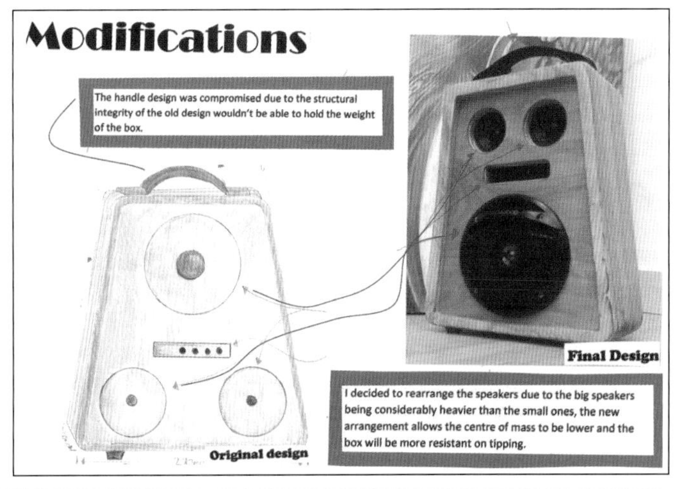

Visual comparison of the final design option drawing with the finished outfit (left) by student Janna Porras, showing modifications; and (right) by Jarrod Murphy for a speaker box

Finished student products

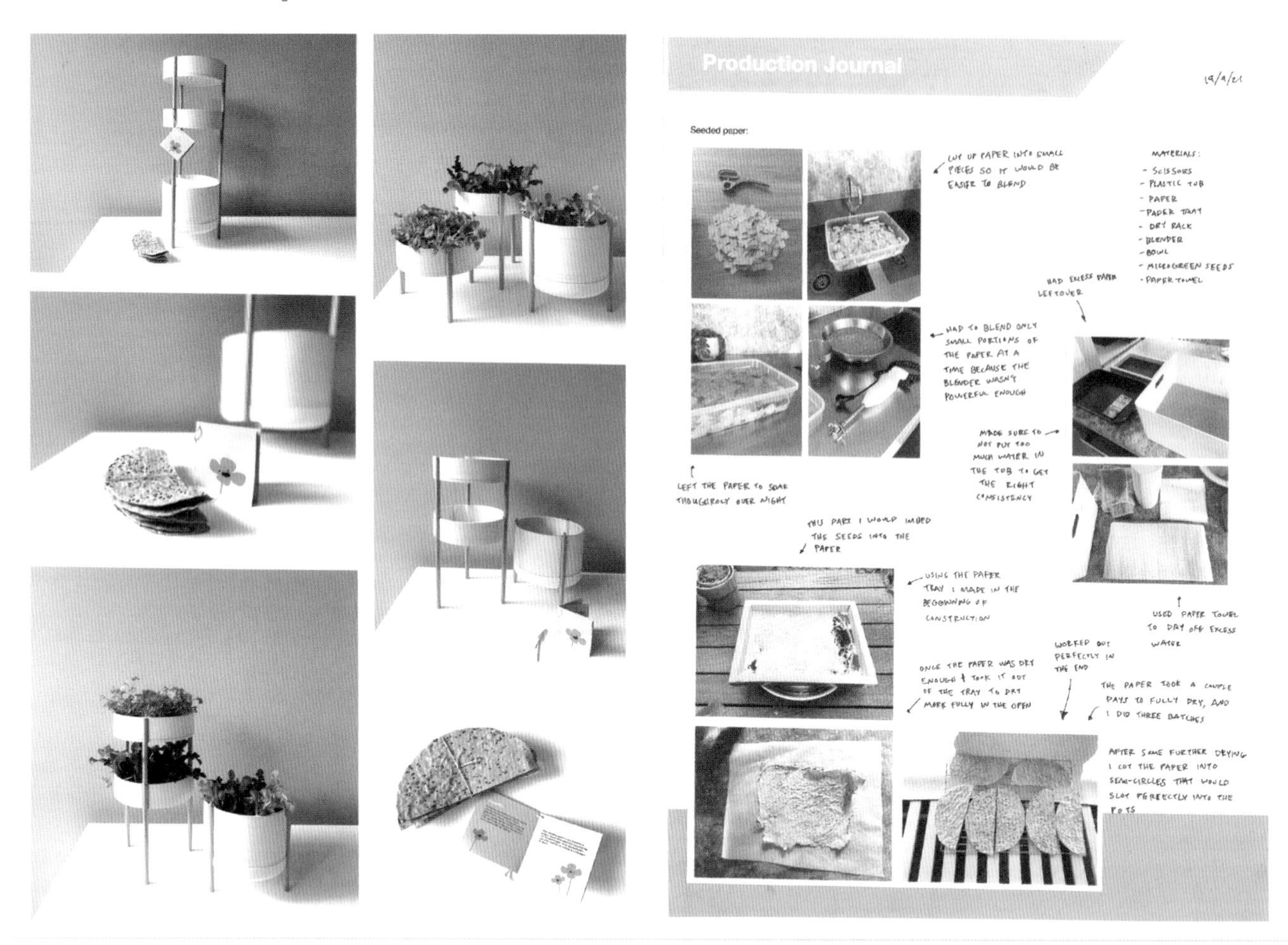

Journal pages by student Joseph Walker, showing how he created seeded paper for his stackable planters, which were 3D printed and joined with wooden dowels. See his technical drawings in Chapter 12 on page 336.

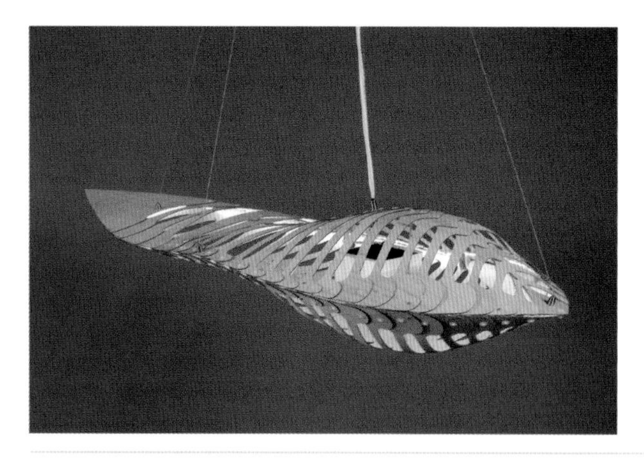

Final suspended lighting/air filter unit and close-up detail by student Ella Costanzo. You can see her developmental work throughout this chapter.

Weblink
Top Designs

To see more examples of student work, refer to the Top Designs exhibition held every year at Melbourne Museum. This exhibition showcases high-quality student work from the previous year. Visit the 'Top Designs' page of the Museums Victoria website and search for this year or previous years to see the works featured.

Plywood model, finished product and two close-ups of a Mid-Century Modern–styled light stand/planter by student Bronte Wiggins

Summary
Chapter 11

Quiz
Chapter 11 revision

In the *VCE Product Design and Technologies Study Design 2024–2028*, the final evaluation of the SAT product is completed in Unit 4, Area of Study 2 (SAC 1 of Unit 4). The SAT product is evaluated in conjunction with a range of products.

Weblink
VCE Product Design and Technologies home page

It is important to check the SAT Assessment Criteria, which are renewed every year and can be found on the VCAA home page of Product Design and Technologies.

DESIGN FUNDAMENTALS

This section covers topics in the cross-study specifications:

- design thinking
- drawing and design – types of drawings
- aesthetics and use of Design Elements and Design Principles.

9780170477499

CHAPTER 12

Design thinking

In this chapter you will learn about:

- design thinking – divergent and convergent thinking and methods for being creative and critical
- being inspired
- types of drawings
- visualisations
- design options
- working drawings
- design aesthetics – the Design Elements and Principles
- using Design Elements and Principles.

Drawing & designing worksheets:

- Thinking and designing creatively **(p.318)**
- Research for inspiration **(p.320)**
- Types of drawings **(p.322)**
- Drawing techniques for resistant materials **(p.330)**
- Drawing techniques for non-resistant materials **(p.331)**

Worksheets:

- 12.3.1 Identifying drawing types **(p.342)**
- 12.5.1 Form **(p.348)**
- 12.6.1 Shape and proportion **(p.354)**
- 12.6.2 Patterns **(p.359)**
- 12.6.3 Inspiration for styles **(p.361)**

Summaries:

- Chapter 12 **(p.361)**

Quizzes:

- Chapter 12 revision **(p.361)**

Case studies:

- Jacki Staude **(p.346)**

To access resources above, visit **cengage.com.au/nelsonmindtap**

Design thinking

Design thinking is used throughout the Double Diamond design process. It is about researching widely and thinking creatively to generate multiple ideas, combined with thinking critically to make decisions, and using speculative thinking, which focuses on the future.

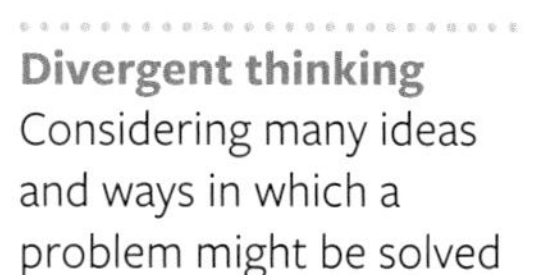

Divergent thinking
Considering many ideas and ways in which a problem might be solved

Convergent thinking
Bringing the most suitable ideas together

Speculative thinking
Considering future possibilities and developing ideas to improve the future

Creative thinking is considered **divergent thinking**; it allows one's thoughts and ideas to go wide. It comes from open-mindedness, flexibility, explorations, investigations and a curious approach.

Critical thinking is considered **convergent thinking**; bringing one's thoughts and ideas into focus and eliminating what is unsuitable or impractical. It comes from questioning, clarifying, planning, analysing, examining and testing information and ideas.

Speculative thinking considers the future and the impact that commercial product design has on humans and the planet. It considers what could be ... Read more on pages 187–92.

The three types of thinking are assisted by research. Each type of thinking involves consideration of important factors such as the end user, the most suitable materials, aesthetics, construction methods, availability of time and facilities and sustainability. Design thinking brings the best ideas to fruition.

Shutterstock.com/NadyaVetrova

Divergent thinking

Divergent, or creative, thinking is used to consider many ideas and a wide variety of ways in which a problem may be solved. 'Divergent' means moving outwards in different directions with an emphasis on quantity – a large amount. At first, divergent ideas can seem to be unrelated but, the more ideas that are expressed, the better the chance there is of finding a highly suitable solution. Divergent thinking is similar to lateral thinking.

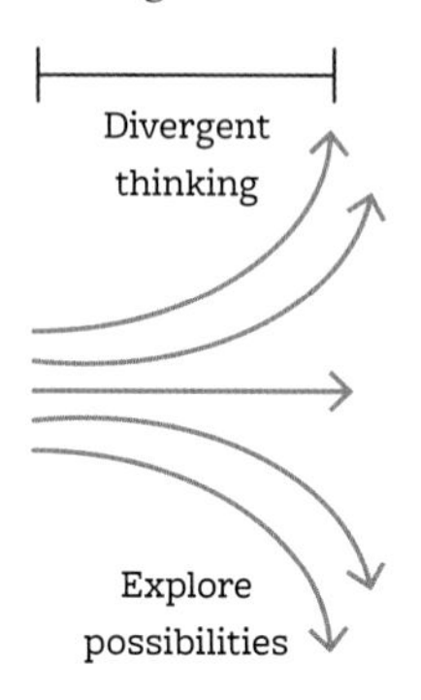

9780170477499

Divergent thinking is conceptual (ideas or theories that are not thoroughly worked out) and problem focused. The Double Diamond process starts with the design problem in the first diamond and asks, 'What is the problem, and how do we explore and understand it?'

Divergent thinking is used again in the second diamond to ask, 'What are the possible solutions to the problem? Can we come up with solutions using different perspectives or from different angles? How can solutions/products be developed and made?'

This quote is attributed to Albert Einstein: 'If at first the idea is not absurd, then there is no hope for it.'

Profile: Edward de Bono

Edward de Bono has spent years researching and writing about how to use the brain creatively. He was the originator of the concept of **lateral thinking** – a term that is now used worldwide and in a wide range of situations. He developed activities and exercises that help people to understand how their minds function, to recognise their strengths and weaknesses, and to work on ways to increase their ability to think laterally, escape habitual mind patterns and open the mind to new connections.

He also identified six modes of thinking and labelled them as different coloured hats. The natural adoption of different thinking methods, or 'hats', by members in a team can create effective solutions when addressing a design problem.

Search out information about de Bono's work and try out some of his activities, such as thinking of a solution to a simple design problem by combining ideas from three completely unrelated everyday objects.

Lateral thinking
'Sideways' thinking – being creative through an indirect approach, rather than an obvious one; using new approaches to solve a problem

Weblink
The de Bono Group

Getty Images/Mint

Edward de Bono

Convergent thinking

Convergent, or critical, thinking involves decisions about bringing the most suitable ideas together. 'Convergent' means coming together from different directions. It refers to the narrowing down of possible ideas to probable or final ones. Convergent thinking is analytical and solution focused.

In the first diamond of the Double Diamond, convergent thinking is used to gather information about the project to understand who the end user is, and what their needs are. This is finalised in the design brief (and analysed further through evaluation criteria).

Convergent thinking is used again in the second diamond to critically evaluate proposals (ideas) against the requirements in the design brief. It can involve considering questions such as:

- 'This is where we need to get to. How do we get there?'
- 'Which is the most effective and viable way to approach the making of the product within the workshop and your skills?'
- 'What is realistic and achievable in the time frame? What are the best materials and construction processes to use?'

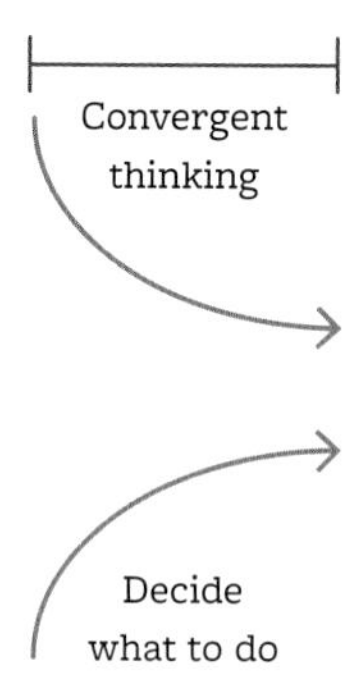

Rational thought and clear judgement are applied to each idea to see how it suits the design situation and the needs of the user. This requires asking the right questions, looking at your designs to consider whether they are too complex or too simple; looking at all the information you have researched, checking facts and then analysing, combining and applying this information to create a rational, well-judged and suitable solution.

 You can read more about convergent and divergent thinking in chapters 1 and 7.

Creative and critical thinking

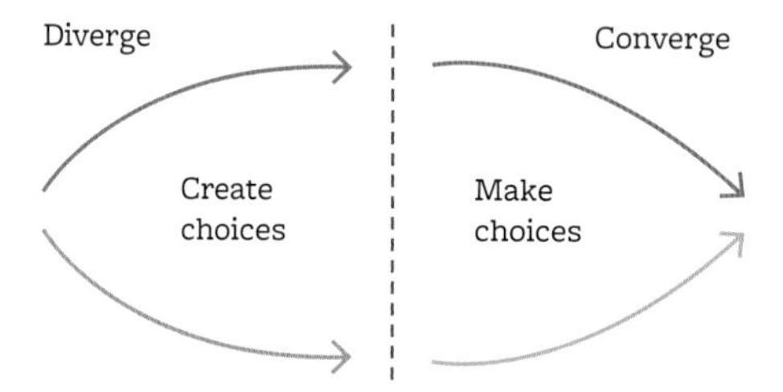

Creative (divergent) and critical (convergent) thinking are sometimes portrayed as opposing things: right- and left-brain thinking; freewheeling and rigid thinking; or 'artistic' and 'scientific'. However, it is important that both forms of thinking are used together to create quality solutions. Look at the techniques in the following table and think about how they can be applied simultaneously to support each other. You may have a natural affinity for a particular type of thinking through a problem; but applying the other type as well is important for developing high-quality solutions.

9780170477499

Creative and critical thinking techniques

Creative thinking techniques	Critical thinking techniques
Brainstorming – useful in a group, many ideas are put forward and all considered, keeping all ideas until the process is over **Mind mapping** – representing and organising ideas visually **Brain writing** – ideas are written down, passed around and extended on **Thinking laterally** and 'outside the square' (see 'Profile: Edward de Bono') **Applying** things from one area to another, unrelated area **Looking,** listening and **feeling** what is around you **Observing** and adapting existing ideas or products and nature **Escape thinking** – a method to challenge conventional thinking **Being curious** – about how things work, what makes people tick, learning about things **Synectic thinking, SCAMPER** or other triggers (see following pages) **Exploring** the Design Elements and Principles (see more later in this chapter) **Synthesising** – putting information and ideas together in new ways **Collecting** objects and materials of interest **Experimenting** with materials, processes and ideas **Expressing**/communicating/recording ideas and **Annotating** them	**Questioning** – asking really targeted questions (How can I do this? How does this idea meet the requirements of the end user?) **Critiquing** – looking at the positives and negatives **Researching** – seeking, sorting, comparing, analysing and using information; filtering what is important from what is not **Analysing** products and behaviour for ideas, problems and successes **Comparing** or contrasting ideas and synthesising the best **Using** logic and reason – thinking clearly; being analytical – breaking things down and examining the parts **Seeking** others' points of view and noting them down **Getting** end user feedback **Using** evaluation criteria to select the most suitable ideas **Noting** your assumptions and habitual thoughts and why they exist **Synthesising** – identifying and sorting common traits; enmeshing, connecting and combining ideas **Checking** facts – using evidence to support your reasoning **Being** precise and specific, not vague **Making** judgements and decisions on what will have the most effect, and putting effort into those things **Deciding** on the best materials, colours, methods and solutions **Using** worksheets such as PMI (Plus, Minus, Interesting) and other graphic organisers (visual methods) for analysis

Being creative and innovative

Creativity and **innovation** are often interlinked, but they have different meanings. Generally, we think about them like this:

Creativity
Approaching, exploring or combining something in new or unexpected ways to create something unique

Innovation
Putting existing ideas, knowledge or concepts together in a new and different way

Breaking down the meanings of 'Creativity' and 'Innovation'

Creativity is creating something through an unusual or unexpected combination – of lines, textures, shapes and colours; of materials and their accepted uses, etc. It is generally linked with the aesthetic or visual aspects of an object or concept, but it is not confined to only that aspect. It can also relate to functional aspects and practical problem-solving. However, a creative work is not necessarily innovative.

Innovation is about being inventive, taking a completely different approach to an existing product or situation, and is accepted by most as an improvement. It might involve applying concepts from one area to another to create something in a previously unthought-of way. It is often linked to the more functional aspects of design. While being creative does not necessarily mean a solution is innovative it is often said that it takes creative thinking for an innovation to occur.

Using triggers to be creative

Synectic thinking technique

Triggers
Words or objects that spark an idea

Synectic
From the Greek, meaning 'bringing different things into a unified connection', synectic thinking focuses on taking things apart and putting them together differently to transform ideas

Synectics is an ordered way of approaching creativity and problem-solving. One method is to twist things around so that what is familiar, typical or ordinary becomes strange, and what is strange, weird or unexpected is looked on as normal. It suggests that if you look for similarity between apparently dissimilar objects, you will look at objects differently and find more ideas. William Gordon, who co-developed this method in the 1950s, expressed his central principle as: 'Trust things that are alien, and alienate things that are trusted.' In other words, don't be afraid to try things that seem strange at first. He said that emotional input into creativity was more important than the intellectual component.

When designing, it can be useful to use **triggers** to open up or spark new ways of perceiving things.

You can take either a new or an existing idea and bounce it around to come up with something original. The list below is part of a **synectic** approach to creative thinking. Use one or many of these, or come up with your own ways of transforming an idea.

Drawing & designing worksheet
Thinking and designing creatively

Synectic approach to thinking

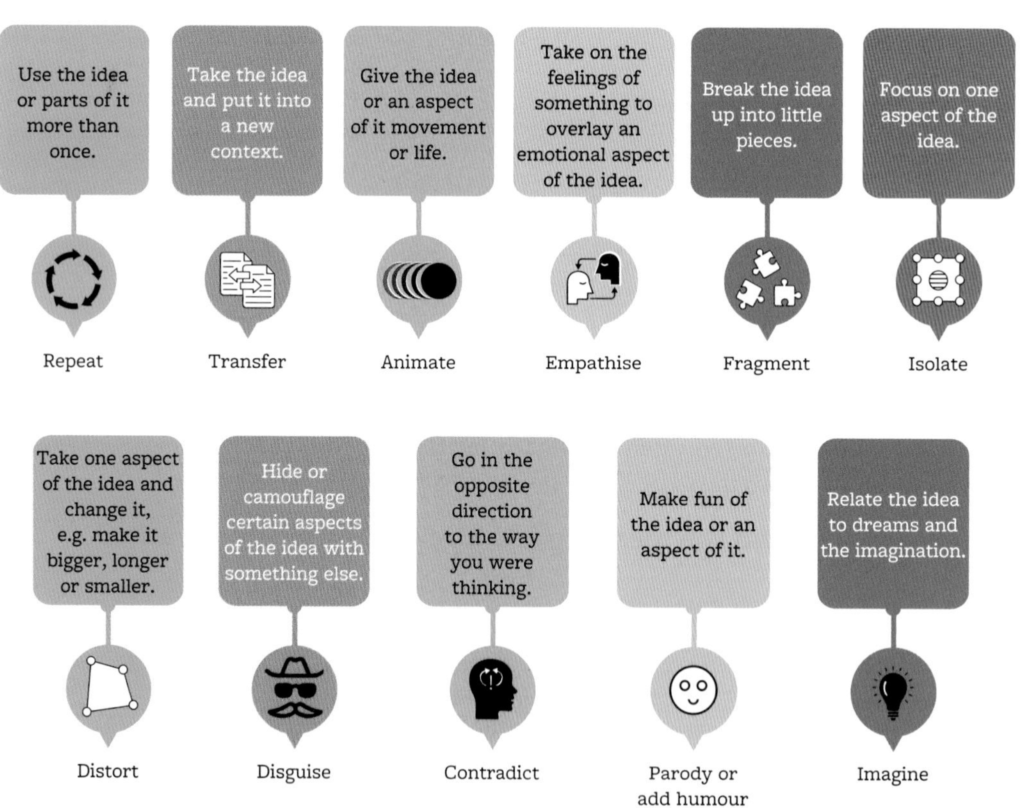

9780170477499

SCAMPER thinking technique

The SCAMPER list of triggers, developed by Alex Osborne in 1953, might also be useful:

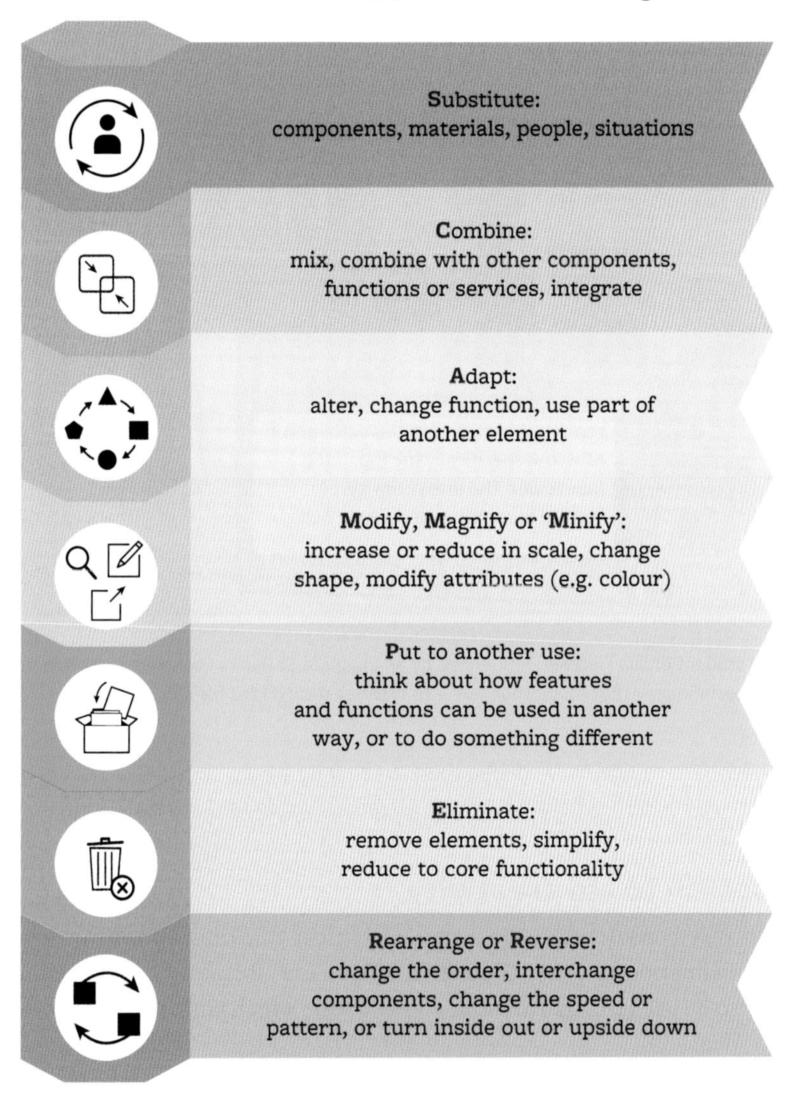

The SCAMPER technique works best when applied to one idea or an existing product (from your research) at a time. Choose the most applicable triggers.

There are many other techniques that can be used to process and transform ideas creatively. Often, they have some crossover, but it is worth checking them out as you might pick up new or different ways of working.

In your folio, choose several of the actions from synectic thinking or the SCAMPER technique, choose an idea or a commercial product to apply them to and draw them. It is better to have visualisations showing your ideas than written text. To explain your work, insert the original product or idea on the page, write the trigger word you are using and annotate your resulting drawings to explain what you have changed.

Visualisations by student Emilia Fletcher, influenced by an image of St Basil's Cathedral in Moscow and using one of the SCAMPER techniques – 'Put to another use' – to create a design idea for a dress

'It is easier to tone down a wild idea than to think up a new one.' (Alex Osborn)
'Creativity demands flexibility and imaginativeness ...' (Taher A. Razik)

Using research to aid creativity

Drawing & designing worksheet
Research for inspiration

Research is mostly considered to be a part of critical thinking, where you check facts or find specific information in order to make a judgement, followed by a decision. However, research can also be used to build a 'catalogue' of ideas, which you can then use as inspiration to create new ideas.

Don't limit yourself to one or two ideas, but research many varied design possibilities and use them to inspire your own ideas, and in that way your design work can evolve and develop. Once you have a multitude of ideas, you will use critical thinking to choose the most suitable ideas to refine.

Some starting points might include looking at a culture, an artist, a historical design movement or an innovative contemporary designer that you are interested in and the design features that make their design work unique or that appeal to you.

9780170477499

Inspiration from a different field

You can research aspects from unrelated fields to give you inspiration. This may prompt you to look at the situation from a different perspective. Designers are often inspired by the work of artists. Many products have been inspired by the painter Piet Mondrian (1872–1944).

In 1965, Yves Saint Laurent, French fashion designer for Dior, created dresses (left) directly influenced by the work of Mondrian, such as the painting *Composition A* (centre). The dresses were made of wool and silk with graphic thick black lines separating blocks of colour. The wooden *Red Blue Chair* (right) by Gerrit Rietveld (1923) was also inspired by Mondrian.

Research to inspire: student examples

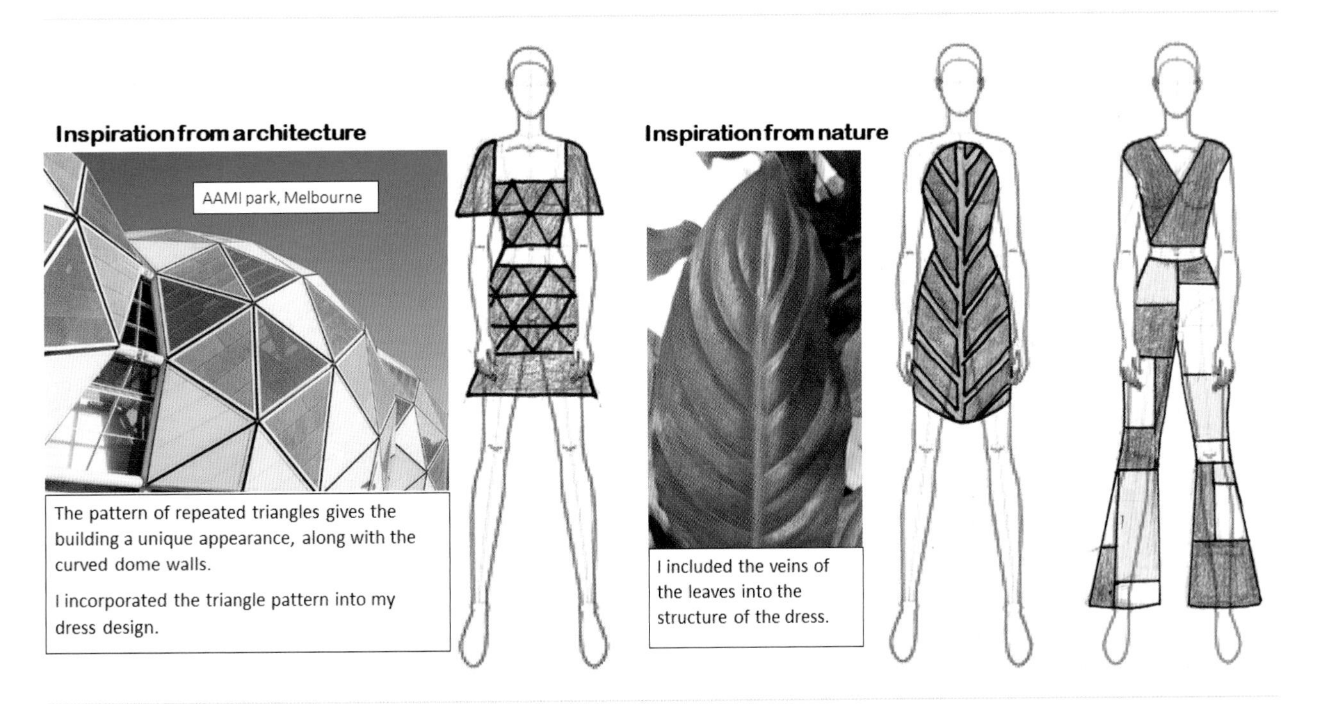

Visualisations by student Emilia Fletcher: drawing (right) inspired by Mondrian's *Composition A*, and other drawings inspired by her own photographs of the Melbourne Rectangular Stadium (AAMI Park) in Richmond and a giant plant leaf

Drawing and design

Types of drawing for graphic product concepts

VCE Product Design and Technologies makes a distinction between the drawings required for visualisations, design options and working drawings. The differences are briefly explained below and throughout this chapter.

Visualisation drawings
Sketches that get multiple ideas down quickly; for exploring and visualising what is in your head. The emphasis is on quantity.

Visualisations are the first step in the designing stage, where you play around with and explore potential ideas for parts and/or the whole of a solution to meet the brief. Graphic organisers, diagrams, mock-ups and rough 3D models can be used in this step. However, mostly you will use **visualisation drawings**, which are quick sketches, called thumbnails, concept or idea sketches. These drawings can be rough or refined; detailed or simple. This step is a record of your design thinking in this early stage, aided by annotations.

Your folio should demonstrate a flow or links between your design brief and research, your visualisations and your design options along with exploration of the Design Elements and Principles.

Drawing & designing worksheet
Types of drawings

Be inspired by researching images other than similar products to your intended product, e.g. from nature or completely different objects. For example, if you are designing a garment, get inspired by kids' toys or architecture; if designing furniture, also get inspired by architecture or microscopic plant/animal details. Think about using different materials or ask questions such as, 'How could you make it more entertaining or fun?', 'How would a tradesperson use it?', 'How would a blind or deaf person use it?' Use some of the creative techniques and triggers outlined on pages 318–19 to explore ideas in new ways. Look at the previous pages, read some suggestions on the next page and look at the student examples in this chapter and Chapter 11.

Design options
Also known as presentation drawings; realistic, detailed 3D drawings that show the whole product and are used to convey a fully worked-out design solution. The emphasis is on quality and detail.

Design options are expected to be in the presentation style. They are realistic (3D for products in resistant materials), show more detail and indicate what the whole product will look like. They are rendered/coloured to represent form and textures. Annotations indicate the proposed materials and construction processes; explain how the product will suit the end user and address the brief; and help to communicate aspects that can't easily be seen in the drawing (unlike labels). The presentation drawing style is expected for the design option question of your VCE Product Design and Technologies exam.

Working drawings
Accurate line drawings that communicate the product's dimensions and components. The emphasis is on accuracy.

Working drawings are needed once the design is finalised – i.e. when the preferred option has been chosen. Working drawings (or technical drawings) are used to show exact measurements, shapes, joins and placement of components for construction or manufacture.

To discover more about the drawings suitable as 'visualisations', design options and working drawings, go to the tables in Chapter 1, on pages 20 and 21, and Chapter 7, on pages 218–19.

9780170477499

Visualisations

Check the Assessment Criteria for the relevant year to see how your visualisations will be assessed for the SAT in Units 3 and 4.

When creating visualisations, it is important to start thinking creatively or divergently. To get started you might do some brainstorming or draw up a concept map and write down some ideas. You can look at your research and 'borrow' ideas and things that interest you. Take your written ideas and create drawings. Get down as many ideas as you can. Quantity is important here!

Drawings can be done in pencil, but different drawing media, such as bold pen and 'splashes' of colour, add visual interest. They also make the work more defined once it is scanned for an e-folio or to turn into a digital file to manipulate or add annotations.

To be truly creative, try some of the following ideas.

3D models and mock-ups are also useful visualisation methods to check ideas or if you don't feel confident with drawing. Be sure to photograph and annotate them for your folio.

Suitable media and design ideas

Using mixed media can enhance your drawings substantially. Instead of only using lead pencil, use pen to create bold outlines and add colour with washes, textas/markers, coloured pencils, textured rubbings, cloth or paper (patterned, shiny, textured, fine, etc.).

Here are some suggested ways to help you get started with sketching objects and design ideas:

- Start with two or three basic shapes or forms, such as circles or spheres, triangles or pyramids, rectangles or blocks.
- Draw lots of quick images – not too large. Play around with the placement, composition, comparative size and order, proportion and linking elements.
- Turn some of the better images into 3D sketches.
- Refine some of the shapes and edges and add some detail.
- Avoid 'stick-figure' representations of objects; add double lines to create the illusion of the material's width.
- Darken some of the lines and add some rendering for depth. To find out more about rendering, go to page 332.
- Highlight the most effective sketches with a black marker background or outline; or with a coloured wash background to add impact.
- Annotate your drawings to explain your ideas and to analyse what does and doesn't 'work' for the design situation.

Software for creating visualisations

Software that can be used to create visualisations includes:

- FireAlpaca
- GoodNotes
- Procreate® for Mac
- Adobe Illustrator/Photoshop
- ibisPaintX.

Weblinks
FireAlpaca

GoodNotes

Procreate

ibisPaint

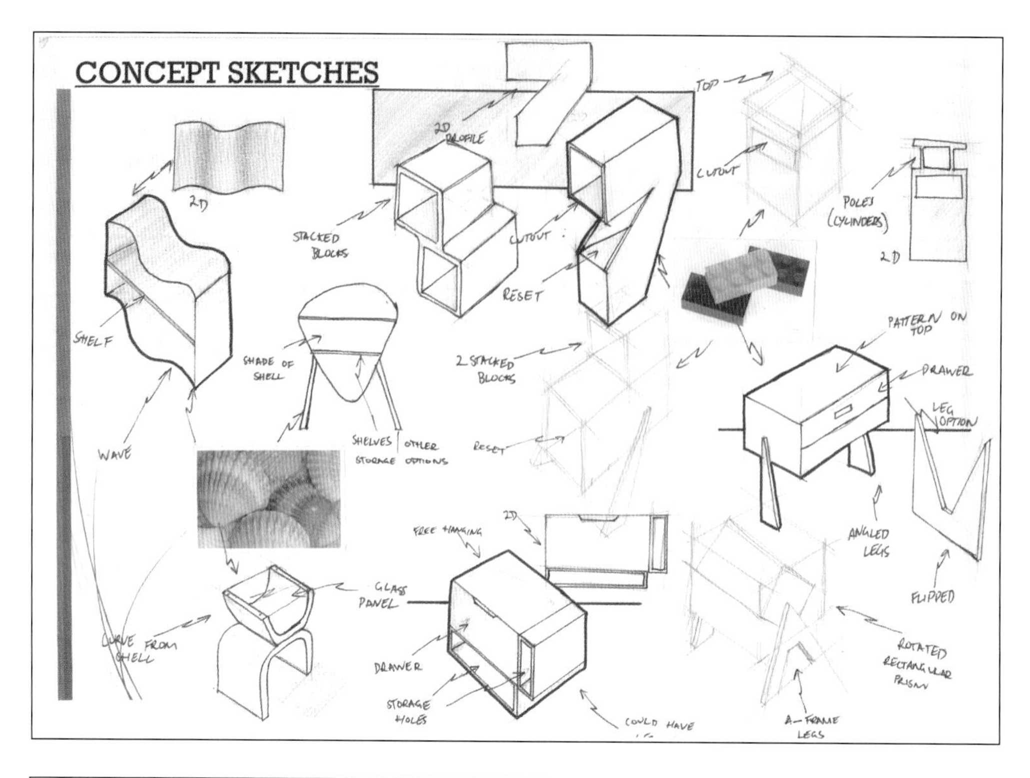

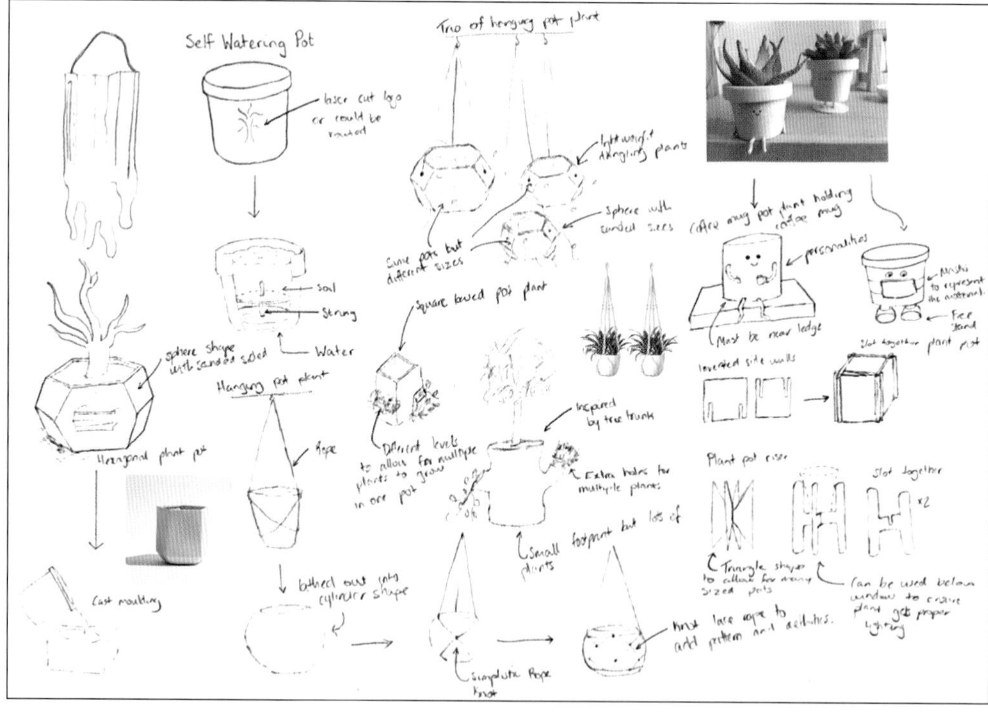

Visualisation drawings for a storage item, inspired by a shell and Lego® blocks, by student Josh Goudge (top); and for using used disposable face masks, by student Hamish Pittendrigh (bottom)

Crating is drawing a 3D 'cage' to assist in creating a 3D image or 3D representation of your idea. Lightly draw a cube or rectangular form, fill it in with your ideas and use the crate as a guide for the edges, as can be seen in some of Josh's work at the top of this page.

9780170477499

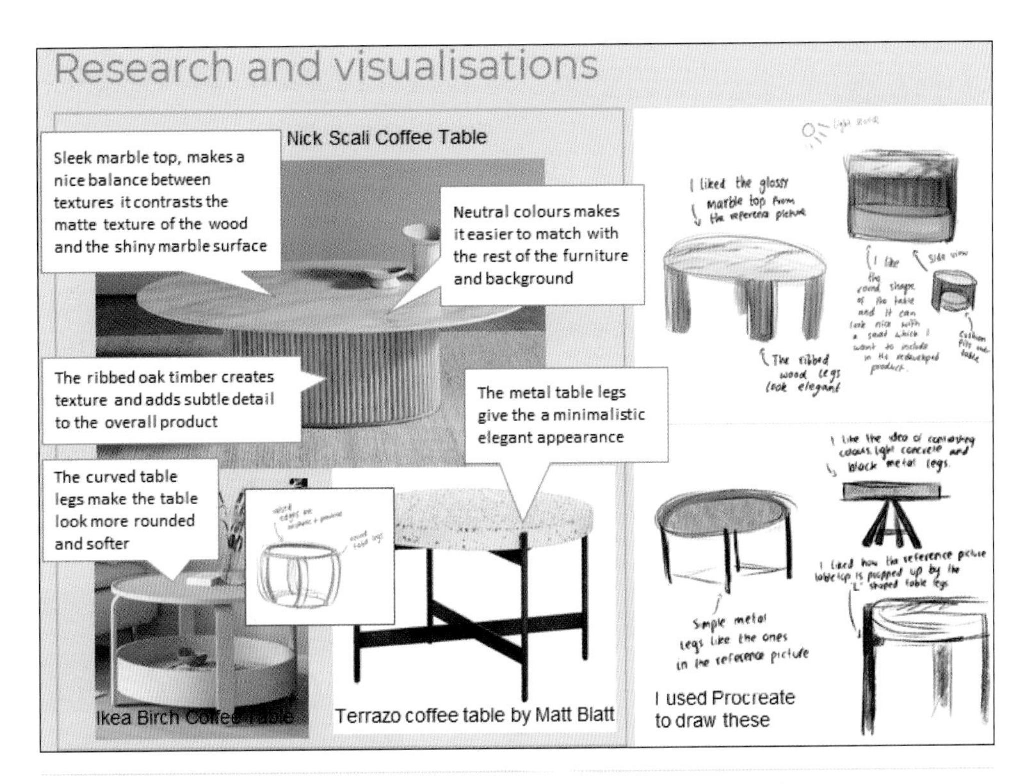

Visualisation drawings for a coffee table, drawn by Unit 1 student Rayan Al-Mosa using Procreate

Note that your visualisation drawings are not expected to be of high quality. Most student examples of visualisations shown in this book are of a much higher quality than expected and were chosen because they communicate clearly and contribute to dynamic folio pages.

Visualisations: student examples

Folio pages by Leya Mackus (left) and Mariah Kehaidis (right), showing visualisation drawings influenced by research of very different objects

Folio drawings by Mariah Kehaidis, displayed at Melbourne Museum in the exhibition Top Designs 2017 (part of the VCAA's VCE Season of Excellence 2017) and featured on pp. 34-5 of the exhibition catalogue Top Designs 2017 (VCE Product Design and Technology, © VCAA 2017).

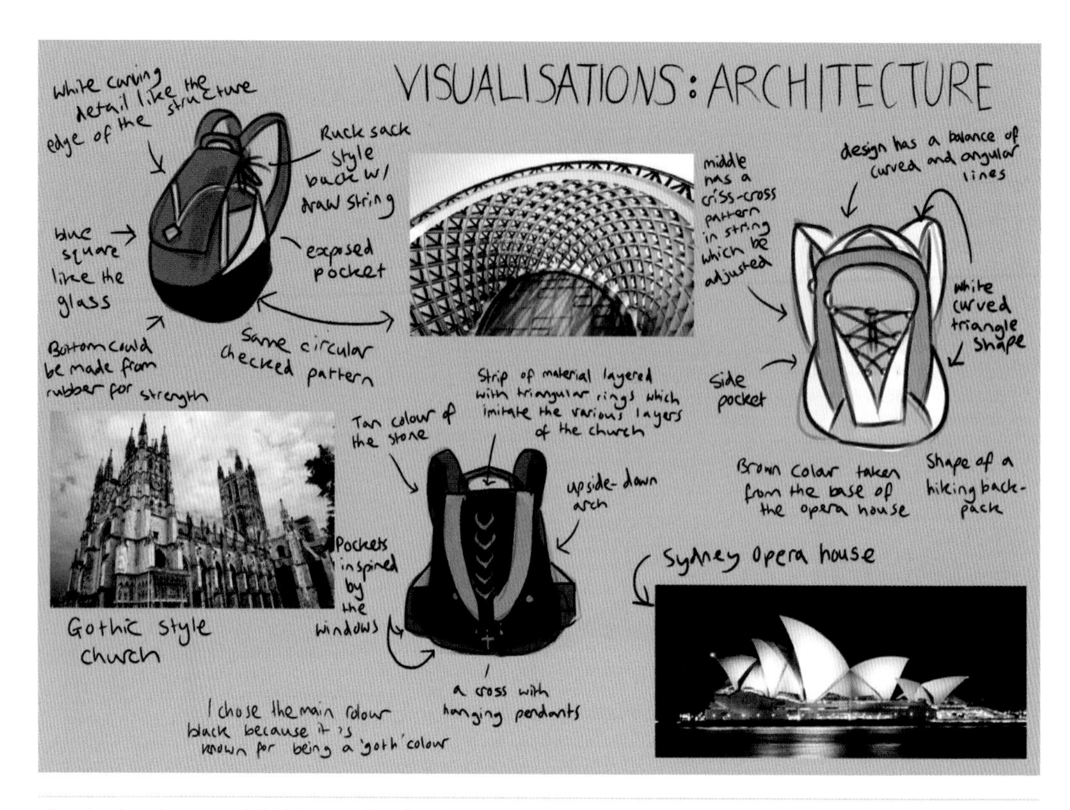

Visualisation drawings (of higher quality than is needed at this stage) for a backpack, inspired by various architecture and drawn using FireAlpaca free software by student Sarah Hall

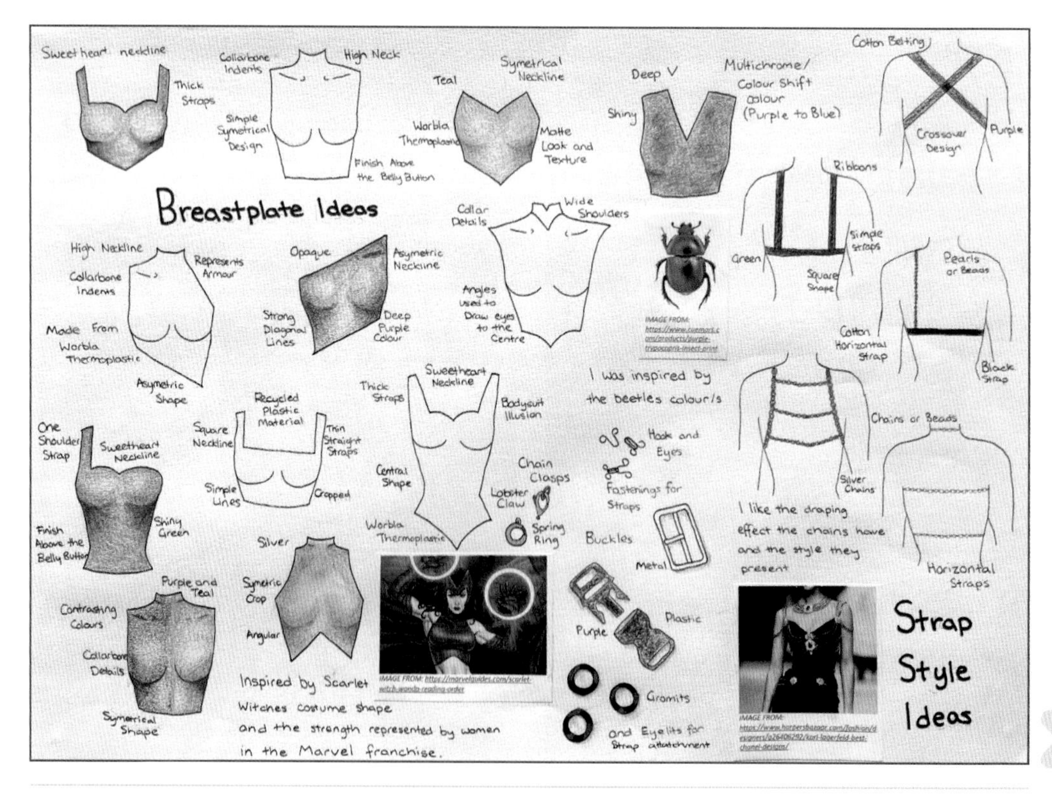

Visualisation drawings for the top half of a gown, inspired by a beetle, cartoons and strap buckles, by student Amelia Sompel

9780170477499

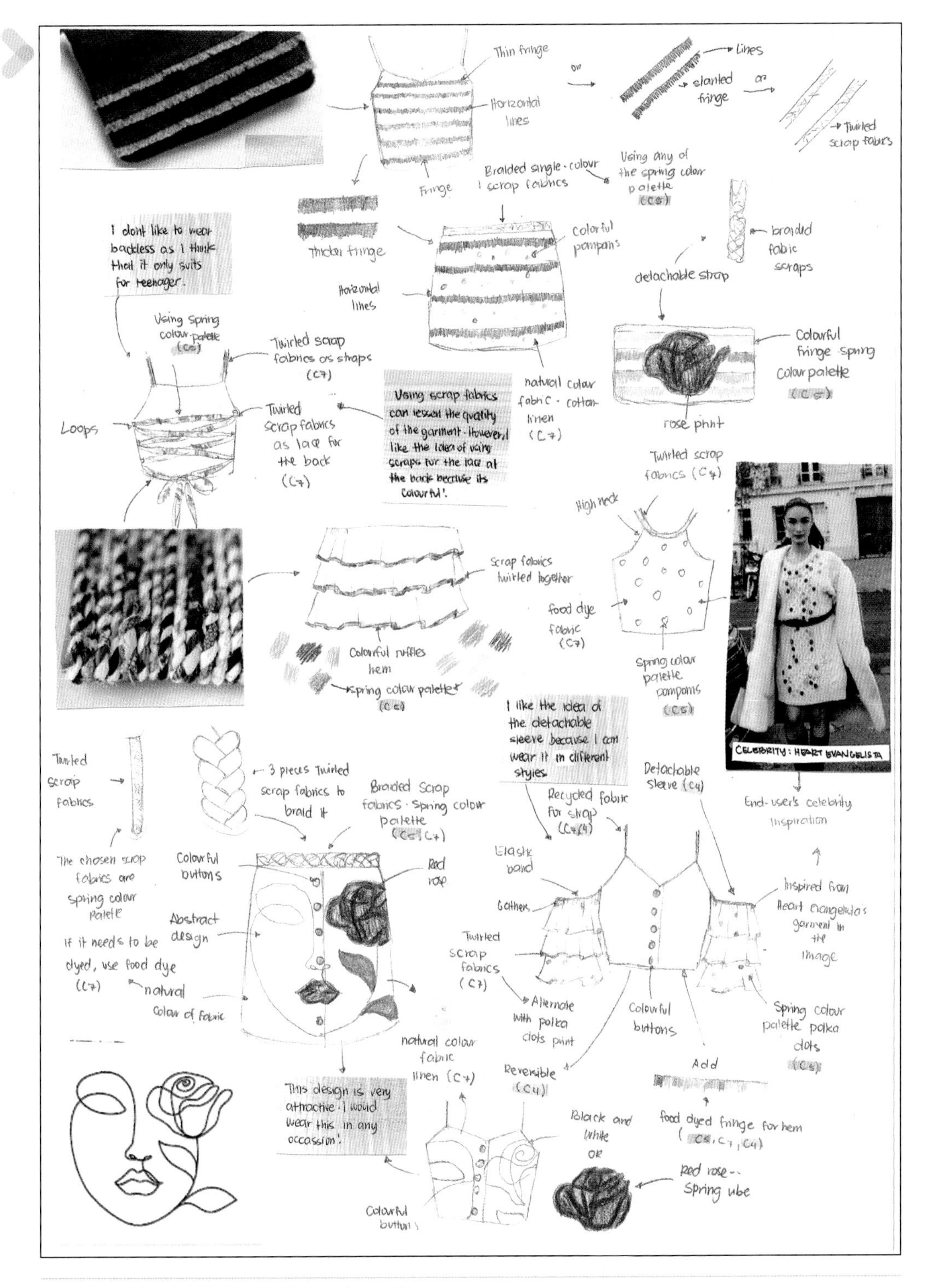

Visualisations inspired by various items, by student Janna Porras

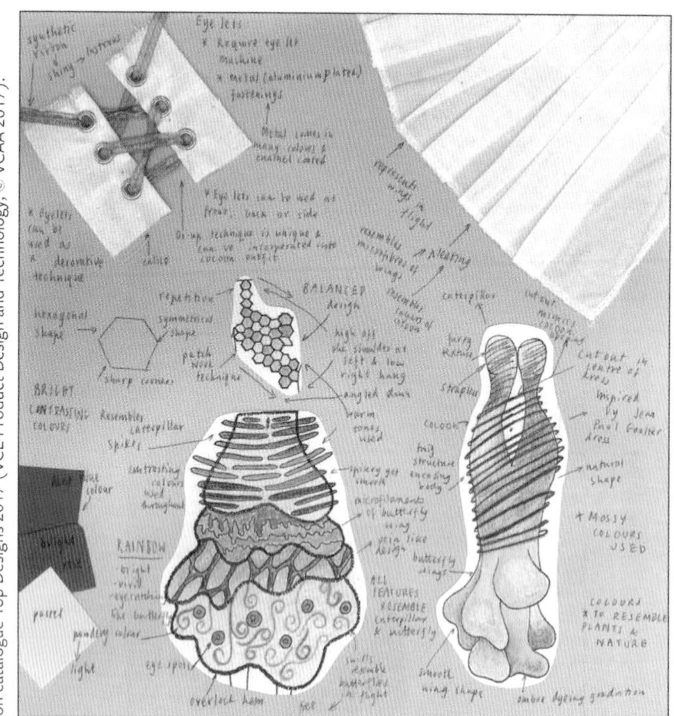

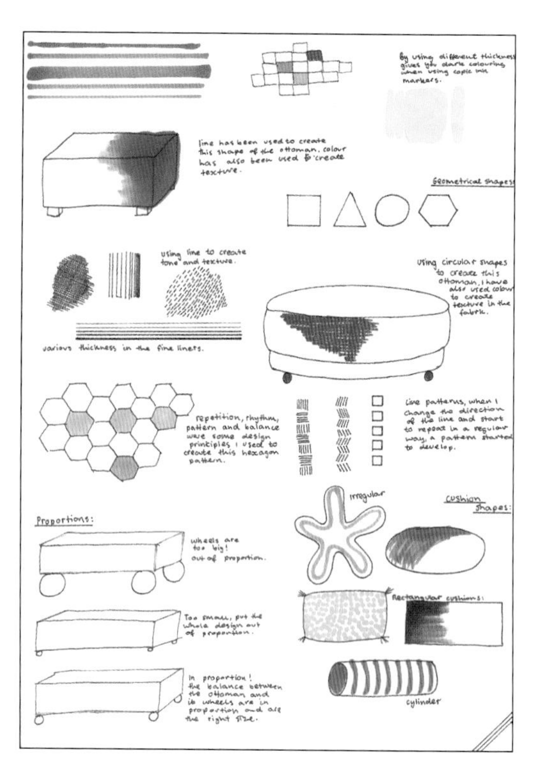

Folio drawings by Mariah Kehaidis, displayed at Melbourne Museum in the exhibition Top Designs 2017 (part of the VCAA's VCE Season of Excellence 2017) and featured on pp. 34-5 of the exhibition catalogue Top Designs 2017 (VCE Product Design and Technology, © VCAA 2017).

Exploring the Design Elements and Principles for a garment by Mariah Kehaidis (left); and for an ottoman by Sophie Allen (right)

Using mood boards

Mood board
A collage of images, design fragments, textures, components, colours, words and phrases, etc., used to provide a reference for visual consistency in designs or as a stimulus for design ideas

Mood boards communicate a mood or emotion in a visual way; by showing the relevant colour, texture, shape, patterns, style and related products or components. They are fun to create, and often stimulate and provide direction for lots of design ideas. They can be crowded with information – clarity of information isn't as important as overwhelming the senses with colour, shape and movement.

Designers use mood boards:

- as a reference point to stimulate ideas related to the set theme
- to set a consistent theme (colours, textures, patterns, components, etc.) to be used in a range.

Mood boards are especially helpful before you create your design options. They can assist you as a reference by defining a suitable style or theme for your end user.

Why have a mood board?

Mood boards are not essential. If you use a mood board, however, it's a good idea to add a heading to communicate its purpose with the viewer – e.g. 'Mood board to define end users' colour palette'. It must be relevant to your folio, with a visual link to your design theme and your design work.

9780170477499

Mood boards can be digital or physical/hard copy, made up of bits and pieces from many different sources, including:

- cuttings and images
- colour and texture swatches of fabric, plastic, metal, etc.
- unusual papers – textured, wrapping paper, etc.
- bits and pieces from existing products
- flat natural objects – feathers, leaves, bark, etc.
- words, phrases or slogans that indicate the needs, desires and aspirations of the users or client.

Where possible, acknowledge the work of others by labelling images with the name of the designer or manufacturer. This can be done directly next to the image or on a separate page that follows the mood board in your folio.

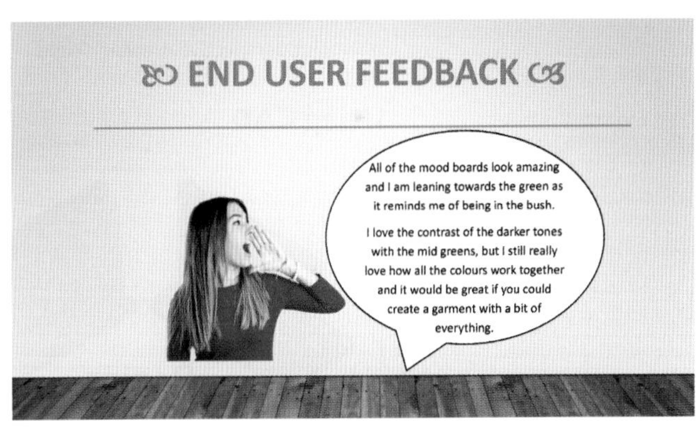

Mood boards by student Sharni Teesdale to present to an end user for feedback and to select the most suitable theme for design options. An opening page explained clearly why three mood boards were created instead of just one. The 'plant' mood board shown was then used as a strong influence on Sharni's design options.

Design options or presentation drawings

Design options need to be of high quality, detailed and realistic.

Check the Assessment Criteria for the relevant year to see how your design options will be assessed for the SAT in Units 3 and 4.

Suitable for resistant materials

Cabinet oblique is a type of drawing in which one view of the object faces the viewer squarely (straight on) and the edges recede to one side at 45°. The front face is drawn to its true size and shape, but the receding edges are drawn to half their true length.

Drawing & designing worksheet
Drawing techniques for resistant materials

Isometric drawing is based on a grid with vertical lines, and lines rising to the left and right at 30° from the horizontal (drawing tools such as a T-square and 30°/60° set square can be used). All lines are completed with reference to these grid lines.

Perspective drawing is used to give a realistic view of the completed article. You can use one-point or two-point perspective.

More information on these drawing systems can be found in the resource on MindTap, on the internet or in drawing books.

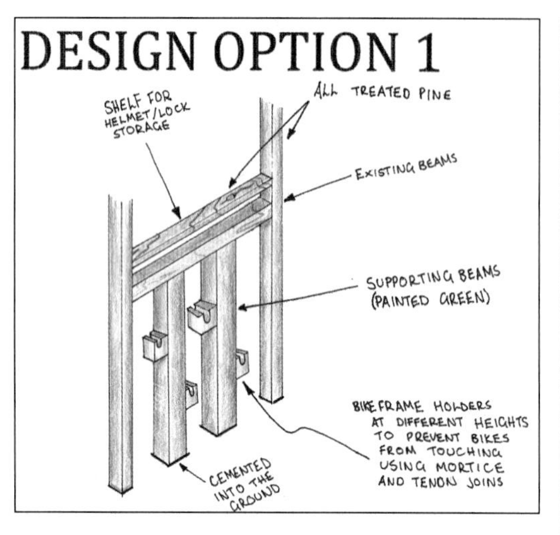

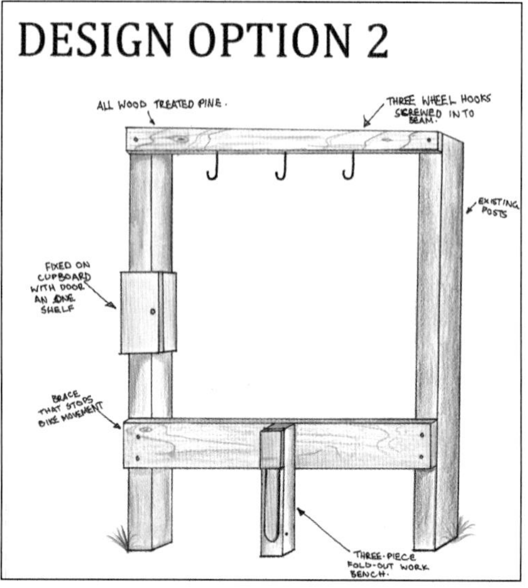

Two annotated design options by student Sam Jeffrey for a bicycle stand – an isometric drawing (left) and a one-point perspective drawing (right)

Using CAD

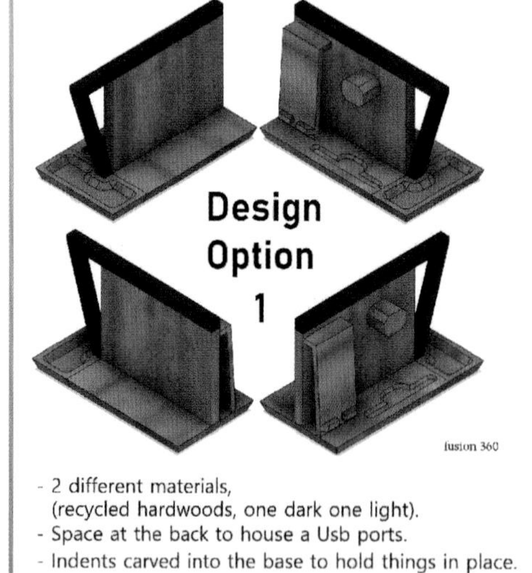

- 2 different materials,
  (recycled hardwoods, one dark one light).
- Space at the back to house a Usb ports.
- Indents carved into the base to hold things in place.
- straight, simplistic lines, emphasising function.
- tapered edges for a futuristic shape
- joins, basic butt joins, glue, nails and dowl joins.

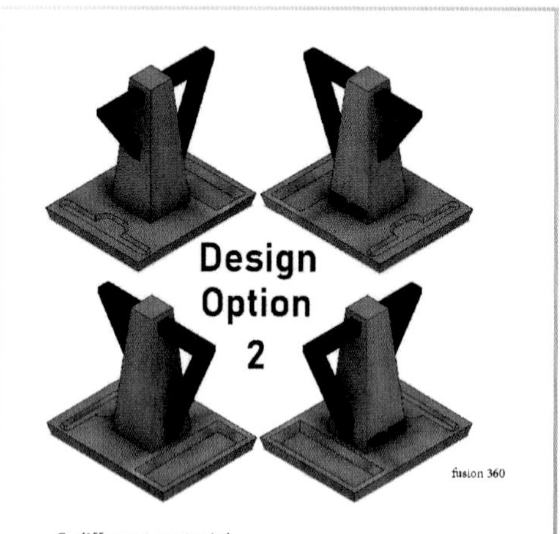

- 2 different materials,
  (recycled hardwoods, one dark, one light).
- Some space at the back for usb ports.
- indents carved into the base to hold things in place.
- straight simplistic shape centered around
  a pillar with tapered edges.
- basic butt joins with nails and glue, as well as
  dowel joins.

Two design options for a desktop cord organiser, with notes, drawn by Unit 2 student Nathan Colbert using Fusion 360

9780170477499

Suitable for non-resistant materials

Descriptive drawings:

- provide clear, detailed information of shape, structure and style, fabric and colour
- show construction details such as darts or shoulder seams
- include notions such as zips, buttons and fasteners, which are drawn or annotated if not visible.

Drawing & designing worksheet
Drawing techniques for non-resistant materials

Fashion illustrations:

- are fluid drawings used to give an impression of line, shape and form
- have a strong sense of atmosphere and individual style.

Body templates (or croquis) are useful when completing fashion drawings, whether they be visualisations or descriptive presentation drawings. Body templates are outline drawings of figures in different poses. These can be enlarged using a photocopier, and then traced and used as a basis to 'dress' with your clothing designs. Note: it is not essential to draw the body in this subject.

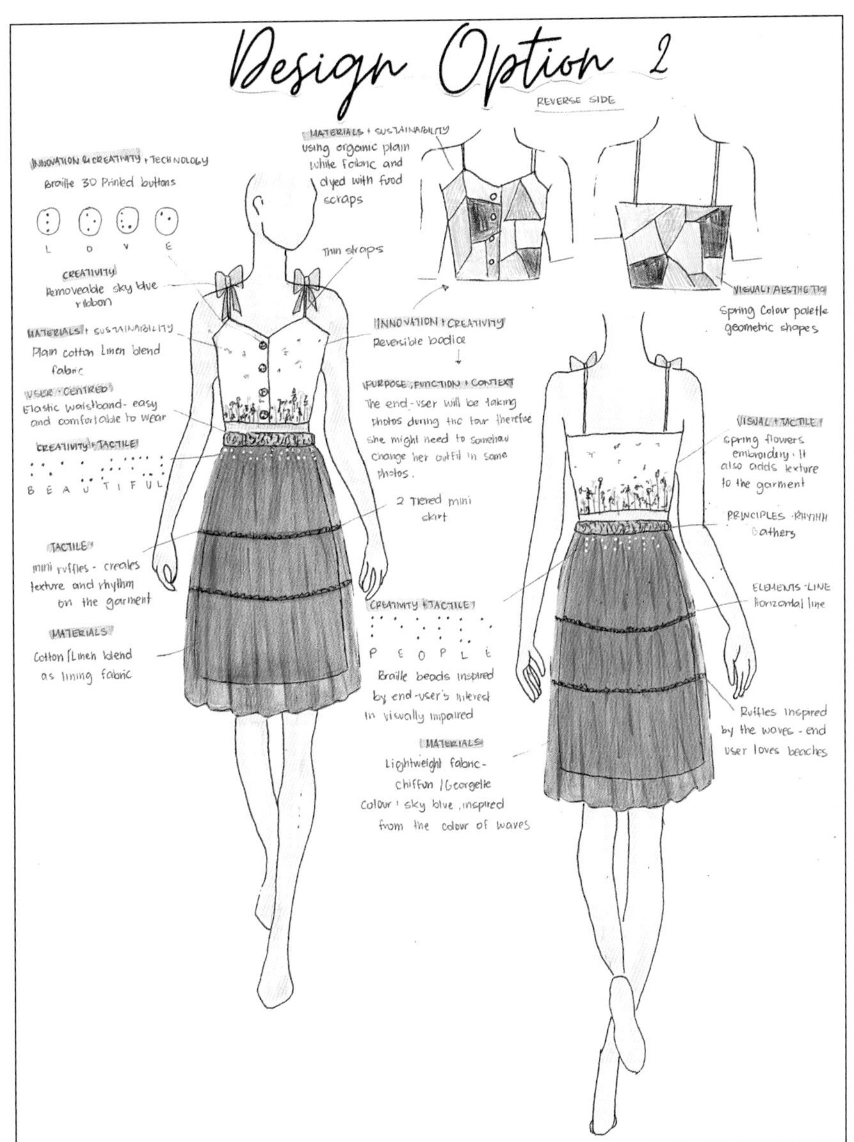

Design option hand-drawn by student Janna Porras. Note that it is not mandatory to present several views; one large view is sufficient. Multiple views can be drawn up for the favoured option.

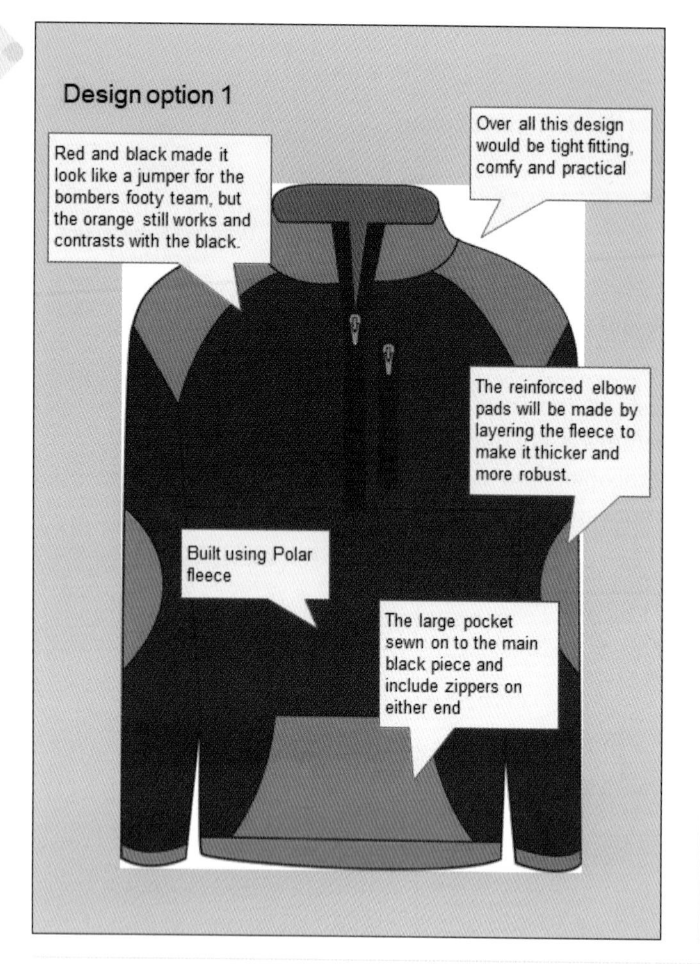

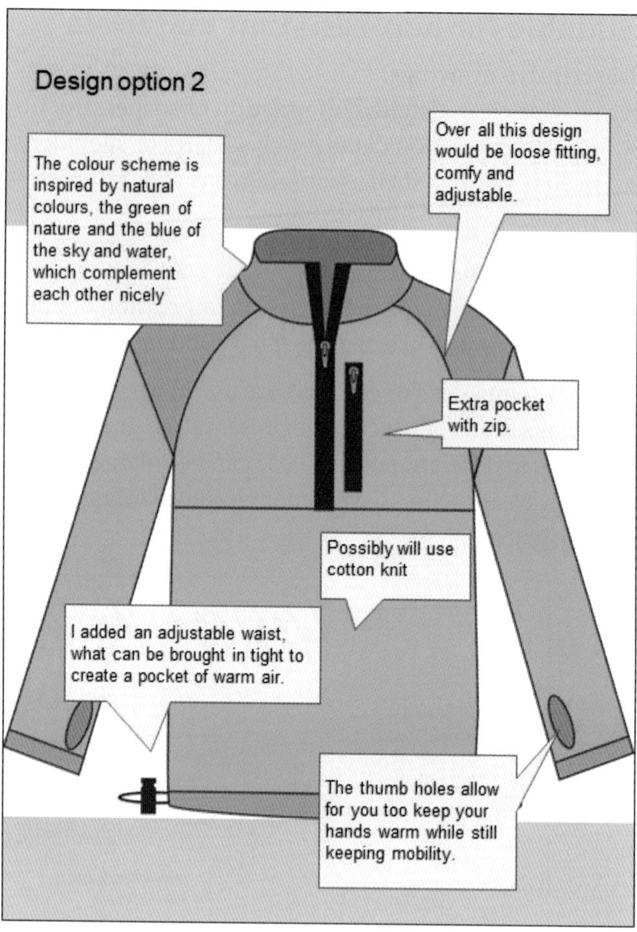

Two design options by Unit 1 student Dylan Edwards, drawn using Adobe Illustrator

Weblink
Designers Nexus

You will find many ideas, techniques, templates and tutorials at the Designers Nexus website.

Rendering

Rendering is used to make drawings more realistic by the careful application of shading, patterning and colour. It is expected in design option drawings. Rendering techniques give form by adding line, colour, texture and/or tone.

For example:

- Colour is usually the most effective way to enhance your drawings.
- Tone shading is used to represent the areas of light and dark on an object if you imagine it being illuminated by a light source coming from a specific direction.
- Texture can be created by lines or patterns that represent the surfaces of various materials, such as wood, metal, brick or glass.

9780170477499

Design options can be drawn by hand and rendered using markers, chalks, poster paints, watercolours, coloured paper, etc.; or with CAD – but remember to render, colour and annotate them.

Read how you can use the Design Elements and Principles to assist in creating aesthetic, creative and interesting designs later in this chapter.

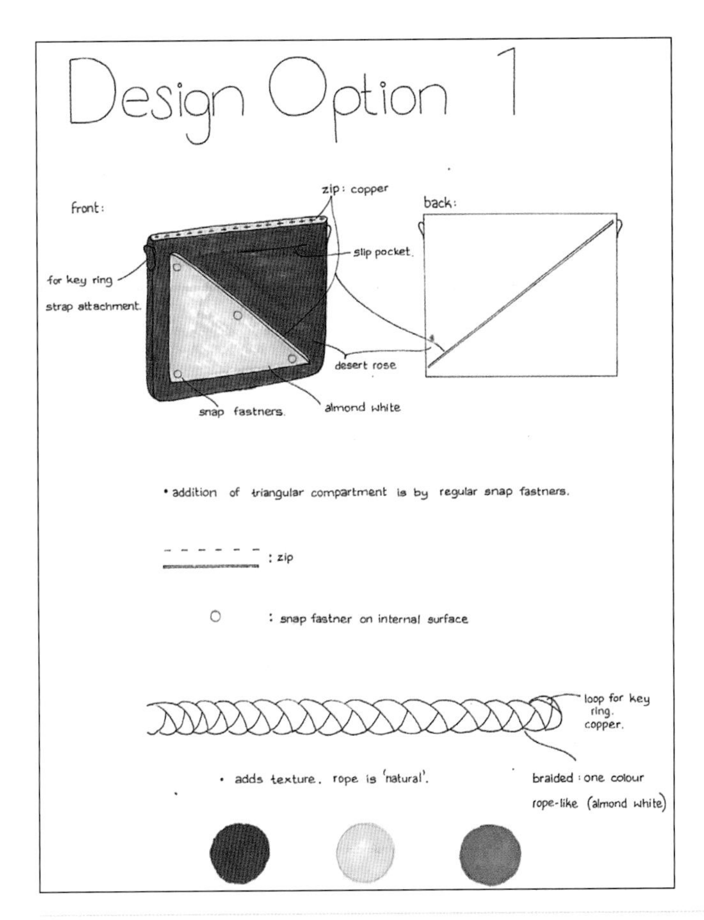

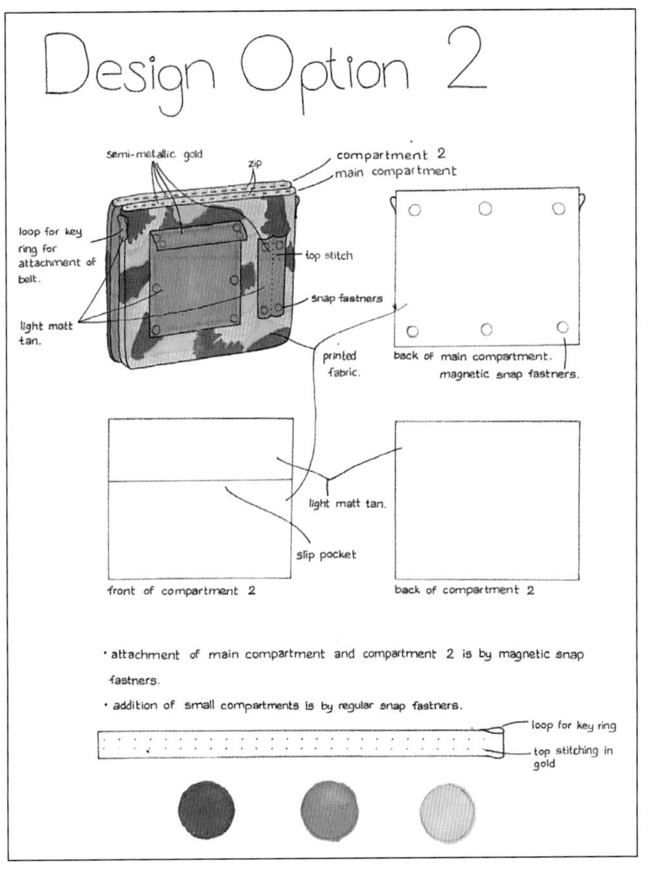

Two rendered design options, hand-drawn by student Sayuri Binaragama. Note that is not necessary to include the technical or working drawings on the same page as the design option.

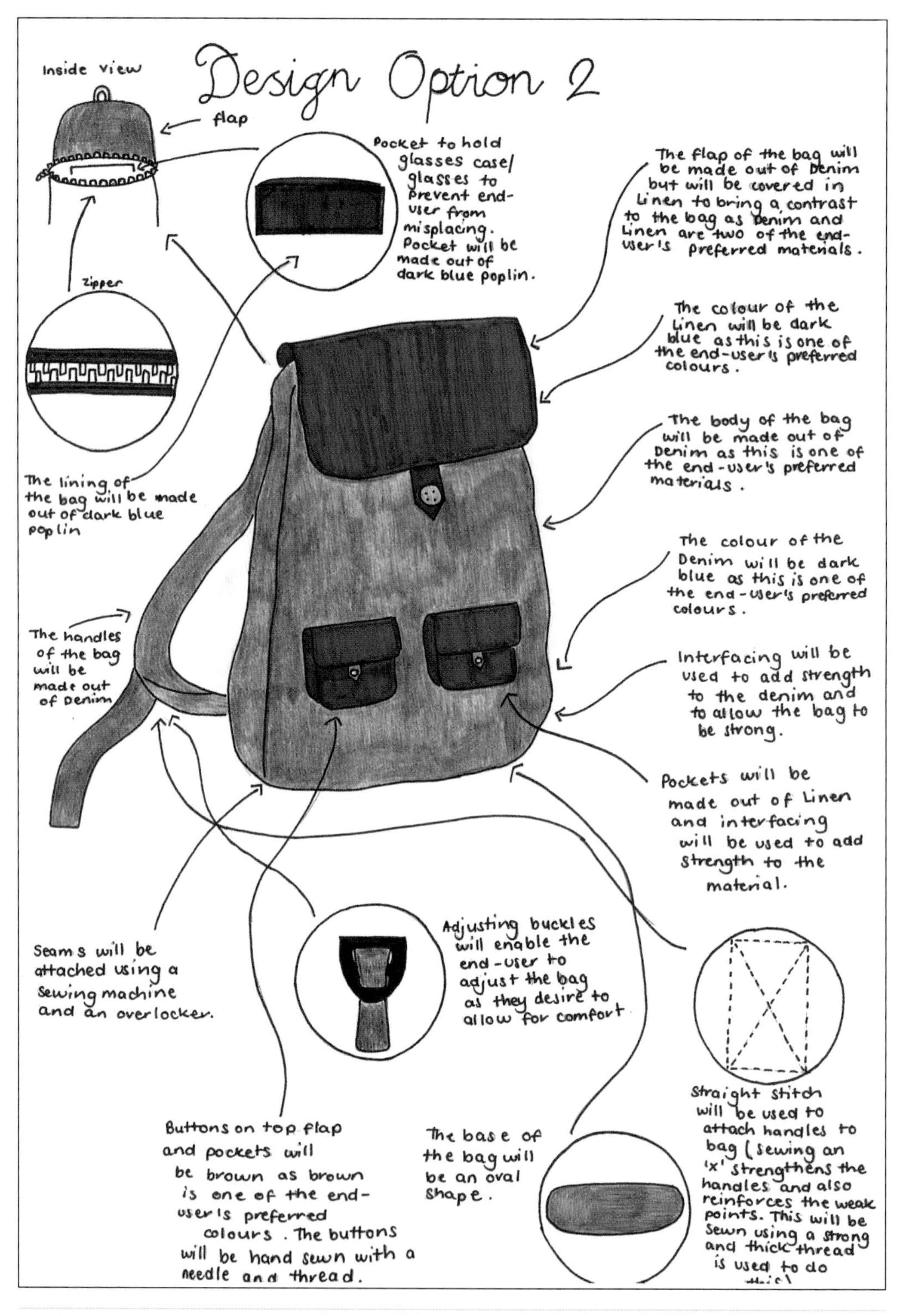

Rendered design option by student Sarah Hampton, showing 'exploded views' of details of the bag

9780170477499

Working drawings

Check the Assessment Criteria for the relevant year to see how your working drawings will be assessed for the SAT in Units 3 and 4.

For resistant materials

Orthogonal drawings

Orthogonal drawings are the most suitable working drawing for resistant materials (wood, metal, plastics). They must be clear, accurate, detailed and to scale, displaying construction detail and all relevant sizes (dimensions). 2D orthogonal drawings show several views as flat 2D shapes aligned with each other according to 'rules' or conventions of Standards Australia in AS 1100: Technical Drawing.

They can be done in CAD. If you are drawing by hand, all lines must be precise, and straight lines must be drawn with a ruler. 3D drawings can be added to aid communication.

You can find many examples and tutorials online to help with technical drawing.

Check with your Visual Communication Design teacher for 'rules' on orthogonal drawings.

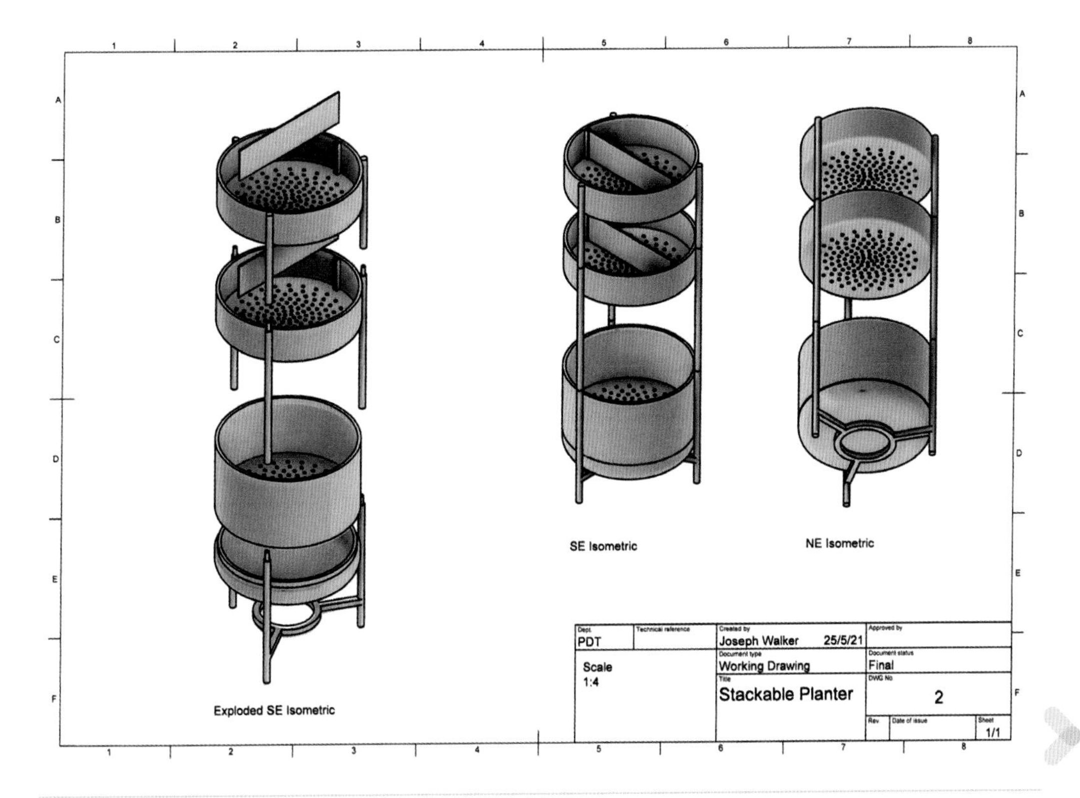

Clear 3D CAD images for a stackable planter by Joseph Walker provide functional information. See his finished dimensioned orthogonal drawing on the next page and his finished work in Chapter 11.

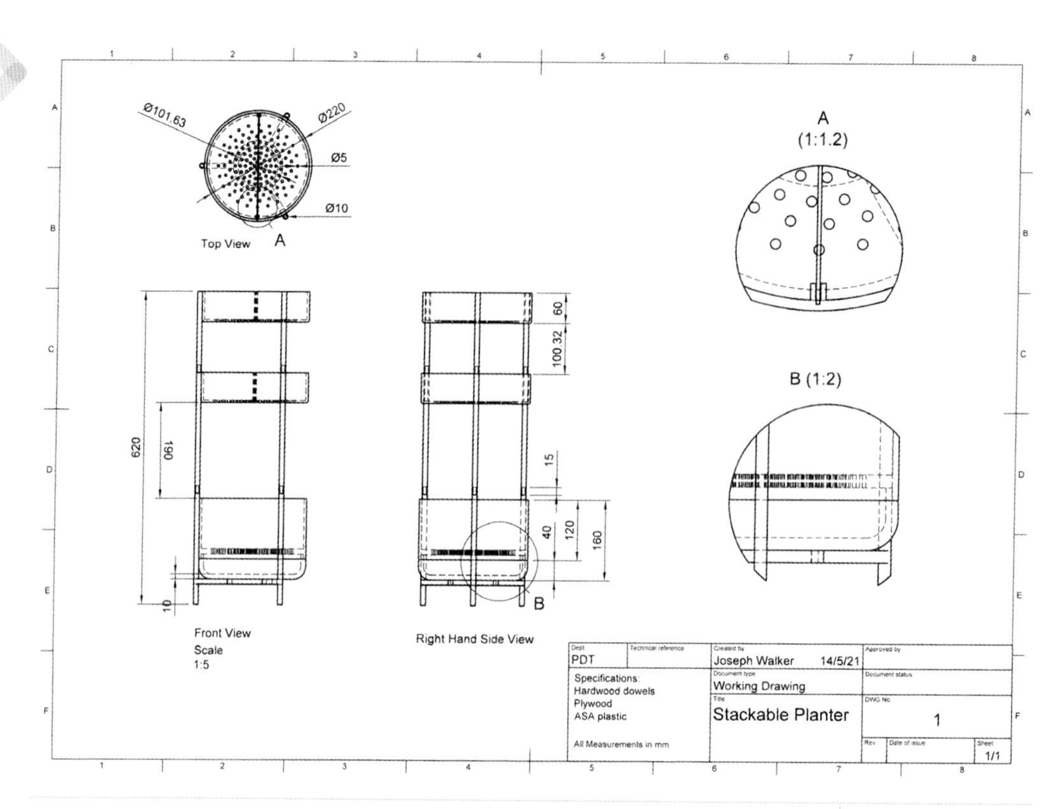

Technical drawings for a stackable planter, by student Joseph Walker

To purchase *Handbook HB1: 1994 Technical Drawing for Students*, go to the Sai Global website and use the search facility.

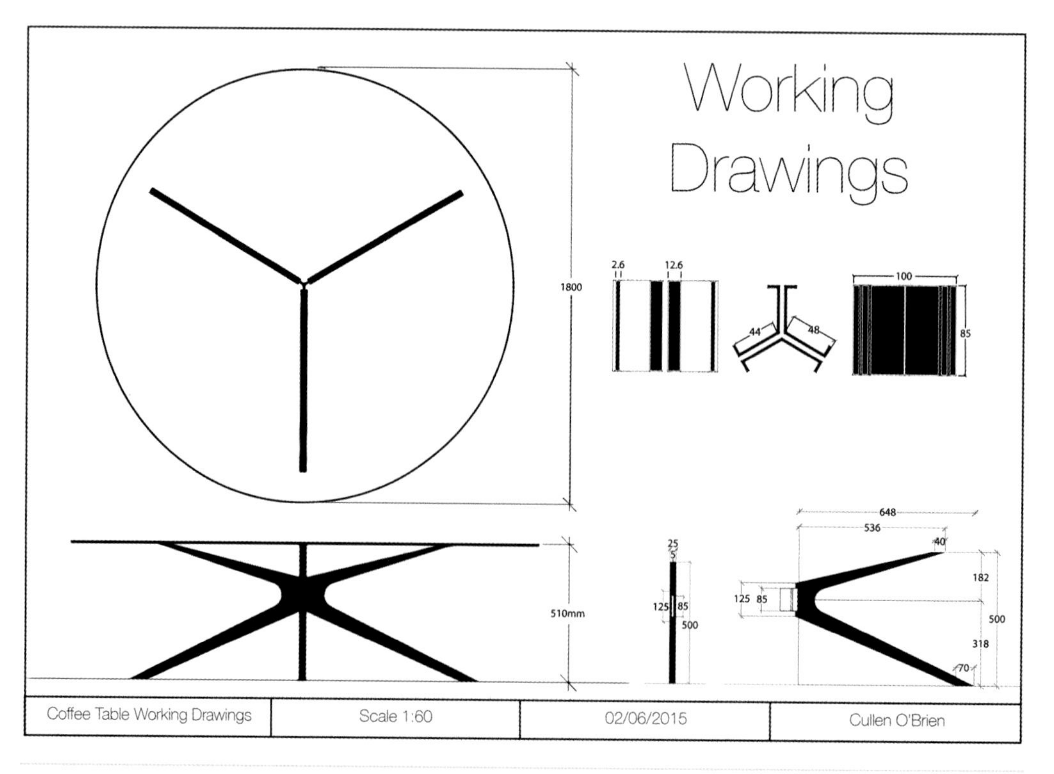

Orthogonal drawing for a coffee table, showing front, top and side views, by student Cullen O'Brien

9780170477499

Detail drawing or exploded view

A **detail drawing** is helpful when a particular section of the main drawing is too small or its details are too complex for accurate representation at the chosen scale. Enlargement (i.e. an exploded view) of that section may become necessary. On a scale drawing, details or exploded views are often drawn at five or 10 times the size of the original drawing, i.e. at a much larger scale.

Exploded drawings can also be drawn in the design option pictorial-style drawing. See the example on a design option for a denim bag by Sarah Hampton on page 334.

A **hidden detail drawing** indicates detail inside an object, or features that may be hidden. It is shown using a thin (0.25 mm) dashed line.

Section drawing

Another method often used to show internal detail is **section drawing**, or **sectioning**. Sectioning requires you to imagine that the object has been sliced open, revealing the internal structure and the thickness of 'walls' or parts. Section drawings are used a lot in architecture. These drawings can be very complicated, so it's important to check the 'rules' in the handbook.

Section drawing/ sectioning
A view of a slice through an object to portray its internal structure

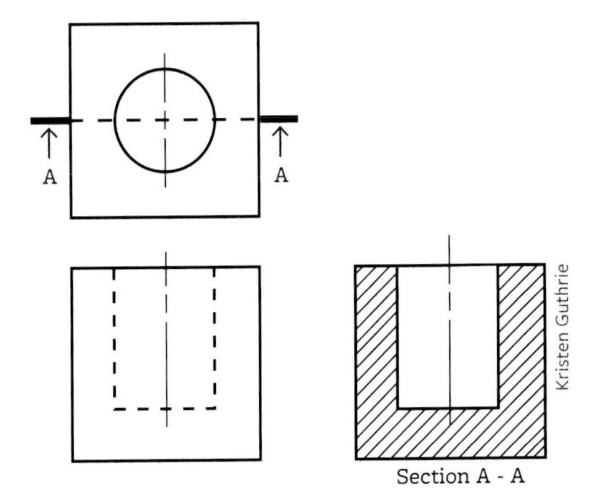

Dimensions

The various methods of indicating dimensions on your drawings are laid out in the Sai Global *Technical Drawing for Students* handbook (see page 336). Wherever possible, they are drawn outside the object.

Extension lines show the start and finish of the distance being dimensioned. They are drawn lightly and precisely. To distinguish them from the object itself, it's important that extension lines do not touch the edge of the object. An arrowed line, called a 'dimension line', is drawn from the start to the end between the extension lines, with the measurement written outside the dimension line. The dimension line is also faint and should be parallel with the edge of the object and drawn 10 millimetres away. Dimension and extension lines should be visibly lighter than the outlines of the object, which are thicker and bolder.

Extension lines
Light lines extending out from the object, defining the distance being dimensioned

Note that:

- all dimension lines require use of a ruler
- all dimensions should be written in the same direction and in millimetres, which must be indicated in the title block
- the shorter dimensions are usually placed closer to the object; the longer dimensions are drawn further away.

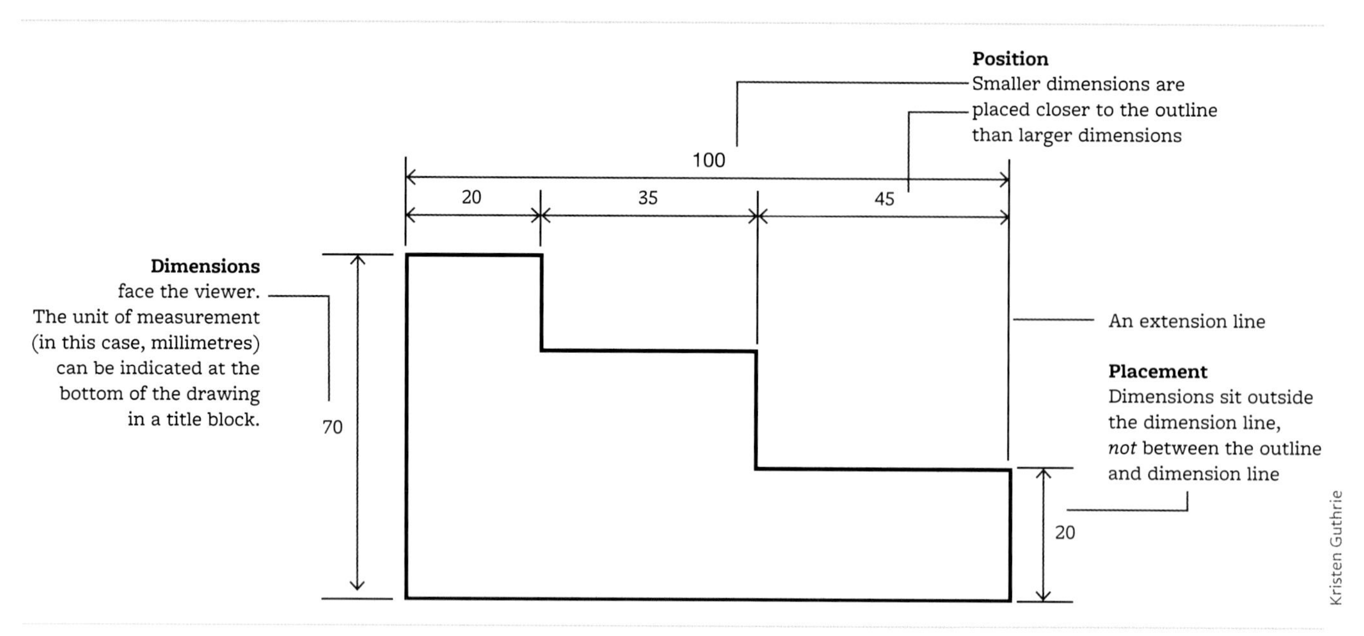

Indicating dimensions (in millimetres). Notice the extension lines that define the start and end points of each dimension.

For non-resistant materials

The most suitable working drawings for non-resistant materials (fabrics, fibres, yarns, etc.) are:

- 'flats' – to show the details of a clothing design as if laid flat
- construction drawings – to show how the pieces are put together.

Flats (trade sketches or technical drawings) show details in outlines, have a front and a back view and are off the body. They are drawn to scale and include measurements.

Designers use them alongside a fashion illustration on a specifications sheet or 'spec' sheet, which explains the details of a product to those who make it.

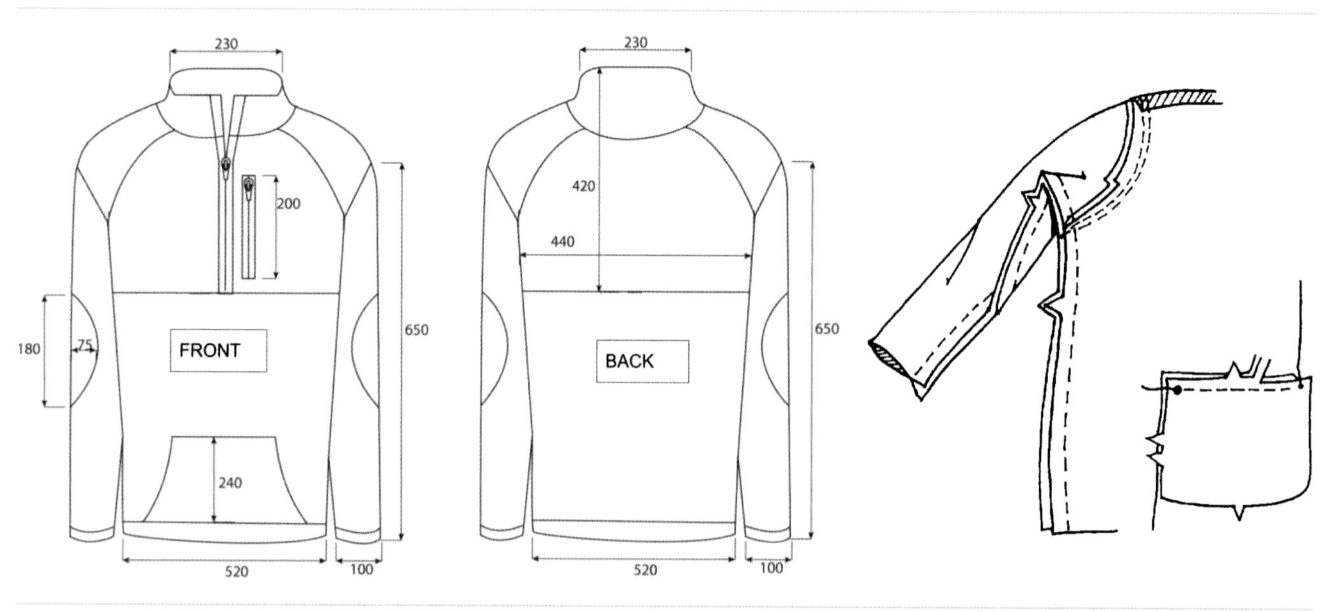

A 'flat' drawing of a sleeved top showing front and back (left), drawn by Unit 1 student Dylan Edwards using Illustrator; and construction drawings (right), showing seam and pocket details

9780170477499

Construction drawings are precise drawings of pattern pieces, or diagrams, that usually show:

- stitching, seams and the placement of zips, buttons, etc.
- enlarged details of difficult or tiny areas
- symbols for the alignment of pattern pieces or the ends of seams.

In construction drawings, the right side of the fabric is shown shaded, the wrong side of the fabric is shown white, and interfacing is shown hatched. Construction drawings are included as part of a sequence of production steps in a commercial pattern to provide instructions to the maker. These drawings may only be required to show modifications if you are working with a commercial pattern. Check with your teacher to be sure of how you will be assessed.

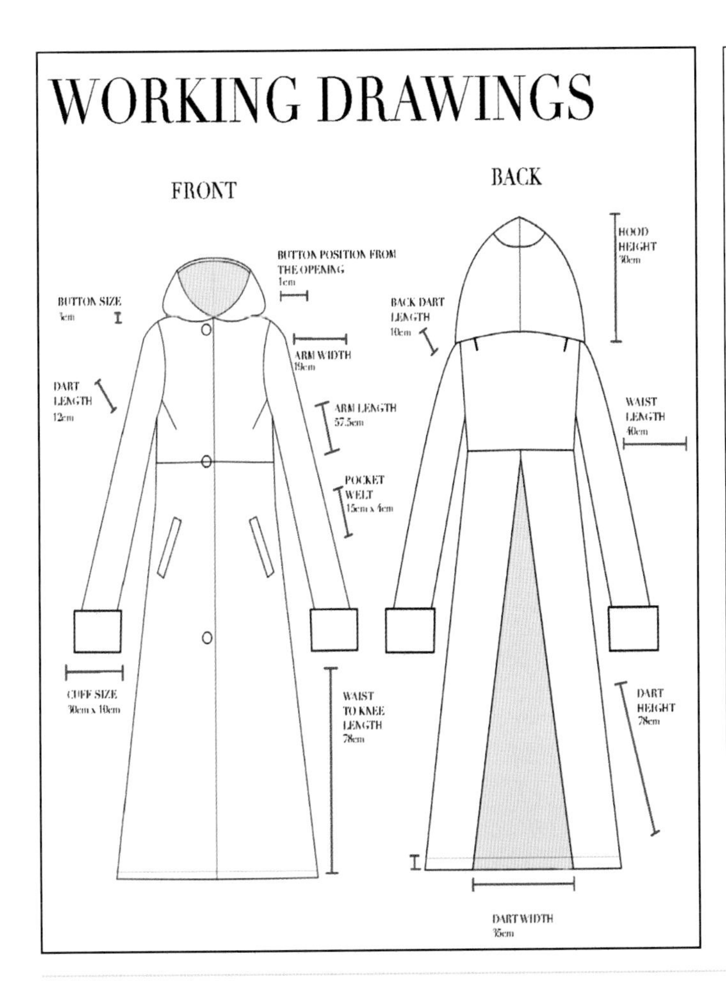

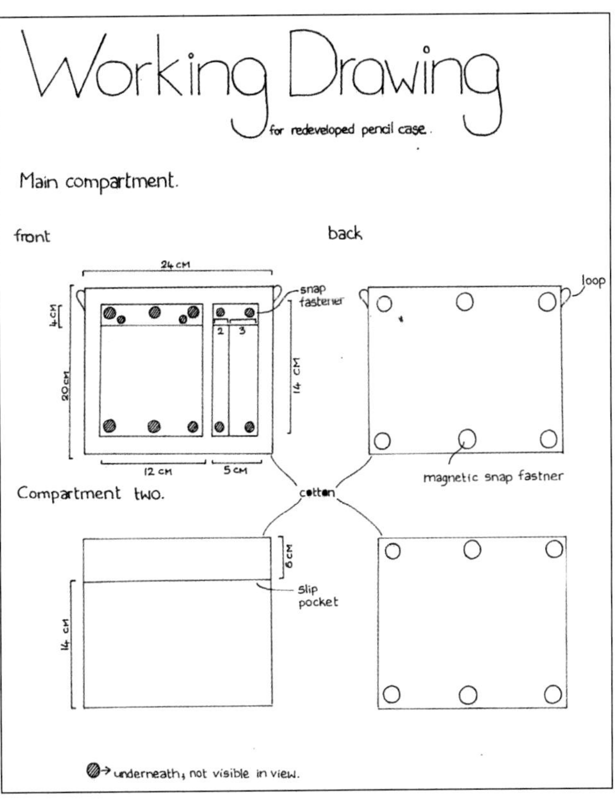

Flat technical drawing for a coat (left) drawn in Illustrator by student Izzy Donat; technical drawing for a pencil case (right) drawn by hand by Unit 1 student Sayuri Binagarama

Modifying a commercial pattern

At VCE level you are permitted to use commercial patterns, but it is expected that you will acknowledge them in your folio and adapt them to reflect your selected design option and to fit the end user. For resistant products, changes can be drawn onto the 'plans' and photographed. If you obtain digital plans, your adjustments and changes should be annotated. Commercial patterns for clothing may be altered by folding, cutting and/or extending pieces to change the garment shape. These can be drawn as pattern modifications that show exactly how the pieces will be altered.

You will also need to consider the layout (the placement of pattern pieces on the fabric) to minimise waste through the careful arrangement of pattern pieces.

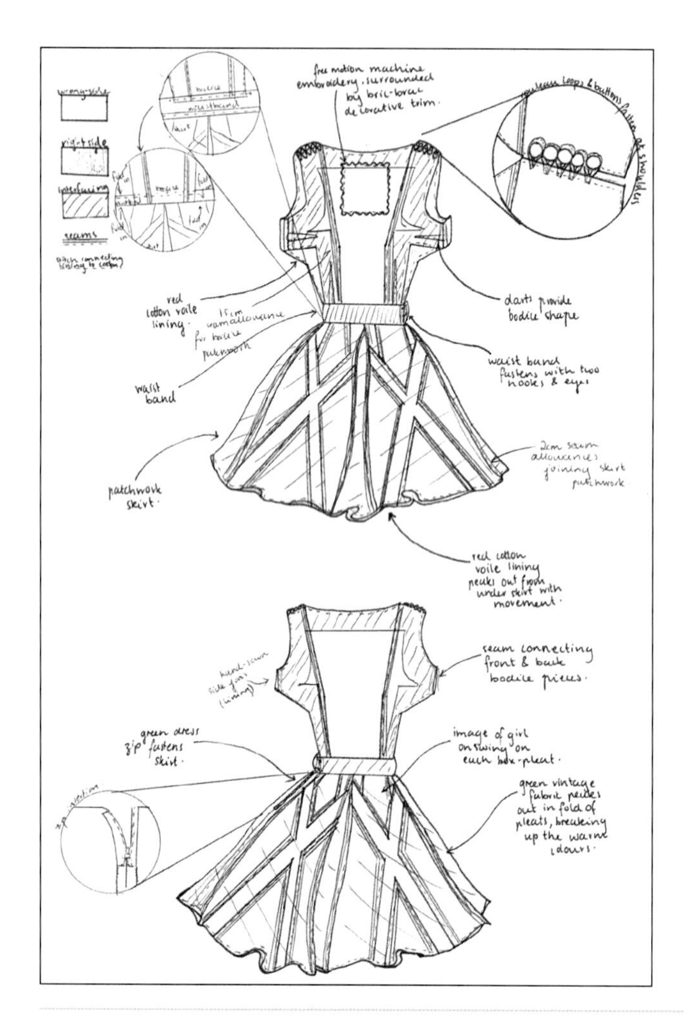

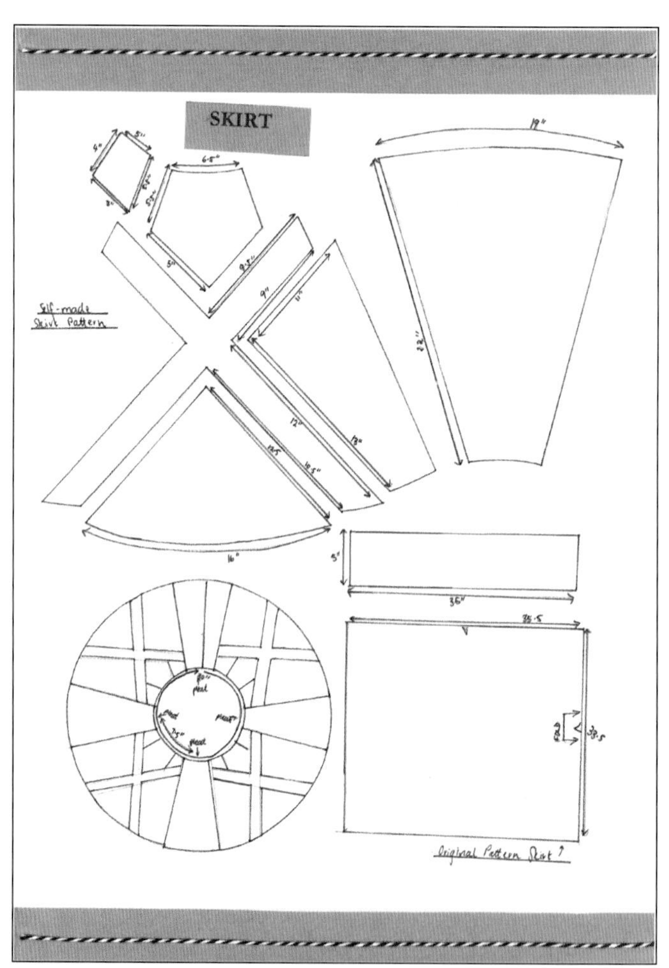

Hand-drawn pattern drafting and adjustments for a dress with circle skirt, by student Laura McKenzie

9780170477499

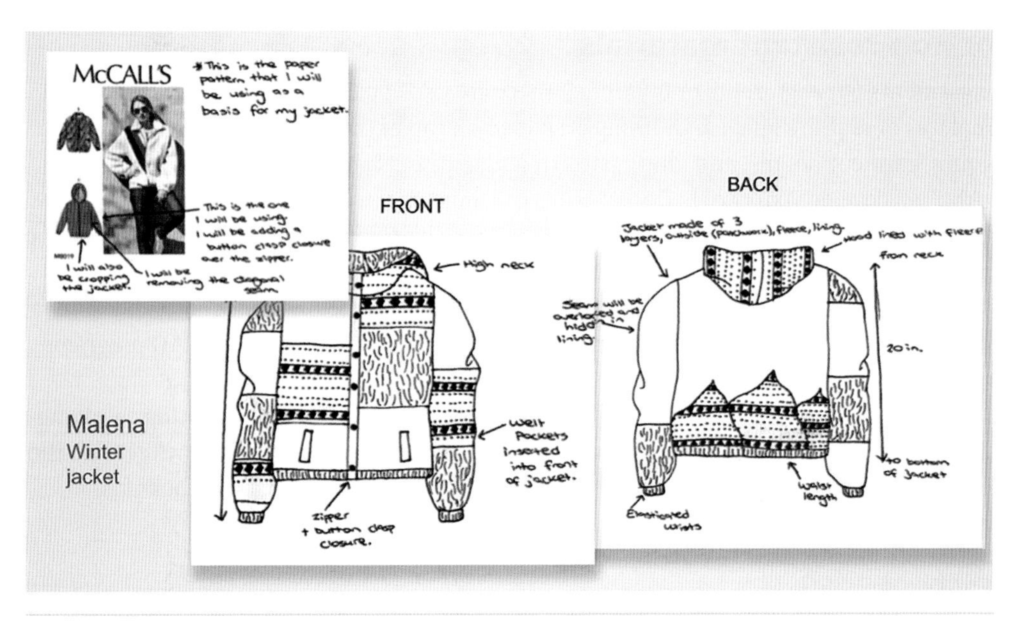

Pattern adjustments and acknowledgment by Malena Lopez, with a hand-drawn working drawing

You will find many 'flat' sketches, and tutorials for adapting them in Adobe Illustrator, at the Designers Nexus website.

Weblink
Designers Nexus

Computer-aided design (CAD)

You can use CAD at any stage in your design process. It can be used for visualisations, design options, working drawings and for 'virtual product concepts', which are digital prototypes. You can use CAD to explore colour combinations and textures along with many other Design Elements and Principles.

For resistant materials

Computer-aided design (CAD) programs are designed to handle objects in 2D and 3D forms. In many cases, this capability has removed the need for physical 3D modelling, particularly in the early stages of design refinement. Structural changes can be made, and surfaces and textures are clearly represented, without committing time or materials. See the student example by Nathan Colbert on page 330.

CAD can be used to create 3D rendered drawings for your design options and 2D (or 3D to aid communication) dimensioned drawings as working drawings. These can also be viewed as 'virtual or digital prototypes'.

The use of CAD is standard practice in industry because of its cost-effectiveness and additional benefits, which include the following:

- Drawings can be easily repeated and errors can be fixed easily.
- Complex shapes can be produced, modelled, show textures and colours and be viewed in both 2D and 3D.
- Production and lead times are reduced.
- It is easily integrated with computer-aided manufacturing (CAM) and/or a 3D printer in your classroom.

It takes many hours of practice for a designer to be able to use a CAD program effectively. Develop your CAD skills using software available in your school, or download and experiment with free drawing software, such as Google SketchUp.

For non-resistant materials

For garments (or bags, tents or toys, etc.) there are CAD programs that store a series of body templates or real-life images. Clothing is shaped on top of these templates. Style, silhouette, colours, garment details, and fabric pattern and textures can be easily manipulated.

You can develop your own body templates in a drawing software package by digitally drawing or scanning a body shape into a program. Overlay the templates with garment shapes and experiment by filling them with patterns, colours and textures (either drawn or scanned).

Search online for tutorials on how to create 'flats' using Adobe Illustrator and other software. Patterns, weaves, knits, textures and prints for fabrics can be planned using CAD programs.

A useful website is the Designers Nexus.

In industry, information about fabric designs is sent directly to printing, weaving or knitting machines via computer. Details can be swiftly changed, and the effects of those modifications can be seen immediately on computer-generated garment mock-ups.

Weblink
Designers Nexus

Patterns, sizing and layouts

In industry, body measurement data is entered into pattern-making CAD programs, or a bank of standard measurements is used to 'grade' patterns. Grading is the process of creating pattern pieces for a garment in numerous sizes.

For examples of how designers use CAD in garment design, see the case study on Rip Curl on page 57.

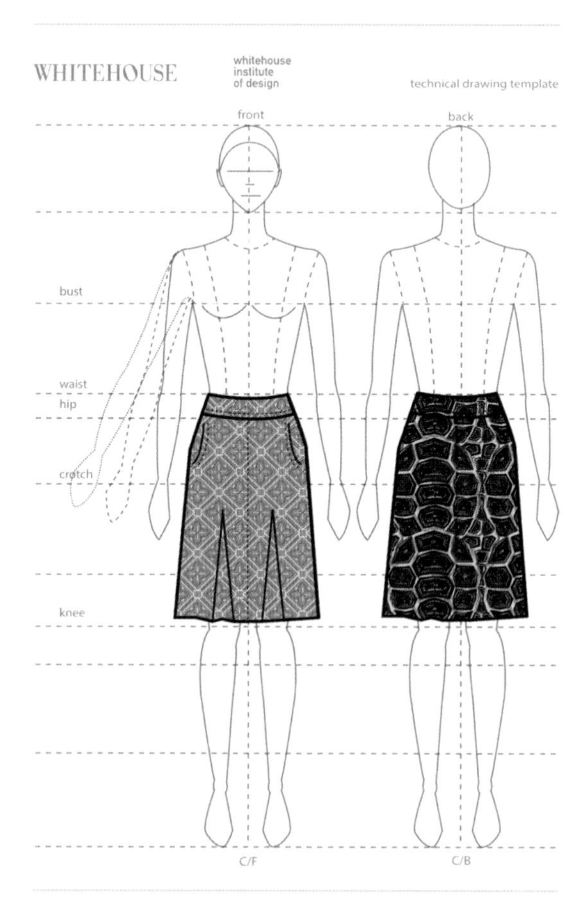

An example of a 'flat' drawing of a skirt with different renderings, done in Adobe Illustrator. Notice the alignment of back and front views.

Worksheet
12.3.1

Use of models and prototypes

Rowena Reed Kostellow (1900–1988) was an influential designer and educator from the USA who believed that three-dimensional designs should be sketched three-dimensionally, not two-dimensionally. Her 'drawing' tools were clay, cardboard, wire, glue and so forth, not paper, pencils and markers. She also believed space to be an important element in 3D design. You can read a lot more about her methodology and the history of the fund created in her name at the tabs across the top of the Rowena Reed Kostellow Fund's home page.

Weblinks
The Rowena Reed Kostellow Fund

Modelling

Using physical 3D models is a very effective way of visualising, communicating and presenting your ideas to others. Designers often use a model in combination with drawings when presenting initial ideas to clients or end users. Changes are easily made if required, and only a small amount of time and resources (usually less valuable) have been committed to their production.

Your first models can be rough to help you explore shape, colour and proportion and gain some initial end user feedback. They can be used to explore and critique ideas and to try out functional aspects with ongoing end user feedback.

Mock-ups or 3D models can be constructed:

- to show the outward appearance and external details only
- from cardboard, craft foam or clay
- from aluminium, acrylic and plywood for more realistic models
- with paper or fabric and glue to try out proportions or combinations.

Later versions of models or prototypes can be of higher quality to give a clearer representation of size, proportion and form as well as texture, surface finish, colour and pattern. From these you will decide on a chosen product concept to become your 'final proof of concept'.

Be sure to photograph and annotate this work to add to your folio. See more in each of the unit chapters 1, 3 and 6.

Student Justin Cauchi building a model on a scaled working drawing

Prototyping

Prototypes are 3D full-scale versions of a product used to test its performance before production of multiples. They are used in mass production to check all external and internal details of a design before production. The prototype is tested, checked and adjusted, and the process is repeated. This is called iteration (repetition). The final prototype is made to the exact specifications of the intended finished product (or parts). From a prototype, a designer can tell:

- whether the design will function as hoped
- how it feels when used
- if there is a need for further development
- the location of potential production problems.

See more suggestions on pages 46–7 in Chapter 2 (Unit 1) and student examples in Chapter 11.

3D printing

Some schools may have access to rapid prototyping equipment for computer-controlled milling (subtracting from a solid block), 3D printing or laser sintering (using polymer powder that is melted with heat from a laser). To create physical objects, CAD files are needed. 3D printing is a great way to check out proportions of the whole product or test dimensions of parts and how they fit together. It can also help gauge functional and aesthetic aspects so details of the design can be finalised before production.

Shutterstock.com/MarinaGrigorivna

These objects were made using 3D printing.

Modelling for non-resistant materials

For most fabric products (other than clothing), you could create versions in paper or fabric offcuts. You could also create miniature bedroom or lounge furniture to trial furnishings for shape, colour combinations and fabric.

9780170477499

The most commonly used form of finalising a garment design is to construct a toile (pronounced twahl), which is a full-sized version of the finished garment made out of an inexpensive fabric, such as calico. The toile can be fitted, checked and altered, and details can be changed, while it is on either a mannequin or a person. Some designers begin by folding and draping fabric directly onto the body form and using this as a stimulus to create their designs (called French draping).

Paper pinned onto a mannequin to check placement, proportions and bulk, by student Michelle Dong

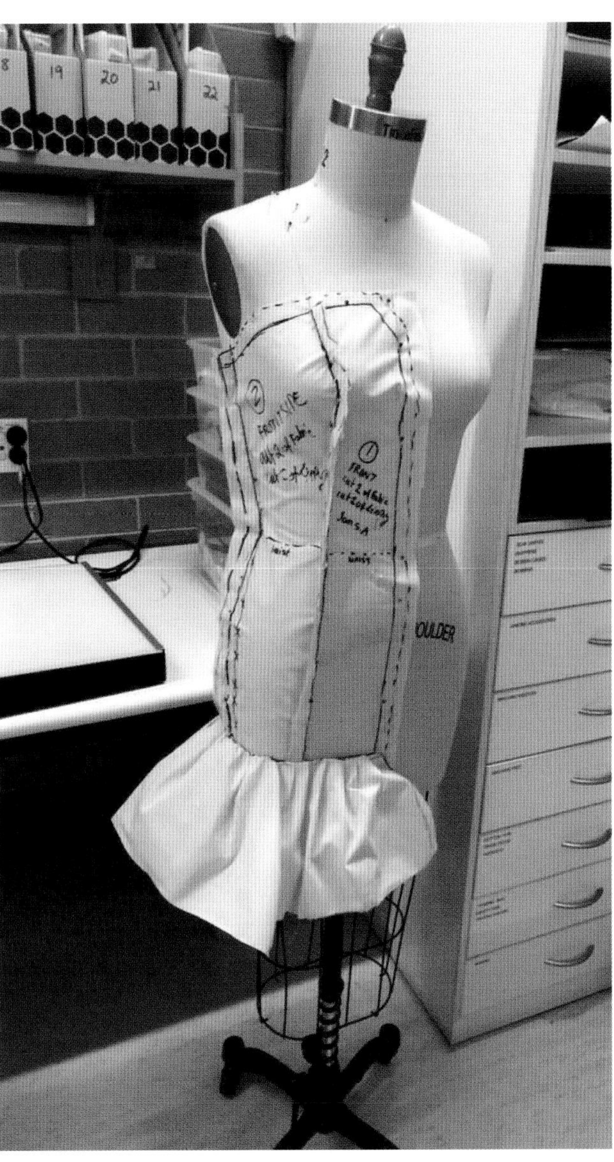

A half toile of calico fabric pinned onto a mannequin with markings for adjustments in black pen, by student Laura Mazzarella

Other ways of trialling/modelling clothing are to:

- drape a small artist's mannequin with different fabrics, of different proportions or lengths, using pins or paper clips, to gain a 3D sense of the finished design
- create models from cut and folded paper
- cut out the negative space of a garment silhouette from thick card and place it over different fabric samples, to get an idea of how the fabric might suit the style, or to check different fabric combinations. Take photos to add to your folio.

Design aesthetics: the Design Elements and Principles

Point
In product design, a particular place or spot that the eye is drawn to that is isolated or by itself

Line
A mark from one point to another

The Design Elements and Principles are the fundamentals that give designs their aesthetic:

- The Design Elements are the building blocks of design. They include **point**, **line**, shape, form, texture, colour, tone, translucency, transparency and opacity.
- The Design Principles are used to structure and order the elements. They include proportion and balance, contrast, symmetry and asymmetry, pattern, movement/rhythm, repetition, positive and negative space and surface qualities.

The Design Elements and Principles are interlinked. The Design Principles refer to the way in which the Design Elements are organised. For example, a discussion of the Design Principle 'pattern' would need to include descriptions of the Design Elements of shape, line and colour used.

Design Elements

Lines

Case study
Jacki Staude

Lines may be thick or thin, straight or curved; they may zigzag or spiral; they may be interrupted or they may flow through the product. Lines may also be used as part of a surface pattern. Designers can use lines within their product to decorate, to emphasise a shape or to create movement and draw the eye around the product or sweep it in a desired direction.

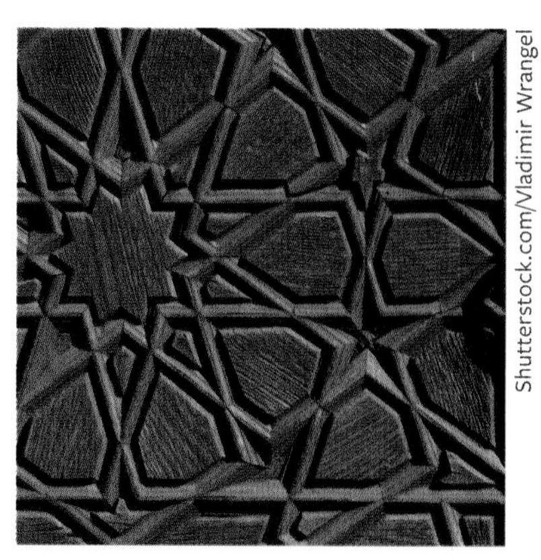

Shutterstock.com/Vladimir Wrangel

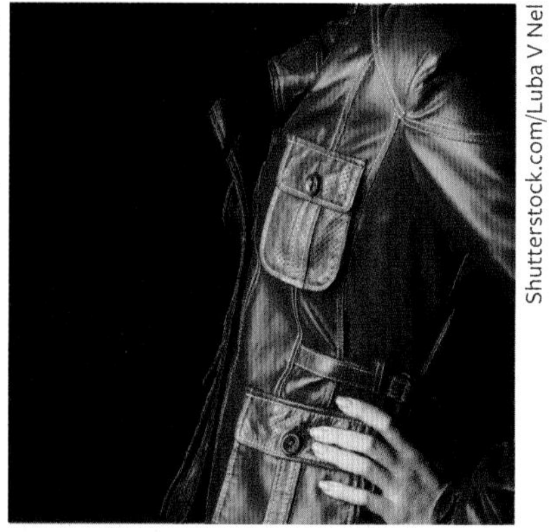

Shutterstock.com/Luba V Nel

Examples of the use of line: geometric lines in carved wood, and top stitching on a leather jacket

9780170477499

Lines in clothing can be formed by seams, the top stitching and folding of fabric and other aspects of the construction. These lines can define the shapes within the garment.

Vertical lines make the body look longer, taller and slimmer, and horizontal lines do the opposite.

Lines can also give character to a garment. Straight lines and clear angles give a sense of structure and order. Curved lines are connected with grace and sensuality. Lots of visual lines can create clutter and distract the eye, while fewer lines or regular, repeating lines create a sense of the garment being streamlined.

Martin Grant, Paris (fashion house), *Look 15*, coat, 2003, autumn-winter 2003–04, Pending gift of the artist, 2023, Courtesy of the National Gallery of Victoria

Top-stitched lines emphasise the simple form of the *Stitch Coat* by Martin Grant.

Shape

Shape is defined as a line meeting itself in two dimensions (2D). A shape is therefore flat.

Shapes can be geometric and regular – e.g. squares, rectangles, triangles and circles – or organic and irregular, like amorphous, squiggly, curvy, plant or animal shapes. Shapes can also be simple or complex.

Shape is one of the first things you notice about a product. It is a Design Element that helps you to visually distinguish one design from another.

Some of the most satisfying shapes for design are based on an understanding and use of both geometric and organic shapes together. A designer can find the simplest geometric basis of an organic shape – e.g. the elegant spiral of a nautilus shell, or the balanced curves of a tulip bud – or they can contrast the rigid stability of some geometric shapes with the flowing unpredictability of organic curves.

Shutterstock.com/Jorge Moro

Shutterstock.com/yod 67

Shutterstock.com/atiger

Organic shapes can be extracted from 3D forms seen in nature.

Form (from shape)

Form is defined as three-dimensional, i.e. not just a 2D shape, but a solid. Form is the overall structure that we refer to in a 3D product.

Forms can be bulky, simple, complex, solid, light, delicate, short, wide, rounded, spindly, spherical, conical, cubic, chunky, convex, concave or bulbous.

When designing a product on a page, 2D shapes need to be turned into 3D forms, such as cubes, cones, spheres, etc. A simple shape or form appeals to some people, but others are attracted by complex combinations of different shapes or forms.

Visual bulk in form can be reduced by the materials used. For example, a table that has a steel frame and a transparent glass table-top tends to have less overall visual bulk because it is delicate and transparent. A larger, thicker opaque top and chunky legs on a timber table give a stronger, more solid appearance.

Using different fabrics in a garment also has a very significant effect on visual bulk. This is partly to do with the transparency or opacity of the fabric, but is also affected by how the fabric falls – its draping quality. For example, a shirt made from a thick, stiff, opaque cotton drill would have a very different visual presence to another shirt made from a flimsy silk organza.

ACTIVITY

Worksheet 12.5.1

Look at the two chairs pictured and discuss words you could use to describe their different forms.

Getty Images/Eric Piermont/AFP

Eileen Gray's *Bibendum* chair (1920s)

Philippe Starck

W.W. Stool (1990) by Philippe Starck

9780170477499

Colour

Colour is a very basic element of design, and many art and design books will explain colour theory in depth. When discussing colour, we talk about:

COLOUR wheel

Where colours are placed in sections of a circle to show their relationship to each other

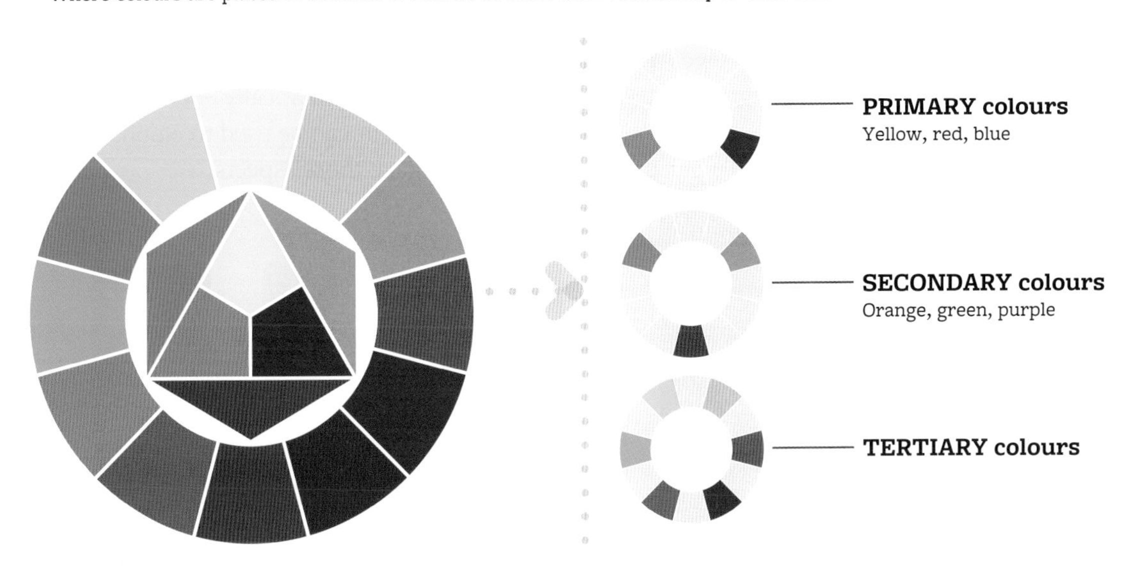

Complementary colours

Colours that are opposite each other on the colour wheel; highly contrasting, but often dynamic when used in design

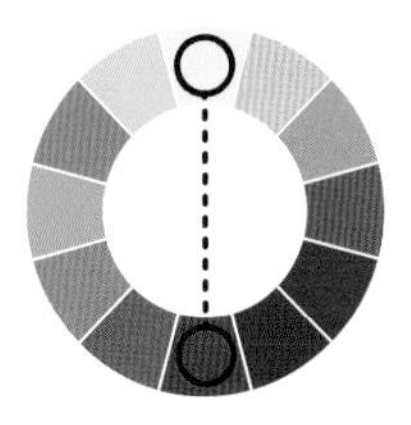

Analogous colours

Colours that sit next to each other on the colour wheel, and are very soothing when combined (blue and purple, yellow and green, etc.)

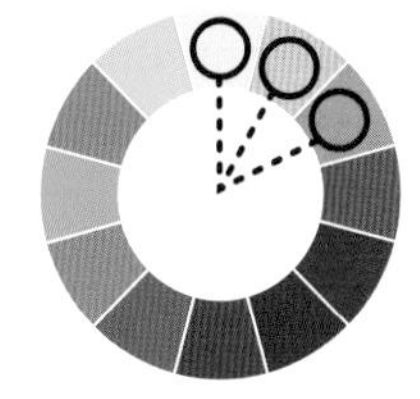

Warm colours

Warm – reds, oranges and some yellows

Cool – blues, greens, some purples, and some metallic colours

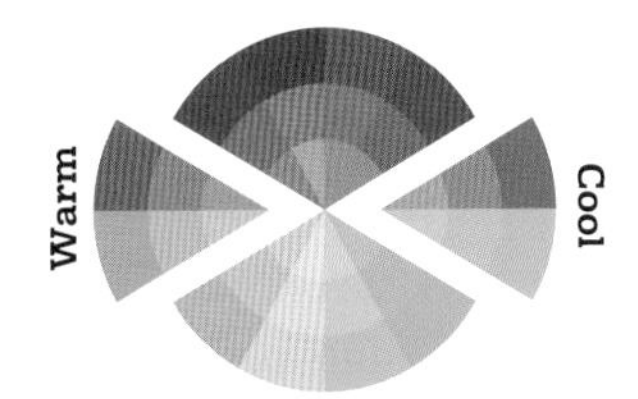

Metallic colours

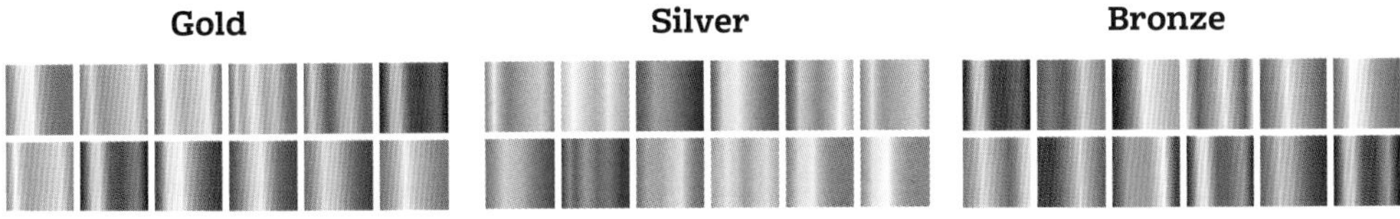

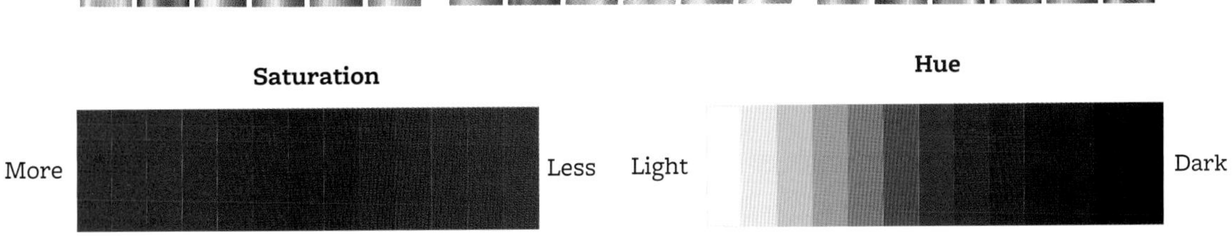

Terms such as hue, tint and shade describe the subtleties of colour mixing and the addition of black and white to make colours lighter and darker.

Colours create mood and evoke emotion, and can transport us to different places (spicy rich reds, oranges and purples may make us think of the heat of India or Morocco, while cool aqua and pale creams/yellow may evoke a holiday on the beach). Some colours are categorised and given a group or theme name, e.g. reflecting different seasons – sombre, dark colours for winter; fresh, bright, 'living' colours for spring; bright, strong colours that reflect the sun and beach colours for summer; and rich, warm colours for autumn.

The suitability of a colour depends entirely on the overall image and purpose of the product being designed. Fluorescent colours are often used for high visibility and project a very different image to beige or pastels. Colour can be used subtly – in light washes, harmonising and blending shades. Or colour can be used dramatically – creating visual tension by combining strong and contrasting colours. Colour can be used to emphasise the shape, proportion and details of a product, or to camouflage those aspects.

Shutterstock.com/Gina Smith

Shutterstock.com/archideaphoto

Examples of vivid colours in clothing, and the use of 'colour blocking' in the chairs

'Colour blocking' is a term used to describe the use of flat, plain colours for the whole or sections (blocks or panels) of a garment. Endless possibilities of colours are available in fabrics. However, in wood, metal, glass and plastic, discussions of colour are driven by the particular material being used. Metal and timber have a limited range of natural colours. However, there are many methods to achieve colour through treatments and finishes – with paints or varnishes, or by treating and adding chemicals. Glass and plastic have a very broad colour range. Sometimes it is important to research the available colours of a material before deciding on an exact colour or hue.

Alamy Stock Photo/WENN Rights Ltd

The Memphis-style *Carlton* bookcase, by Ettore Sosstass

9780170477499

Look at Memphis-style furniture and the work of designers such as Rietveld (see page 321 in this chapter) to see how colours and shape can be used to create strong impact.
Look at other designs throughout this book, and name and describe as many colours as you can see, how they have been combined or contrasted and the feelings they evoke. What combinations and/or contrasting colours could you explore in your own designs that will appeal to your end users?

Shutterstock.com/Natalia Mikhaylova

Striking use of spicy colours and various tones in a Manish Arora collection

Tone

Colour is affected by tone, which refers to how light or dark an object or a colour is. Contrasts or differences in tonal value provide interest and depth in product design. Dark and light tones are often placed next to each other to provide a striking contrast.

Transparency, translucency and opacity

Colour is affected by the degree of transparency of the material. Materials can be transparent (see-through), translucent (allows light through and is partly see-through, or cloudy) or opaque (completely solid; no light passes through). If a material is transparent or translucent, its colour will be affected by the colours behind it. This can be used to great effect – think of layering gauze-like fabrics, or overlapping sections of tinted plastic or glass. French designer Philippe Starck used the quality of transparency dramatically in his *Ghost* series of designs. Dinosaur Designs use translucent materials to give a smoky lightness to their products.

Look at the images of the works on the next page and identify which materials are transparent, translucent and opaque.

Alamy Stock Photo/Avalon.red

Philippe Starck's *Louis Ghost* chair and table

Dinosaur Designs

Dinosaur Designs jewellery, bangles and tableware in resins

Texture and surface qualities

Texture and surface qualities relate to both the look (visual aspects) of a material and its feel (sensual aspects). The sensual cues to texture might be that the surface is rough, smooth, furry, corrugated or slippery. Texture can be represented visually in a drawing by rendering (see page 332) or by gluing actual materials such as paper or fabric onto the drawing.

Sheen
How light is reflected by a surface

When talking about the visual aspect of texture, we often refer to the **sheen** of a material, its finish, or a pattern that is applied to or is part of a material. For example, the heavily cabled pattern of a knitted fabric will give it a variegated texture created by 3D elements and shadows, and a finely etched pattern in metal has a bumpy surface. Patterns also interrupt the play of light on the surface of materials.

Freekpik.com/elements

Shutterstock.com/Oksa

Examples of the use of texture: knitted fabric and textured metal

Fabric texture can be rough, grainy and stiff; soft and fluffy; or it can be smooth and shiny. The texture of a fabric is governed by the way it is constructed (the density and pattern of weaving or knitting), the fibre or yarn it is constructed from (fine or coarse, smoothly spun or knobbly) and whether the fibre or yarn is natural or synthetic. A change to the surface might also give the fabric texture (e.g. corduroy, embroidered, pressed, creased or burnt patterns). Some people talk of the sound of texture – the swish of silk taffeta, or the 'scritching' of a raincoat as you walk.

9780170477499

Corduroy

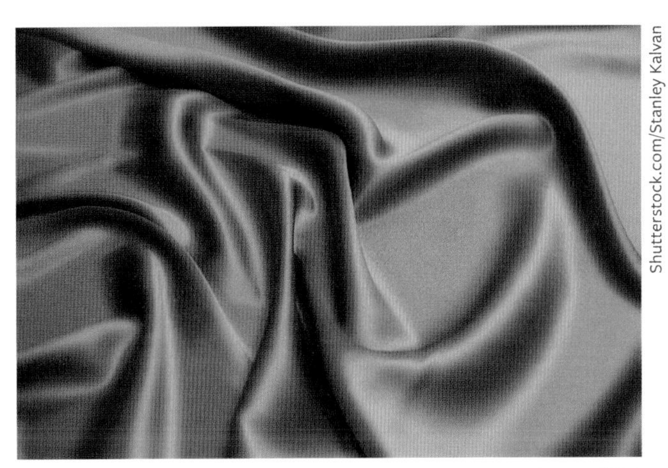

Silk

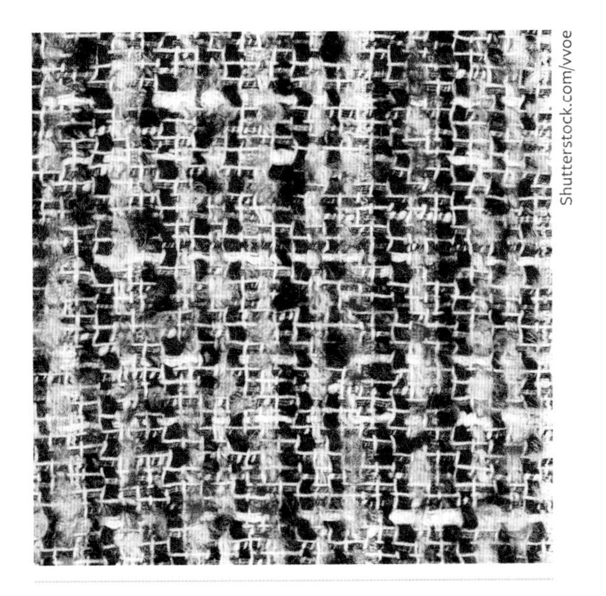

Boucle

See more on texture and patterns on pages 357–60.

Design Principles

The Design Principles describe the different ways that Design Elements are used in a product – how they interact with each other and are combined, or how they are used to provide contrast, drama and style.

Proportion and shape

When working with timber, metal and plastic, it is very easy to limit the shapes used in design to squares and rectangles and to design with standard or normal proportions. However, this limits design choices to a reduced range of geometric shapes and how they are proportioned. If organic shapes are considered, interesting interactions between geometric and organic are possible. Enlarging or exaggerating shapes within a design is a way to change the proportions.

Worksheet
12.6.1

ACTIVITY

Look at the designs here and throughout this book, and name and describe as many shapes as you can see and the proportions in which they are formed. What shapes and proportions can you use in your own designs?

Alamy Stock Photo/Chicade

An interesting interaction between geometric and organic: the Eames *La Chaise* chair

When discussing fashion and clothing, shape can refer to two things:

- the overall or outline shape of the garment, outfit or accessory – its silhouette
- shapes that are used as components of the garment or object (e.g. a sleeve or collar) and that have an impact on the proportions, or that are part of the fabric's graphic design (e.g. as part of a printed pattern).

A silhouette can be described, for example, as tubular, boxy, triangular (wider at the bottom) or hourglass.

The silhouette is affected by the shaping of the waist, the shoulder line and the fullness of the skirt or pants, etc. We often think of fashions of a particular era as using a specific silhouette shape – e.g. the tube-like flapper dress of the 1920s, the full-skirted, bell-shaped ball gowns of the 1950s, and the wedge created by the padded shoulders of 1980s power dressing.

Accessory designers also use shape and proportion in specific ways.

Jacinta O'Leary

Shutterstock.com/Faroe

Simple white shapes to create form and use of line to emphasise the different proportions in a Gucci handbag (left). Men's clothing (right) is often based on boxy, rectangular shapes.

9780170477499

Proportion and balance

Proportion can refer to two related aspects of a design:

- the relationship between the overall dimensions of an object – the length, the width and the thickness of the form
- the relationship between the size of, and space between, the various parts of a design.

Shape and form interact with proportion to give an overall feeling or mood of a product. Short, wide shapes – such as broad-based rectangles or triangles – suggest the concept of solidity, as seen in some men's clothing. Using similar dimensions for component parts can make a product seem balanced or static. Some people find these related proportions satisfying – there is a sense that the object isn't going to fall over easily and is well grounded. Tall, thin shapes tend to indicate delicacy and tension. Some people find these proportions aesthetic and satisfying.

The form a design finally takes is a combination of the designer's preference and the functions the product must perform. Carefully considered proportions create a sense of balance between different aspects or parts of a product.

Many modern designers exaggerate some of the proportions of their designs for the effect that this produces. They might include a section that flies in a particular direction without seeming to be supported, which focuses the viewer's gaze on a particular feature. It also gives the sense of breaking away from expected or conventional forms.

Search for the work of the American industrial design teacher Rowena Reed Kostellow, who developed a very structured process of learning how to use 3D shapes to give visual structure to an object.

When thinking about proportions, it is important to consider the 3D forms that make up an object and how they relate to each other. If they are the same shape and size, this might be boring and repetitive. However, if the forms are of different sizes but with similar length-to-width ratios, this might give a sense of unity while being interesting.

Weblink
The Rowena Reed Kostellow Fund

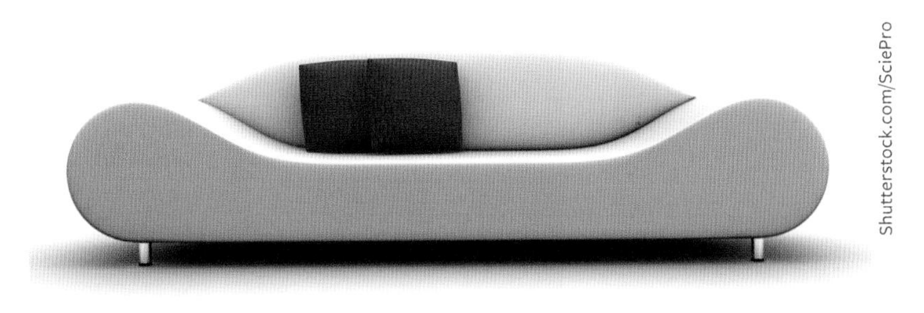

Shutterstock.com/SciePro

Creating an effect with symmetrical, exaggerated proportions and rounded forms

The golden mean

A commonly used series of proportions is based on the 'golden mean', a ratio that was used frequently in Classical Greek architecture. The golden mean relates to a mathematical series that was later defined by the medieval Italian mathematician Fibonacci; successive numbers in the series are in a ratio of 1 is to 1.618. It has been suggested that this numerical relationship also occurs regularly in natural objects.

Find out more about the golden mean in Michael J. Ostwald's article 'The Golden Mean: A great discovery or natural phenomenon' on *The Conversation* website.

Weblink
The Golden Mean

Symmetry, asymmetry and balance

Symmetry is very closely related to balance. When an object is symmetrical, you can place a line through the centre and one side is the mirror image of the other. Some objects or shapes have more than one line of symmetry, and the symmetry lines can be vertical, horizontal and/or diagonal. So you can see that, if an object is symmetrical, it is very strictly balanced.

An object that is asymmetrical has no lines of symmetry. However, the object can still be visually balanced – it might have different shapes with similar volumes to achieve balance, or the designer might use colour to provide balance and unity.

Look at all the products in this chapter and decide which ones have symmetry. Which ones have asymmetry, and which have both?

Positive and negative space

Positive space is the space an object takes up. Negative space is the area around the object, or spaces within or between parts of the object. For example, if you look at the back of a sporting swimsuit, the positive space or shape is the 'X' shape that the straps form across the back. The negative spaces are the elegant, curved shapes formed between the straps, where the wearer's skin shows. In a chair, the negative spaces might occur between the arms and the seat, between the legs, or between sections of a back support, as shown in the image. Designers consider the interplay between the positive and negative spaces.

Negative space – in a swimsuit and in a plastic chair

9780170477499

Contrast

Contrast can be achieved by placing different colours, tones, materials and textures close to each other. The difference between them makes certain elements stand out more. Contrast can be a part of other principles, such as balance or pattern.

Look at all the images in this chapter where contrast exists and try to explain each instance of it clearly to a classmate, to see whether they can identify the images you are describing.

Pattern

Patterns use the elements of line, shape and colour in a repeated way. Designers use patterns to give their designs interest and drama. Many textures on materials form patterns.

Patterns in resistant materials

Timber has inherent patterning of grain lines and **figuring**, including beautiful wave-like and birds-eye figuring. Very thin veneers of these timbers can be attached to the surface of other timber products (such as MDF or chipboard), either in full sheets or in carefully cut interlocking shapes, to make the surface more appealing.

Figuring
Lines created by the grain in timber

Patterns can be added to resistant materials by:

- scratching or etching the surface
- carving, cutting or burning into the surface
- painting a pattern onto the surface
- using techniques such as mokume that layer, press and then grind back to reveal different layer colours
- using chemicals to affect the surface
- adding a painted or enamel patterning to the surface.

Patterns are an important element in clothing or fabric product design.

Shutterstock.com/Tan Yoowang

Shutterstock.com/Passakorn sakulphan

Different types of patterns on metal and wood: mokume and carving

Patterns formed by fabric construction

Patterns can be either formed in the construction of the fabric – the interplay of threads as they are woven or knitted – or printed or embroidered onto the surface.

Patterns that are woven into fabric with different colours often have specific names, e.g. plaids, herringbone and hound's-tooth; while twills and jacquards are fabrics with a pattern formed from threads of the same colour, which are woven using changes in the **warp** and **weft** threads.

Warp
The structural threads of woven fabric that are threaded onto the loom and run along the length of the fabric

Weft
The 'filling threads' that are woven from side to side across the warp

Ikat fabric from Indonesia uses a traditional way of creating patterns – it is formed by a combination of dyeing and weaving.

Transparent and opaque sections of fabric can also be used to create delicate patterns in lace and thermal/chemical burnt-out fabrics (devoré), while pressing into a fabric with a raised pile can create the type of patterning that is evident on crushed or panne velvet.

Knitted patterns and colour

Different colours can be knitted to form traditional 'Fair Isle' patterns or more modern patterns. Threads of the same colour can be knitted to form textural patterns such as 'Aran' and functional ribbing patterns. Complex machine knitting uses colour and texture to create a multitude of pattern options and fabric constructions.

Shutterstock.com/Tania Zbrodko

Dreamstime.com/Butsayapics

Patterns created by knitting (left) and weaving (right)

Printed patterns

Many different shapes can be used in printed patterns, e.g. geometric shapes, such as squares, circles and dots, and organic shapes such as teardrops, animals, plant leaves or flowers. Colours can be striped or checked. Lines can zigzag.

Abstract patterns use non-pictorial, geometric or random elements.

Naturalistic or figurative patterns are based on recognisable, real things – e.g. flowers, fruit, people and figures, or planes and cars. These may be realistic, simplified or stylised.

9780170477499

Shutterstock.com/Igor1503
iStock.com/Bkkweekender

An African semi-abstract pattern and an Indian printing block

Look for examples of patterned fabric and describe the pattern to someone else using any of the language in this chapter on Design Elements and Principles.

Worksheet
12.6.2

Pattern and rhythm

Pattern refers to a surface decoration on materials or a repeated feature that creates a visual pattern. In fabric, pattern is:

- a printed design in various colours and shapes
- a textural aspect, such as a twisted pleat used repeatedly
- repetition of a motif or shape.

Repeated elements in a pattern create a sense of rhythm that is predictable, like the rhythm of a piece of music.

The shapes within an object may be the source of some patterns. These may all be the same shape and size – e.g. the slats on the back of a chair – which will give the object unity and rhythm. Or the forms might gradually change, still being similar in shape or nature, but with the gradual change adding interest and movement.

Some patterns, particularly those that occur in nature, are not regular. The figuring of the grain in a piece of timber, or the irregular markings of fur or some other fabrics, are still thought of as patterns. They have unifying features, but the many parts of the pattern are not identical.

Note that 'pattern' also has a different meaning when it refers to the paper plans available commercially to construct products.

Shutterstock.com/pakww

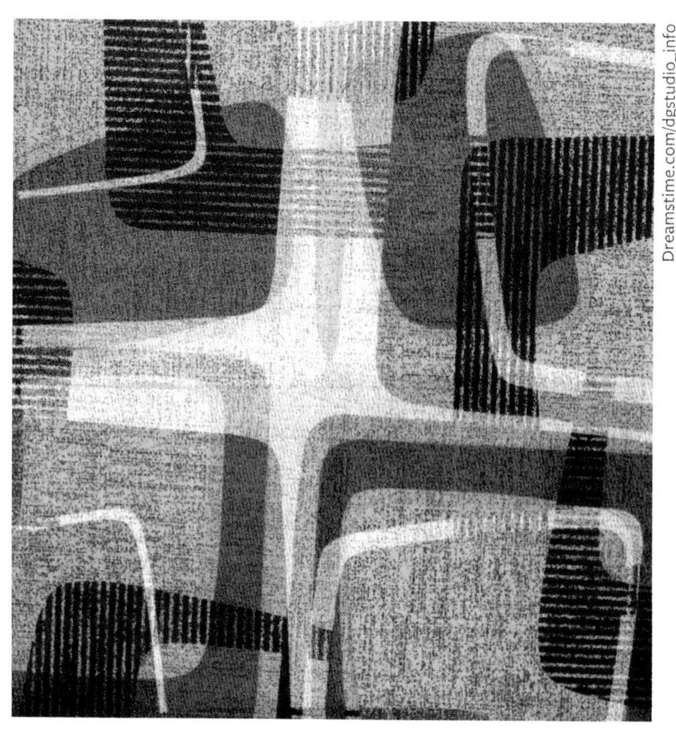
Dreamstime.com/dgstudio_info

Shutterstock.com/hifashion

Examples of the use of regular and irregular patterns: in the slats of a chair, a retro-style fabric and animal-printed fabrics

Movement and rhythm

The word 'movement' means different things when it is applied to different materials and objects. When discussing clothing, the term usually refers to how the fabric flows and drapes. When looking at other objects, such as furniture, it refers to how the eye is drawn across the object by its visual features. In other objects, a repeating pattern, a delicate sense of balance or parts that stretch outwards or upwards can create a sense of movement and/or rhythm.

9780170477499

Using patterned fabric

It is important when designing and making a garment or fabric product out of patterned fabric that you think about the pattern when you are cutting it out. If the pattern has large and/or prominent shapes and colours, pieces need to be placed very carefully before cutting. This is so that parts of the pattern match up at obvious joins in the garment, such as the centre front, back, side and shoulder seams. Having a pattern that can only be used in one direction also affects cutting out: you need to allow more fabric in these circumstances, so that you have the flexibility to move garment pieces around to find a good match for seams.

There used to be strict conventions that limited the mixing of different patterns, but now many contemporary designers use overlapping layers of complex and strong patterns to great effect. This can be seen in some of British designer Vivienne Westwood's work, in which she combined vivid and clashing tartans that previously would not have been seen in the same garment. Much of her work is unavailable for reprint but you can see it in online retail sites.

Style

The style of a product is determined by the designer's use of Design Elements and Principles (shape and proportion, colour, pattern and texture, etc.). Products in a particular 'design style' usually have aesthetic aspects in common – there is a similarity in the way the Design Elements and Principles are applied. This is regardless of the product type (clothing, houseware, cars, etc.). Designers often have a particular 'style' due to their repeated and consistent use of particular Design Elements and Principles. There are also certain styles that relate to a period, or to groups of designers. Groups throughout history have also maintained a style, e.g. Art Nouveau, Victorian, Bauhaus and Memphis style.

For example, the work of Australian designer Martin Grant (now based in Paris) is easily recognised. He has a consistent approach to the way he structures his garments – the simple shapes, the use of line and the range of colours he uses. See the image of his *Stitch* coat on page 347.

Inspiration for styles

Designers sometimes talk about having a 'design story' or stimulus that inspires a fashion range. This could be an exotic place or culture, a past era, or decorative qualities from a different medium. Aspects of these 'stories' will be reflected in most, or all, of the features of design. For example, a designer might be inspired by the clothing and culture of India. This would reveal itself in loose shaping and lots of folded and wrapped construction to reflect Indian clothing styles. Fabrics would be highly decorative and brightly coloured, often in spicy, rich tones.

Summary
Chapter 12

Quiz
Chapter 12 revision

Inspiration for style choices is often researched and put together in the form of 'mood boards'. To revise mood boards, go to pages 328–9.

Research the inspiration and narratives behind the runway shows of designers such as Alexander McQueen, Dolce and Gabanna, Dries van Noten or any of your favourite designers.

Worksheet
12.6.3

CHAPTER 13

Factors that influence product design

In this chapter, you will learn about:

- the importance of the Factors that influence design
- each of the Factors in detail:
  - » what they cover (their scope)
  - » how they influence the work of designers
  - » how to consider them in your design thinking throughout the design process.

Worksheets:
- 13.4.1 Ergonomics and anthropometric data **(p.373)**
- 13.9.1 Intellectual property **(p.390)**

Summaries:
- Chapter 13 **(p.397)**

Quizzes:
- Chapter 13 revision **(p.397)**

Case studies:
- Charlwood Designs **(p.373)**
- Etiko **(p.384)**
- Cantilever **(p.384)**
- Schamburg + Alvisse **(p.385)**
- Valentino's resort range **(p.387)**

To access resources above, visit **cengage.com.au/nelsonmindtap**

Nelson MindTap

9780170477499

What are the Factors?

The process of designing a product is very complex. Designers need to consider a broad range of factors when exploring a product's context and when developing solutions for end users. In our Study Design, these are called the 'Factors that influence design'. They are used as a framework to propose and analyse product concepts, evaluate existing products and as a springboard to developing evaluation criteria. The Factors are referred to throughout Units 1 to 4 and are essential knowledge for all units and outcomes. Almost all the Factors are present in some way in every design project.

In VCE you use these Factors to:

- evaluate existing commercial products
- assist in writing a design brief and the following design process activities:
  - generating related evaluation criteria
  - providing a springboard to direct design and research activities
  - proposing, analysing and finalising product concepts
  - evaluating your own finished product
- understand what designers need to consider when developing products in industrial settings.

Once outlined in your design brief, the Factors will be continually addressed throughout your design process work.

Shutterstock.com/Rawpixel.com

As you become familiar with these Factors, you will see that there is quite a lot of overlap between them – for example, materials can relate to the aesthetics and texture of a product, its function and quality, as well as its sustainability. Mostly, however, you will look at materials in terms of their characteristics and/or properties.

The following pages will explain the Factors in more detail.

Factors that influence product design	Scope of Factor as it relates to product design
Need or opportunity	Identification of the purpose for, or of, a product. Considerations include the context, the purpose for designing and how a product will be used. Needs and opportunities are identified from research and development, feedback from end user/s, new ideas and knowledge and new and emerging technologies (including materials). In VCE Product Design and Technologies, the needs or opportunities that are explored must be real.
Function	The purpose of a product that makes it fit-for-use for its intent.
End user/s	The human and/or non-human 'consumers' of the product for whom, or for which, the product is intended. Consideration of the end user incorporates welfare, which includes quality of life; quality of life encompasses culture and religion, emotional and sensory appeal, universal design, demographics, social and physical needs and trends, safety, accessibility, comfort, ergonomics and anthropometric data.
Aesthetics	Relates to the product's form, appearance and feel. Considerations include Design Elements and Design Principles. Design Elements include point, line, shape, texture, colour (tone, transparency, translucency and opacity). Design Principles of balance, contrast, repetition, movement/rhythm, pattern, proportion, asymmetry/symmetry, negative/positive space and surface qualities are used to combine and arrange the Design Elements. Aesthetics may relate to ethical considerations in design, as aesthetics can influence quality of life.
Market opportunities	Designing innovatively and working entrepreneurially require a creative approach to develop new or improved designed solutions to unsolved problems or new needs or opportunities.
Product life cycle	The resource inputs that span a product's manufacture, which includes sourcing of materials, useful life and the impact of disposal/reuse.
Technologies: materials, tools and processes	Technologies include those materials, tools and processes that are traditional as well as new and emerging. Students should know and experience a variety of materials, tools and processes through making and designing products, as well as through researching designs and the work of designers. Materials are selected for use based on their appropriate properties (their performance and behaviour, both chemically and physically under certain conditions) and desirable characteristics (such as visible features). Examples of appropriate materials, tools and processes are listed on pages 15–17 of the Study Design.
Ethical considerations in design	Ethics in design is concerned with enabling both individual goods and values (such as more time with family and friends) and public goods (such as a fair and just society). This can be realised through products that reflect and enable an end user's values, or work towards social goals such as belonging, access, usability and equity for the disadvantaged. Inclusive design processes can enact respect and concern for humans and non-humans. Ethical considerations encompass sustainability. Sustainability and other ethical considerations are concerned with human and non-human welfare and aim at positive impacts and minimisation of harm with regard to what is made and how it is made, for both present and future generations. Ethics can also involve legal responsibilities. The legal aspects of product design include intellectual property (IP) which … includes copyright, patents, trademarks and registered designs. Australian and International (ISO) standards, regulations and legislation (including OH&S) are other legal responsibilities. Products must be produced safely and be safe for end users.

VCE Product Design and Technologies Study Design 2024–2028, pp. 18–19

9780170477499

Need or opportunity

Need or opportunity	Identification of the purpose for, or of, a product. Considerations include the context, the purpose for designing and how a product will be used. Needs and opportunities are identified from research and development, feedback from end user/s, new ideas and knowledge and new and emerging technologies (including materials). In VCE Product Design and Technologies, the needs or opportunities that are explored must be real.

Source: *VCE Product Design and Technologies Study Design 2024–2028*, pp. 18–9

This Factor is fairly self-explanatory and focuses on what people need or where there is an opportunity to address something through a design solution. Design opportunities can arise for a variety of reasons, such as user behaviour or social trends, new knowledge from R&D and changes in technologies.

Shutterstock.com/Tacar

This product addresses the need to remove eggs safely from boiling water.

The context

The context is the situation, location, circumstances, events or background environment that are relevant to where, how and by whom a product is needed and used. This applies both to existing commercial products and to the product you are designing and making.

Questions to consider are:

- Where will it be used? (city, farm, beach, snow, indoors, outdoors, when commuting, etc.)
- What is the occasion, activity, event or regular occurrence?
- How will it be used? Will it be treated roughly? Used by many people? Barely used or touched? Will it be moved around often?

When developing a design brief

The needs or opportunities decided on must be real and have a functional purpose, i.e. not an expressive artwork or decorative piece. You should not choose a famous person to design for, as you will not be able to collect feedback from them (unless you know them personally). In your design brief, you need to make it clear what need or opportunity you are intending to design for.

When evaluating products

Questions that can be asked about products are:

- Why is it needed?
- What is its purpose?
- What is its function?
- What does it do?
- What is it for?

While these questions are worded differently, they all refer to the same thing: the reasons the product exists.

 Units 1 and 2 students should check with their teacher on what is expected. Unit 3 students can read Chapter 7 (Outcome 2) for more suggestions.

Function

Function	The purpose of a product that makes it fit-for-use for its intent.

Source: *VCE Product Design and Technologies Study Design 2024–2028*, pp. 18–9

The function of a product relates to how it works to address the needs of the end user:
- in broad terms, e.g. how can it solve the problem?
- in specific terms, e.g. how might it work, operate, be useful, how can the parts work together to make the whole product work better? etc.

Often **purpose** and **function** can be expressed in similar ways. When determining function, it is also important to consider all the smaller functional aspects of a product that support its purpose and the main or primary function.

For example, the main function of a bucket is to carry water or objects. Users may have different purposes for a bucket. Functional aspects to support this main function can differ, but most likely there will be:
- a handle to let you carry the bucket
- a lip to enable easy pouring of liquid
- a ridge around the edge of the bucket to provide stability to the rim
- ridges or small 'feet' to prevent water tension sticking the flat bottom of the bucket to a floor.

Shutterstock.com/Emil Timplaru

It is often these smaller functional aspects that make one brand of product preferable over another for certain uses or users. Some functional aspects might be included for specific purposes, e.g. a lid for a bucket would suit specific purposes for some users, yet be unnecessary for others. Some products have only one or two functional aspects: for example, the main function of a pencil is to make marks, and it has a lead to make the marks, a wooden section to hold the lead and perhaps a metal band to hold an eraser. Other products have many functional aspects, e.g. the main function of a car is to move people around, yet a car has so many functional aspects to help it do this that it would take pages to list them. Consumers often choose the brand and the model of a product based on all the functional aspects that will make it more fit or suitable for their use.

Functionality
How well a product can serve its purpose (do its job), its practicality and its range of usefulness

Functionality is another term for the functional aspects of a product.

9780170477499

Quality is an important facet of function. A common way of understanding quality is that a product is 'fit for purpose', i.e suitable for the purpose for which it was purchased. We sometimes incorrectly use the word quality to imply 'good or excellent quality', but it is more appropriate to describe the *level* of quality when making judgements about good or poor quality.

Function and quality also overlap with legal Factors: a product must be safe for consumers to use (i.e. it must work safely). A product's level of quality can be an important factor in safety. Legal responsibilities are discussed on pages 387–94.

STANDARD 100

The OEKO-TEX® label certifies that a fabric is of high quality and doesn't contain any harmful substances

When developing a design brief

For a product to be successful, it needs to fulfil its expected function. The specifics of this expected function need to be clearly defined in the brief. This information will guide your decisions about functional aspects during the design process, and help you to evaluate the product at the end of the process. The context is the circumstances, location, surroundings, environment, background or setting that make up the situation. For example, the context for a design problem related to seating could be an office, a lounge room, an outdoor garden or a kindergarten. To revise the contents of the design brief, go to pages 209–14.

Other aspects relevant to the function could be size or fit, weight, performance, operation, reliability, quality, safety and ergonomics. To find out more about ergonomics, go to page 373.

During the design process

Throughout the design process you will be researching ideas on functional aspects, incorporating ideas into your designs that aid the intended function, and checking that your prototypes and product will function as required.

Test out functional aspects and aim for quality with trials. Get end user feedback. Refer to the context to determine suitability. Create a list of quality measures and check quality in everything you do. See Chapter 2 for suggestions regarding quality measures.

When evaluating products

Questions asked about the function of products are similar to those regarding 'needs and opportunity' – or, more particularly, 'What does it do?'

Analysing products designed by others is an informative process that can add to the quality of your own designs. When analysing products, determine the purpose of the product to suit the context and the end users' needs. Look for the Factors that are important for effective functioning. Identify the main function (or major function) and the smaller functional aspects that support it, i.e. that enable the primary function to perform most effectively and make the design suitable for its context. How well does it work? Does it do what it needs to do?

End users

End user/s	The human and/or non-human 'consumers' of the product for whom, or for which, the product is intended. Consideration of the end user incorporates welfare, which includes quality of life; quality of life encompasses culture and religion, emotional and sensory appeal, universal design, demographics, social and physical needs and trends, safety, accessibility, comfort, ergonomics and anthropometric data.

Source: *VCE Product Design and Technologies Study Design 2024–2028*, pp. 18–9

It is important to note that products are rarely designed for 'everyone' or 'anyone'. They are targeted at people with certain tastes and lifestyles, in different age groups and with different budgets. See the heading 'Deciding on an end user' on page 196 to read more.

Shutterstock.com/Rawpixel.com

Quality of life

Quality of life describes a state of human health and wellbeing. While quality of life is most often influenced by economic and social conditions, it can be impacted by the design of products in several ways. The following aspects are all mentioned elsewhere but are listed here as a brief summary of how they can contribute to quality of life.

- **Functionality** – a well-designed product should be easy to use and perform its intended function well to help improve efficiency and productivity.
- **Safety** – products must be safe to use and designed with safety in mind to reduce the risk of injury or other negative outcomes.
- **Comfort and ergonomics** – these can ensure that a product is designed to suit the human body, is comfortable to use and doesn't cause strain or pain.
- **Accessibility** – designing products to be more accessible to people with disabilities or other limitations can help people who might otherwise struggle to use certain products.
- **Universal design** – well-designed products, particularly public spaces and public utilities, should cater for all.
- **Sustainability** – products designed sustainably can help reduce waste and minimise environmental impact. This can help create a healthier and more sustainable future for everyone.
- **Aesthetics** – visually appealing or emotionally satisfying products create a sense of joy or satisfaction when being used.

9780170477499

Demographics

Demographics is information used to name, describe or delineate a population or group. It defines them by characteristics such as age, gender, race, ethnicity, income, education level, marital status, occupation, special interests and other social and economic indicators. Demographic data is often used in research, marketing, and other fields to understand and segment a population into different groups based on these characteristics. Information is often collected through surveys, censuses, and other forms of research. Businesses and organisations use the information to better understand and analyse a particular group of people (called the 'target market') and to direct a product or service towards them. Demographics can also be used to identify trends and changes in a population over time.

Weblink
Australian Bureau of Statistics

 You can find many statistics on populations on the Australian Bureau of Statistics website.

Culture and religion

Cultural and religious beliefs (often arrived at through history and traditions, ethnicity, location, the climate and economics) make an important contribution to human behaviour and therefore needs.

Many religions and cultures have rules by which people live their lives: the way they relate to one another, to other living things and to the environment. This influences their thinking, designs, architecture and use of technology. All of this affects or determines the way people behave in many contexts, which in turn determines how they use products and how they want those products to be.

Aboriginal and Torres Strait Islander society has specific needs, traditional customs and very different beliefs from those of the predominantly Eurocentric Australian society that has taken shape in the last 250 years. Australian Aboriginal people are currently some of the poorest, most marginalised and most incarcerated (jailed) people in the world, and many have very poor health outcomes. Many designed solutions have failed to cater in a holistic way for Aboriginal people and have not been successful in meeting their needs. This is due largely to a failure on the part of others to understand or take into consideration traditional beliefs and customs when trying to remedy their situation.

 For information about cultural influences, particularly those related to Aboriginal and Torres Strait Islander culture, read Chapter 5.

Emotional and sensory appeal

Emotion plays a huge role in guiding a consumer's purchase, and it can be both conscious and unconscious. Products can be bought because of sentiment, nostalgia, desire for status, the need to 'fit in', laziness or simply newness or novelty. Companies often market their brand to tap into these emotional responses.

Designers can tap into people's emotional needs for a product by making it appeal viscerally, behaviourally and reflectively. A part of this is sensory appeal. The way a product looks, smells and can be heard, felt and touched also affects a consumer's willingness to purchase it.

As a designer, consider how people will react to your product emotionally and with their senses.

Accessibility

In this subject the term 'accessibility' refers to the design of products, devices, services or environments for people who experience disabilities or who are disadvantaged.

To improve accessibility, designers need to consider the physical needs of users, in particular those with disabilities or mobility, vision or hearing issues. This is of particular importance for public and community spaces and facilities, and public transport, but it also applies to products purchased for individual use.

Accessibility is also an important consideration when designing for different types of assistive technologies, such as screen readers or voice assistants.

Shutterstock.com/Chansom Pantip

A refreshable braille display is designed to make text output accessible to people with impaired vision who cannot use a regular computer monitor.

Universal design

Inclusive design – also called 'universal design' or 'design for all' – is about designing products for use by a broad range of people, especially those who are sometimes excluded from using mainstream products because of their age, gender, ability or economic status. Universal design has a close connection to accessibility. However, where accessibility looks at the personal (making something accessible for a person with specific needs), universal design aims to broaden access to a product, space or experience to the widest range of people possible. Like accessibility, universal design is very important for public spaces or products that will be purchased by users with a wide variety of abilities and levels of mobility.

9780170477499

Bellemo & Cat

Urban Nodes: this design for a public area in Reservoir (Melbourne) by Bellemo & Cat is an example of inclusive design as it caters for elder pedestrians needing a rest, for shoppers wanting to chat, and has bike racks for cyclists and paving for easy access of trolleys, wheelchairs and pushers. It also promotes a feeling of community.

People are living longer, and often an increase in age brings diminished capabilities, such as reduced or impaired strength, dexterity, eyesight, hearing and mobility. These are the important capabilities for interacting with products. Inclusive/universal design aims to cater for both the elderly and those with disabilities.

When designing products, strategies to increase the range of users could include making sure that:

- visual aspects are large and bold, with different textures and colours to indicate different parts
- universal symbols are used for controls
- mechanical parts can be moved easily
- the product is lightweight and/or easy to lift or, conversely, is immovable for safety purposes
- all parts of the product can be easily reached
- the components of the product are reliable
- the size of the product and its parts caters for a wide range of users.

Such approaches consider the ways people interact with products through vision, hearing, thinking, instinct or reflex action, reach and stretch, and dexterity.

Social and physical needs and trends

As humans, we have a strong need to be social, to interact with others. Many products cater for this need by connecting us, allowing us to share and providing reasons to communicate with each other.

After our basic physical needs have been met, almost every product we buy satisfies a social need, such as the desire for new clothing to wear to a party.

Some of the social and physical needs that guide purchases are:

- to belong
- to feel safe
- to have status
- to feel good
- to release energy
- to be comfortable in hot/cold conditions
- to eat well
- to keep active
- to achieve and do well
- to earn a living
- to look our best.

People like to keep up with trends, and this can affect whether they will purchase and use a product. Designers need to produce products that appeal to this. Fashion and trends could also be considered a social need – to be seen as being up to date, or part of the fashionable set or 'in crowd'.

Trends are also a factor in the design of our communication products and technology, our vehicles and our houses.

Designers take into account social and physical needs and trends by conducting research and analysis – for example, by researching the social and physical needs of the users who will interact with the design (or potential product) through interviews, surveys and observation to understand their behaviour, preferences and cultural background. This can include diverse elements such as colour, symbols, language and interactive behaviour. Designers can analyse emerging technologies, cultural shifts, and lifestyle changes that can impact designs. This can help them anticipate and respond to future user needs and preferences. Designers will also consider people's concern about the environment.

Safety

Safety is important for any product, no matter what its purpose. It is a legal requirement in Australia that any product sold is safe for the user. However, this becomes even more important if the product is required to meet government-regulated safety standards, as is the case for items such as life jackets, seatbelts or items intended for use by children.

Following ergonomic principles (see page 373) or complying with the relevant Australian Standard (see page 392) can help to ensure the user's safety.

stock.adobe.com/Alla

Children's toys are required to meet stringent safety standards.

9780170477499

Comfort, ergonomics and anthropometric data

Comfort refers to how well the product fits the user or functions for them. This is linked closely to safety and **ergonomic** factors. To make products physically suitable for people, ergonomic principles are applied.

Ergonomic
Relating to the scientific study of the relationship between people and their living and working environment

Ergonomics is about designing equipment and products to match the human body and the way it moves – to address comfort and ease of use, and to reduce strain, injury and fatigue. This includes the tools, materials and equipment that are used.

The study of ergonomics uses data from several disciplines:

- **anthropometry** – body sizes and shapes (see more detail following)
- **biomechanics** – body movement, muscles and strength (in relation to levers and forces)
- **environmental physics** – noise, light, heat, cold, radiation, vibrations that affect body conditions, hearing, vision and sensations
- **psychology** – perceptions, skill/experience, learning and communication that affect behaviour or contribute to errors.

Case study
Charlwood Designs

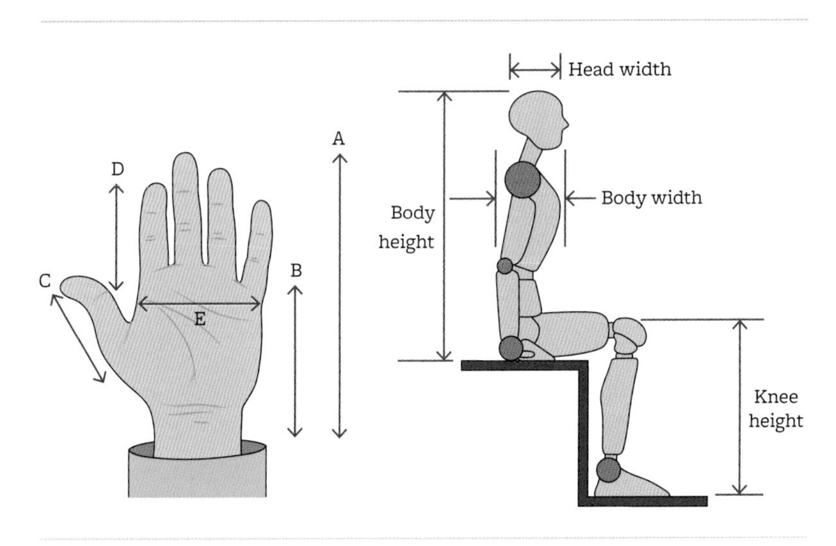

Anthropometric data: average hand measurements (left) can be important when designing handles; and body measurements (right) are important for seating.

Anthropometry

Anthropometry is the measurement of the human body. It covers all the aspects of body size: height, length of limbs, circumferences, shape, weight and proportions.

Anthropometric data is gathered from the measurement of populations and the variations within these populations. It is used in ergonomics to design products and spaces to suit the human body in areas such as size, reach, etc.

All human beings have the same basic form, but we each have our own unique set of dimensions and proportions. For mass production, designers need to know the range of sizes into which most people will fall. Anthropometric data is used to determine these limits. This is based on scientifically collected measurements from a wide variety of people to provide us with a range of sizes for various parts of the body. These sizes will be suitable for use with the majority of our population. However, some products will need to refer to data in the 5th (the smallest) or 95th (largest) percentile, e.g. if designing specifically for very short or very tall people.

Worksheet
13.4.1

Anthropometric measurements could range from the length of an index finger to the vertical reach of a person standing upright. They could be for a particular group of people – for example, children 3–6 years old, nurses, or the elderly – or for a specific situation or environment, such as public transport or schools.

Designers use anthropometric data in various ways:

- products for the average user – using measurements in between the largest (95th) and smallest (5th) percentiles
- products for extreme sizes – selecting measurements for either the largest or smallest percentiles
- products to fit most sizes – selecting measurements including those from both the largest and smallest percentiles
- products with adjustable sizing – where possible and cost-effective, using the highest and lowest measurements to accommodate most individuals, e.g. in safety helmets, car seats, office chairs, stadium seats or life jackets.

Weblink
Anthropometric data

For detailed anthropometric data, download the 52-page extract from a 1988 survey of US Army personnel.

Men's size chart

Shirt sizes								
Australia	XXS	XS	S	M	L	XL	XXL	XXXL
Europe	36	37	38	41	42	44	45	46
UK/USA	14	14.5	15	16	16.5	17.5	18	18.5
Neck (cm)	36	37	38	41	42	44	45	46
Chest (cm)	81-86	86-91	91-96	96-101	101-106	106-111	111-116	116-121

Trouser sizes													
Australia/ UK/USA	28	29	30	31	32	33	34	35	36	37	38	39	40
Italy	39.5	40	40.5	41	41.5	42	42.5	-	43	43.5	44	44.5	45
Japan	XS	S	S	M	M	L	L	L	XL	XL	XXL	XXL	XXXL
Waist (cm)	71	73.5	76	78.5	81	83.5	86	88.5	91	94	96	99	101.5

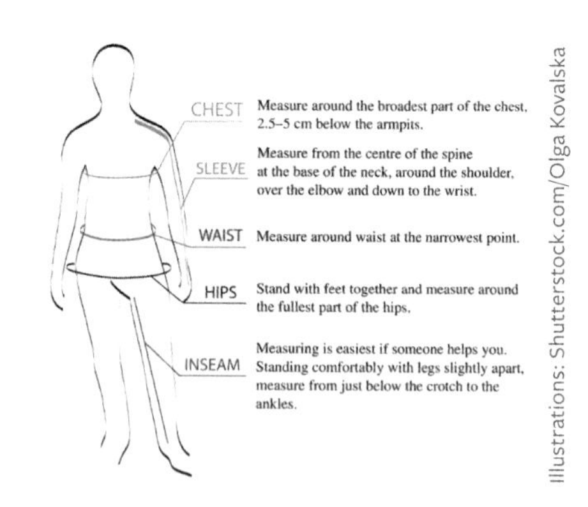

Illustrations: Shutterstock.com/Olga Kovalska

Women's size chart

Australia/ UK	4	6	8	10	12	14	16	18	20	22	24
China	S	S	M	M	L	L	XL	XL	XXL	XXXL	XXXL
USA	1	2	4	6	8	10	12	14	16	18	20
Bust (cm)	82.5	85	87.5	90	92.5	95	99	103	107	112	117
Waist (cm)	64	66	68.5	71	73.5	76	80	84	88	91.5	95
Hips (cm)	89	91.5	94	97	99	101.5	105.5	109	113	118	123

SLEEVE Measure from the centre of the spine at the base of the neck, around the shoulder, over the elbow and down to the wrist.
BUST Measure around the fullest part of the bust.
WAIST Measure around waist at the narrowest point.
HIPS Stand with feet together and measure around the fullest part of the hips.
INSEAM Measuring is easiest if someone helps you. Standing comfortably with legs slightly apart, measure from just below the crotch to the ankles.

Clothing size charts are constructed using anthropometric data.

9780170477499

When developing a design brief

In your design brief, include a profile of your end user. In Units 1 and 2, this can be a brief statement, outlining your end user with some background and demographic information, including their lifestyle and/or their aesthetic taste. In Unit 3 it can be a little more descriptive or detailed – you can also add visual images related to the end user, their needs and relevant interests/influences. It can also include numeric/quantitative data collected from research you have carried out about end users' needs and opinions. In all units, your design brief should include any specific requirements for your end user that need to be considered in the solution. Read through the end user Factor details to determine what should be included.

Focus on the user by:

- considering their specific needs
- clearly defining the context and what is special about it. Give a (brief) story about the situation and why a new solution is needed
- examining existing solutions and identifying problems with them, i.e. aspects that don't quite perform suitably for your intended user, to include in your design brief.

During the design process

Throughout the design process, seek end user feedback and insert it into your folio where appropriate.

Consider how you can:

- tap into people's emotional responses to your ideas and attract or draw them to your design
- include ergonomic theory into your design and use anthropometric data to give the user less strain, fatigue or chance of injury
- focus on the specific needs of your user, rather than just adapt existing solutions
- incorporate connections to the end user's background and values (e.g. related to culture, traditions or religion)
- research widely to be sure you are abreast of developments or fashion and trends related to the use of your intended product; and understand any cultural sensitivity around the activity to be performed with your product.

When evaluating products

When analysing designs developed by others, aim to identify the user's need and how it has been catered for in the product. Consider how deeply the designer looked into the specific issues or needs of the user. Questions that can be asked are:

- Who exactly is the end user? What separates them from other people or groups? What specifically do they get with this product? Does this product improve their quality of life?
- How do some products take into account issues for particular end users?
- Has the product been designed ergonomically? Is the product a better solution for a particular group of users than other products?
- Does the product appeal to its intended user on an emotional level, whether this be for prestige, 'cuteness' or some other reason?
- How well do you think the product contributes to the user's wellbeing or improves their quality of life?

Aesthetics

Aesthetics	Relates to the product's form, appearance and feel. Considerations include Design Elements and Design Principles. Design Elements include point, line, shape, texture, colour (tone, transparency, translucency and opacity). Design Principles of balance, contrast, repetition, movement/rhythm, pattern, proportion, asymmetry/symmetry, negative/positive space and surface qualities are used to combine and arrange the Design Elements. Aesthetics may relate to ethical considerations in design, as aesthetics can influence quality of life.

Source: *VCE Product Design and Technologies Study Design 2024–2028*, pp. 18–9

Aesthetics
The study of beauty and style, or what different people see, feel or experience as beautiful

The idea of **aesthetics** relates to the visual or sensory appeal of art, nature and other objects and how we see or react to them, or what each of us considers to be pleasing to our senses. Most often, this is determined by our past experiences and knowledge or exposure to various aesthetics, who influences us, our values and our culture.

Aesthetics appeal to the senses (sight, hearing, smell and touch) and are involved in the creation of objects, spaces, surfaces and experiences to engage those senses. Aesthetics are important because products we use every day affect our wellbeing. There are, however, no fixed rules on what defines beauty or makes something attractive. Whether you find an item attractive or not can depend on a range of factors, such as your age, your cultural background or peer group, or the current fashion.

Designers use the Design Principles (such as proportion, space, balance and repetition) to organise the Design Elements (such as shape, form, colour and texture) which are combined to create a particular 'aesthetic' or style. To revise the Design Elements and Principles, go to Chapter 12.

Aesthetics is also about how things feel. Certain materials have 'tactile' qualities that may attract us to a product. We want a winter jacket that feels soft, warm and a bit furry, a floor covering that will not let us slip, a swimming costume that is very smooth and will not irritate us when it gets wet, a handle surface that provides good grip or a tabletop that is smooth to touch. Words such as furry, shiny, smooth, rough, corrugated, pitted, silky, spongy and prickly can all be used to describe how something feels or its tactile qualities.

Aesthetics of materials

Materials are not only chosen for cost and function; often, they are selected for visual appeal. Materials have certain qualities that contribute to the 'look' or style of a product, such as:

- colour, and colour variation
- texture and 'feel'
- level of sheen and ability to reflect light
- markings or figuring
- transparency (something transparent is see-through, e.g. clear glass), translucency (allows some light through but images cannot be seen clearly, e.g. glass bricks) or opacity (nothing can be seen through it).

9780170477499

iStock.com/Anton Minin

A handbag design may use hardware such as screw heads for aesthetic effect.

Aesthetics and styles

There are many different design styles – they may relate to a design period or movement, or the work of a particular designer. A style usually has a set of aesthetic features in common, and products within a style show a similar use of some Design Elements and Principles. If you are creating a product for a context that has a specific style, or if your end user prefers (and has asked that the product reflect) a style, you will need to research and identify the specific features that are common to the style, and then work out how to reinterpret the features, or apply them in a new way.

When developing a design brief

In your design brief, describe the aesthetics that appeal to your end user. Use the language of the Design Elements and Principles to help you. Include any relevant tactile requirements. If the product is required to be in a particular style, then define the chief Design Elements and Principles that make up that style.

During the design process

In the design and development stage, employ different Design Elements and Principles to enhance and enliven your ideas. Use nature, geometry and mathematics to inspire and create interest or uniformity in your designs. Other creative techniques, such as SCAMPER, can also be used to extend your use of the Design Elements and Principles and to stretch your ideas.

When evaluating products

Questions that can be asked are:

- What visual and tactile features appeal to the target market, demographic or end user of the product? In other words, in what way would the target market find this product beautiful?

- How have aesthetics been incorporated for touch and sight or experience?
- How have the Design Elements and Design Principles been combined to create this aesthetic?

When analysing designs by others, you can deconstruct the aesthetics by identifying and describing the Design Elements and Principles. For example, 'the small, pale yellow squares repeated down the left side in a random manner create contrast with the rust-coloured front surface of the bag'.

Market opportunities

Market opportunities	Designing innovatively and working entrepreneurially requires a creative approach to develop new or improved designed solutions to unsolved problems or new needs or opportunities.

Source: *VCE Product Design and Technologies Study Design 2024–2028*, pp. 18–9

Market opportunities are areas where existing products do not already cater for a need, or where suitable products haven't been developed and a new product can fill a gap in the market.

When developing a design brief

In your design brief, you may have identified a particular gap or opportunity in the market that you intend to develop a solution for. This needs to be clearly described.

For example, many city buildings do not currently have secure spaces where visitors to the building can park a bicycle. This could be a market opportunity as it is not a product that commonly exists, yet could be in high demand as more commuters choose bicycles as a means of transport over private vehicles or public transport.

When evaluating products

Questions that can be asked are:

- How does this product provide an innovative solution?
- What creative ideas have gone into the product?
- What new need does this product satisfy?

iStock.com/JennaWagner

There could be a market opportunity here for a bike parking solution.

9780170477499

Product life cycle

Product life cycle	The resource inputs that span a product's manufacture, which includes sourcing of materials, useful life and the impact of disposal/reuse.

Source: *VCE Product Design and Technologies Study Design 2024–2028*, pp. 18–9

Product life cycle is what is analysed in a life cycle assessment (LCA). It involves consideration of the main stages in a product's creation. They are:

- sourcing/extracting raw materials and the environmental impact of this
- processing of raw materials into a usable form and the resources required
- product manufacture and supply chains
- distribution and the transport required (through all its life stages)
- product use, including cleaning, repair and maintenance, plus any waste or emissions created
- product disposal – whether it be landfill, incineration, reuse or through biodegradation.

  You can read a lot more detail on these in Unit 3, Outcome 1 in Chapter 6 on pages 146–8.

When developing a design brief

In your design brief, you will consider the life cycle of your intended product and its sustainability. This may not be explicitly written, but you should give some thought to your intended product's impact on the environment.

During the design process

Consider the life-cycle stages (and their impacts) when researching and developing your product. Research in depth and make choices around the overall design of your product, the materials you choose to use (check the source location and the supply chain if possible), waste minimisation, efficient production processes, quality construction methods, the resources used in the 'product use' phase, and disposal or reuse of the materials in your product (are they reusable, recycled, recyclable or biodegradable/compostable, or will they most likely end up in landfill or be incinerated?).

When evaluating products

Questions that can be asked are:

- What material is the product made from?
- How is this material sourced, i.e. extracted and processed for use?
- Does it require a lot of energy to make it usable from the raw material?
- What happens to the product's material when the product is used? When it is washed, does it shed microfibres?
- Does it use electricity?
- Does it create waste chemicals or pollute the air?
- Does it last a long time before it gets disposed of?
- Does it end up in landfill after one use or a very small amount of use?
- What will be the most likely method of disposal – landfill, incineration, reuse or letting it biodegrade naturally? Will it require industrial methods to make it biodegrade?

Technologies: materials, tools and processes

Technologies: materials, tools and processes	Technologies include those materials, tools and processes that are traditional as well as new and emerging. Students should know and experience a variety of materials, tools and processes through making and designing products, as well as through researching designs and the work of designers. Materials are selected for use based on their appropriate properties (their performance and behaviour, both chemically and physically under certain conditions) and desirable characteristics (such as visible features). Examples of appropriate materials, tools and processes are listed on pages 15–17 of the Study Design.

Source: *VCE Product Design and Technologies Study Design 2024–2028*, pp. 18–9

iStock.com/Laurence Dutton

Technologies used in designing

Designers can use artificial intelligence (AI) for coming up with ideas, drawings, models and 'photos'. These can be used in the initial brainstorming or idea creation stages. A suitable idea can be selected, worked on and details finalised in drawings created using computer-aided design (CAD).

Many technologies used in the designing of products are inextricably linked to manufacturing. For example:

- CAD is used initially for designing where virtual prototypes can be generated, but it has the bonus of being able to send digital files straight to computer-aided manufacturing (CAM).
- 3D printed pieces or rapid 3D prototypes rely on files created in CAD. Prototypes are commonly used to check product details before manufacturing. However, 3D printing also has a role in the manufacturing of parts for some products, which in turn is reliant on CAD.
- Lasers are commonly used in manufacturing but can also play a role in designing as a checking method before finalising a prototype.

9780170477499

Technologies used in manufacturing

This term refers to the tools, machinery and other equipment used in factories for production.

Refer to Unit 3, Outcome 1 in Chapter 6 to review technologies used in designing and manufacturing. For the end-of-year exam, you need to know these technologies as they are listed in the Study Design for Product Design and Technologies.

Technologies used by you

In some ways you are constrained by:

- what is on offer at your school
- your skills and knowledge of how to use technologies safely
- the relevance or suitability of the available technologies to your project.

Refer to Unit 3, Outcome 1 in Chapter 6 and the new and emerging technologies covered. Choose those from Chapter 6 that are available at your school and review them and/or conduct research to find out more about them and their uses. Make sure you are familiar with any hazards associated with them and precautions required to prevent possible injuries or harm.

List any technologies in your workshop that aren't listed in Chapter 6. Research how to use them and how to use them safely by following teacher instructions, reading the manuals or researching the manuals online.

Substances

New products for finishing, protecting or embellishing materials are constantly arriving on the market. If you are using any substances, it is important that you follow the instructions on the containers to the letter and, where necessary, obtain the safety data sheet (SDS).

Materials

When making decisions about materials, it is important to consider how their **characteristics** and **properties** will suit the design situation. You will find many different definitions of 'characteristics' and 'properties' in different texts, depending on the field.

There is so much overlap between characteristics and properties that in this text, and for the purposes of this Factor, we will group them together.

Characteristics and properties can include words such as those listed here (and hundreds of others):

- conductivity
- corrosiveness
- density
- durability
- elasticity
- feel
- hardness
- impermeability
- insulation
- permeability
- rigidity
- shine
- strength
- taste
- thermal resistance.

Note that colour is not considered a property or characteristic of fabric unless it changes the feel, but it can be attributed to timber and metal.

Refer to Chapter 14 for more detailed information and definitions relating to material characteristics and properties.

Characteristics
Visible features of a material; something typical of that material or special to it

Properties
How a material performs and behaves – chemically, physically or mechanically – under certain conditions

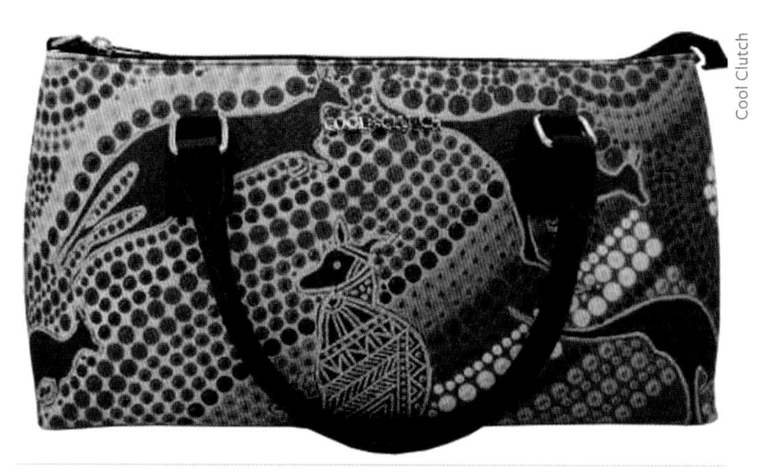
Cool Clutch

This *Cheree Cool Clutch* is a fashionable bag designed to be used to carry food and drink and keep them cool. The design of the fabric is based on a painting by Aboriginal artist Cheree Stokes. The bag has an insulated interior and waterproof aluminium foil lining that will keep its contents cooler or warmer as needed – the material to suit the job.

Compatibility of materials

When using more than one material in a product, you must consider their compatibility, whether it relates to appearance (whether they make an aesthetic match), joining methods or their physical proximity.

The dictionary meaning of 'compatible' is 'things that can exist in the same place and at the same time without harming each other'. If the materials are not compatible, the viability of the product may be compromised.

When joining different materials, a designer must ensure that joins are strong and durable and that any adhesives used are not going to 'degrade' one of the materials. When materials touch – for example, where fabric rubs against brick – neither one should deteriorate. Dust and particles from the brick would settle on the fabric and contribute to the wear and tear of the fabric.

Other situations in which materials 'touch' each other include:

- Ferrous (iron-based) metals in close proximity to wood can put rusty stains on the wood.
- Wire can push holes through fabric when put under pressure, such as in a mattress.
- Tent frames make contact with the fabric in all weather conditions and must not cause damage.
- Different fabrics in clothing may be problematic if the dye runs in one and is absorbed by the other, or if they have different washing constraints.

When developing a design brief

In your design brief, you need to outline the characteristics and properties that would make materials suitable for your intended product. It is best to avoid naming specific materials. Your evaluation criteria also need to avoid specifying particular materials, and should focus on what is required of the materials, in areas such as:

- the characteristics and properties needed
- how the materials perform in the product's particular context/location
- required aesthetic features.

9780170477499

At this stage, you may not write explicitly about the technologies you will use but you will need to keep in mind what is available to you along with the time, access to the technology and your level of skill in using it.

During the design process

Throughout the design process, you will be researching and testing materials, tools and processes, and demonstrating your knowledge of them and their suitability for your product.

When developing your scheduled production plan, you will plan out the production steps (i.e. the construction processes and the technologies needed) that you expect to follow. These should be written as instructions to yourself.

Almost every product goes through the main construction processes of:

- measuring and marking out the pieces
- cutting the pieces to shape or 3D printing of any parts (some materials may require refining of the shapes)
- joining pieces using specific types of joins suitable for your material. These joins should be named in your production steps
- assembling or constructing all pieces into the final product
- surface treatment such as embellishment, decoration, cut-outs and finishing treatments.

The order of these production steps will depend on the needs of your product, and aspects such as the availability of equipment, etc. Your production steps plan will have a timeline and should aim to:

- list the steps in the order you expect them to occur
- give an estimated time required for each step
- set a date when you expect to complete that step.

Note that your production steps plan is just that – a *plan*! It will not necessarily reflect what you actually do. Any changes during production can be recorded as modifications in your journal, or record of progress. However, it is important to create plans before you start production. Doing so will cement your 'ownership' of the process, give you some control and independence from your teacher and may prompt you to check ahead for availability of facilities that you intend to use. This plan should also clearly identify any processes or technologies you intend to 'outsource', i.e. to get completed outside the classroom.

See more details in Unit 1 on pages 36–8, Unit 2 on page 90 and Unit 3 on pages 231–3.

Use the critical thinking technique of asking questions when researching, analysing and selecting the most suitable material. Ask what characteristics or properties are required of the product's materials (related to the product function and form, the user's needs and the context) as well as how to cut, manipulate, form or process the material to suit.

Examples include:

- What are the strength requirements? Is the material flexible? Can it be shaped or bent? Does it return to its original shape or dimension? Does the material need to be rustproof? Will it scratch or dent easily?
- Are there special conditions under which the materials must perform (outdoors, under rough treatment, underwater, in heat, humidity, etc.)?
- Does the design require tolerances for material movements? Will expansion/contraction limit the performance of the material?
- Does the product require materials with particular thermal or conductive properties?

Other aspects are also important for any materials-related decision:

- cost of the material – 'Is it affordable?'
- environmental impact of the sourcing, production, use and disposal of the material
- aesthetic qualities of the material – 'How do I want it to look?'
- ability of the material to be processed – 'Can I work with it?'
- availability of the material – 'Can I get it?'

When evaluating products

Questions that can be asked when evaluating existing products are:

- What technologies (or particular software) would most likely have been involved in the *design* of this product?
- What technologies would most likely have been involved in the *manufacturing* of this product? What technologies does one of your favourite designers use?
- What materials have been used? What characteristics or properties of the material used make it suitable for this product? Have any substances been used to enhance, finish or embellish the materials?

When evaluating your product, ask:

- Were the technologies chosen for the product the most suitable? Would other technologies have resulted in a better made and more effective product?
- Were the best, most suitable materials chosen for the product? Did the properties and characteristics of the materials suit the needs of the end user and the design situation?

If the information is not readily available, you may need to research an alternative, similar type of product.

Ethical considerations in design

Case studies
Etiko

Cantilever

Ethical considerations in design

Ethics in design is concerned with enabling both individual goods and values (such as more time with family and friends) and public goods (such as a fair and just society). This can be realised through products that reflect and enable an end user's values, or work towards social goals such as belonging, access, usability and equity for the disadvantaged. Inclusive design processes can enact respect and concern for humans and non-humans.

9780170477499

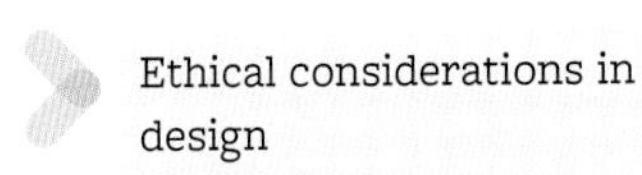

Ethical considerations in design	Ethical considerations encompass sustainability. Sustainability and other ethical considerations are concerned with human and non-human welfare and aim at positive impacts and minimisation of harm with regard to what is made and how it is made, for both present and future generations. Ethics can also involve legal responsibilities. The legal aspects of product design include intellectual property (IP), which refers to creations of the mind through intellectual or creative activities. Intellectual property includes copyright, patents, trademarks and registered designs. Australian and International (ISO) standards, regulations and legislation (including OH&S) are other legal responsibilities. Products must be produced safely and be safe for the end users.

Source: *VCE Product Design and Technologies Study Design 2024–2028*, pp. 18–9

Case study
Schamburg + Alvisse

Ethics in design

Ethics relate to our values – what we believe is good and bad, right and wrong, what is important to us. In product design, we consider the impacts of our products. Ethical designers aim to:

- reduce negative impacts – the things that might change our lives or the environment for the worse
- create products that have a positive impact – that improve health and wellbeing, that increase people's access and opportunities, and that are sustainable.

Designers can use their 'ethical compass' or sense of **ethics** to meet end users' values. This Factor has a lot in common with the 'end user/s' Factor. Products can be designed to help improve the quality of life of all humanity by using a 'speculative design approach' or by considering social needs such as belonging, access and/or usability.

Review speculative design thinking on pages 266–7.

Ethics
A branch of philosophy; moral principles concerned with human conduct, which identify right and wrong behaviour

Ethics and sustainability

Sustainability has environmental, economic and social dimensions. To be sustainable, designers consider the environmental impact of design, how it affects people, and how the economy can be healthy without negative impacts on the other two dimensions. A healthy economy is often said to be about growth in wealth creation and distribution. However, there is a mounting realisation that economies cannot continue to depend on growth that is based on endless consumption of finite resources. This century, it has become clear that expanding material wealth for the eight billion people who live on Earth is limited by the **biophysical** world from which all resources come and within which all economic activity takes place.

Weblinks
Design Sustainable Products

Design Sustainable Fashion

Go to the Business Victoria website for information on how to design sustainable products and sustainable fashion. Refer to Chapter 6 on sustainability and pages 142–4 to read about the triple bottom line and the three Ps: planet, profit and people.

Biophysical
Relating to the physical world and all the living things in it

You can equate the sustainability dimensions with the three Ps as follows:

- environmental dimension = planet
- economic dimension = profit
- social dimension = people.

Planet – looking after our natural resources:

- Choosing materials that come from an ethical source or that can be reused or recycled later, significantly reducing the use of non-renewable resources
- Reducing energy and water use where possible by using new and efficient technologies and efficient methods to distribute products (e.g. by train or boat rather than by air)
- Managing waste and pollution – following and increasing environmental regulations

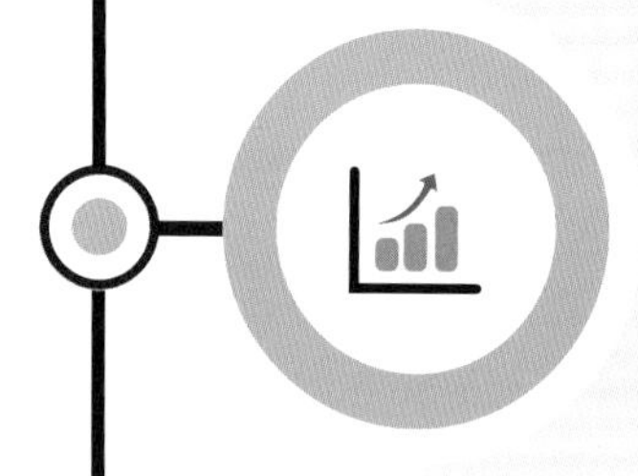

Profit – staying in business ethically:

- Managing and taking responsibility for long-term financial viability
- Avoiding unethical or unfair methods of making profits
- Honestly and accurately calculating and predicting overheads (e.g. operating costs, transport) and limiting financial waste
- Paying bills, taxes, costs and wages fairly and in reasonable time frames
- Ensuring quality and price are consistent across all regions
- Connecting with other 'ethical' businesses in the supply chain
- Investing in research and development that contributes to sustainability for the benefit of all
- Paying for the environmental and social 'costs' of products

People – supporting the health of all people, workers and communities:

- Considering, and improving where possible, the living conditions, health (mental and physical) and social impacts on consumers, societies and workers throughout the whole life cycle of a product, from extracting raw materials, processing and manufacturing to use and disposal of the product
- Giving back to the community where possible with grants or public infrastructure, by investing in environmental programs or by being philanthropic (donating to causes)
- Ensuring workers receive a living wage, don't experience undue pressure to work long hours or in unsafe conditions, and have opportunities to be trained – which all contribute to creating loyalty and therefore retaining workers, and reduce the cost of hiring and training new employees
- Ensuring products are safe and not damaging to consumers' mental or physical health.

9780170477499

Legal responsibilities

In any society citizens must obey the law and should expect to be penalised if they do not. There are several areas of legal responsibilities relevant to VCE Product Design and Technologies. They are intellectual property, occupational health and safety (OH&S) and mandatory Standards.

Case study
Valentino's resort range

Religious or cultural misappropriation

Appropriation is the act of borrowing elements from a religion or culture, usually one that is seemingly exotic or marginal, for purposes unrelated to their original significance.

Aboriginal and Torres Strait Islander Peoples have different secret stories and rituals, songs, dance and symbols that are sacred to specific members of the group, such as women, children or young men. Traditionally, artwork and markings on artefacts were a way of passing on this knowledge about Country and culture to specific people, and all people were restricted in what they could depict without causing offence.

Today, art and the use of symbols is really important for Aboriginal and Torres Strait Islander people in Australia; not only does it tell their 'stories', but it is closely linked to employment and livelihoods. There are many instances where Aboriginal artwork is misappropriated: it is 'taken' and reproduced on cheaply made, mass-produced 'tourist artefacts' by people with no understanding of the symbols they are using in this way. The use of unlicensed imitations of Aboriginal art on products may deeply offend Aboriginal people, harm their livelihoods and infringe copyright and moral rights. Ethical approaches require designers to collaborate with, or seek permission from, Aboriginal and Torres Strait Islander artists to use cultural or religious artwork, and to avoid 'misappropriation'.

Shutterstock.com/krithnarong Raknagn

Only if a product is made by an Australian Aboriginal person can it be called 'Authentic Aboriginal Art'; otherwise, it has to be called 'Aboriginal Styled Art'.

Intellectual property (IP)

Intellectual property (IP) is the property created by your mind or intellect – your knowledge and ideas. It also includes the means used to express ideas, such as writing, artwork, music, designs and inventions. Protection of IP is about the right to own, and earn income from, your creations.

Types of IP

Types of IP relevant to VCE Product Design and Technologies are:

- copyright and registered designs – the most relevant
- patents and trademarks (more relevant to industry).

Other types of IP that are not relevant are: circuit layout rights, plant breeders' rights and confidentiality/trade secrets.

Each area of IP protects a different aspect of a product. It is important to note that copyright differs from the other types in that it is free and automatic. It is necessary to apply and pay a fee for the other types of IP, and the protection is of much shorter duration.

Intellectual property categories most relevant to Product Design and Technologies

Copyright	Design	Patents	Trademarks
Free and automatic right	Need to be applied for, entail a cost and are administered by IP Australia		
Protects the physical expression of ideas, not the ideas themselves; rewards intellectual skill and effort in works such as books, films, music/sound recordings, newspapers, magazines and one-off artistic or craft works such as painting, sculpture, drawing, engraving, photographs, one-off fashion garments, furniture pieces or jewellery.	Design registration protects the overall appearance of a functional product (not the way it works). Design refers to the features of shape, configuration, pattern or ornamentation. It must be new and distinctive from other products on the market. It protects fashion, jewellery, furniture, homewares, toys and other accessories that are to be mass-produced.	Patents protect how an invention works or functions. They are for devices, substances, methods or processes that are new, inventive and useful. The types of products that can be patented are extremely varied. 'Innovation patents' are suitable for an innovative step, rather than an invention.	A trademark protects a letter, number, word, phrase, sound, smell, shape, logo, picture, colour, aspect of packaging or any combination of these – i.e. branding – that is used to distinguish the goods and services of one trader from those of another.
Protected for 70 years after death of creator	Protected for 5–10 years if fees are paid	Protected for 20 years, with fees payable every 5 years (8 years for an innovation patent)	Protected indefinitely if fees are paid every 10 years

Copyright and designs overlap

Once a 2D design or artistic work becomes a 3D product and is produced commercially (more than 50 articles are produced), it needs to be registered as a design in order to receive protection. It is then considered to be a commercial product and will no longer be granted the free and extended time protection of copyright awarded to one-off pieces.

- Copyright ownership can be sold or transferred to others.
- Design registrations or trademarks can belong to companies, not to individuals.

Weblink
Avoiding the Gaps in the Copyright/Design Overlap

You can read more in the article 'Avoiding the Gaps in the Copyright/Design Overlap' on legal firm Sharon Givoni Consulting's website.

9780170477499

3D printing

Both design and copyright laws can apply to 3D printing, so users need to find out who created any existing design being printed to reduce the risk of infringement.

What constitutes a design for registration?

Fashion	Industrial design
A fashion design can be registered for its visual appearance, such as: • a skirt with ruffles • the cut of a shirt or dress • an embroidered or textured surface • a decorative 3D pattern made with studs, sequins, ribbons, etc. • the shape of a purse • special pocket designs on the garment. Fashion companies may have both trademark and design protection on a product such as a bag or a shoe. Patents may be granted to clothing if there is a new device added, such as temperature-detecting technology or medical identification tags. They can also be granted to textiles such as fabrics treated with conductive polymers to transfer heat.	'Industrial design' and 'design objects' are terms that refer to functional articles such as furniture, homewares, toys, machinery or fashion accessories. Such articles can be registered for their visual appearance, such as: • a pattern woven into carpet • a chair with unusual legs or back support • the arrangements of shelves and compartments inside a fridge • the shape of the body, spout and handle of a kettle.

Examples of patents

Patents protect the idea of how something works or functions. The patent itself is a document containing technical drawings, measurements and written descriptions – it covers the **scientific and/or technical knowledge** contained in its descriptions. Patents can be granted for products such as bicycle helmets, garage doors, wheelchairs and air conditioners. Some recent Australian patents have been in the surf industry for products such as fin plugs, foot strap systems, removable and adjustable surf fin systems, a storage compartment within a surfboard, heated wetsuits and surfing sunglasses.

Shutterstock.com/Wright Studio

Multiple registrations

It is possible that one product may require several forms of protection.

Worksheet
13.9.1

Protecting your own IP

IP protection gives the owner the right to:

- exploit (use, license or sell) their work commercially for a given time
- prevent its commercial use by others.

If you want to commercialise your 'creation', it is important to seek legal advice on the best protection against your design being copied or misused. You are also responsible for monitoring the market to check for infringements.

If using another person's IP, you must seek permission. Legal action can be taken against those who infringe (see below). Note that the creator is not necessarily the IP owner: the creator may be working under contract, in which case the contract will state who owns the IP.

What is IP infringement?

Using another person's intellectual property commercially, i.e. copying and selling it, without permission or acknowledgement of the IP owner is considered an infringement and is against the law. For example, you could not use the Nike 'swoosh' on your own products and sell them commercially. If you were discovered to be doing so, you could be issued a warning letter; a request for a licensing arrangement (granting permission, usually with associated fees); an out of court settlement (i.e. money); or court action.

Moral rights

The concept of moral rights was introduced into Australian law in 2001, in the *Copyright Act 1968 (Cth) Part IX*. The original creator of a work has moral rights. They may not own the IP (if the work was created on behalf of their employer, or due to contractual obligations or on-selling of their IP) but they have non-economic rights.

These include the right:

- to be attributed as the creator of the work in a clear and reasonably prominent way so that it is obvious to the viewer
- not to have work falsely attributed (i.e. attributed to another person) or to have their work altered and be attributed as unaltered
- to take action if their work is treated in a derogatory way (mutilated, distorted or displayed in a context that affects their reputation).

Moral rights: attributing others' work in your folio

When using images of work created by others in your folio, you must observe their moral rights by displaying their name in a clear way so that it is obvious to the viewer. As a student, you are entitled to use others' IP for educational purposes, but you must still acknowledge them. However, if you are displaying your work in public in a commercial way, such as for sale or in a public exhibition, you do not have the same allowance and must seek permission. (Schools pay fees for licences to use copyright works in the course of educational instruction.)

The best way to acknowledge the IP or moral rights associated with images, ideas and other forms of 'borrowed' content in your folio is to identify the creator of the work (the designer, manufacturer, etc.) alongside the image or work. If you don't know the creator, identify the source from which you borrowed the work. Note that a bibliography (which is *not* required in this subject) at the end of your folio would not always correctly and clearly acknowledge moral rights. This would require you to number every image in your folio and

9780170477499

include corresponding numbers in the bibliography. You should also remember that long URLs copied from online retail sites can be very unstable, are sometimes withdrawn after a short time and very often do not give the creator's name (i.e. the name of the designer or the brand name).

When developing your folio, review the information provided about IP recognition for Unit 3, Outcome 2 in Chapter 7. It explains how to make a clear distinction between the work done by you and any work done by others.

Australian and International (ISO) Standards

What is a standard?

Standards are documents that set out technical specifications or other criteria necessary to ensure that a product, service or system will consistently do the job it is intended to. Their purpose is to ensure:

- safety
- reliability
- consistency
- quality.

They are living, changing documents that reflect progress in science, technology and systems.

Australia has close to 7000 standards and, wherever possible, they mirror International Standards. The International Organization for Standardization (ISO) is responsible for developing worldwide standards.

International Standards ensure that products manufactured in one country can be sold and used in another. Standards reduce technical barriers to international trade, increase the size of potential markets and position Australian firms to compete in the world economy.

Standards are identified by a number and the year they were last amended. Australia and New Zealand have common standards, hence you will see many standards with the prefix 'AS/NZS', followed by their number and the year.

Note that not all products have standards, and that most standards are adopted voluntarily.

Mandatory standards

Mandatory
Required by law

Some products have **mandatory** standards.

Many of these standards relate to products for children, including:

- a wide range of children's toys (AS/NZS 8124.1: Year updated 2018)
- children's nightwear (AS/NZS 1249: Year updated 2014)
- bunk beds (AS/NZS 4220: Year updated 2010)
- cots (AS/NZS 2172 and 2195: Year updated 2013)
- beanbags (Safety Standard 2014).

Weblink
Product Safety Australia

Standards Australia

A standard becomes mandatory when the government identifies it or refers to it in legislation. Find out more about mandatory standards, and about safety with beanbags in particular, at the government website Product Safety Australia. Go to 'Product Safety Laws' and select 'Mandatory Standards'. Read more about Australian Standards at the Standards Australia website.

International Standards

Weblink
ISO Standards

Many International Standards relate to how a product is produced, rather than to the product itself – they outline management systems such as Quality Management Systems (ISO 9000) and Environmental Management (ISO 14001). They are generalised and can be applied to all companies, whether they are making cheese or supplying counselling services (for example). ISO does not certify products; each country has its own organisation to do this.

You can read more and watch a short video at the ISO Standards website.

Regulations and legislation (including OH&S)

Occupational Health and Safety (OH&S) laws stipulate that all workplaces must be kept safe for employees, customers and the general public. Workplace illness and injury can cost a business a great deal, both financially and in reputation.

All businesses in Australia are legally obliged to provide:

- safe premises
- safe machinery and materials.

Employees have a responsibility to follow any safety directions and guidelines and to inform their employers if there are any safety issues.

In Victoria the *Occupational Health and Safety Regulations 2017* (made under the Occupational Health and Safety Act) specify details to support the Act (e.g. requiring licences for specific activities, keeping records and notifying certain matters). The regulations are closely aligned with the *Model Work Health and Safety (WHS) Act*, published by Safe Work Australia, which aims to harmonise safety regulations across Australia. Safety regulations are often changed, so it is best to check the relevant websites for updates.

Weblinks
Safe Work Australia

Australian Government Business

WorkSafe Victoria

Look at the websites of Safe Work Australia, Australian Government Business and WorkSafe Victoria to find out more about OH&S regulations.

9780170477499

Product safety

Products must be produced safely and be safe for end users.

In Australia, consumers, businesses, government agencies, safety experts, standards writers and consumer advocates work together to maximise the safety of all goods sold and used. A combination of measures exist to promote product safety. They include:

- voluntary actions by suppliers
- government laws that give incentives for making safe products
- restrictions on selling unsafe products
- information and education that enables consumers to choose safe products and use them safely.

Shutterstock.com/Green Mountain Montenegro

When developing a design brief

- Include requirements (constraints) relating to sustainability or other ethical values.
- Check whether there is an Australian Standard that is relevant for your intended product. If your intended product is regulated by a mandatory standard, then you will need to obtain the standard to find out the requirements. Examples are bunk beds, beanbags, basketball rings, cots, nightwear for children, and toys. Standards can be very expensive, but you may only need to purchase relevant parts; alternatively, your school could buy these as a resource.

During the design process

- Research in depth and make ethical and sustainable choices around the overall design of your product, the materials you choose to use (check the supply chain), waste minimisation, efficient production processes, quality construction methods and disposal or reuse of the materials in your product.

- Attribute IP of any 'work' to its creator in your folio pages by making their name clear and obvious.
- When designing your product, think about the safety of the end user (e.g. in terms of protruding sharp edges, and the way it is to be constructed for reliability) and consider that your product may be used in an unintended way. To revise risk management, go to pages 48–9.
- During production, be sure to follow OH&S guidelines. Plan for safe production by doing a risk assessment of the environment, for handling materials and chemicals and using machinery, and consider modifications to ensure that your finished product is safe for users.

When evaluating products

Questions that can be asked are:

- What materials and construction methods were used, where was it made and how is it transported to consumers? (Many companies now have websites that give details of their entire supply chain.)
- What information does this give you about the impacts of the materials, how far they have travelled, and the types of environmental and worker protections that may have been in place?
- How could this product be more sustainable? Your suggestions could be for the product itself or the way it is distributed.
- Does the company put back into society by donating to research or by running community programs? Research will be required, and you will have to investigate claims made by the company.
- Is the product protected by one or more types of intellectual property?
- Are there any safety issues with the product during production and use? How have they been addressed?
- Do you believe the factory provides a safe environment for the workers in it? Explain or give evidence.

When analysing designs by others, acknowledge the IP (attribute the creator), check if there is a relevant standard and research its number, and determine what aspects of the product contribute to its safety.

Design specialisations

This section explains **some of the new terminology** under the heading 'Design specialisations' starting on page 15 of the *VCE Product Design and Technologies Study Design 2024–2028*.

Interdisciplinary and transdisciplinary nature of design

A discipline is a practice you are skilled at (or in which you are building your skills) or an area of knowledge that you have cultivated. The term 'interdisciplinary' can refer to the use of hand skills along with many technologies in a particular material category. 'Transdisciplinary' refers to the knowledge or skills that you may have from working in one area, such as a category of materials or the use of CAD, and how you can share or transfer that knowledge or those skills to other areas and other people, while learning new things from them in turn.

9780170477499

Technacy

This term refers to the ability to work across many technologies with many materials and with a knowledge of the user and the context, to solve problems effectively. It also covers all the knowledge and skills required to achieve an innovative and creative solution.

Assistive products

Assistive products are those products that provide accessibility to people with disabilities or limitations. Products are specifically developed to assist these people to perform daily tasks and to function independently. They can help by improving mobility (e.g. wheelchairs, prosthetic limbs), communication (e.g. hearing aids, screen readers for the blind), accessibility (e.g. clothing or utensils) and overall quality of life. These products are also known as assistive technologies or assistive devices.

Biophilic design

Biophilic design is a design philosophy, used within the building industry, that seeks to connect people with nature and to incorporate elements of nature. It builds on the concept that humans have a deep-seated and innate connection to nature. Incorporating elements of nature into our built environments creates spaces that look and feel natural. This can positively impact our health, wellbeing and productivity. Research has shown that biophilic design can have various benefits, including reducing stress and anxiety, improving cognitive function, boosting creativity and productivity, and enhancing overall wellbeing. It is even said to reduce crime.

Biophilic
Demonstrating a love of nature and living things

Biophilic design can take many forms, including incorporating natural light, using natural materials, using plants and greenery, using natural patterns and shapes, and providing views to the natural environment. Biophilic design can be applied to a wide range of settings, including homes, offices, schools, hospitals and more.

Wearable technology

The term 'wearable technology' refers to electronic devices or gadgets that can be worn on the body or incorporated into fabrics from which clothing or accessories are made. Wearable technology is possible due to advancements in miniaturisation of electronics, sensor technology and connectivity (the ability to connect to other devices or networks).

Common examples of wearable technology include smart watches, fitness trackers, smart glasses, smart clothing and wearable medical devices. Most of these provide functionalities such as tracking biometric data, monitoring posture or controlling body temperature. Wearable medical devices, which include heart rate monitors, insulin pumps and prosthetic limbs with embedded sensors, are designed to monitor vital signs, deliver medication or assist in rehabilitation.

You may not have the capacity to design the electronic aspect of a wearable technology, but you could consider the idea of designing the casing or 'container' for such a thing.

Lapidary

Lapidary refers to the art and craft of working with gemstones, minerals and other hard materials to create decorative objects. Lapidary work involves cutting, shaping, polishing and engraving these materials to enhance their beauty and create items such as jewellery, sculptures, ornaments and decorative pieces. Lapidaries, or lapidary artisans, use specialised tools and techniques to transform rough stones into finely crafted and polished pieces, often highlighting the natural colours, patterns and textures of the materials. Lapidary work requires skill, precision and an understanding of the properties and characteristics of different gemstones and minerals. It is a creative and meticulous craft that combines artistic expression with the appreciation and manipulation of natural materials.

Fibres such as roving or yarns

Fibres are the small, thread-like structures that are the fundamental building block used to create various textile materials. They are processed, spun into yarns, and then woven, knitted or otherwise formed into fabrics. The selection of fibres depends on factors such as the desired end use, comfort, durability and cost.

Fibres can be either natural or synthetic.

- **Natural fibres** are derived from plant or animal sources. Examples of natural fibres are cotton, wool, silk, linen, jute and hemp.
- **Synthetic fibres** are manufactured fibres created by using a chemical process to transform raw materials such as petroleum-based chemicals or cellulose from wood pulp into polymers that can be extruded into long filaments or spun into staple fibres. Common examples are polyester, nylon, acrylic, rayon and spandex.

Textile manufacturers often blend different fibres to achieve specific performance or aesthetic qualities. Blending fibres can combine the desirable properties of different materials, such as the strength of polyester and the comfort of cotton. Fibre blends can exhibit enhanced fabric durability, breathability and other functional characteristics.

Roving is a type of fibre preparation commonly used in spinning and felting processes. It consists of long, loosely aligned fibres that are typically drawn from a fibre preparation or carding machine. Roving fibres are generally thicker and less processed than yarns, making them suitable for creating bulkier and textured yarns or for felting projects.

Yarns are continuous strands made by twisting or spinning fibres together. Yarns can be created from a wide range of fibres, including natural materials like cotton, wool, silk and linen as well as synthetic materials such as polyester, acrylic or nylon. Yarns can vary in thickness, texture and composition, offering different qualities and characteristics for various applications. They are used for knitting, crocheting and weaving.

Biomaterials

Biological materials, or biomaterials, may not be easy to obtain as many are experimental or in limited supply. You can read more about **mycelium** and how it can be used to create new materials on pages 161–2 of Chapter 6 (Unit 3, Outcome 1).

If you wish to use biomaterials, you will need to research their availability and cost. You could do your own experiments, but you must also take into account how long it might take you to get the material in a usable form.

Some biomaterials are used in **medical and healthcare applications**. Scientists and engineers are exploring innovative approaches, such as 3D bioprinting and nanotechnology, to develop biomaterials with enhanced properties, improved biocompatibility, and more precise control over tissue regeneration.

9780170477499

- **Natural biomaterials** are derived from organic sources, such as plants, animals or their byproducts. Examples include collagen, alginate, chitosan, silk and cellulose. Natural biomaterials often possess biocompatibility, meaning they are well tolerated by the body and do not cause adverse reactions.
- **Synthetic biomaterials** are manufactured materials designed to mimic or enhance the properties of natural tissues. Examples include biodegradable polymers, such as polylactic acid (PLA) and polyglycolic acid (PGA), as well as hydrogels and ceramics. Synthetic biomaterials offer advantages such as tunable properties, controlled degradation rates and tailored functionalities.

Summary
Chapter 13

Quiz
Chapter 13 revision

CHAPTER
14

Materials and testing

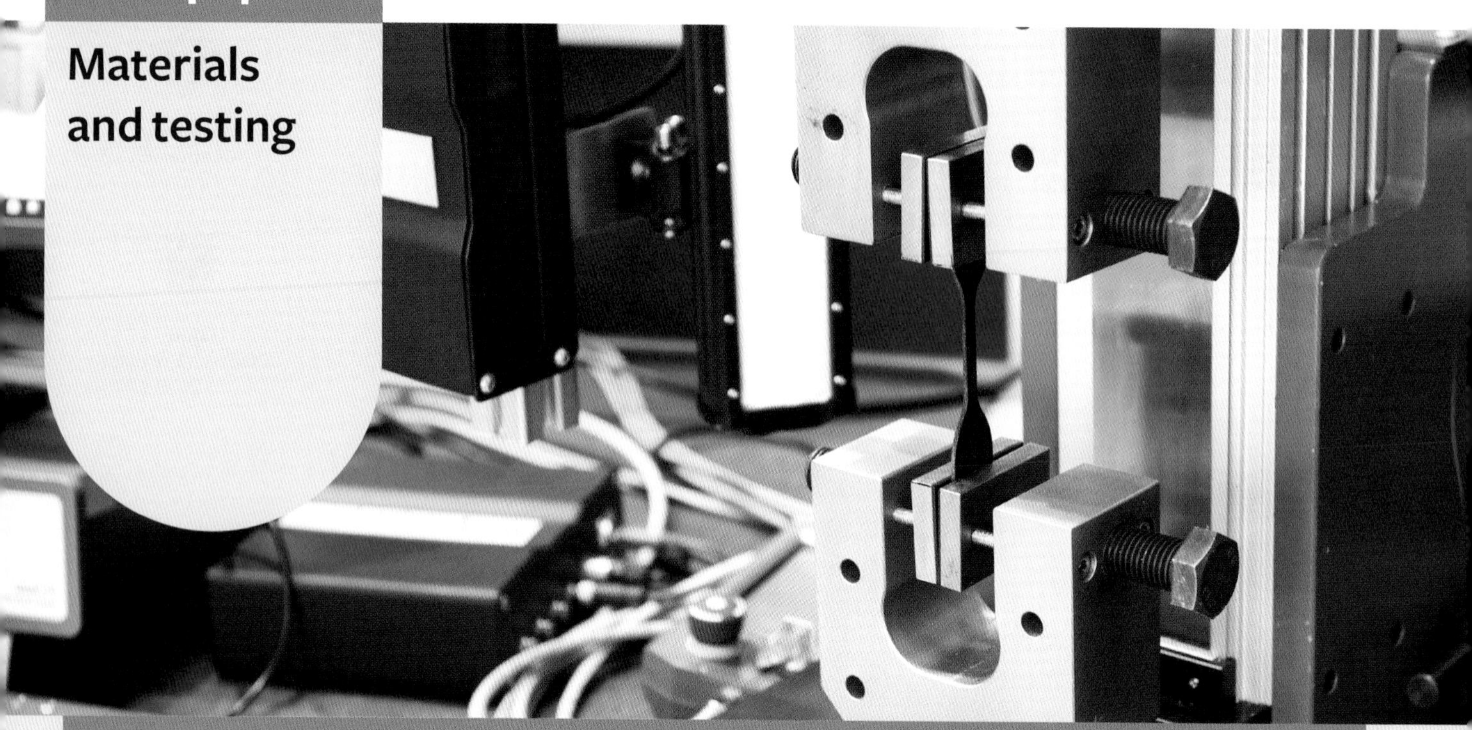

In this chapter, you will learn about:
- materials categories
- the characteristics and properties of materials
- methods of testing, and to report on tests
- classification, materials processing and production, and issues related to:
  - » wood
  - » plastics
  - » metal
  - » textiles.

Materials worksheets:
- Materials research
- What materials could you use?
- Materials testing
- Research on sustainability
- Construction process trials

Summaries
- Chapter 14 **(p.427)**

Quizzes:
- Chapter 14 revision **(p.427)**

Resources:
- Materials and processes

Case studies:
- Bryan Cush **(p.409)**
- Cantilever **(p.409)**
- Jackie Staude **(p.410)**
- Marcos Davidson **(p.410)**
- Bodypeace **(p.416)**
- Emily Barrell **(p.416)**
- FOOL clothing **(p.416)**
- Chapter 14 case study report activity

Nelson MindTap

To access resources above, visit **cengage.com.au/nelsonmindtap**

9780170477499

Knowledge of materials

A broad knowledge of materials helps designers to make decisions about the most appropriate materials for a product. So an important aspect of design thinking is researching, testing or experimenting with materials and analysing information to help select the best material solution for each design situation. Materials testing gives you a greater understanding of the properties and characteristics of materials – and usually involves a comparison of two or more materials.

The section that follows provides you with guidelines for developing, carrying out and analysing materials tests. You are also given suggestions for tests on pages 422–7 that might be useful for resistant and non-resistant materials.

Resources
Materials and processes

Understanding materials

Materials worksheet
Materials research

This chapter covers many different aspects related to materials. It covers:

- materials categories
- characteristics and properties
- testing of materials – including materials testing suggestions
- general information about material groups – wood, metal, plastics and textiles
- classification
- sourcing and processing
- quality
- health and safety issues
- environmental issues.

Resistant and non-resistant materials

There are many ways of classifying materials, depending on the purpose or reason for your classification. One simple way is to define the material according to its resistance, i.e. what happens when it is pushed, pressed or bent? Does it move easily or resist?

Examples of non-resistant and resistant materials

Non-resistant materials	Resistant materials
Materials that are flexible, easily moved, bent and shaped; materials that can be used to cover surfaces: • fibres, yarns and fabrics • synthetic/plastic and bio-membrane • rubber and leather.	Materials that are generally solid in form (at room temperature) and, because they show only a limited amount of flexibility, are often used for structural purposes: • wood • metal • glass • ceramics • solid plastic • stone • concrete.

Materials worksheet
What materials could you use?

Material categories

In VCE Product Design and Technologies, materials are classified in broad categories and subcategories. It is helpful to be aware of the classification of all materials, and it is critical to know about the classifications of the materials you are using.

Each of the major material categories (wood, metal, fabric, plastic, etc.) is broken into a number of subgroups. Placement of a material within a subgroup is usually determined by its source or its chemical structure. Materials in these subgroups often have similar characteristics or properties. Some of the main subgroups are then further broken into minor groupings.

Material category	Major classification subgroup		Examples
Wood	Softwoods		Pine, cedar, fir, spruce
	Hardwoods		Eucalypts, acacias, beech, mahogany
	Manufactured boards		Plywood, particle board, MDF
Metal	Ferrous		Mild steel, stainless steel, cast iron
	Non-ferrous		Copper, tin, aluminium, zinc, silver, gold
	Alloys		Bronze, brass, pewter, stainless steel
	Coated metals		Galvanised iron, powder-coated metal
Plastics	Thermosetting		Resin, epoxy, silicon, urethane
	Thermoplastic	Petrochemical	ABS, PVC, polypropylene, polystyrene, acrylic
		Bioplastic (bio-based polymer)	PLA, cellophane, bio-polyethylene
	Composites		Fibreglass , carbon fibre
Fabric, fibres and yarns	Natural	Protein (animal)	Wool, silk, angora, alpaca
		Cellulose (plant)	Cotton, linen, jute, hemp
	Manufactured	Regenerated (cellulose-based)	Viscose, acetate
		Synthetic (petrochemical)	Nylon, acrylic, polyester
	Blended and laminated		Polycotton, Gore-Tex™
	Skins, membranes and bio-materials		Leather, rubber, latex, mycelium, Scoby

Resistant materials: characteristics and properties

Appropriate use of a material depends on knowing its particular characteristics and properties – that is, how it will behave in different circumstances.

Read the sections on specific categories of materials later in this chapter for more detailed information about their characteristics and related issues.

Mechanical properties

- **Strength** is one of the most important properties since it determines a material's ability to hold loads without failure. Strength can vary greatly depending on how the forces of a load act on the material. Cast iron, for example, is strong in compression strength but comparatively weak in tensile strength.
- **Hardness** is the ability of a material to resist scratching, abrasion or indenting, and is closely related to strength. Sunglass lenses, for example, need to be scratch-resistant.
- **Elasticity** is the ability of a material to return to its original shape and dimensions when a deforming load is released. Bungee jumpers, for example, depend on the elasticity of

9780170477499

the cords to absorb shock when they jump, and to return to their original shape for use by the next jumper.

- **Stiffness** or **rigidity** is the ability of a material to resist deflection under load. It is important in the design of products such as shelving or sailboard fins.
- **Plasticity** is the ability of a material to be permanently deformed without breaking. It is important when shaping processes are being considered, as the material may have to be heated or worked in special manner.

Bookshelves shouldn't deflect or bow under a heavy load of books.

- **Malleability** is a material's ability to manage all types of deformation without damage. Gold, silver and other precious metals are well known for being malleable. They can be readily hammered into thin sheets or drawn into fine wire.
- **Toughness** is the ability of a material to absorb energy when being deformed, and therefore to resist deformation and failure. Steel, for instance, would be considered tougher than window glass, particularly where suddenly applied impact loads are a concern.
- **Ductility** is similar to malleability. In particular, it is the ability of a material to be drawn out (stretched longitudinally), yet remain intact, e.g. copper and gold can be pulled into thin 'wires' without breaking.

Physical properties

- **Density** is the mass per unit volume of a material. Materials of high density are comparatively heavy for their volume, e.g. gold and lead. Low-density materials seem light for their volume, e.g. balsa wood.
- **Porosity** depends on the number of holes or pores present in a material; it is usually expressed as a percentage of the total volume.
- **Moisture content** of a porous material can have significant effects on its strength properties. 'Green' timber, for instance, has different properties to seasoned wood. Many plastics also absorb water – enough to change their levels of strength and electrical insulation. Some 3D filaments absorb moisture over time and become less reliable. The volume of a material may also be altered by moisture content. Changes in the moisture content of timber through absorption or seasoning (drying) may result in the timber swelling or shrinking.
- **Thermal conductivity** is the ability of a material to transmit heat. Metals are by far the best conductors because of their atomic structure. As the temperature of a material is raised, it normally expands (increases in volume).
- **Heat resistance** is determined by the melting point of a material, and its chemical stability and strength at high temperatures. Most heat-resistant materials are ceramic-based, but these have a poor resistance to thermal and mechanical shock (rapid changes in temperature, or being struck). This limits their application as moving components. On the other hand, metals such

Copper effectively conducts electricity.

as tungsten, titanium and molybdenum retain their strength and toughness at high operating temperatures and remain heat- and shock-resistant.

- **Electrical conductivity** depends on the material's structure and bonding, as flow of current relies on the movement of charged particles (usually electrons) throughout the material's atomic structure. All metals and carbon (in the form of graphite) are good conductors.
- **Corrodibility** refers to the vulnerability of a material to corrosion, or destruction due to chemical or electrochemical attack. Metals corrode due to oxidation (e.g. rust), however, non-metals, such as concrete and stone, are also susceptible to atmospheric and chemical corrosion.

Shutterstock.com/Oleksii Fedorenko

Shutterstock.com/CHAIYA

Corrosion of metal and concrete

- **Durability** refers to how much a material can take before it breaks or is damaged beyond usefulness. It depends a lot on the forces applied, exposure to weather, chemicals and time, and the way it is treated.
- **Ease of machining** is not essentially a property but is related directly to how a material responds to being machined. For example, ironbark is an extremely dense, heavy and hard timber, which makes it difficult to machine. Ease of machining is definitely something to consider in the choice of materials for your projects.

Shutterstock.com/ben44

Fine, even-grained timbers are easier to turn on the lathe.

Non-resistant materials

It is important to understand how non-resistant materials (such as fabrics, fibres and yarns) will perform under different circumstances in order to ensure that suitable materials are chosen. The characteristics and properties of non-resistant materials are, however, difficult to compare directly with those of resistant materials. They are defined and tested quite differently.

9780170477499

Characteristics and properties

When considering the characteristics and properties of a fabric, the following will have an effect on the characteristics of that fabric and how it responds in different situations:

- the structure of the fibre
- the physical form and chemical components of the fibre
- the method and tension of twist in the yarn
- the construction of the fabric.

For example, woollen fibres have a scale-like structure and are usually crimped. These characteristics contribute to the thermal insulating property (when the fibres are processed into fabric, the unevenness of the fibre creates little air pockets). Cotton has an uneven and twisted fibre that does not reflect light easily, so it has a low sheen. Most silk and most synthetic fibres have a regular and smooth surface. They reflect light, and therefore have a higher level of sheen.

Yarn is traditionally made from fibre. The characteristics of yarns depend on the type of fibre and the physical structure of the yarn – the direction of twist, the amount of twist, and the number and twist direction of the threads that form the yarn. The blending of fibres in a yarn will also change its properties. For example, blending a small amount of lycra with cotton gives it more elasticity, while blending a natural fibre with a metal will add a stiffness.

Cloth fabric is traditionally made from yarn or thread, though some types of cloth, such as felt, are made directly from fibre. The characteristics and properties of a fabric relate to the type of fibre used and the construction of the fabric – whether it is woven, knitted or felted, whether it is densely or loosely constructed, and whether layers are laminated or bonded to form the fabric. The way a fabric is finished will also affect its properties. For example, the brushing of a fabric will increase its ability to insulate, and a plastic coating will make it waterproof.

Note that there is a lot of overlap between characteristics and properties, and the terms are sometimes used interchangeably. These terms have specific definitions in particular technical fields. In this study, however, we usually group them together for ease of understanding.

Strength usually relates to the ability of the fibre, yarn or fabric to withstand tension without breaking. The strength of some fibres changes when wet – e.g. wool loses strength and cotton gains strength when wet.

Durability is a more general property that takes into account the material's strength, elasticity and resilience. The level of durability will indicate how the fabric will withstand everyday wear and tear. A fabric's durability may be tested by repeated washing and/or through abrasion testing.

Elasticity is the ability of a fibre, yarn or fabric to stretch without breaking, and also its 'stretch recovery' – whether the material returns to its original shape after stretching. Elasticity of a fibre usually affects its 'wrinkle recovery' – whether it stays creased after crushing or appears 'uncreased' – and its ability to retain shape. Wool, for example, has elasticity, retains its shape and has good wrinkle recovery. Elasticity in fabric is also important for comfort and fit.

Moisture absorbency relates to how much water the fibre, yarn or fabric absorbs. When considering absorbency, the terms 'hydrophilic' (water-loving) and 'hydrophobic' (water-hating) are often used. Fabric construction and surface finishes may affect the ability of a fabric to absorb or repel water. The ability of a fabric to dry quickly may also be related to its level of absorbency, e.g. cotton absorbs moisture and takes a comparatively long time to dry. It is important to think about comfort when analysing absorbency, as some fabrics may absorb much moisture but feel cold and clammy next to the skin (e.g. cotton), whereas others absorb moisture well but still feel warm (e.g. wool).

Moisture-wicking fabric allows moisture to travel easily from the inner to the outer surface of the fabric. Such fabrics are often used in activewear or sportswear as the fibre and fabric structure draws moisture away from the body to the outer surface of the clothing, where it can evaporate quickly and easily.

Getty Images/FatCamera

Some sports clothing is made from absorbent material, such as cotton, for comfort; these uniforms are made from synthetics to wick moisture away from the body.

Shutterstock.com/Getmilitaryphotos

Good thermal insulators keep the body at a regular temperature.

Thermal properties refer to the ability of a fibre, yarn or fabric to insulate against, or conduct, heat or cold. If a fabric is a good thermal insulator, then it will keep the body at a regular temperature and will not be affected by outside conditions. If a fabric is a good thermal conductor, it will allow the body to heat or cool quickly according to the outside temperature.

Some fabrics, fibres and yarns have adverse **reactions to chemicals** of different types, such as acids, alkalis and bleaches. Understanding the effect of these chemicals on a material may determine how it should be treated and washed. For example, wool is not affected by acids but is weakened by alkalis and bleaches, and therefore should be washed with specially formulated soap and not bleached. Conversely, cotton is resistant to alkalis but is weakened greatly by acids.

9780170477499

Some fabrics, fibres and yarns are greatly weakened by prolonged **exposure to sunlight**. Understanding these effects is important, particularly when making choices about materials that will be used outside.

Shutterstock.com/Stephen Denness

Children's beach clothing should give a high level of UV protection.

Related to sunlight, the level of **protection against ultraviolet (UV) light** may be determined by the fibre (some fibres are naturally protective; others may have protective treatments added during processing), the construction of the fabric (its density) or a finishing process (a protective coating may be added to the fabric surface).

Fabrics, fibres and yarns vary in their **resistance to biological attack**. Fungi, mildew and cloth moths may damage some materials, whereas others are resistant. An understanding of this is important when making choices of material for a product used in humid, damp conditions, or when considering how to treat clothing for storage. Bamboo fibre is thought to have this resistance.

The **ability to take and retain dye** also varies. Different fibres absorb and retain dyes in different ways. Some natural fibres will absorb a range of dyes, whereas others – such as flax – cannot be dyed without weakening the fibre. Generally, hydrophilic (water-loving) fibres can be easily dyed. Many synthetic fibres that are hydrophobic (water-repelling) may need to have the colour added during fibre production.

Shutterstock.com/Paul Prescott

Some natural fibres will easily absorb a range of dyes.

Shrinkage is what happens when a fibre, yarn or fabric becomes smaller after washing. The washing temperature and washing solutions may affect the level of shrinkage.

Drape is a term that refers to the flexibility of a fabric – how it falls and folds. Drape is determined by both the stiffness/elasticity of the fibre, and the density/weight of the fabric construction. A fabric is said to drape well if it falls softly over forms, is flexible and folds easily.

Shutterstock.com/Luba V Nel

Fabric that drapes well falls softly over forms.

Wrinkle resistance and recovery relate to a fibre's elasticity. If a fibre is elastic (i.e. it bends and returns to shape), then fabric made from it is less likely to be affected by handling. Some fibres become more elastic (e.g. wool) and others less elastic (e.g. cotton) when wet. This means that cotton fabrics need to be handled appropriately to reduce washing wrinkles. The

plasticity of a fibre (its ability to be moulded and changed) may also affect its wrinkle resistance and recovery. A fibre that is considered 'plastic' (e.g. silk) can wrinkle severely, but can be easily returned to its wrinkle-free state.

Wood

Timber classification: hardwood and softwood

Timber is classified according to its botanical features as either hardwood or softwood.

There is a general misconception that the term 'hardwood' means these timbers are harder than softwoods. Many hardwoods are fairly strong and dense, but the group also contains such timbers as balsa wood. Hardwood trees usually have broad leaves, flowers and fruit. Softwoods come from trees with needle-like foliage and cones.

 Bamboo is a plant that is increasingly being used for furniture and other products. It is difficult to place in timber categorisation as it is technically a grass.

Shutterstock.com/Harm Kruyshaar

Examples of commonly used hardwoods and softwoods

Hardwoods		Softwoods		Manufactured boards
Balsa Beech Blackwood Jarrah Mahogany	Meranti Merbau Redgum Teak Victorian ash or Tasmanian oak (eucalypt)	Cedar (excluding Australian red cedar) Cypress pine Douglas fir Hoop pine Huon pine	Queensland and NZ kauri Radiata pine Redwood Rimu Spruce	Blockwood Masonite MDF Mouldings Particle board Plywood

9780170477499

Timber production

Both hardwoods and softwoods are obtained and prepared for use by the processes of logging, milling (cutting from the log) and seasoning (drying). When timber is first logged, it has a high moisture content. Seasoning removes moisture from the timber until it reaches a stable and usable level. This can be done either in the open air or in a kiln (kiln drying is faster, more controlled and more stable).

Moisture content and milling methods

Different trees have a different moisture content when they are newly cut down – radiata pine can have a moisture content of up to 180 per cent (there is more water in the cells than there is plant material) but trees that grow in a drier climate, such as cypress, may have a moisture content of only 45 per cent. After timber is seasoned, it should have a moisture content of about 15 per cent. As the moisture evaporates, the cells shrink, which means that radiata pine will shrink a lot more than cypress during seasoning.

The way timber is milled also has an impact on how the timber moves and shrinks as it loses moisture. The cells in the outside layers of a tree have a higher moisture content and will shrink more. If a piece of timber is cut radially towards the centre of the log (quarter-sawn), the part that is furthest from the centre will narrow a little. However, if the timber is cut across the log (through, live or back sawn), the sides and ends closest to the outside will shrink more – cupping or warping the plank and causing more distortion.

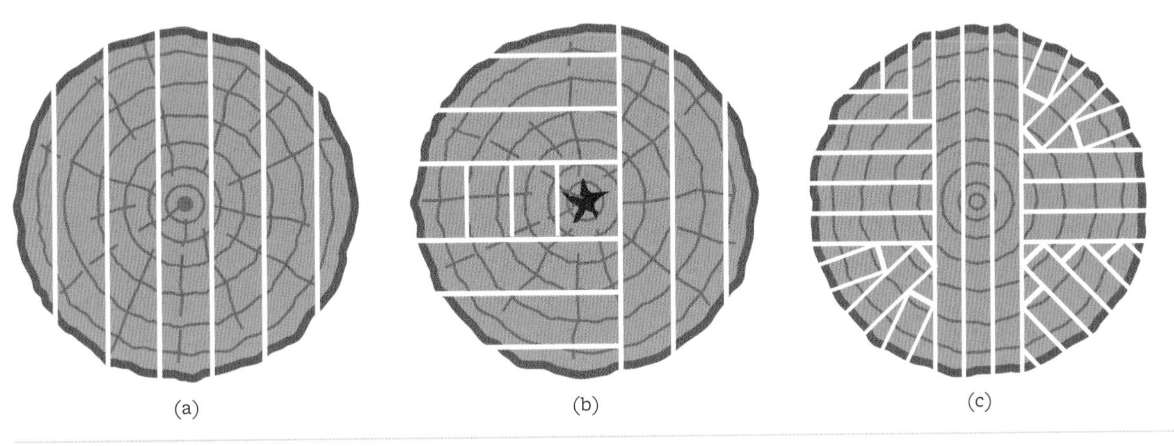

Different methods of sawing: (a) plain, through or live sawing, (b) back sawing, (c) quarter sawing

Manufactured boards

Manufactured boards use wood as the raw material, usually in veneer, chip or pulp form. These are processed to make new materials, usually in sheets, boards or moulded shapes. Manufactured boards are made in various thicknesses, textures and qualities, and are useful for constructing a wide range of products for different contexts.

Commonly used manufactured boards include:

- plywood – made from thin layers of timber veneer, with the grain of each layer turned at 90 degrees and crossed. It is very strong for its weight, and versatile
- chipboard or particle board – made from larger timber chips bound with glue and pressed
- medium-density fibreboard (MDF) and highly moisture-resistant (HMR) fibreboard – made from finely processed timber fibres bound with glue (not recommended for use in schools)

- blockwood – a core of solid timber strips covered on both sides with layers of veneer (often used for bamboo sheets)
- masonite (a hardboard) – made from finely processed timber fibres processed with high-pressure steam, and without glue or binder.

Shutterstock.com/NataVilman

Manufactured boards are made from processed raw wood.

A thin layer of quality veneer is often used to cover the outside surface of manufactured boards. This conserves the highly valued timber and improves the appearance of the boards.

General properties of wood

Seasoned timber is a good thermal and electrical insulator, and is stronger in tension and compression than when it is subjected to shear forces. Wood is combustible and suffers from biological attack. Some timbers, however, have antibacterial properties.

Health and safety issues related to wood

In its solid form timber poses few, if any, dangers apart from those that can occur through poor lifting and carrying procedures.

Timber dust can be an irritant and long-term exposure to it can cause sinus and respiratory health problems. Certain timbers can bring on an allergic reaction in some people, either in their nasal passages through the inhalation of dust, or on the surface of their hands through handling.

Machining operations on timber and timber products pose hazards through the creation of dust and noise, along with the associated dangers of machine operation, such as flying debris and machine contact. Adequate ventilation and eye and ear protection are vital.

The application of glues and finishes by spraying or other methods can create mist, vapour and fume hazards. Use appropriately filtered masks to counteract these problems. Instructions on safe use of finishes can be found on packaging and on a Safety Data Sheet (SDS) – this should be available in your school.

Some chemicals used in treating timber can be hazardous.

Safety concerns related to manufactured boards

Manufactured boards that are made from very fine wood fibres, such as medium-density fibreboard (MDF), can cause significant illness. The dust caused by shaping and sanding MDF can be very harmful when inhaled. The fine particles are not filtered out by the upper

9780170477499

respiratory system and can lodge in the lungs and initiate cancerous growths. MDF should not be used in schools without highly efficient, industrial-level dust extraction.

The timbers, processes and glues used in some manufactured boards can cause the release of gases called volatile organic compounds (VOCs) that are considered to be harmful. Many boards are now made with lower levels of these compounds ('Low VOC') and are much safer to use.

Case studies
Bryan Cush

Cantilever

Chapter 14 case study report activity

Quality of timber

Timber, and timber products, are graded for their strength and their resistance to rot and moisture, as well as the quality of their grain or surface. Strength gradings for timber are specified in a range of Australian Standards. Materials receiving the highest ratings cost more but deliver superior levels of performance. Kiln-dried timber is usually more expensive than air-dried timber, but is more stable and less likely to warp. Some specialty timbers are very costly due to their unique characteristics, and because they are often difficult to source and process.

Environmental issues

The major environmental issue associated with timber is the logging of forests and the associated destruction of habitat. Of particular concern is the destruction of complex and irreplaceable rainforest ecosystems through clear-felling. The logging of other types of forest is also of concern, as it also reduces or removes habitats for native birds and animals. Clear-fell logging of any forest is much more destructive than selective logging, causing the destruction of habitat and degradation of land.

Alamy Stock Photo/Mark Boulton

Erosion caused by removal of forest

Some timbers are rare and the tree species they are sourced from are under threat of extinction. Avoid using any timber that is on the red lists of either the Convention on International Trade of Endangered Species (CITES) or the International Union for the Conservation of Nature (IUCN) – check the Wood Database or the Greenpeace Good Wood Guide.

Weblinks
The Wood Database

Greenpeace Good Wood Guide

Forest Stewardship Council

Erosion is another consequence of logging forests. It occurs when the removal of a stabilising root structure leads to the loss of topsoil and a greater likelihood of flooding.

Plantation timber is usually a more environmentally sustainable option. However, plantations of exotic species can be problematic as the local flora and fauna cannot exist among these 'foreigners'. Even a plantation of a single type of native tree creates a disturbed environment where other plants, birds, animals and insects find existence difficult.

When selecting timbers that are more environmentally sustainable, it is critical to look for Forest Stewardship Council (FSC) certification – this means that the forest from which the timber is sourced is independently accredited as being sustainably managed. For more information about the accreditation process, go to the Forest Stewardship Council of Australia's website.

Forest Stewardship Council

The FSC logo indicates that timber and furniture products come from sustainably managed forests.

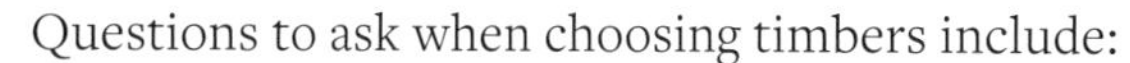

Questions to ask when choosing timbers include:

- Where was the timber sourced?
- Was it locally grown and processed? This limits impacts related to transport, use of petrochemicals and pollution.
- Was it sourced from rainforests, other naturally growing forests or plantations?
- Is the source FSC certified?
- Is the timber on the CITES or IUCN red lists?
- Does the timber or manufactured board create a possible health risk when machined?
- How durable is the timber? Will it help the product to last well?

Metal

Origins and properties

In comparison with other naturally occurring materials, humans have only used metals for a relatively short time. Gold, silver and copper were the first metals used by humans because they were reasonably available in their free state (i.e. they didn't have to be processed much from their raw state to be usable). It is also possible that small quantities of iron from meteors were used. By about 3000 BCE, gold was being refined and worked into jewellery on a regular basis, and iron production was well established in Egypt and Syria by 1500 BCE.

Today, metals of all kinds have become very important in our lives. They are one of the most versatile and widely used groups of structural materials.

Materials worksheet
Research on sustainability

General properties of metals

Most metals appear in nature as ores and are greyish in colour when refined, except for bismuth, gold and copper. Their melting temperatures range from −39°C for mercury to 3410°C for tungsten. Metals are generally strong and resistant to different types of stress, although there is considerable variation between metals. They are marked by properties such as hardness, tensile strength and some degree of elasticity. They can also be deformed (bent) without breaking, and can be shaped and moulded when heated. Many are also prone to corrosion or chemical attack.

Case studies
Jackie Staude

Marcos Davidson

Chapter 14 case study report activity

Shutterstock.com/RHJPhtotos

Metal ore

9780170477499

Metal classification

Metals are divided into two major classifications based on their constituents: ferrous metals, which contain iron, and non-ferrous metals. Another commonly referred-to group of metals is alloys – these are simply mixtures of different ferrous and/or non-ferrous metals. Alloys often have qualities quite different to those of their constituent metals, and are an extremely important area of metallurgical research.

Despite being prone to corrosion, ferrous metals are used extensively because of their mechanical properties and low cost. Surface coatings such as paint are often used as a protective covering. Non-ferrous metals are not prone to rust, which makes them extremely useful in wet environments. Their main disadvantages are that they are usually higher in cost and generally have lower strength than ferrous metals.

Classification of metals

Ferrous metals	Ferrous alloys	Non-ferrous metals		Non-ferrous alloys	Coated or plated metals
Cast iron	Corten steel	Aluminium	Nickel	Brass	Anodised aluminium
Low-, medium- and high-carbon steel	Stainless steel	Chromium	Platinum	Bronze	Corrugated iron
Mild steel		Cobalt	Silver	Soft solder	Galvanised iron
		Copper	Tin		Powder-coated steel
		Gold	Titanium		Tin plate
		Magnesium	Zinc		

iStock.com/lucentius

Different metals: (from left) tin, copper, brass, brass, aluminium, lead, iron and zinc

Health and safety issues related to using metal

Most metals melt at a high temperature, and in liquid form can cause serious burns. However, metals for the most part are relatively safe in their solid form – dangers come from sharp edges and incorrect lifting.

Machining operations create hazards associated with flying particles, heat and machine contact. Welding and forming operations create radiant heat, fume, vapour and fire hazards. Welding has the added possibility of eye damage and burning skin from a flash.

Quality of metals

Steel in all its variations is by far the most widely used metal due to its strength and ease of machining, but its durability is lessened due to corrosion. However, there are many alloys and special-purpose products that have mechanical qualities far exceeding steel's. A metal may need to resist corrosion for a marine use, be lightweight for a bike frame, or maintain its strength at high temperatures for an engine. The materials available that provide these properties are quite expensive, and are difficult to work because of their physical properties.

Environmental issues

For metals, the most pressing environmental issues relate to sourcing them. Mining may involve:

- the destruction of forests and rivers, causing the loss of habitat and biodiversity
- erosion due to land clearing and disturbance, and land degradation due to mine dumps and tailings dams
- the significant use and contamination of water (e.g. by arsenic and other heavy metals and acids, which are sometimes used in the extraction of metals and can leak into waterways and the water table).

Mine abandonment is an issue when the land and waterways are not rehabilitated after a mine is closed.

There are significant health issues related to mining metals. Inhalation of fine dust particles and contamination of heavy metals cause the premature death of many mine workers and those who live around mines, particularly mines with poor OH&S practices. Mining in some developing countries has also been connected to the use of child labour.

Some environmental problems are caused by the processing of metals. The smelting of some metals produces high levels of air and water pollution – e.g. the smelting of sulfide ores contributes to acid rain, while the smelting of copper can release particles of copper and selenium into the air. The production of some other metals uses large amounts of electricity – e.g. the conversion of bauxite into aluminium.

iStock.com/Douglas Sant Anna da Cunha

The collapse of this tailings dam in Brazil caused a toxic flood that killed 20 people and millions of freshwater fish, degraded local land and polluted a turtle-nesting area.

On a more positive note, many types of metal can be successfully recycled, reducing the need for further amounts to be extracted.

Questions to ask when choosing metals include:

- Is the metal recycled and/or recyclable?
- Where was the metal sourced and processed?
- Is it sourced from a country that has strong environmental protection laws and OH&S protection for workers?
- Does the mining of the metal cause significant environmental damage?
- Is the metal ore mined and processed in Australia? How great is the distance that the ore and processed metal has travelled?
- Does the company producing the metal have a history of environmental accidents/incidents?

Materials worksheet
Research on sustainability

9780170477499

- Does the processing of the metal use large amounts of energy? Can you find out the energy policy of the processing company (e.g. does the company use energy from renewable or non-renewable sources)?
- How durable is the metal? Will it help the product to last well?

Plastics

Origins and properties

Plastic is a term used to describe the long-chain organic polymers that are formed by extrusion, moulding, spinning or casting processes. The development of plastics began in about 1860, when a manufacturer of snooker and billiard balls offered a $10 000 prize to anyone who developed a satisfactory substitute for natural ivory. John Wesley Hyatt developed a way to form celluloid. Although highly flammable and light-sensitive, celluloid remained in popular use for shirt collars, dental plates and other products until 1920. Table-tennis balls are probably the only celluloid products readily available today.

Plastics are used for a huge range of products every day and come in a wide variety of forms. Many types of plastic are replacing traditional materials (e.g. alternatives for timber and metal in seating, for ceramic in crockery, for glass in spectacles, as a fibre in threads, yarns and fabrics and as a finish for timber).

General properties of plastics

Plastics generally show high strength-to-weight ratios, are excellent thermal and electrical insulators, and have a good resistance to attack from acids, alkalis and solvents. They have few directional properties (unlike timber, where grain orientation is critical to strength) and withstand many types of stress. However, the strength of plastics is generally lower than that of metals.

Even though there is an almost indefinite range of plastics available, most can be classified as either **thermoplastic** or **thermosetting** polymers:

- When thermoplastic polymers are heated, they become flexible and can be easily shaped and reshaped. If you look at the diagram below, you can see that there are no cross-links in thermoplastics and that the chains of molecules can slide over each other.
- Thermosetting polymers do not soften when heated because the chains of molecules are cross-linked together and remain rigid. They cannot be reshaped once set.

Thermoplastic
Plastic that becomes flexible and can be shaped when heated; can be re-formed many times

Thermosetting
A plastic that cures (sets), usually through heat or chemical reaction, and then cannot be re-formed

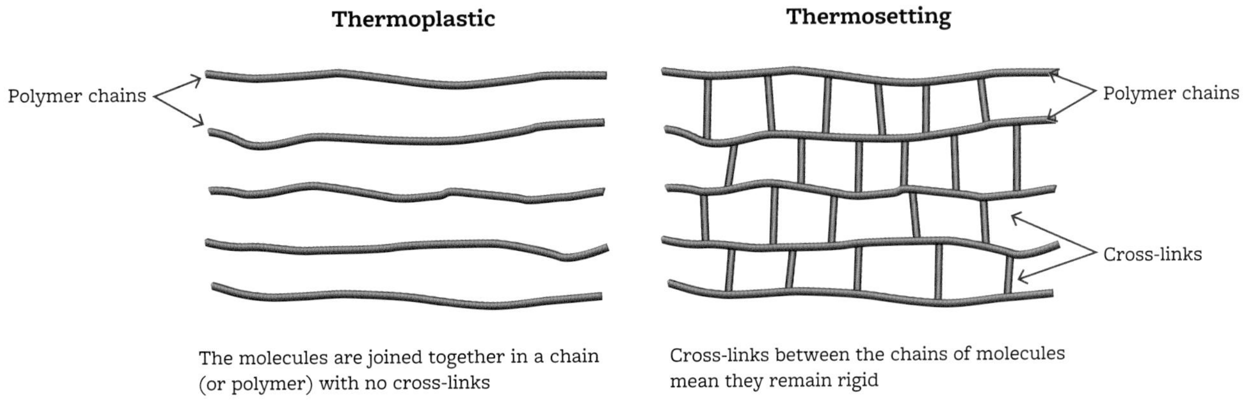

The molecular difference between thermoplastic and thermosetting polymers

Classification of plastics

Thermoplastic		Thermoset
Petrochemical	**Bioplastics**	
ABS	Cellulose-based plastics	Bakelite
Acrylic	Lignin-based polymers	Epoxy resin
PET	PHAs	Melamine
Polyethylene (PE)	PLA	Polyester resin
Polypropylene (PP)	Starch blends	Polyurethanes
Polystyrene		Silicone resins
		Urea formaldehyde
		Vulcanised rubber

Quality of plastics

There are many plastics, all with different characteristics and properties. Plastics need to be carefully selected, as an inappropriately chosen plastic will reduce the quality of the end product. Plastic that is contaminated with impurities (possibly through poor recycling processes) will have reduced quality, and may only be appropriate for low-grade products. The quality of plastics can be improved through the addition of specific chemicals during processing, e.g. chemicals can provide resistance to UV damage.

The creation and processing of plastic materials and components is highly mechanised and standardised. Quality is uniform and consistent when compared with natural materials, such as timber, that have considerable variation in their structure.

Health and safety issues relating to plastics

Plastics can be among the most volatile materials. They require extreme care and should be treated as any other type of synthetic chemical. For example:

- Although sheet materials are relatively inert, cutting and sanding operations produce dust hazards. In schools, hot-wire cutting should only be carried out on polystyrene in controlled conditions with adequate ventilation.
- When they are combined, two-part plastics (resins that use a chemical catalyst) often produce fumes that can cause nausea, while skin contact can cause dermatitis. These should be used in a fume cupboard wherever possible.
- Some solvent glues that react with the surface of acrylics are not to be used in schools because of the dangerous fumes produced and dangers of direct skin contact.
- Fire hazards can arise from incorrect storage and from mixing incorrect ratios of catalyst to resins.

Environmental issues

The sourcing of many plastics involves the use of petrochemicals. The processing of these materials causes environmental damage through the use of non-renewable materials and the production of pollutants. In addition, some plastics emit hazardous fumes during their useful life (off-gassing), and the common use of plastics in the home and work/school environment increases our exposure to those fumes. The disposable nature of many plastic products increases waste problems, particularly as many commonly used plastics do not decompose, but last for hundreds of years. These problems include the increased need for landfill and the damage caused to sea life by plastic rubbish getting into waterways.

9780170477499

Bioplastics are now replacing some plastics from a petrochemical source. These plastics may be less damaging to the environment – they come from renewable sources and decompose a little more readily.

Shutterstock.com/Mohamed Abdulraheem

The spread of plastic rubbish in landfill

Many plastics can be recycled, but for this to be successful, effective recycling systems need to be more fully implemented and consumers need to improve their level of recycling. Currently, less than 10 per cent of plastics are recycled; most still go to landfill. Because most plastics last a long time, they should be used for long-lived products rather than disposable products. Products need to be thoughtfully designed so that they can be easily dismantled and the recyclable content can be retrieved, sorted and reused. It is important that plastics are labelled for effective recycling, as contaminants reduce the quality of recycled plastics. Most recycling collections handle plastics labelled with numbers 1 to 6.

1 PETE	2 HDPE	3 PVC	4 LDPE	5 PP	6 PS	7 Other
Polyethylene terephthalate	High-density polyethylene	Polyvinyl chloride	Low-density polyethylene	Polypropylene	Polystyrene	Others (not readily recyclable)
Bottles and containers for soft drinks, mineral water, fruit juice, cooking oil	Containers for milk, cleaning agents, laundry detergents, bleaching agents, shampoo, washing and shower soaps	Trays for fruit Plastic packaging Bubble foil Some cling film to wrap foodstuffs	Shopping bags Highly resistant bags Some cling films	Furniture Luggage Toys Car bumper bars	Toys Hard packaging Food trays Make-up bags Costume jewellery	Acrylic Fibreglass Nylon Polycarbonate Polylactic fibres

Shutterstock.com/esbeauda

Symbols for different types of plastics

Questions to ask when choosing plastics include:

- Is the plastic recycled and/or recyclable? Is the plastic labelled with a recycling number? Are there systems put in place for recycling?
- How was the plastic sourced and processed?
- Does it come from plant-based materials or is it predominantly from a petrochemical source?
- Is the plastic made/processed in Australia? How great a distance has it travelled?
- How quickly and well does the plastic decompose? If it breaks down into small particles, will these become an environmental danger?
- How durable is the plastic? Will it help the product to last well or will it become an environmental problem when the product has reached the end of its useful life?

Materials worksheet
Research on sustainability

Fabrics, fibres and yarns

The general heading 'fabrics, fibres and yarns' covers all the elements of the textiles area, including:

- natural and synthetic fibres, filaments and yarns
- woven, knitted, tufted, braided, bonded and embroidered fabrics
- non-woven fabrics, consisting of mechanically and chemically bonded fibres.

Case studies
Bodypeace

Emily Barrell

FOOL clothing

Chapter 14 case study report activity

Textile fibres and fabrics have an extremely broad range of properties. Some fibres are efficient conductors of heat, transferring heat away from the body quickly, others are good insulators, keeping hot or cold air trapped. Some fabrics are hydrophobic and don't absorb moisture, others absorb moisture very effectively. Synthetic fibres have good strength characteristics, while natural fibres are weak in comparison.

The strength and directional properties of a fabric are directly related to its method of manufacture. Knitted fabrics tend to be stretchy and flexible, woven fabrics are stronger and less stretchable, and some non-wovens are stiff and even less stretchable in comparison. The level of drape/stiffness, permeability and other properties are also affected by the density of the fabric construction.

Fibre classification

The two main categories of fibres, natural and synthetic, relate to the source of the fibre:

- Natural fibres occur, ready be to processed and used, in their natural form. They can be split into two further groups:
  - those coming from animals (protein)
  - those coming from plants (cellulose).
- Synthetic (or manufactured, man-made) fibres can also be broken down into two further groups:
  - regenerated fibres, which are based on a natural fibre that has been chemically altered
  - fully synthetic fibres, which are formed from a molten polymer.

A third category covers skins and membranes from both natural and manufactured sources.

Types of fibres and fabrics

Natural		Synthetic		Skins and membranes
Protein	Cellulose	Regenerated	Fully synthetic	
Wool	Cotton	Rayon	Nylon	Leather and suede
Mohair	Hemp	Bamboo	Polyester	Latex and rubber
Silk	Jute	Tencel	Acrylic	Mycelium
Alpaca	Linen	Acetate	Lycra	Scoby

The physical structure of a fibre has an effect on the fibre's properties and appearance. At a microscopic level:

- Woollen fibres have an overlapping, scale-like structure, which increases their insulating quality and their tendency to felt. They have a soft, non-shiny appearance.
- Most synthetic fibres have a very smooth shaft, which increases the lustre or sheen of fabric made from those fibres as the fibre is uniform and reflects light.
- Microfibres are extremely fine synthetic fibres. They have distinct characteristics: they are less absorbent and have a wicking quality (i.e. they draw moisture from the body); fabric made from these fibres can have a higher density, and is usually very soft and drapes well.

9780170477499

Science Photo Library/EYE OF SCIENCE

Electron micrographs of various fibres

Fibre blends

Different fibres can be blended to combine fibre characteristics or to overcome weaknesses. For example, most T-shirts, hoodies and sheets are made from poly-cotton (a mixture of polyester and cotton), which combines the moisture absorbency of cotton fibres with the strength, wrinkle recovery and fade resistance of polyester.

Yarns

The degree and type of twist can affect the properties of the yarn and the fabric it is made into. Yarns are usually hard or soft spun.

- Hard-spun yarns have a tight twist.
- Soft-spun yarns are lightly twisted. The yarn is more absorbent, more flexible and has good wrinkle recovery. Examples of the use of soft-spun yarns are woollen fabrics, fabrics with a nap, knitwear and knitting yarn, and brushed fabrics.

Fancy or novelty yarns use threads of differing thickness and tension in construction, and can be heavily textured.

Skins and membranes

Skins such as leather and sheepskin are hard to categorise in the textile area. They are used like a fabric but are not constructed. Membranes such as latex and rubber are also used like a fabric. New developments in this area have led to the creation of materials made from the mycelium of fungi (the root-like, underground part of the fungus), which is able to digest and bond substrates, and leather made from Scoby (**s**ymbiotic **c**ulture **o**f **b**acteria and **y**east), a living culture that forms a rubber-like mat.

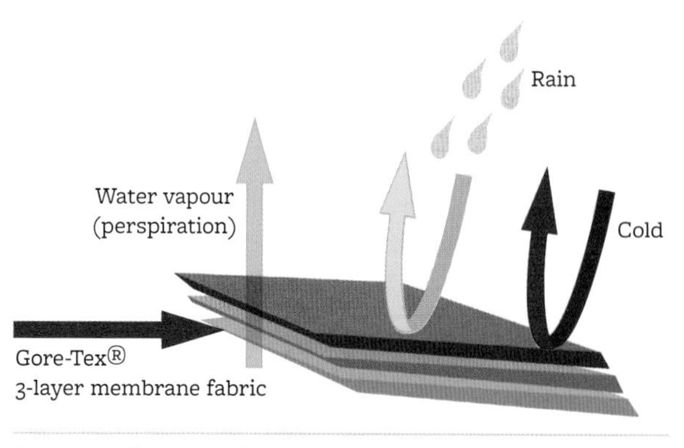

Structure of Gore-Tex®

Technical fabrics

Laminated or bonded fabrics

To create a laminated or bonded material, fabrics of different physical characteristics or fibre content are bonded or fused in layers. Many high-performance sports fabrics are laminated or bonded fabrics, which have been developed after research into the needs of the body during intense activity in different environments. Laminated fabrics include Gore-Tex® and Mexican oilcloth.

Geotextiles

Geotextiles are extremely hardwearing and durable fabrics and membranes that are used in agriculture, road-building and construction. They are often used to stabilise soil, provide a waterproof barrier or act as a membrane for other construction purposes.

Treated fabrics

Fabrics can be treated to give them special properties required for particular applications – to make them:

- stain- or water-repellent
- flame-retardant
- antistatic
- antibacterial
- sound-absorbing
- thermally insulating
- protective against UV
- insect-repellent.

Smart yarns

'Smart yarns' are constructed from hollow fibres filled with chemical substances. Fabrics constructed from these yarns are primarily used for:

- pharmaceutical purposes – for protecting the skin, distributing medication to the skin (e.g. nicotine patches), drawing contaminants away from the skin and sensing biomedical conditions
- children's clothing – the hollows may be filled with synthesised fragrance (e.g. chocolate or banana) to make the clothing more appealing.

For more information about new developments in fibre technology, go to the CSIRO website and search for 'textiles and fibre technology' or 'advanced fibres'.

Smart electronic fabrics: wearable technology

Smart electronic fabrics have electronic components incorporated into the structure of the fabric. Examples are:

- clothing for high-performance athletes – fabric that monitors heartbeat, tracks movement, etc.

9780170477499

- electronically heated fabric
- fibres that monitor blood health
- solar collector and device powering.

Hardware and software that run or monitor the smart components are developed together as an integrated system. A 'smart yarn' has been developed that uses carbon nanotubes and conducts electricity and heat, and is strong enough to protect the wearer from bullets.

For more information, go to the Science Daily website and search for a recent article on 'smart yarns'.

MYZONE

The OMSignal smart sports bra, which monitors heart rate, breathing, muscle tension and stress levels

Weblinks
CSIRO

Science Daily

Quality of materials

The quality of a natural fibre depends on the circumstances under which it was grown (e.g. the quality of wool depends on the sheep's breed and the care of the sheep, as well as the climate). For all fabrics, the quality and functional characteristics depend on the quality and properties of the original fibres, and the quality of processing during manufacture. For instance, the quality of dye used and the physical characteristics of a fibre affect colour fastness, while the tightness and evenness of weave or knit and the strength of a fibre will determine the level of resistance to abrasion and stretching.

Health and safety issues

There are a number of health and safety issues related to the production and use of textiles:

- Farmers and workers who tend and harvest fibre crops on which a lot of fertilisers and/or pesticides have been used often suffer multiple health problems.
- Health concerns related to textiles also occur during the yarn and fabric production stage, where chemicals (for cleaning, degreasing, dyeing, etc.) and heavy machinery are extensively used.
- During manufacture, hazards include the chemicals used in dyeing, treatments to distress denim, fumes from heat-sealing joins in synthetics, unsafe factory buildings, long working hours and the machines used in construction.
- There are health risks associated with the 'off-gassing' of volatile organic compounds (VOCs) from some fabrics and fabric treatments.

Environmental issues

Many people assume that the environmental impacts of natural fabrics and fibres are low. However, the pesticides used in the production of cotton have a significant environmental impact (although the quantity of pesticides used has been reduced in the last 10–15 years). In Australia, marginal land that is used for raising sheep for wool production has been irreparably damaged. On the other hand, the production of synthetic fibres uses petrochemicals and other non-renewable materials, and produces pollutant by-products during processing. There is a lot of concern about micro- and nanofibres – tiny fibres of microscopic size – that are released into the environment in waste water when synthetic clothes are washed. These fibres are now being found in human body systems and those of animals, fish and birds. We do not know the long-term damage these fibres will cause.

Materials worksheet
What materials could you use?

As clothing becomes cheaper to buy and is driven by a shorter fashion cycle (the fast-fashion phenomenon), more is thrown away before it wears out. Clothing made from synthetic fibres takes thousands of years to break down in landfill, causing a rapidly increasing problem. Even clothing made from natural fibres takes a long time to break down in an airless (anaerobic) landfill environment.

Some moves have been made to lessen the impact of textile production. Cotton plants have been genetically modified to make them resistant to insects and disease, reducing the need for chemical sprays. The demand for, and production of, organic and sustainably produced cotton is steadily growing, and cotton can now be certified through:

- Global Organic Textiles Standards (GOTS)
- Fairtrade certification.

OEKO-TEX®

The OEKO-TEX® label certifies that a fabric is of high quality and doesn't contain any harmful substances

Another form of textiles certification is OEKO-TEX® STANDARD 100, a standard that indicates that the textile product has been extensively tested for harmful substances (e.g. dyes, finishes) and is safe for human use (STANDARD 100, certified by the independent international OEKO-TEX® organisation).

Hemp, bamboo and soybean fibre are increasingly being used to replace cotton, as their growth is more environmentally friendly. However, bamboo and soy have their own environmental impacts due to land clearing and chemicals related to fibre processing. Fabrics such as the original Polartec polyester fleece reduce waste, as 60 per cent of the fibre used is sourced from recycled plastic drink bottles. However, like most synthetics, this fabric when washed still sheds microfibres into waterways.

Shutterstock.com/rj lerich

Outdoor clothing can be made from fleece with recycled PET (polyethylene terephthalate) content.

Questions to ask when choosing fabrics, fibres and yarns include:

- Does the fibre come from a natural or a synthetic source?
- If natural:
  - Is it 'organic'? Or were significant amounts of pesticides or herbicides used in its growth?
  - Did it use a lot of water to grow and/or be processed?
- If synthetic:
  - Did the materials for creating the fibres mainly come from petrochemical sources?
- Were toxic chemicals or heavy metals used in the processing of the material?
- What forms of pollution were created in the processing of the material (e.g. degreasing, dyeing)?

Materials worksheet
Research on sustainability

9780170477499

- Does the fabric or fibre have any Fairtrade, environmental or organic certification?
- Is the fabric or fibre recycled? Can the fabric or fibre be recycled?
- Is the material locally sourced? Has the fabric or fibre travelled many kilometres for sourcing and processing?
- How durable is the fabric or fibre? Will it help the product to last well?
- Does the fabric shed lots of microfibres when washed? Can this be reduced?

Materials testing

The reliability of all products depends on the quality and appropriateness of the materials used in their production. Materials testing will help you choose the best and most appropriate materials. This usually involves comparing two or more materials in a planned test or experiment. A materials test is effective if the test procedure is:

Materials worksheet
Materials testing

- **well planned** – identify the one aspect you want to test and plan a test procedure that will isolate for that. Make sure that you have all of the equipment organised before you start the test, and plan a clear method for measuring and recording your results
- **consistent and controlled** – ensure that the only thing that changes in your test is the material. You need to be able to repeat the test procedure in exactly the same way for each material sample
- **repeatable** – carry out the test on each material a number of times to check whether your process and results are reliable
- **accurately measured and recorded** – source the most accurate measuring equipment that is available (you might need to borrow equipment from your Science Department).
- **safe** – follow all safety procedures relevant to your material. If your test involves heat or the use of chemicals, your teacher may ask for a Risk Assessment before you start.

Materials test report

The following elements need to be included in an effective materials test report:

- aim – a description of the characteristic or property being tested and why the information will be useful (its relevance and context, which are related to the design situation)
- a description of the materials being tested
- a prediction of results – based on your knowledge of the properties of the materials to be tested
- testing procedure – step by step with a diagram and/or photos of the testing set-up and procedure
- accurately measured results (in table and graph form) and observations
- analysis and interpretation of results (What do the results tell us about these materials? Do the test results differ from your predictions or expectations?)
- consequences for material use (How will this affect the decisions you make about the use of these materials?)
- an evaluation of the testing procedure (Was this an accurate test? How could it be improved? What sort of equipment would improve the accuracy of your results? How might the test be carried out in industry or in scientific laboratories?).

Consistency in testing

The first step is to decide:

- What (which characteristic or property) are you testing for?
- Which materials will you compare?

It is important to isolate just one thing for each test. For example, if you are comparing the moisture absorbency of two materials, you need to control all the variables that might affect the sample pieces. In other words, start with samples of the same size, expose all sample pieces to the same conditions, apply the same amount of moisture and use the same equipment to measure your results. Avoid the temptation to add another characteristic (such as shrinkage, for example) in the same test.

Record results

Record your results in a methodical way, to assist with clear presentation in your report.

Safety

Follow all safety procedures relevant to your material. Your teacher may ask for a risk assessment before you start.

Writing your report

Include the following components in your test report:

- an accurate identification of the materials being tested
- a description of the characteristic or property you are testing for and how it relates to your product
- a clear description of the test and how you are measuring differences (including photos of your samples before and after testing)
- the results of the test (presented in numerical form, and noting your observations)
- an analysis of your results
- your conclusion, explaining the choices you will make as a result of your testing.

Test suggestions for resistant materials

Property tested	Description of test	Diagram
Hardness (checks dents and/or scratches)	Drop a 12 mm ball bearing or a heavy pointed object onto the material samples you wish to compare from a specified height (e.g. 800–1000 mm). Measure and compare the size of indentation on both samples. You could also measure the height to which the ball rebounds. Harder surfaces will rebound further and not mark from the impact. Or do a scratch test. Gather a series of material samples. Scratch one against each of the others. The material that marks the other sample is the harder material. Rank the samples from softest to hardest.	SAMPLE
Strength and durability	Hang a load from the material sample, increasing the load until failure (cracking, splitting or breakage) occurs. Record the load weight required for failure. Repeat on each sample.	LOAD SAMPLE Clamp

9780170477499

Property tested	Description of test	Diagram
Elasticity (ability to bend and return to original shape)	Subject the material sample to loading until it bends. Release the load and check to see if the sample returns to its original position or measure the distance that it returned to from the original position (you will need a reference point to measure from). Repeat on each sample.	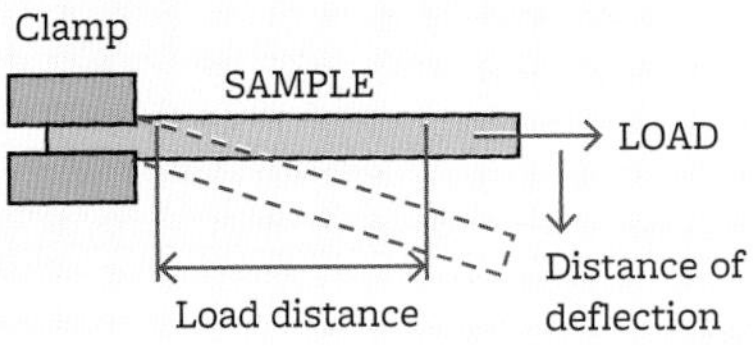 Clamp, SAMPLE, LOAD, Load distance, Distance of deflection
Ductility (ability to be stretched, bent or 'spread' until it deforms or shows damage) or tensile strength (ability to be stretched until it breaks completely)	Accurately measure the diameter or cross-section of the sample. Clamp one end and apply a load (tensile force) to the lower end for a set number of minutes. Release the load and remeasure the diameter or cross-section area to determine if it is thinner (appropriate for metals) or observe and note any deformation. A ductility test will only give tiny changes – very accurate measuring equipment and acute observations are required. For a tensile strength test, record the weight applied to break the piece and the time it took. Repeat on each sample.	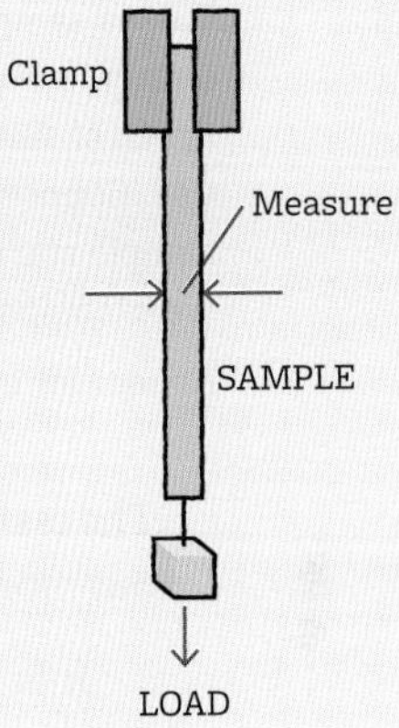 Clamp, Measure, SAMPLE, LOAD
Corrosion resistance (ability to avoid damage from oxygen, water or chemicals)	Place each material sample in exactly the same intended work environment (e.g. water, boiled water, water with salt, washing powder, vinegar) and observe over an extended period of time. Consider using separate containers to avoid any crossing over of molecules. Photograph and note the degree of degradation (rate of change of colour, surface damage, etc.) in each sample.	 SAMPLE
Ease of machining	Place a material sample in the selected machine (e.g. lathe, milling machine, router). Starting with a shallow cut, increase the depth on successive cuts and note the ease or difficulty the machine has in cutting through the depth. Note observations – effort required, sound of machinery while cutting, difficulties due to material variation, and the 'cleanness' of the cut on each sample etc. (This test doesn't give easily measurable objective results.) Repeat for each sample.	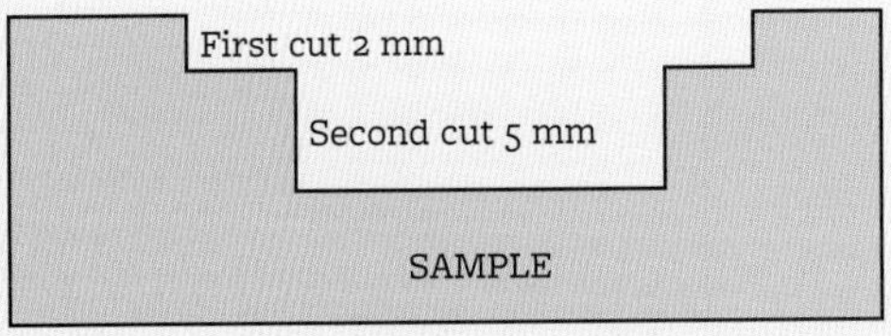 First cut 2 mm, Second cut 5 mm, SAMPLE
Thermal conductivity (ability to hold or lose heat)	Attach a thermometer to one end of the sample material and note the rise in temperature when the other end is heated for a specified time. The thermometer should always be the same safe distance from the heat source, and the heat source should be at a constant temperature. Repeat for each sample.	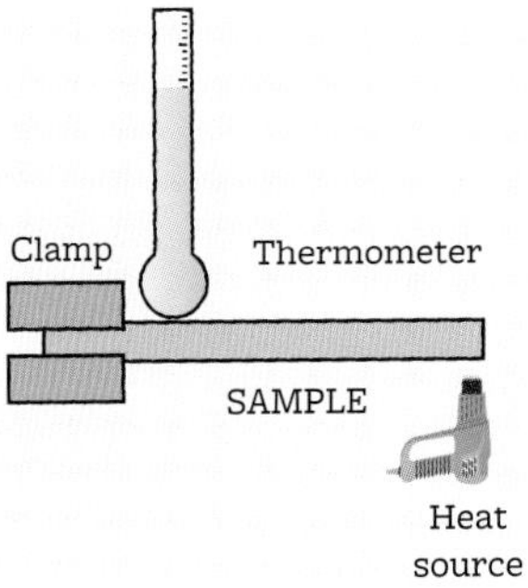 Clamp, Thermometer, SAMPLE, Heat source

Property tested	Description of test	Diagram
Electrical conductivity	Obtain a multimeter and set it to ohms or resistance scale. Place one probe at each end of your sample. Materials that are conductive will give a reading while non-conductors will not. Repeat for each sample.	Multimeter SAMPLE
Wear resistance	Take a sample of the material and subject it to continuous contact and movement from the types of surfaces it will experience when in use. Count the number of movements until the sample starts to show wear, or observe the wear compared with an untouched sample. Repeat for each sample.	Test surface SAMPLE

The test descriptions in the table are very brief, and you will need to conduct research to find out more details or ask your teacher for help.

It is important to know:

- that some tests are more suitable for one material area than others (e.g. hardness for timber, thermal conductivity for metals). You need to select your tests carefully
- the correct and full names of the materials you are testing and using in your product. Ask your teacher or get the information from the place where you purchased the material. Write the full name each time to help you remember it
- that some tests on materials (particularly plastics) may not be safe. You may need to limit yourself to secondary research (collating information about the materials that other people have researched).

9780170477499

Test suggestions for non-resistant materials

Property tested	Description of test	Diagram
Strength	Attach a fabric sample to a wooden frame (carefully plan the method of attachment so that the attachment itself doesn't damage material or tear during testing). Load the sample with weights, and measure the weight needed to tear or damage the fabric. Repeat for each sample, using the same weight.	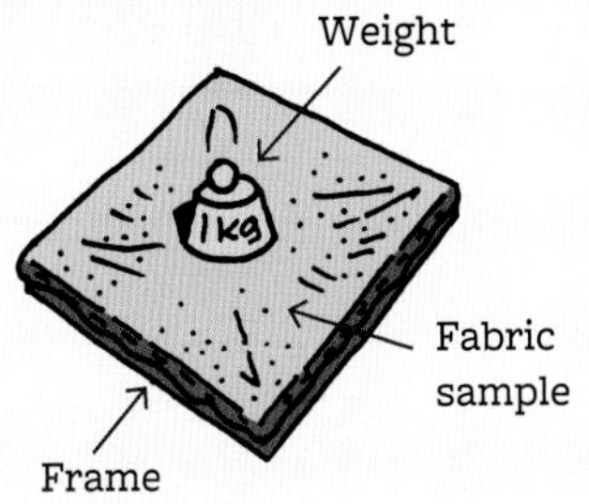
Durability	Cut several samples and keep one of each material intact for comparison later. Repeatedly wash fabric samples, noting shape change, size, colour change and strength (or do simple strength tests similar to above). Abrasion tests can also be done (e.g. rubbing surface with sandpaper block) in the same manner for each sample.	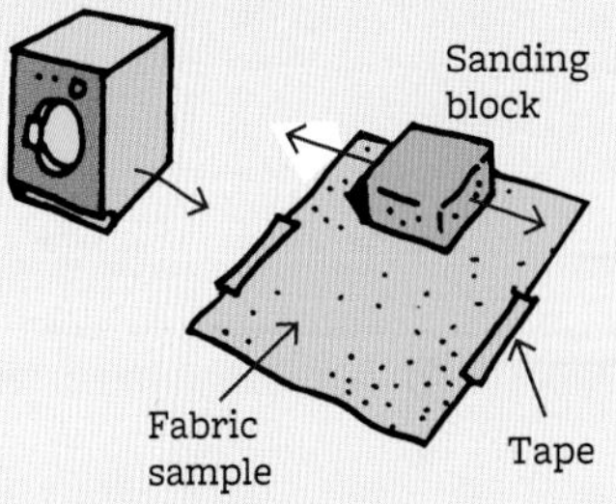
Elasticity or stretch	Measure how far a fabric will stretch (start with samples of different fabrics the same length and record this length). Hang each sample securely and attach the same weight. Record distance of stretch. Measure fabric after weight is removed to record level of recovery. Yarns or fibres can also be tested in this way if appropriate to your product, e.g. for knitting or weaving.	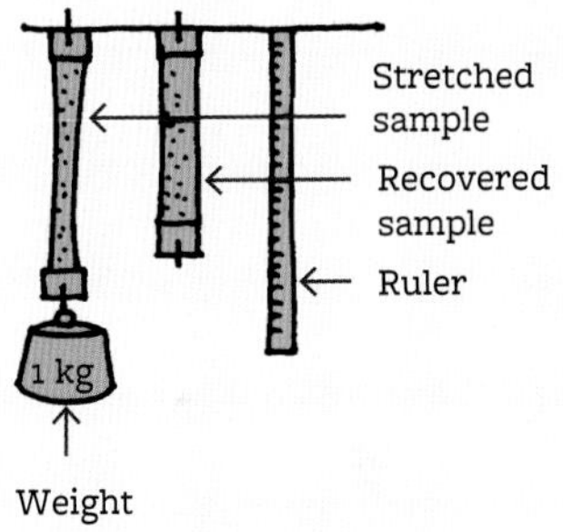
Moisture absorbency	Either drip a controlled amount of water onto a fabric sample and measure the spread of moisture, or dip the ends of fabric samples into water and measure how far the moisture travels. Make sure both samples are placed on the same or identical surfaces. Repeat on samples of a different fabric.	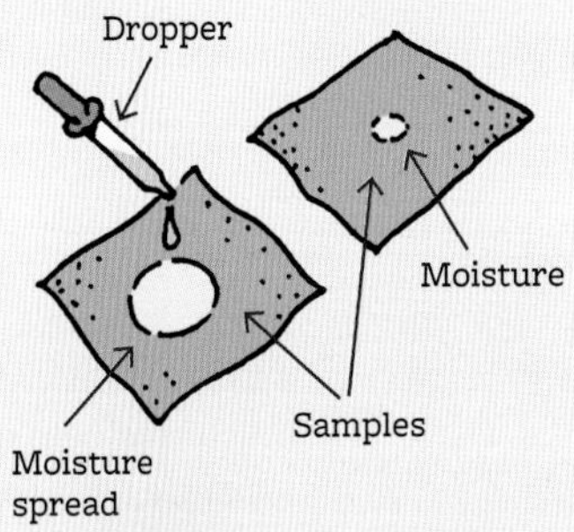
Thermal insulation (ability to hold or lose heat)	Completely cover a beaker of hot water (containing a thermometer) with a fabric 'bag' (made from a sample of one fabric) tied tightly. At regular spaces of time, record the temperature change in the water. Repeat in the exact manner with a different fabric sample.	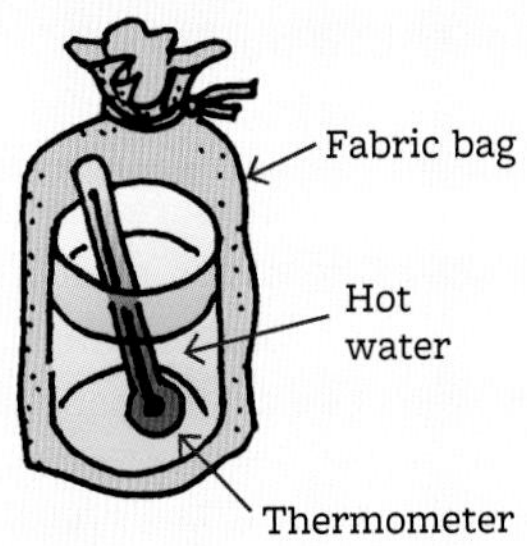

Property tested	Description of test	Diagram
Reaction to chemicals (acids, alkalis, bleaches)	Cut several samples of identical size of each fabric. Submerge fabric samples in dilute forms of acid (e.g. vinegar, lemon juice), alkali (detergent) or bleach. Remove after a length of time and test for strength. Repeat with samples of a different fabric.	
Effects of sunlight	Leave fabric samples in direct sunlight over an extended period of time (e.g. two weeks). Record observations of changes in the fabric (e.g. colour) and test for strength.	
Resistance to biological attack	Insects – collect cloth moths and/or silverfish and trap with samples in a jar (with ventilation). Record the amount of holes (if any) after defined periods of time. Fungi and mildew – leave samples in an appropriate place (bathroom) for an extended period of time. Record level of fungal/mildew growth.	
Colour fastness (ability to take and retain dye)	Place samples in dye and record level of dye intake (devise a 1–10 grading system). Use a range of different dyes and colouring agents. Colour fastness can be tested by washing a sample with another sample of white absorbent fabric and recording the level of colour that is transferred.	
Shrinkage	Measure and precisely cut 4 samples of the same size from each material. Keep a sample of each fabric unwashed for comparison later. Wash and record changes in size for cold, warm and hot water. (Large samples show greater change of size and are easier to measure.) You may also do the same tests with different detergents. Measure, label and photograph as evidence.	

9780170477499

Property tested	Description of test	Diagram
Drape	Place a large circle of fabric (the diameter needs to be twice the glass height plus the glass diameter) over a glass and trace the shape of the fall of cloth. The greater the number of folds and the smaller the shape, the softer the drape. Repeat with a sample of a different fabric of the exact same size.	Paper Glass Draped fabric
Wrinkle resistance/ recovery	Scrunch fabric sample into a ball and place a weight on it for a set period of time (this can be done wet or dry). Unfold, and record or photograph the number of folds/creases in a small section (use a cardboard frame with inner measurements of 50 mm × 50 mm). Rate the depth of creases (shallow, medium, deep). Hang samples for a set time and record level of recovery. Iron samples at cool, medium and high heat (if appropriate for fibre) and record recovery. Repeat with a sample of a different fabric of the exact same size.	Weight Scrunch fabric Wrinkled sample Cardboard frame 50 mm × 50 mm

The preceding test descriptions are very brief, and you will need to conduct research to find out more details or ask your teacher for help. If you are using any chemicals or heat testing, check with your teacher and be sure to do a risk assessment first.

It is important to know:

- the correct and full name of the fabric (weave and fibre content, e.g. 100 per cent silk satin or 100 per cent polyester satin) that you are testing and using in your product. Ask your teacher or get the information from the place where you purchased the material. Write the full name each time to help you remember it.
- that you will need to create your own tests or do your own research on suitable tests for any new materials that are not covered in the tables.

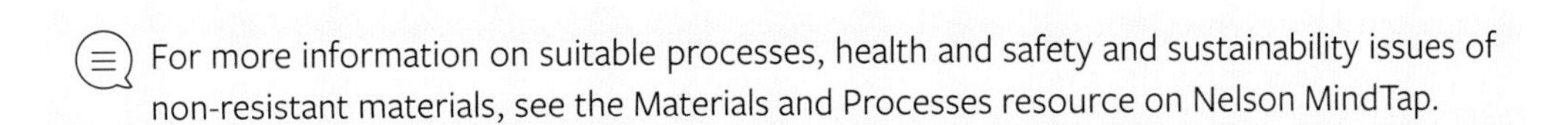
For more information on suitable processes, health and safety and sustainability issues of non-resistant materials, see the Materials and Processes resource on Nelson MindTap.

Summary
Chapter 14

Quiz
Chapter 14 revision

Resources
Materials and processes

Materials worksheet
Construction process trials

GLOSSARY

Aesthetics The science or philosophy of beauty and taste, which relates to visual and tactile aspects of a product. It is explored through the design elements and principles.

Bespoke or **custom-made** Designed and made to suit a customer for a specific situation

Biophilic Demonstrating a love of nature and living things

Biophysical Relating to the physical world and all the living things in it

Brainstorming Working as a group to share ideas and explore new ideas, encouraging creativity and problem-solving

Characteristics The visual and textural aspects that are inherent in or typical of a material

Collaborate To work together cooperatively for a common goal or task

Convergent thinking Bringing the most suitable ideas together

Couture The French word for tailoring and dressmaking

Creative thinking Thinking to come up with new and original ideas and explore new possibilities

Creativity Approaching, exploring or combining something in new or unexpected ways to create something unique

Critical thinking Thinking that questions, researches, analyses and makes judgements/decisions

Critique To review and appraise; to provide an analytical assessment

Design options Also known as presentation drawings; realistic, detailed 3D drawings that show the whole product and are used to convey a fully worked-out design solution. The emphasis is on quality and detail.

Detrimental Negative and destructive

Disparities Great differences

Divergent thinking Considering many ideas and ways in which a problem might be solved

Economies of scale Cost savings made when the per unit cost is low as materials can be purchased in bulk, and discounts and manufacturing set-up costs are shared over many products

End user Someone or something that might use the product or solution. Can be a person or a creature (human/non-human).

Enhancement Something that will increase or improve quality, visual appeal, functionality or value

Entrepreneur A person or group that starts their own business and takes on financial risks to bring a product to market in the hope of profits

Ergonomic Relating to the scientific study of the relationship between people and their living and working environment

Ethics A branch of philosophy; moral principles concerned with human conduct, which identify right and wrong behaviour

Ethnographic Related to a branch of anthropology that studies peoples, cultures and their customs, habits and mutual differences

Extension lines Light lines extending out from the object, defining the distance being dimensioned

4IR The Fourth Industrial Revolution, where technologies and data are interconnected and responsive to each other

Fasteners or functional components Zippers, buttons, velcro, hinges, latches, belt buckles, loops, press studs, snap/lock fit, handles

Figuring Lines created by the grain in timber

Fossil fuel A carbon-containing compound, formed over millions of years, that releases greenhouse gases such as carbon dioxide when burnt

Functionality How well a product can serve its purpose (do its job), its practicality and its range of usefulness

Greenhouse gas Carbon dioxide, methane or nitrous oxide, which are all naturally present in Earth's atmosphere and trap the sun's heat to sustain life, but which have increased out of proportion over the last two centuries, increasing global temperatures

Hazard A thing, action or behaviour that could cause harm

Incident An event that comes close to causing injury and, in a workplace, should usually be reported to management

Inextricably In a way that is impossible or difficult to disentangle or separate

Injury The actual harm or damage that could occur to a person, ranging from scratches and bruises to broken limbs, head injuries, entanglement injuries, electrocution or death, etc.

9780170477499

Innovation Putting existing ideas, knowledge or concepts together in a new and different way

ISO International Organization for Standardization, which develops and promotes global standards

Iterate Repeat, adjust and refine through small improvements

Last A solid form in the shape of a human foot, around which shoes are made

Lateral thinking 'Sideways' thinking – being creative through an indirect approach, rather than an obvious one; using new approaches to solve a problem

Likelihood The probability that a given event will happen; often rated low, medium or high

Line A mark from one point to another

Low-fidelity prototype A basic prototype that has just enough detail to communicate the essence of an idea

Mandatory Required by law

Maquette A scale model, made from substitute materials or using less material than the final product

Mood board A collage of images, design fragments, textures, components, colours, words and phrases, etc., used to provide a reference for visual consistency in designs or as a stimulus for design ideas

Multimodal Using multiple methods or modalities of presenting work, e.g. text, images and audio

Non-resistant materials Materials that are soft and that move when touched, in particular textiles (fabrics, fibres and yarns)

Occupational health and safety (OH&S) Programs that aim to ensure workplaces are safe for all who enter them

Outsource To hand over to a more experienced person/group or to pay for a technology that is not available to you

Personal protective equipment (PPE) Equipment such as safety glasses, shields, masks and gloves, needed for protection when using a machine or equipment

PET Polyethylene terephthalate, a recyclable plastic widely used in drink bottles and other products

Plant Any machinery, equipment, appliance, implement or tool; any component of any of those things; anything fitted, connected or related to any of those things

Point In product design, a particular place or spot that the eye is drawn to that is isolated or by itself

Primary research Information that is created first-hand

Primary sources Sources of information that is obtained first-hand

Properties How a material performs and behaves – chemically, physically or mechanically – under certain conditions

Resistant materials Materials that are hard when knocked, such as wood, metal and plastics

Risk The severity or degree of harm arising from an accident, ranging from minimal (requiring first aid) to serious (requiring hospitalisation), and the likelihood of an incident happening

Safety control A precaution you will take or put in place to remove or reduce a risk. You must demonstrate that you know exactly what to do to remain safe. List the exact PPE or set-up required.

Secondary research Information collected and published by others

Secondary sources Published accounts of information supplied by others

Section drawing/sectioning A view of a slice through an object to portray its internal structure

Sheen How light is reflected by a surface

Speculate To form a theory, ideas or opinions about something without all the facts or information; to invest in the future in the hope of gain; to think and make guesses about what might happen or how something might work; to make a prediction based on evidence that is followed by actions to find the outcome

Speculative thinking Thinking ahead about new ways of doing things to benefit all of humanity

Substance Any natural or artificial material, which might be in the form of solid granules, liquid, gas or vapour

Supply chain Every step involved in getting a finished product to consumers, from sourcing raw materials, outsourcing of components, labour, transportation at all stages and finally to retail outlets

Synectic From the Greek, meaning 'bringing different things into a unified connection', synectic thinking focuses on taking things apart and putting them together differently to transform ideas

Synthesise To research, collect and make sense of information in a way that has meaning to you, connecting it with information you already have to create something new

Technical Relating to the way things work in practice in industry, science or art

Thermoplastic Plastic that becomes flexible and can be shaped when heated; can be re-formed many times

Thermosetting A plastic that cures (sets), usually through heat or chemical reaction, and then cannot be re-formed

Toile A toile (pronounced twahl) is a prototype or trial version of a garment. A toile is made from cheaper, plain fabric for testing and fitting. Toile is also a translucent linen or cotton fabric originally used for fittings. Calico is a common fabric choice for toiles as it is inexpensive, is a stable, woven fabric and you can draw lines and write notes on it. If you intend to make a garment in a floaty silk or jersey knit, choose a fabric for your toile that has similar drape.

Triggers Words or objects that spark an idea

Visualisation drawings Sketches that get multiple ideas down quickly; for exploring and visualising what is in your head. The emphasis is on quantity.

Warp The structural threads of woven fabric that are threaded onto the loom and run along the length of the fabric

Weft The 'filling threads' that are woven from side to side across the warp

Working drawings Accurate line drawings that communicate the product's dimensions and components. The emphasis is on accuracy.

Worldview issues Things that a person may believe to be fundamental aspects of life, such as what is good or right and what concerns different groups of people

9780170477499

INDEX

A

Aboriginal and Torres Strait Islander people 104–115, 192, 387
- assessment 115
- contemporary Indigenous designers (case studies) 107–113
- cultural appropriation 105
- culture and connection to Country 104–105
- impact of culture on choices 113–115
- influence of culture on product design 105
- possum-skin cloaks 106–107
- traditional Indigenous Australian design 105–106

accessibility 369, 370
accessible design 66–67
accurate measurements 49
accurate measuring (in testing) 41
adaptive clothing: Christina Stephens (case study) 76
additive manufacturing 132
aesthetics 13, 30, 369, 376–378
- Design Elements and Principles 346–361
- factor 29
- of materials 376–377
- and styles 377

agile manufacturing 135
AI *see* artificial intelligence (AI)
alternative materials 167–170
- bamboo 169–170
- Fungi Solutions (case study) 174–175
- impacts of 173
- vegan leather 167–169

animals (non-humans), design for 68–69
annotations 24, 26, 223
anthropometric data 43, 373–374
anthropometry 373–375
appropriation 387
Art Deco 30
artificial intelligence (AI) 126, 219, 380
- automation and 127–128
- for virtual product concepts 234–235
- in manufacturing 128
- in products 127
- use of AI in your work 176

artificial reefs (case study) 79
assessment 56, 115, 283
assistive products 395
asymmetry 356
audio recording apps 82
Australian and International Indigenous Design Charters 105
Australian and International (ISO) Standards 391–394
Australian Bureau of Statistics (ABS) 262
Australian Government 266
Australian Nuclear Science and Technology Organisation (ANSTO) 266
Australian Tax Office 263
automation 126, 172
- and artificial intelligence 127–128

automotive industry 264

B

balance 356
bamboo 169–170
- as fibre 170
- or hardwoods 169

Bamboo Labs (case study) 73
bar graph 204–205
batch production 121–122
bespoke product 120
bibliography 390
- in design folio 200–201

'bill of materials' 50
bio-compatible polymers 163
biodegradable plastics 165
biodegradable polymers 161, 163
biological cycle 138
biomaterials 149, 396–397
biomechanics 373
bioproducts 161
body templates 331
bonded fabrics 418
bottom line 142
brainstorming 13, 16, 288–289
Braungart, Michael 148
Browne, Nikki (case study) 109–110
Brundtland Report 136
business, quantitative data in 205

C

cabinet oblique 329
CAD *see* computer-aided design (CAD)
carbon dioxide 137
carcinogen 169
case studies
- adaptive clothing: Christina Stephens 76
- artificial reefs 79
- Bamboo Labs 73
- Browne, Nikki 109–110
- Cobalt Design 177–181
- contemporary Indigenous designers 107–112
- Fungi Solutions 174–175
- Havaianas 274
- Kami project 233
- lean manufacturing at Zara 136
- Lego Braille machine 75
- Life Saving Dot 78–79
- LifeStraw 74
- A Liter (Litre) of Light 72
- Monks, Nicole 107–109
- *Oi* bike bell by Knog 15
- OXO kitchen tools: design that grew out of experience 89
- of positive impact 80–82
- products for those with autism 77–78
- Profile: Edward de Bono 315
- Radical Yes 139–142
- Rip Curl: wetsuit development 57–61
- Rothy's: sustainable shoes and accessories 274–276
- The Shoe That Grows 71
- Stories from Layers of Blak 109–112
- Wise, Tracy 111–112

case study investigation and report 82–85
characteristics 37, 381
Choice magazine 139
chosen product concept *see* preferred product concept
circular economy 138, 155
- and clothing 139
- Product Stewardship and 154

climate change 190
cloth fabric 403
clothing 139, 342
- adaptive 76, 77–78
- CAD and 342
- circular economy and 139
- colour in 350
- drawing for 38, 39, 331–332, 338–341
- environmental issues 419–421
- lean manufacturing and 136
- lines in 346–347
- modelling for 28, 345–346
- modifying commercial patterns 340–341
- movement in 360
- obsolescence in 159–160
- patents for 389
- patterns in 357–358
- properties of 403–405
- regenerative design and 190
- shape in 354
- size charts 374
- wearable technology 395, 418–419

cloud-based sharing platform 11
CNC routing 41
Cobalt Design (case study) 177–181
collaboration 9–14
collating and forming information 206
colour 348–351, 358
'colour blocking' 350
comfort 43, 369, 373
commercial pattern 340–341
Commonwealth Scientific and Industrial Research Organisation (CSIRO) 266
competitive edge 264
competitive environment 135
components, products 44

composite metals (metal matrix composites) 166–167
composite timber (engineered wood) 167
computer-aided design (CAD) 11, 28, 97, 130–132, 234, 341–342, 380
  non-resistant materials 342
  resistant materials 341
computer-aided manufacturing (CAM) systems 123, 130–131
computer drawing program 28
computer numerical control (CNC) 131–132
concept/idea sketches 24, 25
concept map 36, 177, 289, 294, 323
concepts *see* product concepts
conducting technical research 36–37
conductive polymers 163
connection to Country 104, 105, 110, 111, 387
consistency (in testing) 41, 421–422
constraints 20, 133, 212
construction drawings 338, 339
construction processes 53
construction process trials 40–41
constructive comments 48
constructive feedback 184
consumers, impacts of technologies and alternative materials for 171–173
contemporary Indigenous designers (case studies) 107–113
continuous production 123–124
contrast 357
controls, safety 244, 256
Convention on International Trade of Endangered Species (CITES) 409, 410
convergent thinking 8–9, 14, 186–187, 218, 231, 314, 315–316 *see also* critical thinking
copyright 199, 201, 265, 388
Copyright Act 1968 (Cth) 390
corrodibility 402
couture 120
Couzens, Debra 106
Couzens, Vicki 106, 107
COVID-19 pandemic 125, 174, 191, 270
'Cradle to Cradle' concept (C2C) 148–149, 155
'cradle to grave' impact 148
Craiyon 219
Creative Education Foundation 187
creative thinking 7–8, 185, 187, 219, 231, 314, 316–317
  brainstorming 13
  idea sharing 13
  knowledge and skills 13
  techniques 220, 317
creativity 317
critical feedback 14
critical thinking 7, 9, 185, 187, 231, 314, 316–317
  checking and evaluating 14
  critical feedback 14
  decision-making 14
  process management 14
  techniques 220–222, 317
critiquing 30, 82–84
  useful questions for 83
  using Factors that influence design 82
cultural and religious beliefs 369, 387
cultural appropriation 105, 387
cultural influences on design 102–115
cultural knowledge 104
cultural misappropriation 387
culture, Aboriginal and Torres Strait Islander 104–115
currency 188
custom-made product 120

D

Darroch, Lee 106, 107
data
  collect and present 278
  collection and communication tools 82
  interpretation 278
  presentation 283
  qualitative 16, 202–203, 231
  quantitative 16, 203–205, 231
data storage 171
de Bono, Edward (case study) 315
decorative and finishing methods 39
deep pressure touch (DPT) stimulation 77
define phase 7, 186
deliver phase 7, 187
demographics 369
density 401
Department of Climate Change, Energy, Environment and Water (DCCEEW) 155
descriptive drawings 25, 331
design
  aesthetics 346
  for animals (non-humans) 68–69
  collaboration in 9
  cultural influences on 102–115
  development 96
  drawing *see* drawings
  Elements and Principles 29–30, 346–361
  ethics in 385–386
  Factors that influence 20, 362–97
  and prototype development 268–269
  sustainable 69, 71–73, 76, 106, 107–110, 136–138, 142–156
design brief 6, 14, 19–23, 56, 90, 101, 177, 186, 209–214, 293, 365, 367, 375, 377, 378, 379, 382–383, 393
  design ideas 92–93
  design process using 177
  for different team approaches 20–22
  elements of 210–212
  and evaluation criteria 22–23, 92
  formulate 213
  materials 94–95
  for product range 21–22
  requirements of 55
  research and testing 213–214
  sample 213
  SAT folio 291–294
  team/individual 22
  writing 19–20, 291–294
Design Council (UK) 7, 8
design, developing and conceptualisation 4–33
  Double Diamond design process 6
  generating and designing 7
  investigating and defining 7
  producing and implementing 7
Design Elements 346–353, 376
  colour 348–351
  form (from shape) 348
  lines 346–347
  shapes 347
  texture and surface qualities 352–353
  tone 351
  transparency, translucency and opacity 351–352
designers working
  research methods 81
  in specialisations 80
  together 81
design folio, bibliography in 200–201
Design for Disassembly (DfD) 149–152, 155
  extracting materials 150–152
  framework 232
design for recycling or recovery (DfR) *see* Design for Disassembly (DfD)
design ideas 323
  developing 178
  generating and recording 92–93
design technologies 121, 122, 123, 125–133, 271–273, 280, 380–381
design options drawings 23, 24, 92–93, 217–219, 222–225, 286, 299–302, 309, 322, 329–334
  analysing 33
  developing 31–32
  evaluation 32–33
  non-resistant materials 331–332
  rendering 332–334
  resistant materials 329–331
  in SAT folio 222–224
Design Principles 346, 353–361, 376
  asymmetry 356
  balance 356
  contrast 357
  movement 360–361
  pattern 357–360
  positive and negative space 356
  proportion and balance 355
  proportion and shape 353–355
  rhythm 359–361
  symmetry 356
  using patterned fabric 361
design problem 186
design process 6–33, 36–56, 57–61, 184–192, 256, 367, 375, 377, 379, 383, 393
  case studies 57–61, 139–142, 177–181

9780170477499

design solution 187
design specialisations 394–397
  assistive products 395
  biomaterials 396–397
  biophilic design 395
  fibres 396
  interdisciplinary and transdisciplinary 394
  lapidary 395–396
  technacy 395
  wearable technology 395
design team 9–14, 47
design thinking 230–231, 314–317
  models and prototypes 343–346
  speculative 266–267
detail drawing 337
develop phase (divergent) 7, 186–187
digital/CAD skills 41
digital pattern grading 59
digital prototypes 234, 341
digital technologies, in team 11
digital visualisations and modelling 28
discover phase (divergent) 7, 186
disparities in access 272
dispersion-strengthened composite metals 166
disposal methods 144
divergent thinking 8, 13, 179, 186–187, 218, 231, 314–315 *see also* creative thinking
  creative and innovative 317–321
  definition 314
diversity 66, 67, 105, 275
documenting production 257–259
Double Diamond design process 6–9, 84–85, 90–91, 184–187, 256, 288, 315
  activities in 185–186
  beginning 15–23
  components 84
  core design principles 8–9
  creative thinking 187
  critical thinking 187
  critiquing designer's use of 84–85
  design ideas, generating and recording 92–93
  drawings *see* drawings
  exploring, researching and defining 90–91
  ideas during 23
  phases of 6
  school-assessed task and 184–185
  stages 84
downcycling 148
drape 405, 416
  testing 426
drawings *see also design options drawings, visualisations, working drawings*
  computer programs 28
  and design 322–342
  design options 24, 92–93
  forms of 92–93
  graphical product concepts and 24–25
  for graphic product concepts 322
  isometric 330
  lack confidence in 220
  perspective 330
  product concepts *see* product concepts
  for product design 23–25
  for textiles 25
  types 217
  visualisations 24, 92–93
  for wood, metal and plastics products 24–25
  working 24, 49–50, 92–93
ductility 401
  testing 423
durability 402
  testing 422

E

ease of machining 402
economic issues 272
economies of scale 122
elasticity 401, 403
  testing 423
electrical conductivity 402
  testing 424
Elkington, John 142
Ellen MacArthur Foundation 138
emotional and sensory appeal 369–370
end user profile 19, 21, 201, 210–212, 291–292
  intended function 212
  using icons 212
end users 6, 55, 64–85, 86–101, 139–40, 178, 368–375 *see also* end user profile
  accessibility 370
  anthropometric data 373
  anthropometry 373–375
  comfort 373
  cultural and religious beliefs 369
  cultural influences on 115
  deciding on 196–198
  defining 197–198
  demographics 369
  emotional and sensory appeal 369–370
  ergonomics 373
  feedback 30–31, 231, 282, 308–309
  needs 15–16, 90–91
  non-human 196
  positive impacts for 64–85
  of products 21
  profile 19
  quality of life 368–369
  research 19, 201, 290
  safety 372
  social and physical needs and trends 371–372
  and target market (case study) 139–140
  universal design 370–371
enhancements 282
enlargement 337
entrepreneurial activity 264–266, 267, 279, 284
environmental impact 69, 147–148, 155–156, 173
environmental issues 272
  fabrics, fibres and yarns 419–421
  greenwashing 157
  metal 412–413
  obsolescence 158, 160
  plastics 414–415
  timber 409
Environmental Management (ISO 14001) 392
environmental physics 373
environmental standards and regulations 280
ergonomics 15, 17, 43–44, 89, 369, 373
erosion 409, 412
ethical considerations 231, 271–273, 280–281, 364, 384–386
ethical designs, production for 254–259
ethical product design 182–227
ethical research 198–201
ethics 385
  considerations 231
  in design 384–394
  design problem 194–196
  gathering end user research 90–91
  legal responsibilities 387–390
  in primary research 17
  in secondary research 17
  and sustainability 385–386
ethnographic research techniques 201
evaluation
  assessment 283–284
  data 278–283
  materials tests and trials 306–307
  range of products 262, 278–283
  SAT folio 305–307
  speculative design thinking 266–267
  your own product 282
evaluation criteria 22–23, 48, 55, 94, 214–216, 282, 291–294
  for graphical concepts 215–216
  SAT folio 291–294
experimental materials 161–167
  bioproducts 161
  composite metals 166–167
  composite timber 167
  Fungi Solutions (case study) 174–175
  impacts of 173
  innovative polymers for 3D printing 163–164
  mycelium 161–162, 174–175
  mycelium composites 162–163
  plastics 164
  repurposed plastics 165–166
exploded view 334, 337
Extended Producer Responsibility (EPR) 152–155
  government role 155
  strategies for producers 154
extension lines 337

F

fabrics 416–421
- characteristics and properties of 39
- components for 44
- construction 358
- design options drawings for 331–332
- disassembly 150
- geotextiles 418
- laminated 418
- modelling product ideas in 46, 344
- smart 418–419
- technical 418–419
- texture 352–353
- treated 418
- types of 416
- working drawings for 338

factors that influence design 362–397
- aesthetics 376–378
- design specialisations 394–397
- end users 368–375
- ethical considerations in design 384–394
- function 366–368
- market opportunities 378
- need or opportunity 365–366
- product design 363–364
- product life cycle 379
- technologies 380–384

fashion illustrations 25, 331
fasteners or functional components 151, 223
feedback
- critical 14
- end user 231, 308–309
- evaluation criteria and 100
- market research and 267
- team 61
- team and end user 30–31

ferrous metals 411
fibre-reinforced polymers 163
fibres 39, 396, 416–421
- bamboo 170
- blends 417
- types of 416

figuring 357
final proof of concept 228–251, 307
- *see also* proof of concept
- definition 233
- design thinking 230–231
- prototyping 233–238
- risk assessment 247–249
- risk management 244–247
- scheduled production plan 241–244
- testing materials, tools and processes 231–232

finite resource 137
5 Ps 269–271
flats 21, 93, 97, 219, 226, 338–341
- using CAD 342

flexible and responsive manufacturing 135
Flourish 191
flow chart 99
fluorescent colours 350
form (from shape) 348
formaldehyde 169
fossil fuels 137
Fourth Industrial Revolution (41R) 126
Frascati Manual 2015 262
French draping 345
functionality 366, 368
functional obsolescence 158–159
function 366–368
Fungi Solutions (case study) 174–175

G

Gensler 189
geotextiles 418
Global Organic Textile Standards 76, 170, 420
goal setting 257
good planning (in testing) 41
Google Slides 11
Google Trends 195
Gordon, William 318
GPS tracking devices 82
graphical product concepts 24–25, 217–219, 236, 294–302
- design options 299–302
- drawing for 322–342
- evaluate and critique 225

greenhouse gases 137, 152, 169, 171, 280
greenwashing 157

H

Hamm, Treahna 106, 107
hand sketches 58 *see also* visualisation drawings
hard-spun yarns 417
hardwoods 169, 170, 400, 406–410
harm 244
Havaianas (case study) 274
hazards 52, 244
- types of 52

heat resistance 401–402
Herman Miller company 152
hidden detail drawing 337
high-volume production 122, 123–124
'human-centred design' 191

I

ibisPaint X 220
iconic design styles 30
ideas *see* creative thinking, divergent thinking, visualisation drawings
incidents 247
incineration 137
Independent newspaper 278
Indigenous Australian *see* Aboriginal and Torres Strait Islander people
Indigenous knowledge 104
influencers 271
informed consent 17
injection moulding 133
injuries 244–245
- guidelines to avoid 245–246
- types of 52

innovation 159, 171, 284, 317
- of products 279–280
- R&D's importance for 264–266

innovative polymers, for 3D printing 163–164
inputs 147
inspiration 195, 321
intellectual property (IP) 265, 388
- infringement 390
- multiple registrations 390
- protection 390
- types 388

intended function 212, 293
interdisciplinary nature of design 394
International Organization for Standardization (ISO) 144, 391
International Standards 391, 392
International Union for the Conservation of Nature (IUCN) 409
investment 265–266
ISO 14040 and 14044, for LCA 147
isometric drawing 330
ISO *see* International Organization for Standardization (ISO)
iteration 8, 179, 187, 233, 256

J

job losses 172, 272
job-lot production 121–122
journal 54–55, 251
- production 257–259

K

Kami project (case study) 233
KeepCup *Brew* 81, 179–180
knitted fabrics 352, 358, 416
- patterns in 358

Kostellow, Rowena Reed 343

L

labour costs 59, 122, 127
labour management 280
laminated fabrics 418
landfill 125, 137, 144, 415
lapidary 395–396
laser cutting 41, 128
laser technology 128–129
last 141
lateral thinking 315
Layers of Blak design program (case study) 109–112
lean manufacturing 133–136
- approaches/methods 135
- flexible and responsive 135
- plan-do-check-act method 134
- wastes of 133–134
- at Zara (case study) 136

legal responsibilities 387–390

9780170477499

Lego Braille machine (case study) 75
life cycle analysis/assessment (LCA) 146–147, 155, 379
ISO 14040 and 14044 for 147
Life Saving Dot (case study) 78–79
LifeStraw (case study) 74
lightweight packaging (Mycofoam) 162
lines 346–347
A Liter (Litre) of Light (case study) 72
long-term memory 176
low-fidelity maquette 239
low-fidelity prototype 236
low-volume production 121–124, 141
costs of 121
technologies used in 122

M

machine learning (ML) 126
machinists 59–60
malleability 401
mandatory standards 392
manufactured boards 407–409
manufacturing 120–124 *see also* production
additive 132
artificial intelligence in 126–128
automation in 127
laser technology in 129
lean 133–136
R&D role in 264
robotics in 129–130, 135
scales of 120–124
subtractive 132
sustainability and 136–156
technologies in 125–133, 381
maquettes 236
marketing
elements 269–271
sustainability and 156
market-led designs or opportunities 15
market opportunities 195, 378
market research 195–196, 201–206, 269–271, 284
in product development process 269–271
Martinuzzo, Steve 177–181
mass (high-volume) production 122–124
materials 94–95, 105, 351
alternative 167–170
categories 399–406
characteristics and properties 37–40
choice of 179–180
classification 37, 399–421
compatibility 382–383
experimental 161–167
fabrics 416–421
fibres 416–421
knowledge of 37–40
metal 410–413
non-resistant 39, 331–332, 338–341, 399, 402–406
for physical product concepts 235
plastics 413–415
research 40
resistant 38, 329–331, 399
testing 231–232
testing and/or process trials 40–42
understanding 399–406
wood 406–410
yarns 416–421
material safety data sheets (MSDS) *see* safety data sheets (SDS)
materials cutting and costing list 50–51, 249–250
materials testing 94, 421–427
consistency 421–422
non-resistant materials 424–427
resistant materials 422–424
and trials 306–307
McDonough, William 148
medium-density fibreboard (MDF) 408
membranes 417
Memphis style 30, 350–351, 361
metal matrix composites (MMC) 166–167
metals 150, 410–413
classifications 411
environmental issues 412–413
general properties of 410
health and safety issues 411
quality of 411
microfibres 416
microplastics 164
mock-up *see* prototyping
models/modelling 44–46, 93, 304, 343 *see also* prototyping
for non-resistant fabric products 46, 344–346
for resistant materials 45
Model Work Health and Safety (WHS) Act 392
modern-day slavery 280
moisture absorbency 403–404
moisture-wicking fabric 404
Mondrian, Piet 321
Monks, Nicole (case study) 107–109
mood boards 27, 218, 328–329
moral rights 390–391
Morgan, Amanda 174
Morris, William 92
Moscicki, Kerryn (case study) 139–142
movement 360–361
MSDS *see* safety data sheets (SDS)
multimodal 192
multimodal record of progress 63, 101, 184, 192–193, 257–258
mycelium 161–162, 396
for packaging 163, 174–175
mycelium composites 162
myco *see* mycelium

N

National Clothing Stewardship Scheme 139
natural and man-made disasters 125
natural biomaterials 397
natural fibres 396, 416
need/opportunity 365–366
context 365
design brief 365
design to aid 67–68
end users 68
factors 365–366
investigating and defining 15–23
end user 15–16
ethics 17
identifying 15
to improve something ergonomically 15
perceiving through personal experience 15
primary and secondary research 16
quantitative and qualitative research 16
negative space 356
neoprene 60
neural networks 127
new and emerging technologies 125–133
considerations in 271–273
economic issues 272
environmental issues 272
social issues 272
worldview issues 272–273
new technological developments 15, 41–42
non-ferrous metals 411
non-magnetic materials 150
non-resistant materials 37, 39, 232, 331–332, 338–341, 399, 402–406
computer-aided design (CAD) 342
design options drawing 331–332
fabric products 46
materials testing 424–427
modelling for 344–346
working drawings 338–341

O

obsolescence 157
costs of 160
functional 158–159
planned 157–161
reasons for 157
researching types of 161
style 160
technical 159
occupational health and safety (OH&S) 247, 392
Oi bike bell by Knog (case study) 15
one-off manufacturing 120–121
online AI software 234
online tools 195
open-ended questions 16
open-source methods 232
opportunity factors 365–366
organic waste 137
Organisation for Economic Co-operation and Development (OECD) 262

orthogonal drawings 335–336
Osborne, Alex 319
OTO chair 77
outputs 147
outsource 41
OXO kitchen tools: design that grew out of experience (case study) 89

P

patents 388, 389
pattern drafting 57
patterned fabric 361
patterns 357–360
  colour 358
  by fabric construction 358
  drafting 57–58
  grading 59, 342
  knitted 358
  printed 358–359
  in resistant materials 357
  and rhythm 359–360
people 270
personal protective equipment (PPE) 245
perspective drawing 330
physical models 28
physical product concepts 37, 44–48, 217
  developing and refining 233–236
  suitable materials for 235
physical properties 401–402
physical prototype 132
physical 3D models 343
place 270
plan-do-check-act method 134
planned obsolescence 157–161
  benefits and issues with 158
  on sustainability 160–161
planning
  production 98–100
  SAT folio 308–311
planning research 36–37
plant 248
plantation timber 409
plasma cutting 41
plastic filament 132
plasticity 401, 406
plastics 150, 164, 413–415
  classification 414
  environmental issues 414–415
  health and safety issues 414
  precious 232–233
  properties 413
  quality of 414
  repurposed 165–166
Plus, Minus, Interesting (PMI) activity 18
plywood model 45
point 346
pollution 172
polyethylene terephthalate (PET) 233
polylactic acid (PLA) 163
polymer-bonded rare-earth magnets 163
polystyrene, for packaging 163
porosity 401
positive impact design
  assessment 101
  design ideas, generating and recording 92–93
  evaluation 100–101
  gathering end user research 90–91
  by identifying a need 88–101
  on lives and wellbeing 100
  materials 94–95
  mock-up/prototype 97
  other research 95–96
  production planning and safety 98–100
  proof of concept 97
  solution-focused technical/practical research 94
  working drawing 97–98
  working technologically 91–92
positive impacts, opportunities for 64–85
  animals (non-humans), design for 68–69
  case studies of 80–82
    investigation and report 82–84
    research and report 85
  data collection and communication tools 82
  designers working in specialisations 80
  design to aid specific needs 67–68
  products 70–79
  research methods to investigate 81
  sustainable design 69
  universal and accessible design 66–67
  working together 81
positive space 356
possum-skin cloaks 106–107
practical research 94
precautions 244
precious plastics 232–233
preferred product concept 48–49, 236
  justification 49
  proof of concept 49
presentation drawings 329–334
presentation style 218
price 270
primary research 16, 289
primary sources 36, 198
printed patterns 358–359
privacy laws 82
privacy, respect for 17
'problem space' 185
process management 14
process trials 40–42
producers, positive impacts for 171, 173
producing technologically 91–92
product concepts 217–226, 233, 268
  chosen 236
  drawings and graphical 236
  graphical 294–302
  physical 235
  virtual 234–235, 303–304
product design 105
  advantages and disadvantages 18
  drawing 23–25
  influence 20
  pros and cons of 18
  purpose/function of 19
  range 21–22
  and team evaluation report 55–56
  visualisations 24–30
product development process 267–269, 284
  design and prototype development 268–269
  market research in 269–271
  product concepts 268
  product evaluation and modification 269
  production and distribution 269
  research 268
  retail and consumer use 269
production 54–55, 100, 256 *see also* manufacturing
  comparison of scales 123–124
  designs/plans during 259
  and distribution 269
  for ethical designs 254–259
  evaluation 55–56
  and implementation 50–54
  journal documents 257–259
  low-volume 121–122
  mass (high-volume) 122, 123
  materials cutting and costing 50–51
  one-off 120–121
  planning 50, 54
  planning and safety 98–100
  planning steps 241–243
  quality and degrees of difficulty 256
  quality measures 53–54
  recording 54–55
  record/journal 100
  risk assessments 52–53
  risk management 52, 256
  risk matrix 53
  stages, risk control 249
  steps 241–243
  steps and tools/equipment 51
  team members and 51
  textiles 243–244
  time management 256–257
production log 54–55
productivity 171
product life cycle 379
products 270
  artificial intelligence in 127
  case studies examples of 71–79
  culture influences on 113–114
  environmental impact on 147–148
  ethical design of 206–209
  evaluation 368
  evaluation and modification 269
  innovation of 279–280
  key factors 276–277

9780170477499

life stage 146–147
positive impacts 70–79
SAT 279
success or failure of 273–276
unethical 209
product safety 393–394
products for those with autism (case study) 77–78
Product Stewardship *see* Extended Producer Responsibility
Product Stewardship Centre of Excellence 152
Profile: Edward de Bono (case study) 315
project scope 19–20
promotion 270–271
proof of concept 49, 97, 217
final *see* final proof of concept
properties 37, 381
proportion
and balance 355
and shape 353–354
prototyping 33, 37, 46–48, 97, 133, 185, 233–238, 344
drawings product concepts 236
examples 236–238
graphical product concepts 236
iterative 179
materials 47
models and 44
physical product concepts 233–236
purpose of 46
rapid 128
rapid 3D 132–133
working drawings and 303–304
psychology 373
public spaces 66

Q

qualitative research methods 16, 201–206, 231, 278
collecting data 202–203
in school-assessed task 205–206
quality 367
of metals 411
of plastics 414
timber 409
Quality Management Systems (ISO 9000) 392
quality measures 53–54, 98, 250–251, 256
quality of life 368–369
quantitative research methods 16, 201–206, 231, 278
in business 205
collecting data 204–205
in school-assessed task 205–206

R

range of products 262
and data 278–283
rapid prototyping 128, 132, 135
rapid 3D prototyping 132–133
reality, research based on 202
recording 206
recording production 54–55
record of results (in testing) 41
recycling composite metals 166
refined ideas 222–224
regenerative design, as speculative thinking 189–192
registered designs 388
regulations and legislation 392
reinforced composite metals 166
rendering 332–334
renewable resources 161
repeatability (in testing) 41
repurposed plastics 165–166
research 198–201, 289–291
to aid creativity 320
conducting technical 36–37
end user 201, 290
end user needs 15–16
ethics 17, 198–201
into existing products 18
for inspiration and ideas 293–294
materials 40
other 95–96
planning 36–37
plastics 166
polystyrene and mycelium 163
possible areas 43–44
primary and secondary 16, 198
projects 209
qualitative and quantitative methods 201–206
quantitative and qualitative 16
reports on tests and trials 42
for school-assessed task 207–209
tasks and evidence 36
and testing 213–214
research and development (R&D) 60–61, 262–266
forms of 263–264
for innovation and entrepreneurial activities 264–266
in manufacturing role 264
research methods 81
Research Service Providers (RSP) 266
resistance to biological attack 405
resistant materials 37, 38, 53, 232, 399–402
computer-aided design (CAD) 341
design options drawing 329–330
materials testing 422–424
mechanical properties 400–401
physical properties 401–402
working drawings 335–338
'restorative' design 189–192
rethink, refuse, reduce, reuse, repair, recycle (6Rs) 144–146
rigidity 401
Rip Curl: wetsuit development (case study) 57–61
risk assessments 52–53, 99, 247–249, 256
chart 247
hints for risk control 248
quality measures 250–251
safety data sheets 248–249
risk control, hints for 248
risk management 52, 244–247, 256
process 54–55
production 256
risks 244
investment and 265–266
levels 52
matrix 53
robotics 122, 129–130, 135
Rothy's: sustainable shoes and accessories (case study) 274–276
roving fibres 396
Royal Melbourne Institute of Technology (RMIT) 200

S

safety 368, 422
end users 372
in testing 41
materials and costing list 249–250
product 393–394
risk assessment 247–249
risk management 244–247
safety controls 98, 243
general 245–247
safety data sheets (SDS) 248–249
Safe Work Australia website 52
Safe Work Procedures (SWP) 248
SAT folio 192–193
design brief 291–294
design options in 222–224
evaluation 305–307
evaluation criteria 291–294
examples of items 288–311
generating and designing 294–304
investigating and defining 288–291
planning and managing 308–311
visual checklist for 286–287
working drawings in 226
scales of manufacturing 120–124
aspects of 124
for chair 125
computer-aided design (CAD) 130
computer-aided manufacturing (CAM) 130–131
computer numerical control (CNC) 131–132
continuous production 123–124
laser technology 128–129
low-volume production 121–122
mass (high-volume) production 122
new and emerging technologies 125–133
one-off manufacturing 120–121
robotics 129–130
traditional technologies 125–133

SCAMPER 220
  thinking technique 319–321
scheduled production plan 241–244, 308
  implementation 256
scheme 152
school-assessed task (SAT) 184
  Double Diamond design process and 184–185
  ethical design problem 194–196
  evaluation criteria 214–216
  folio 192–193
  further research 216
  products 279
  qualitative and quantitative research in 205–206
  recording, collating and forming information 206
  research for 198–201, 207–209
'science fiction design' 231
scientific thinking 14
seasoned timber 408
secondary research 16, 289
secondary sources 36, 198
section drawing 337
sectioning 337
self-healing polymers 163
separate timeline 243
shape memory polymers 163
shapes 347
sheen 352
The Shoe That Grows (case study) 71
shrinkage 405
silhouette 354
6Rs 144–146, 155
skins 417
smart electronic fabrics 418–419
smart yarns 418, 419
social and physical needs 371–372
social cost 160
social issues 272
social practices 105
soft-spun yarns 417
software 126, 134
  for visualisation drawings 323–325
softwood 406–410
solidity 355
solid metal 132
solution-focused technical/practical research 94
'solution space' 185
specialisations, designers working in 80
Speculative and Critical Design 188
speculative design thinking 266–267, 284
speculative thinking 185, 187–192, 231, 264, 314
  example scenario for 188–189
  how to use 192
  regenerative design as 189–192
standards 391
Starck, Philippe 273, 351
Statista 195
stems 169
Stephens, Christina (case study) 76
stiffness 401
Stories from Layers of Blak (case study) 109–112
style obsolescence 160
styles 361
  aesthetics and 377
  inspiration for 361
substances 248, 381
subtractive manufacturing 132
success criteria 22
suitable media 323
Summary of the Occupational Health and Safety Act 2004: A handbook for workplaces (Victoria) 247
supply chains 280
  defined 126
  world events on 125
surface coatings 411
sustainability 280–281, 284, 369
  'Cradle to Cradle' concept (C2C) 148–149
  defined 136
  Design for Disassembly (DfD) 149
  ethics and 385–386
  Extended Producer Responsibility (EPR) 152–155
  frameworks 138–139
  influence of 155–156
  life cycle analysis/assessment (LCA) 146–147
  manufacturing 136–156
  materials and processes 232
  planned obsolescence on 160–161
  rethink, refuse, reduce, reuse, repair, recycle (6Rs) 144–146
  strategies 138–139, 156
  terminology linked with 137
  triple bottom line (3BL) 142–144
  waste hierarchy for 138
sustainable design 69
Sustainable Development Goals (United Nations) 70
symmetry 356
synectics 318
synectic thinking technique 318
synthesis 187
synthetic biomaterials 397
synthetic fibres 396, 416

T

'take, make, waste' linear system 138
target market 139–140, 369
team 140–141
  activity 31
  advantages of 9–10
  approaches to 12
  convergent/critical thinking 14
  design brief *see* design brief
  digital technology use 11
  divergent/creative thinking 13
  and end user feedback 30–31
  evaluation report 55–56
  feedback 48, 61
  needs 17
  visual mood board for 27
  working as 11
teamwork
  benefits 9–10
  in design 9
  Rip Curl 10, 57
technacy 395
technical cycle 138
technical drawing 97
Technical Drawing for Students 337
technical fabrics 418–419
technical knowledge 13
technical materials 149
technical obsolescence 159
technology/technologies
  comparing 174
  in designing 380
  factors 380–384
  in manufacturing 381
  materials 381
  materials, tools and processes 380–384
  negative impacts of 171–172
  new and emerging 125–133
  positive impacts of 171
  substances 381
testing materials 40–42, 231–232
textile production 244
textiles 39
  drawings for 25
  fibres 416
  guidelines 246
  production steps for 243
  products, quality measures for 53
texture and surface qualities 352–353
thermal conductivity 401
thermal properties 404
thinking *see also* specific types
  and planning technologically 91
  types of 6, 314
3D printing 28, 41, 132–133, 344, 389
  innovative polymers used for 163–164
  for SAT 164
thumbnails 24
timber
  classification 406
  environmental issues 409
  production 407–408
  quality 409
timeline 243–244, 256, 257

9780170477499

timeline table 99
time management 256–257
toiles 235, 304, 345
tone 351
total quality management (TQM) 136
toughness 401
toxic chemicals 137, 161
Toyota Production System (TPS) 136
trademarks 388
traditional Indigenous Australian design 105–106
traditional patterns 105
traditional technologies 125–133
transdisciplinary 394
translucency and opacity 351–352
transparency 188, 351–352
transportation 156
trends 371–372
trials 306–307
  process 40–42
  tests and 42
triggers 318
triple bottom line (3BL) 142–144

U

Ubuntu 192
understanding materials 399–406
unethical products 209
unfair distribution 172
universal design 66–67, 370–371
  principles of 67
UN Sustainable Development Goals 70
upcyling 148

V

value 136
VCE Product Design and Technologies 214, 322
Victoria, WorkSafe 246
virtual product concepts 217, 303–304, 341
  artificial intelligence for 234–235
visual bulk 348
visualisation drawings 23–30, 92–93, 217–219, 294–298, 322–329
  advantage of 28
  annotations 26–27
  design elements and principles 29–30
  digital visualisations and modelling 28
  examples 325–328
  in folio 219–222
  software for 323–325
  suitable media and design ideas 323
  using mood boards 328–329
volatile organic compounds (VOCs) 409

W

warp 358
waste 136, 137
  concept of 148
wearable technology 395, 418–419
weft 358
well-made model 45
Western and traditional construction methods 104
wetsuit development 57
  case study 57–61
Wise, Tracy (case study) 111–112
wood 406–410
  general properties 408–410
  health and safety issues 408–409
  manufactured boards 407–408
woollen fibres 416
working digitally, advantages of 193
working drawings 23, 24, 49–50, 92–93, 97–98, 217–219, 239, 322, 335–341
  in folio 226
  for models 33
  non-resistant materials 338–341
  orthogonal drawings 335–336
  and prototyping 303–304
working technologically 91–92
  producing 91–92
  thinking and planning 91
worldview issues 158, 272–273, 284
wrinkle resistance and recovery 405

Y

yarns 39, 396, 403, 416–421

Z

Zara, lean manufacturing at (case study) 136
zipper 53